THE NERI OXMAN MATERIAL ECOLOGY CATALOGUE

PAOLA ANTONELLI
WITH ANNA BURCKHARDT

THE MUSEUM OF MODERN ART, NEW YORK

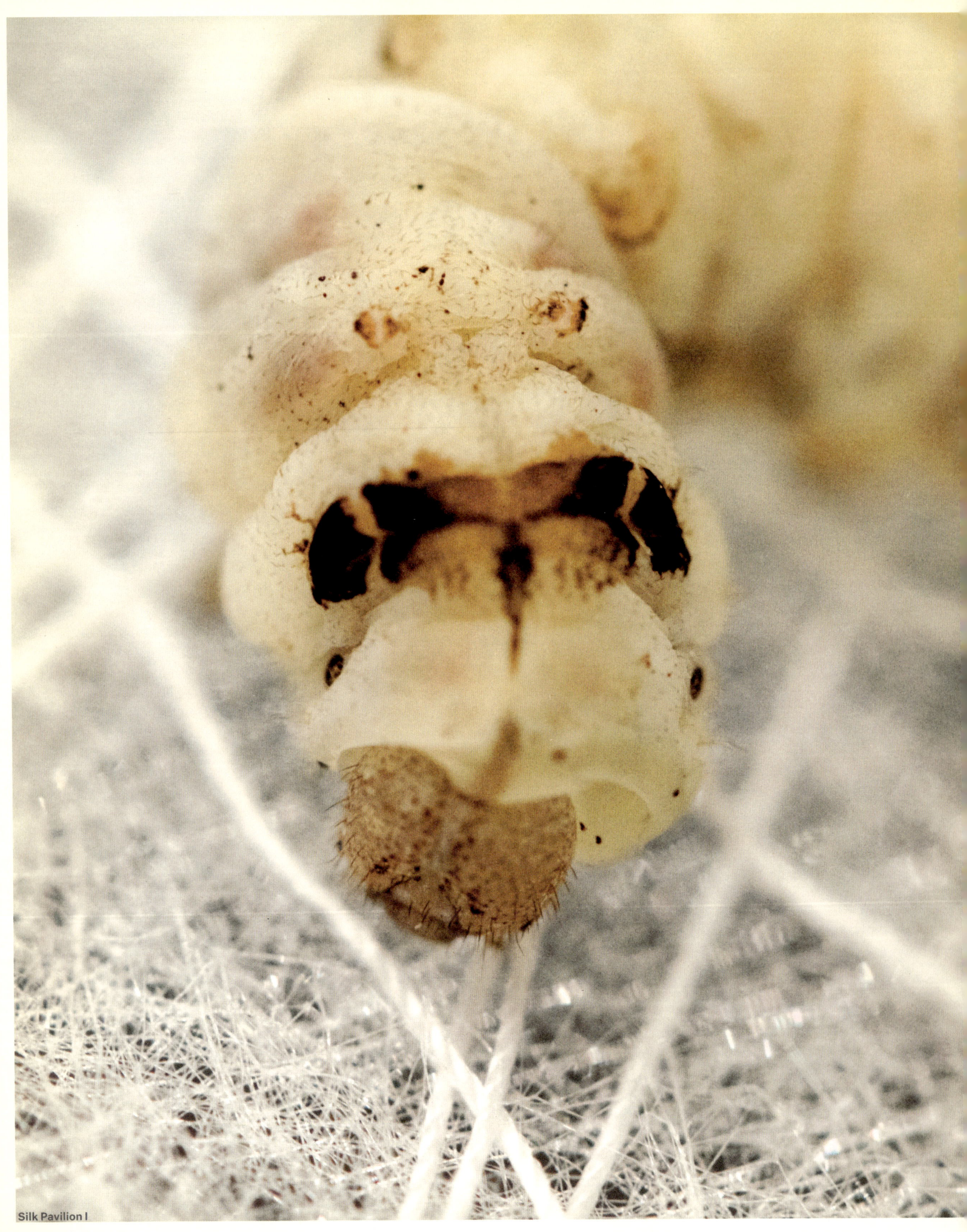
Silk Pavilion I

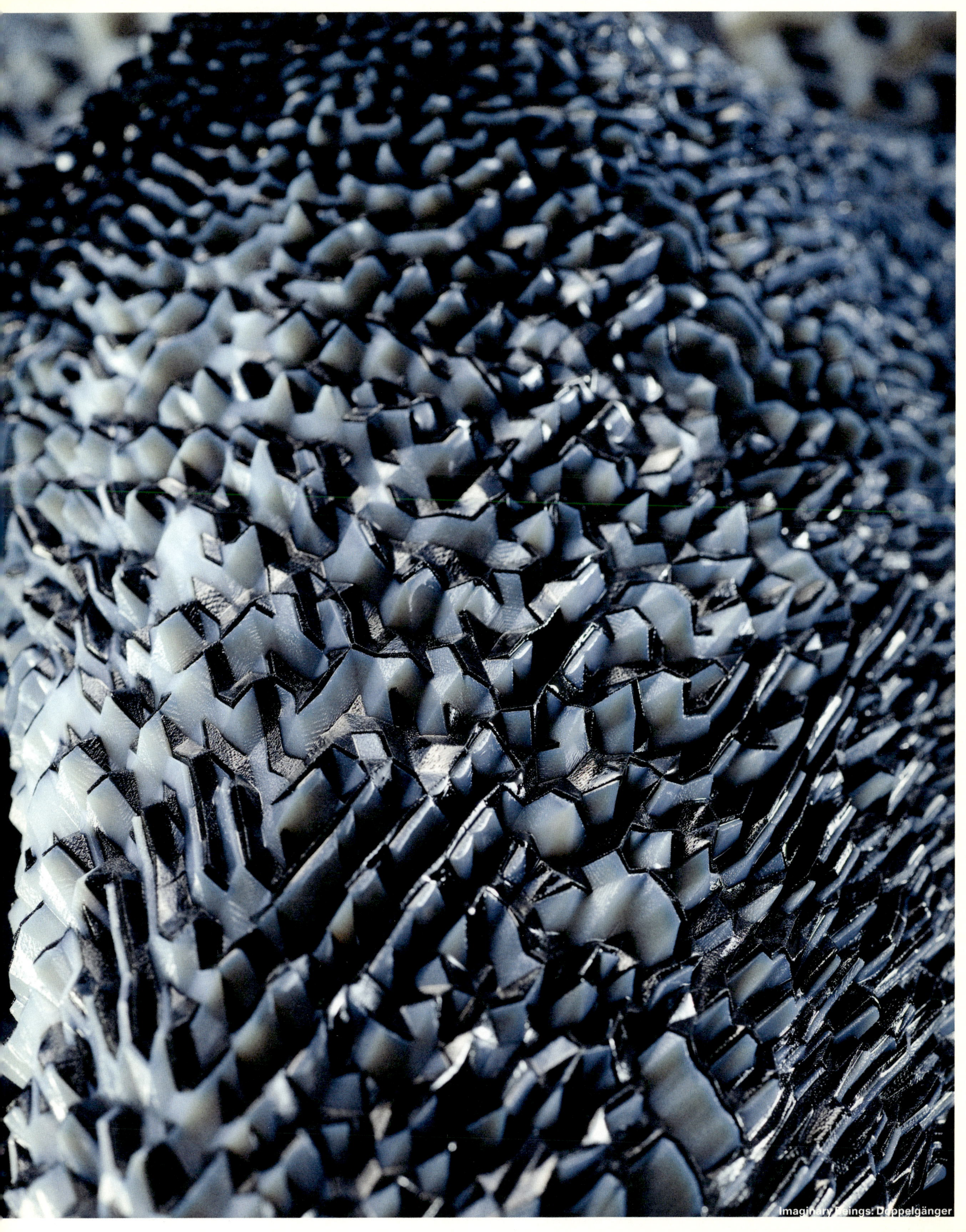
Imaginary Beings: Doppelgänger

Published in conjunction with the exhibition
Neri Oxman: Material Ecology, at The Museum of Modern Art, New York, February 22–May 25, 2020.
Organized by Paola Antonelli, Senior Curator, Department of Architecture and Design, and Director, Research and Development; and Anna Burckhardt, Curatorial Assistant, Department of Architecture and Design

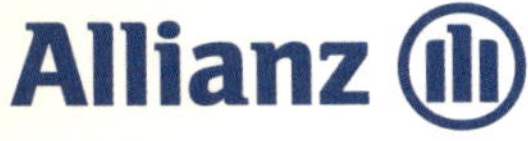

The exhibition is made possible by Allianz, MoMA's partner for design and innovation.

Generous support is provided by The International Council of The Museum of Modern Art and The Modern Women's Fund.

This publication is made possible by the Nancy Lee and Perry Bass Publications Endowment Fund.

Produced by the Department of Publications, The Museum of Modern Art, New York
Christopher Hudson, Publisher
Curtis Scott, Associate Publisher
Hannah Kim, Business and Marketing Director
Don McMahon, Editorial Director
Marc Sapir, Production Director

Edited by Emily Hall and Jennifer Liese
Designed by Irma Boom
Production by Marc Sapir
Printed and bound by Ofset Yapımevi, Istanbul

This book is typeset in Neutral.
The paper is 115gsm Munken Print White.

Library of Congress Control Number: 2020930268

ISBN 978-1-63345-105-6

Published by
The Museum of Modern Art, New York
11 West 53 Street
New York, New York 10019
www.moma.org

© 2020 The Museum of Modern Art, New York
Certain illustrations are covered by claims to copyright cited on page 177.

All rights reserved

Distributed in the United States and Canada by
ARTBOOK | D.A.P.
75 Broad Street, Suite 360
New York, New York 10004
www.artbook.com

Distributed outside the United States and Canada by
Thames & Hudson Ltd.
181A High Holborn
London WC1V 7QX
www.thamesandhudson.com

Printed in Turkey

Front and back covers:
Neri Oxman and The Mediated Matter Group.
Silk Pavilion I. 2013 (see page 98)
Image modified by Irma Boom

CONTENTS

Project texts by Neri Oxman (Materialecology and Imaginary Beings),
and by Neri Oxman and The Mediated Matter Group (all others)

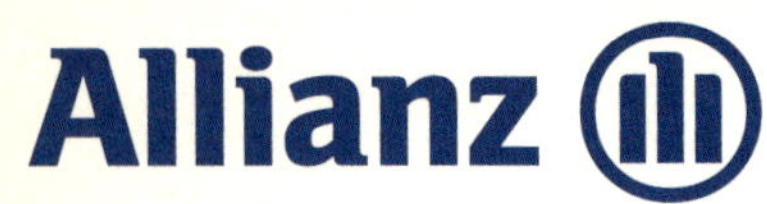

Allianz is delighted to support the inspiring works of Neri Oxman as part of our renewed partnership with MoMA. As the largest sustainability investor in the world, we see in this exhibition a reflection of our commitment to the creation of new narratives through a forward-looking fusion of the computing, ecological, and engineering disciplines.

Oxman envisions a future in which these disciplines work together to directly address the existential challenge of our time: the securing of a sustainable future for us all.

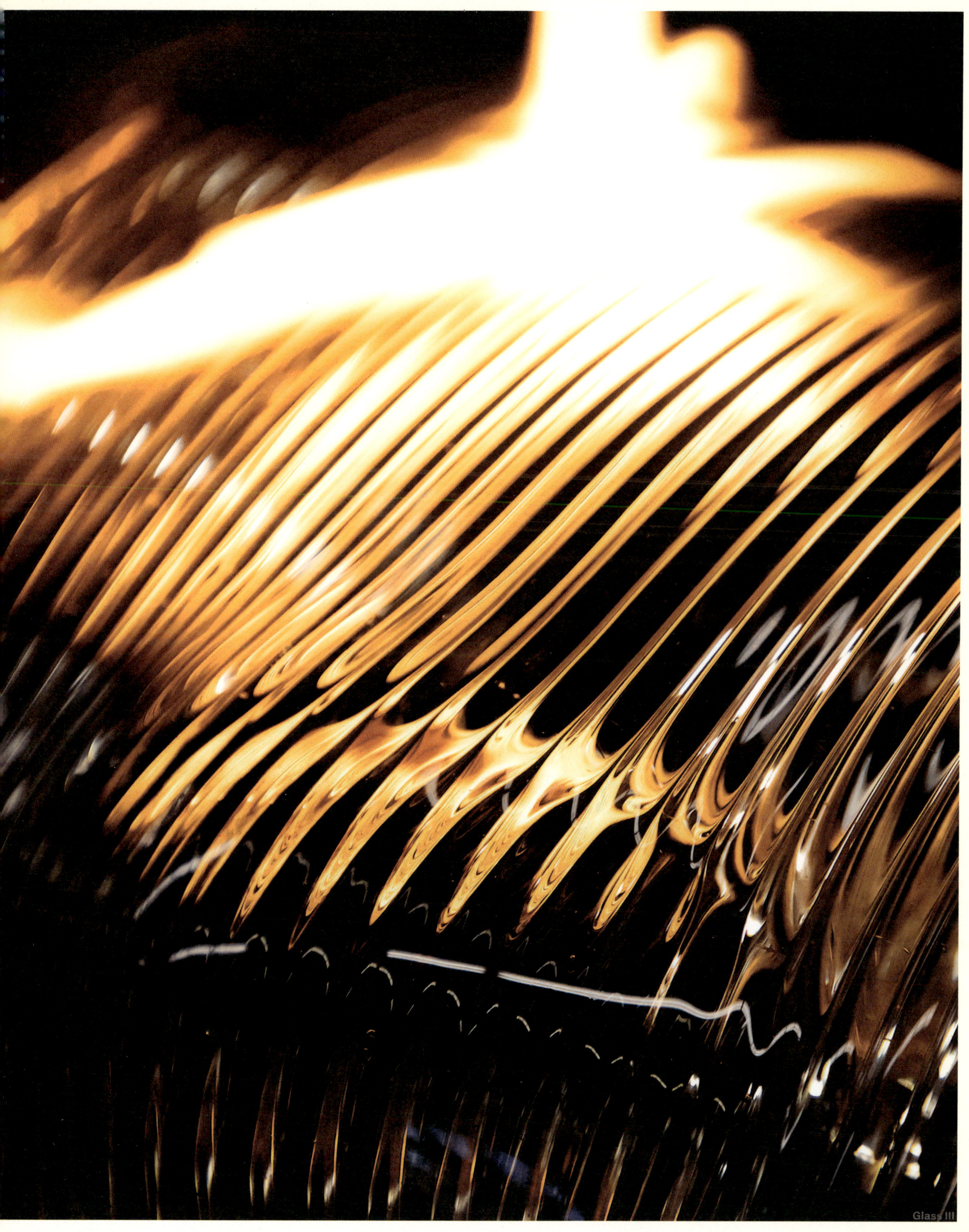

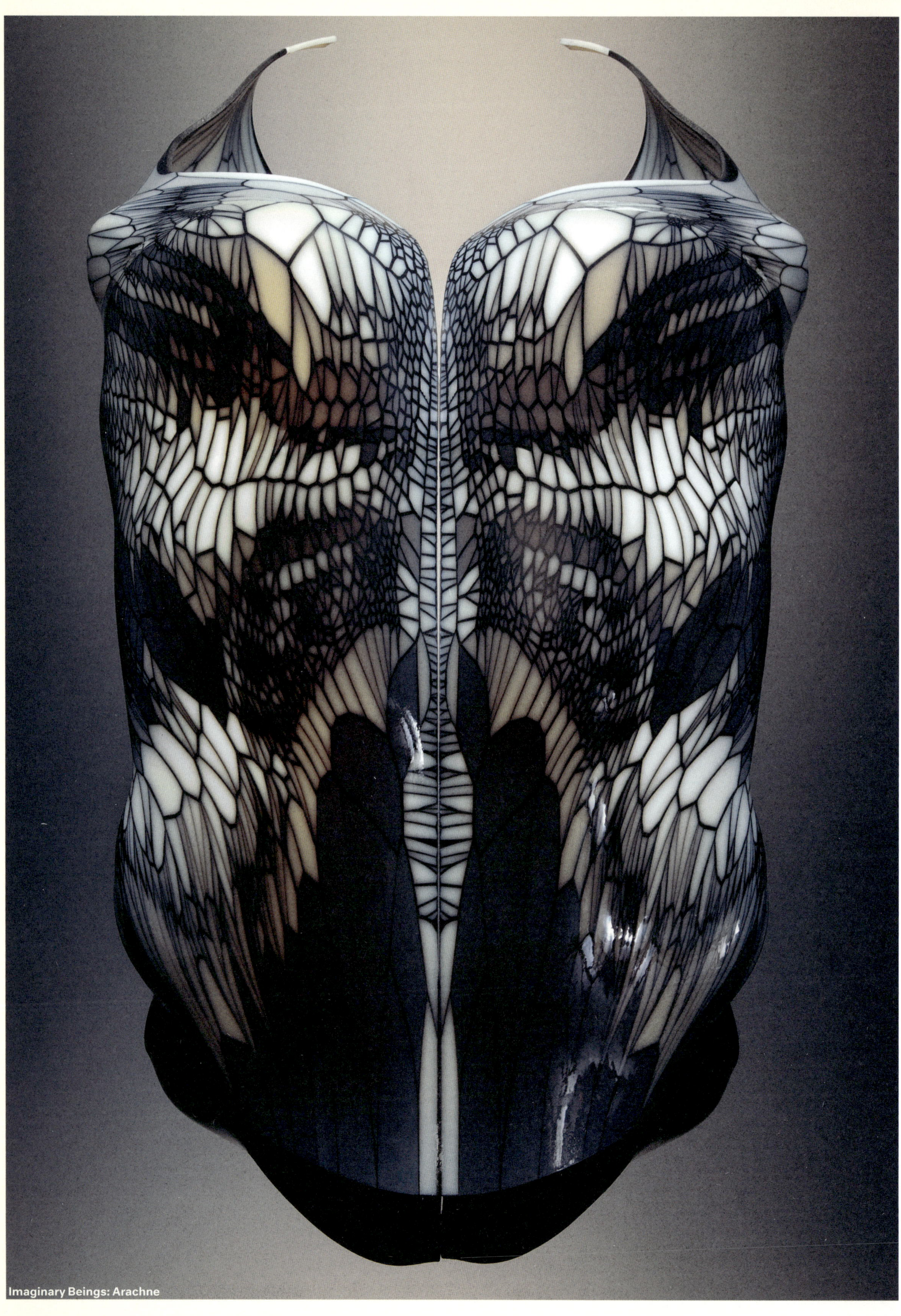

Imaginary Beings: Arachne

Foreword

With the founding of the Department of Architecture and Design, in 1932, The Museum of Modern Art began its positioning of design as a means for building a better present and future. The department's influential programs and displays have highlighted new techniques, typologies, and technologies, and have explored how they might change all of our lives for the better. Early exhibitions such as *Modern Architecture: International Exhibition*, in 1932, and *Machine Art*, in 1934, introduced innovative architects and buildings from all over the world and presented the elegant precision of industrial objects by showcasing them reverentially on pedestals "like Greek sculptures," according to the press release for *Machine Art*. This was just one of the ways in which Philip Johnson, who organized both exhibitions (the former in collaboration with Henry-Russell Hitchcock), used the language of art to elevate the role of architecture and design in cultural discourse.

Neri Oxman's work honors Johnson's vision and turns it on its head. At first glance, her arresting artifacts could easily be confused for sculpture. Their aesthetic elegance, however, is only a function of the processes they embody: the advanced science and technology—including synthetic biology, digital computation, and additive manufacturing—through which she and her collaborators are designing new ways of building. They envision dynamic materials and techniques that produce objects that behave as if grown in response to their context and environment—customizable, intelligent, and specific. Her first appearance at MoMA was in 2008, in the exhibition *Design and the Elastic Mind*, Paola Antonelli's comprehensive foray into design's relationship with science. Over the past twelve years, her work has been an important part of MoMA's evolving vision of architecture and design's present and future roles—a vision that extends beyond buildings to include processes, materials, and strategies privileging an ecosystemic view. Now, with *Neri Oxman: Material Ecology*, our investigation into the relationship between design and science continues, in an exhibition that surveys a selection of Oxman's work since 2008, some of it now in MoMA's collection, as well as a new commission, the site-specific installation Silk Pavilion II.

The term Material Ecology, coined by Oxman around the same time as *Design and the Elastic Mind* was conceived, encapsulates one of our priorities in collecting and exhibiting design in the twenty-first century. Material Ecology is a pragmatic philosophy: a method of design and production that brings together humans, automated processes, and nature to transform architecture into a hybrid act of building and growing. The striking objects included in the exhibition and in this publication are not sculptures, but neither are they architectural fragments or design items—at least not in the way we are accustomed to think of architecture and design. They are demonstrations of the kinds of tools that should and will be available to architects and designers, perhaps sooner than we think; displayed with videos of the experiments that generated them, they paint an incisive picture of a possible future.

I thank Paola Antonelli, Senior Curator, and Anna Burckhardt, Curatorial Assistant, in the Department of Architecture and Design for organizing this extraordinary exhibition, which brings to MoMA the work of an architect and designer who is paving the way in her field, proposing a path forward in this era of anxiety and uncertainty, galvanizing her community, and stimulating cross-disciplinary collaborations in order to advance positive change. I would also like to thank the many colleagues who have collaborated with the curators to bring this complex exhibition to life.

I am also deeply grateful to Allianz, whose generosity has made this timely project possible, and to The International Council of The Museum of Modern Art and The Modern Women's Fund for their support.

Glenn D. Lowry
The David Rockefeller Director
The Museum of Modern Art, New York

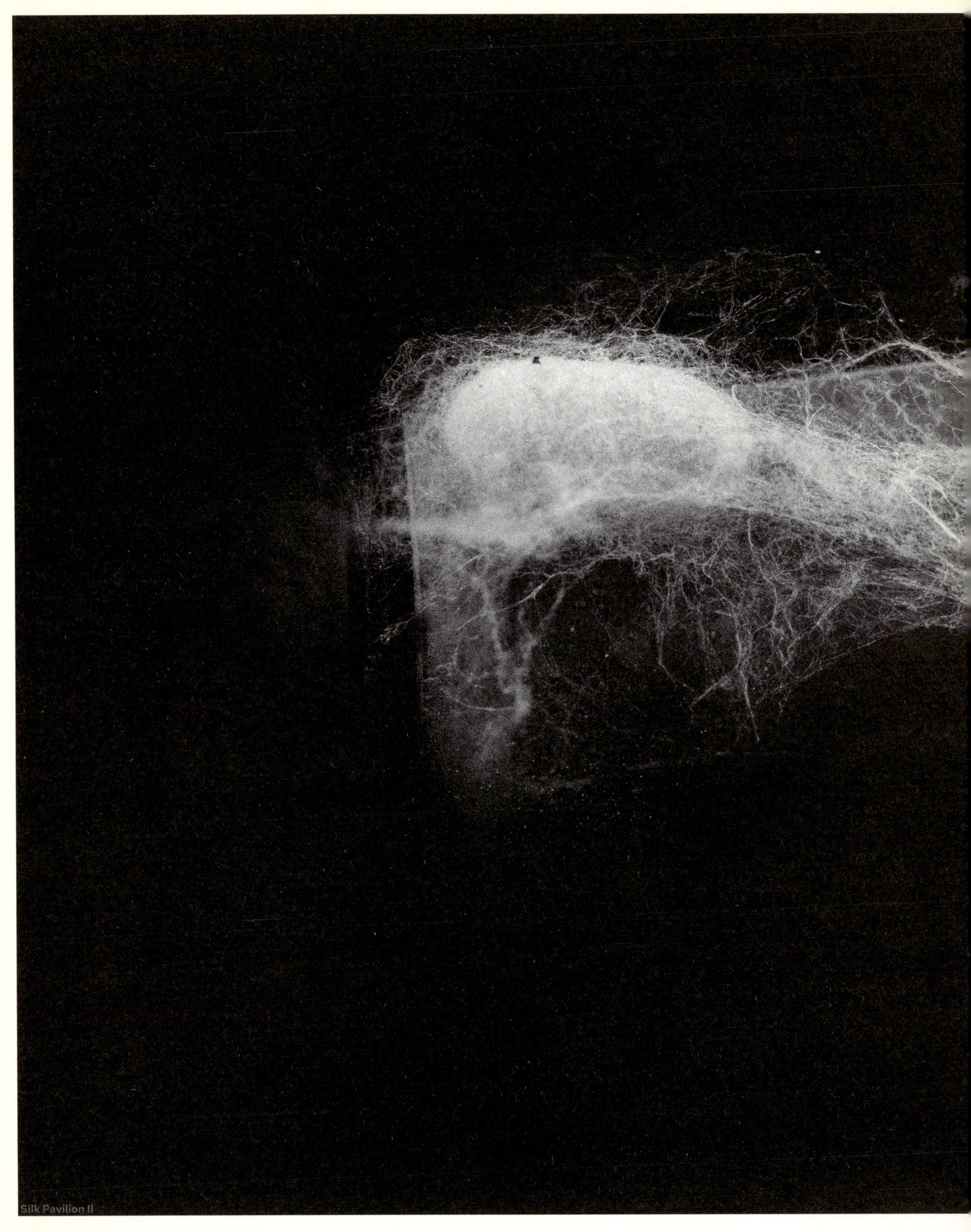
Silk Pavilion II

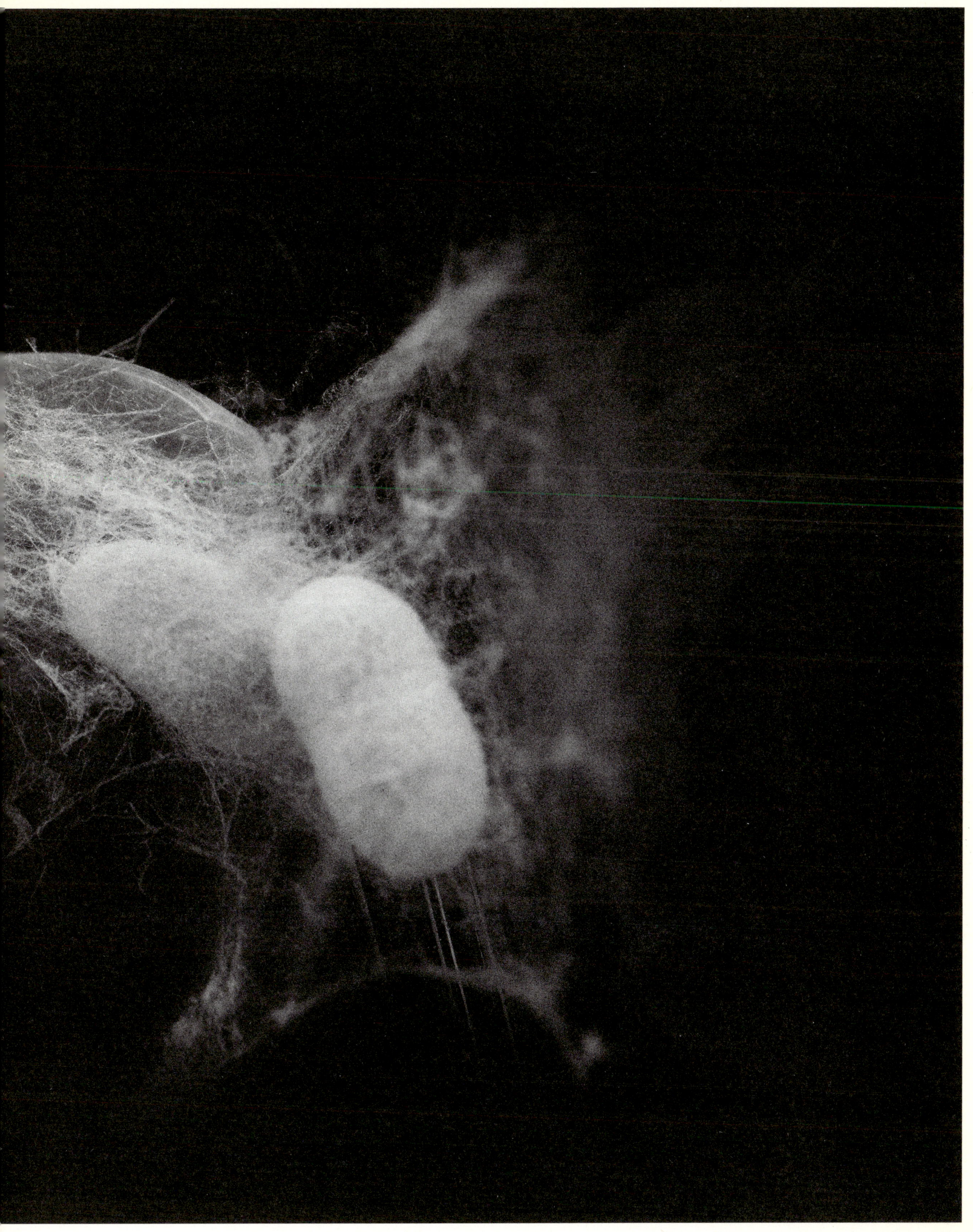

THE NATURAL EVOLUTION OF ARCHITECTURE

PAOLA ANTONELLI

EVERYTHING FLOWS, AND NOTHING STAYS THE SAME. —HERACLITUS[1]

One of the most distinctive characteristics of the human species is a fraught relationship with change. Inescapable, change touches each creature, community, and system uniquely; to each, it manifests at distinct speeds and scales and in different cycles. Most entities—glaciers, plankton, clouds, tigers, or dandelions, for instance—go with the flow, adapting and evolving over time to accommodate change and accept its aftermath, however unfortunate. Not humans. Except for the faithful or the wisest among us, most human beings either resist, pursue, seek to control, or amplify change. We take pride in our ability to interfere with and even manipulate the flow. In so doing, we create consequences—not only for us, but for all species. So much have we tinkered that we seem to have lost control of the mutation, which now ever accelerates, like a cancerous growth.

Indeed, humans themselves are like cancerous cells, so self-absorbed and single-minded about their survival and predominance that they have invaded and destroyed much of what's around them. Neri Oxman, by contrast, seems in no rush. Her practice is a powerful anticipation of a better possible future, and it is projected toward that future without apparent anxiety, patiently weaving connections between disciplines and between species, slowing down the pace of making by marrying the latest technologies with the most ancient and deliberate of tempos—those of silkworms, bees, and microbes. She engages change using change's own momentum.

I met Oxman in 2006. We were introduced by the architect Enrique Norten, who thought she would be a great fit for an exhibition I was preparing at the time. *Design and the Elastic Mind* addressed the unexplored affinities between design and science and their ability to communicate with each other without relying on technology—a well-meaning but sometimes misleading interpreter. Indeed, technology was played down in the exhibition as an instrument with which scientists and designers could realize their joint visions and ventures. On display were mighty examples of scientists' and designers' unbridled collaboration, organized according to scale—albeit a new set of dimensions that took the digital universe into account, one based not only on physical size but also on complexity.

Norten was right: Oxman and her work were the embodiment of the exhibition's thesis. An architect by training, Oxman started medical school before following the family tradition (both parents are architects), attending Tel Aviv's Technion and London's Architectural Association. Her belief in science did not wane, however; she learned its language and dialects in order to be able to engage scientists in productive conversations and collaborations. She also immersed herself in technology, understanding its syntactic power to enact those collaborations. She chose the Massachusetts Institute of Technology (MIT) for her doctoral studies, but even at that forge of so many twentieth- and twenty-first-century revolutions—the wind tunnel, radar, the PET scan, and e-mail among them—Oxman tried her best to avoid being gratuitously seduced by technology, trying instead to look at it as a means she could prod and shape to her ends.

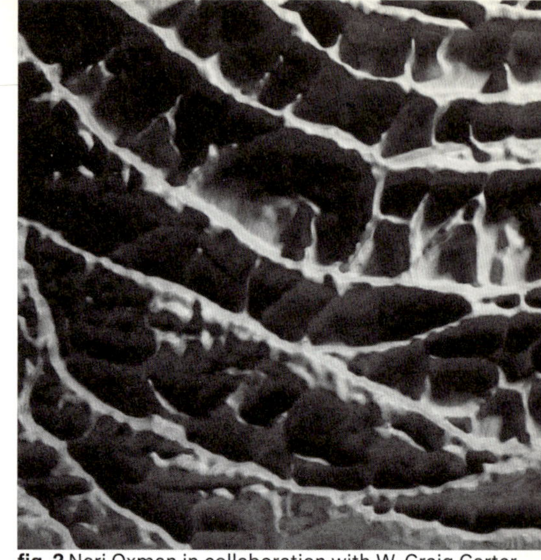

fig. 2 Neri Oxman in collaboration with W. Craig Carter. Subterrain. 2007. Wood composite, 20 × 20 × 2 ¼ in. (50.8 × 50.8 × 5.7 cm). The Museum of Modern Art, New Yo Gift of The Contemporary Arts Council of The Museum of Modern Art

Micrograph image of scorpion-claw tissue. The image was analyzed and reconstructed in three dimensions using a CNC mill and wood composite.

At the time of *Design and the Elastic Mind*, Oxman was also ambitiously contemplating working across scales. She framed the issue clearly a decade later in her tenure statement: "To date, designers invariably encountered a dimensional mismatch between the 'environment space' and the 'object space.' In principle, this mismatch entailed a loss of information when a higher dimensional environment is projected or mapped onto a lower dimensional object."[2] Her overarching goal in response was to detect and

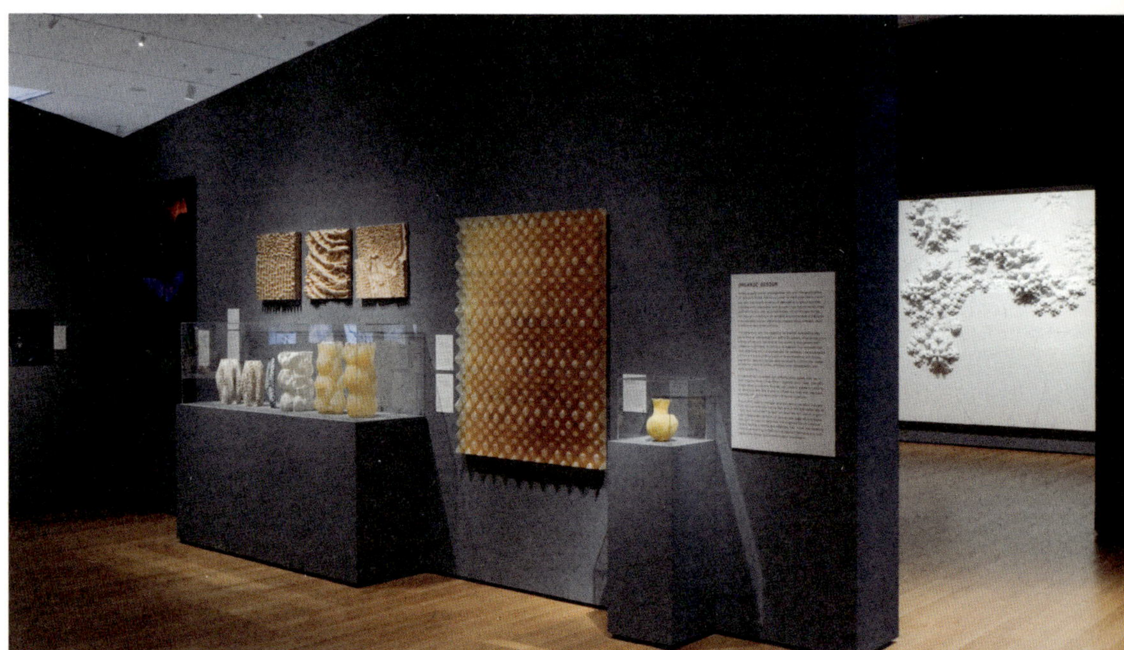

fig. 1 Installation view of *Design and the Elastic Mind*, The Museum of Modern Art, New York, February 24–May 12, 2008

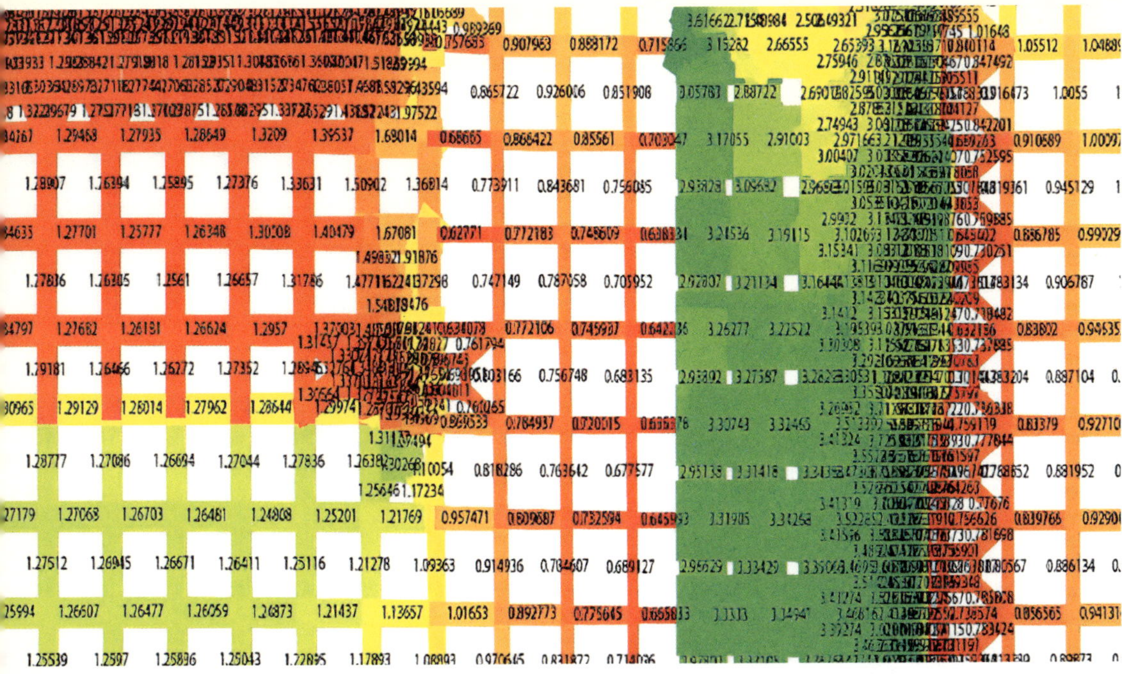

fig. 3 Subterrain. Diagram detail.

fig. 4 Subterrain. Micrograph image of a butterfly wing (left); analysis of material behavior according to stress, strain, heat flow, stored energy, and deformation from applied loads and temperature differences (middle); reconstructed tissue (right).

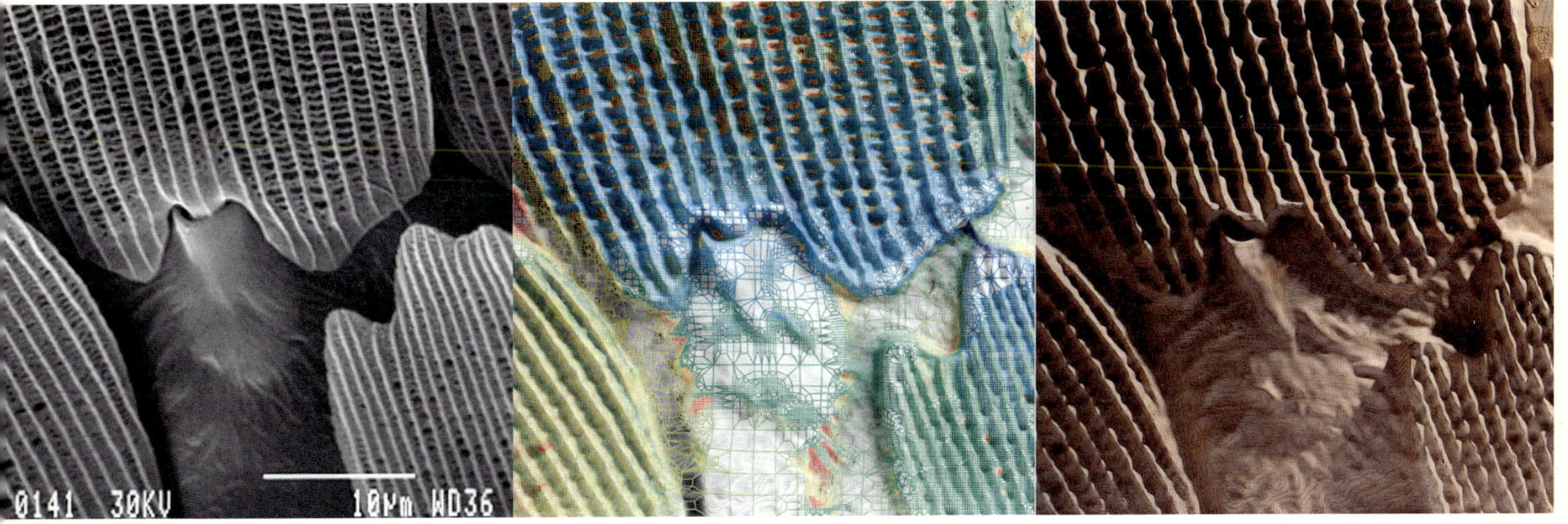

decipher nature's myriad structural and essential design lessons and render them digitally for future application *at all scales*, for the benefit of architects and designers. To steer the relationship with science and modulate fluctuations in scale, she mastered the most expedient ingredient of all contemporary technologies: computation. To me, intent on finding new and innovative ways to conceive organic design, Oxman's approach seemed most timely and effective, visionary yet pragmatic, augmented by the extra gear that only digital technologies and morphologies could offer. She called it Material Ecology: "An emerging field in design denoting informed relations between products, buildings, systems, and their environment. Defined as the study and design of products and processes integrating environmentally aware computational form-generation and digital fabrication, the field operates at the intersection of biology, material science and engineering, and computer science, with emphasis on environmentally informed digital design and fabrication."[3]

Design and the Elastic Mind featured four works from Oxman's Materialecology project—Subterrain, Raycounting, Monocoque, and Cartesian Wax (all 2007; **fig. 1**). Each explored natural and biological phenomena and the potential to use computation to reconstruct them at larger scales, demonstrating how this new technology could inform the future of designing and making objects.[4] In Subterrain (**figs. 2–4**), three tissues (a leaf section, a butterfly wing, and a scorpion claw) were analyzed at the microscale and reconstructed in macroscale using a digitally controlled, very fine mill to create three-dimensional wood prototypes. Raycounting (page 52) generates accurate three-dimensional replicas of objects by measuring the intensity and orientation of light rays—a tool that could hypothetically mimic natural behaviors and apply them to facade treatments, for instance. In three prototypes built using the Monocoque technique (page 64), "veins" on the skin, rather than an internal structure, carry the loads. In Cartesian Wax (page 58), the object's surface is thickened where it is structurally required to support itself, and its transparency also modulates according to light conditions and heat flux. Together, the experiments shown in *Design and the Elastic Mind* highlighted Oxman's early preoccupation with architecture and design's ability to be fluid, to adapt to changing environmental conditions, to contextual, functional, and infrastructural configurations, as well as to levels of occupancy. Computation, digital fabrication, material science, and biology enabled her aspirations. Further, Oxman's research was then already focused on the centrality of individual experience and the ability of architecture and design to respond to local and personal conditions. Oxman's disciplinary breadth, depth, and curiosity made me recognize in her a kindred spirit thirteen years ago. She has since willed dreams into experiments, and experiments into prototypes.

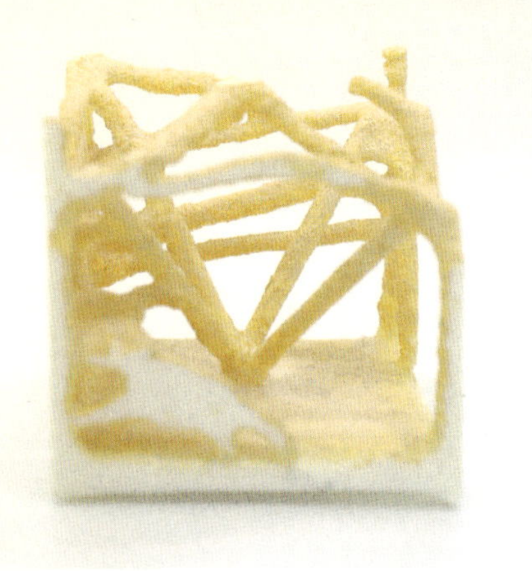

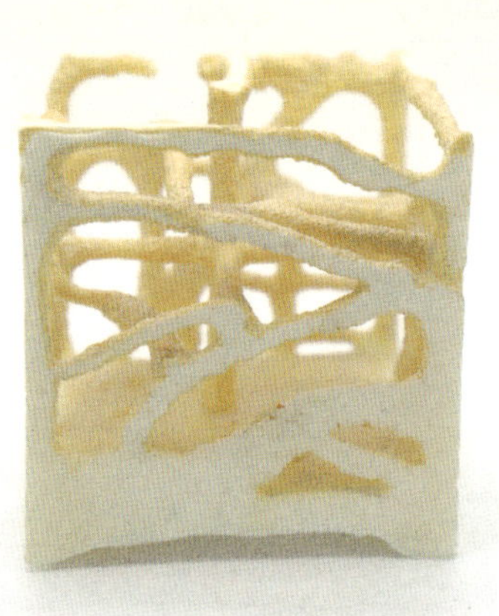

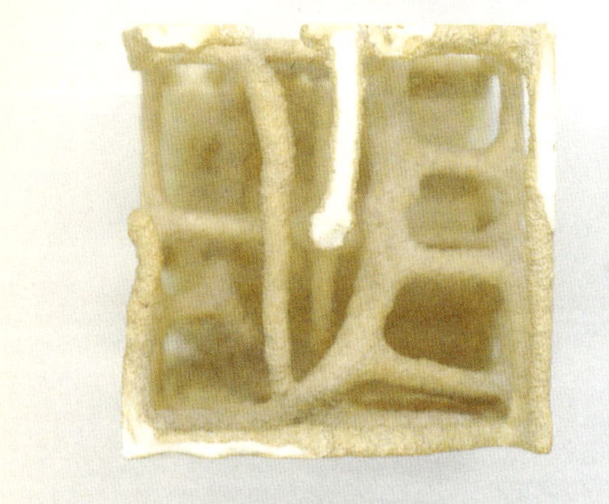

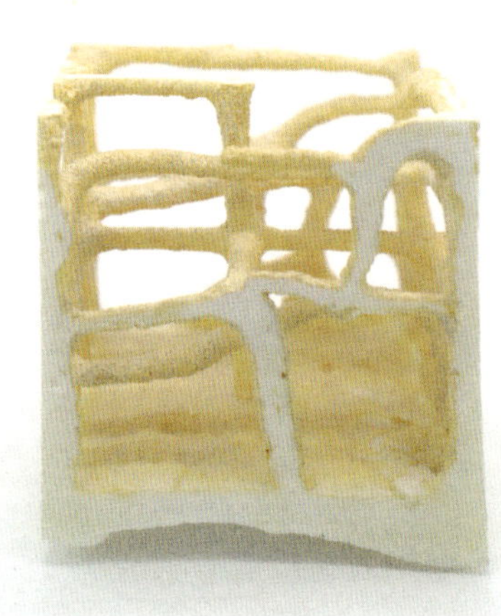

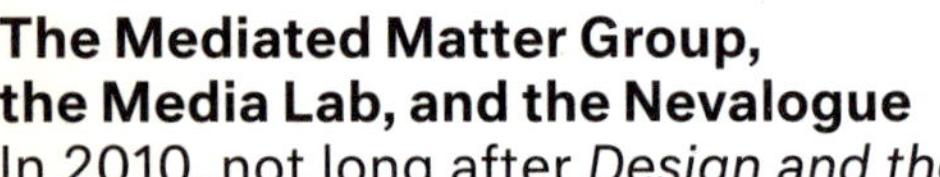

figs. 5–8 Neri Oxman and The Mediated Matter Group. Swarm Fabrication Studies. 2017. Polyurethane foam, 6 × 6 × 6 in. (15 × 15 × 15 cm)

Structures formed by a swarm of ants directed with UV lighting.

The results are complex, articulated, fine-tunable demo objects made with one process and often a single material.

One principle emerges from MMG's motley array of techniques: the process and the materials are the object—of research, of investment, of passion—or, put simply, the end shapes the means. Oxman draws on Gottfried Semper's ideas on this subject: "That matter is secondary to shape constitutes the fallacy of design after craft. By nature, and in its rite, the material practice of craft is informed by matter, its method of fabrication, and by the environment."[8] The ends, one could add, are shaped by culture, one that Oxman has designed for herself and for MMG to be attuned to the systems of nature, a great architect and builder that can be engaged as a powerful partner. Engaging nature, however, requires all hands on deck, a new creative culture that foregrounds symbiotic collaboration among disciplines and practices. If a Venn diagram, with its eureka moment of encounter and superimposition, sufficed to describe the tools necessary to define a new way to design and make, it is inadequate when the goal is to distill a new interdisciplinary approach to applied creativity—one that might also entail unmaking. That purpose is better served by the complexity of the Krebs metabolic cycle (pages 16–17, **figs. 12–14**).

The Mediated Matter Group, the Media Lab, and the Nevalogue

In 2010, not long after *Design and the Elastic Mind*, Oxman established The Mediated Matter Group (MMG) at the MIT Media Lab. MMG has been the center of her collaborative practice ever since. The Media Lab, founded in 1985 by Nicholas Negroponte and Jerome Wiesner, is organized in several interdisciplinary—or, as its former director Joi Ito would say, *antidisciplinary*—research groups that bring together advanced technology and engineering with science, art, and design. The groups' names—Tangible Media, Sculpting Evolution, Fluid Interfaces, for example—combine the familiar and the cutting edge, suggesting their future-facing and yet still human-scale research strands.[5] The MMG team currently comprises two computer scientists focusing on computational design and artificial intelligence (Christoph Bader and Jean Disset), a multimedia designer (João Costa), a product designer (Felix Kraemer), three architects (Nic Lee, Joseph H. Kennedy Jr., and Ramon Weber), a biologist (Sunanda Sharma), a biomedical engineer (Rachel Soo Hoo Smith), a mechanical engineer (Michael Stern), an artist (Ren Ri), a marine scientist (James C. Weaver, as a research affiliate), and a weaver (Susan Williams). The studio includes a biosafety level 2 wet lab (allowing work with moderately hazardous pathogens)—a first in a design studio—and the group has worked with tools, species, and materials as diverse as silkworms, incandescent glass, ants, bitmap printers, bacteria, robotic arms, and bees (**figs. 5–10**), to name just a few. Broadly stated, as Oxman has described it, the team of researchers is focused on "inventing and developing new design tools, techniques, and technologies that have the potential to redefine the way we make things."[6] Oxman also often refers to a Venn diagram (**fig. 11**) that shows the intersection of four research areas—computation, additive manufacturing, materials engineering, and synthetic biology. "Computation allows us to design complex forms with simple code," she has explained. "Additive manufacturing lets us produce parts by adding material rather than carving it out. Materials engineering lets us design the behavior of materials in high resolution. Synthetic biology enabled us to design new biological functionality by editing DNA."[7]

Oxman's polymathic background made her an ideal fit for the Media Lab, a unique academic sandbox mixing art, design, technology, engineering, science, and entrepreneurship. Ito and the distinguished alumnus John Maeda espoused a theory, based on a book favored by MIT students and professors, Rich Gold's *The Plenitude: Creativity, Innovation, and Making Stuff*, that they called the Creative Compass.[9] The compass, rendered in four quadrants, includes art and design in addition to engineering and science. Ito argued that the Media Lab needed more teachers and students who could integrate all four quadrants. Maeda, while the president of Rhode Island School of Design, from 2007 to 2013, lobbied the United States government to add art to the traditional STEM (science, technology, engineering, and math) curriculum, making it STEAM.[10] Such interdisciplinary integration was uncommon at the time in U.S. academic circles, which were steadfast in mimicking the division between sides of the brain and espoused a contrived hierarchization of the disciplines (science on top of engineering, art on top of design). Their integration was instead natural to architects, like Oxman, who were steeped in European education, more theoretical and open to contamination.

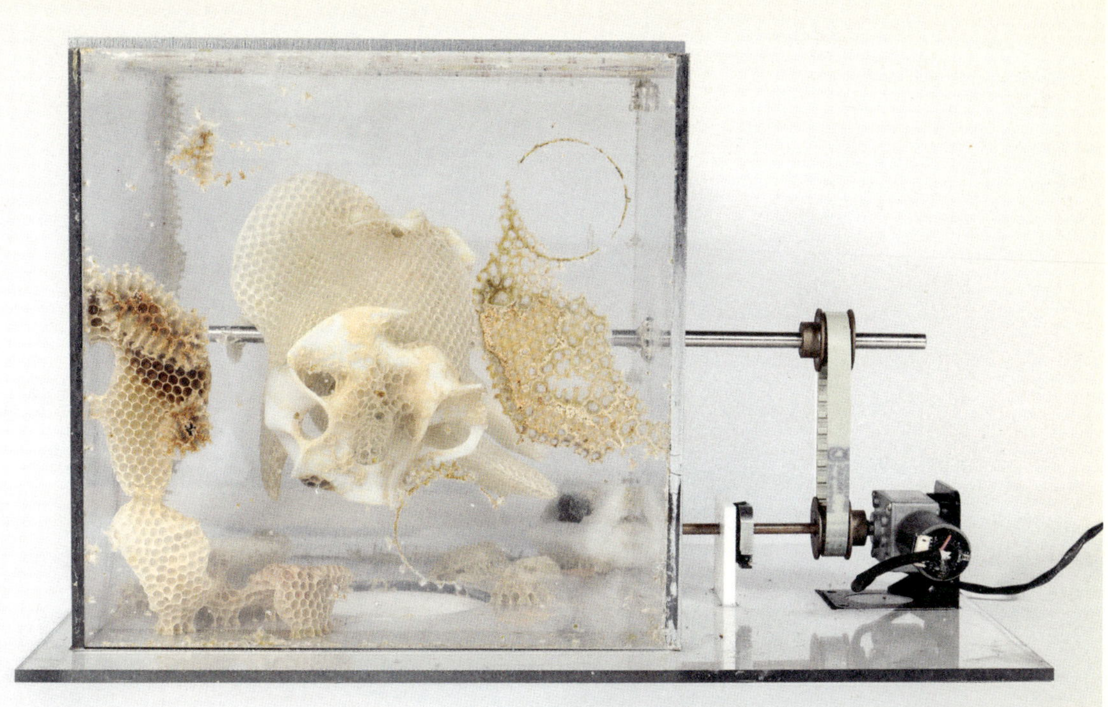

ig. 9 Neri Oxman and The Mediated Matter ìroup. Beehive Studies. 2018. Natural eeswax, 3D-printed resin, stainless steel, olycarbonate, and prepared DC motor, 5 ½ × 19 ½ × 11 ¾ in. (40 × 50 × 30 cm)

ìemains of beehive generated by the nteraction of honeybees and a rotation nechanism.

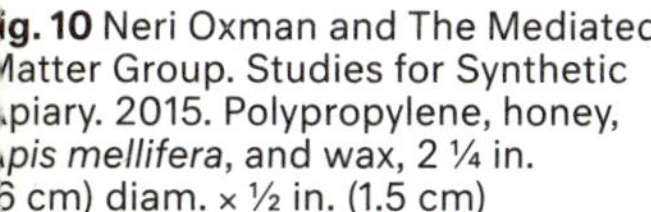

ig. 10 Neri Oxman and The Mediated ⁄latter Group. Studies for Synthetic piary. 2015. Polypropylene, honey, *pis mellifera*, and wax, 2 ¼ in. 5 cm) diam. × ½ in. (1.5 cm)

small sample from the first hive, vhich was installed on the roof of he MIT Media Lab.

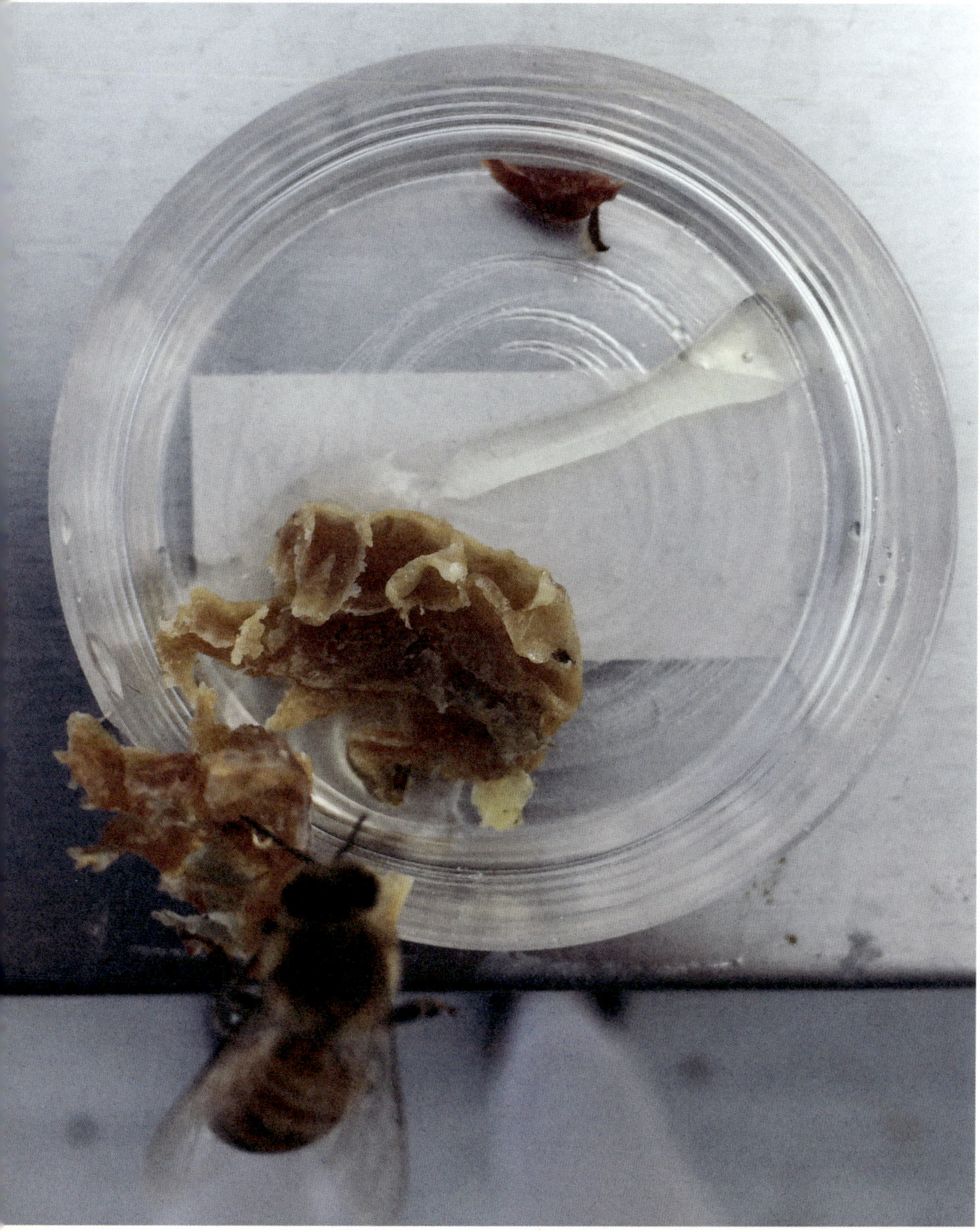

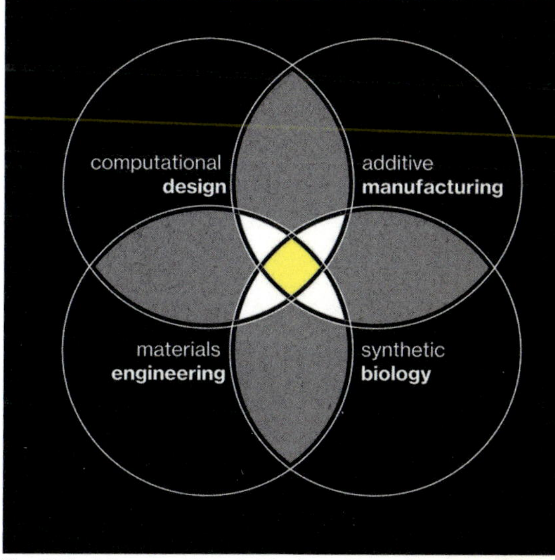

fig. 11 Neri Oxman. Material Ecology Venn diagram. 2015

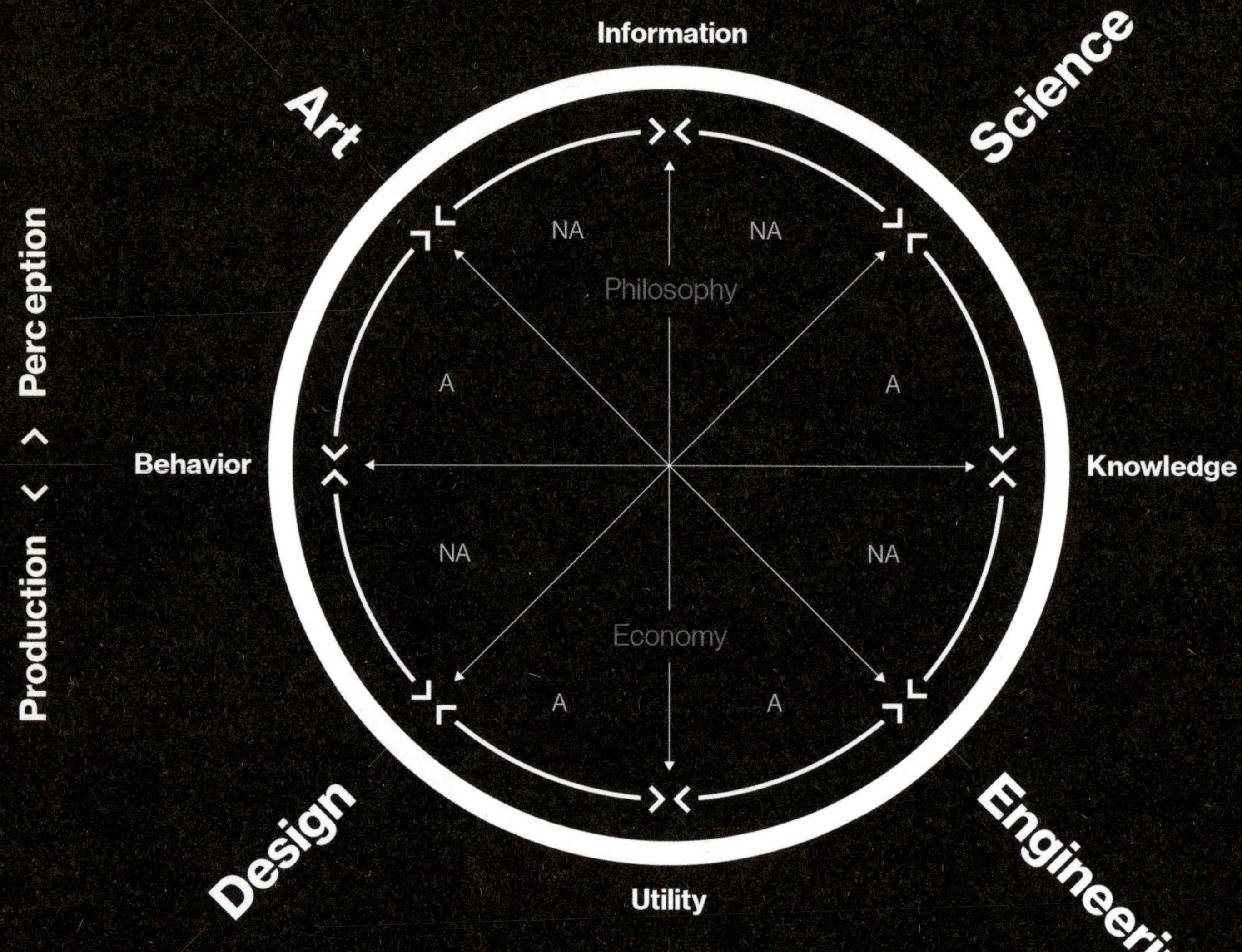

fig. 12 Neri Oxman. The Krebs Cycle of Creativity I (Domains). 2016

The Krebs Cycle of Creativity is a framework that considers the domains of art, science, engineering, and design as synergetic forms of thinking and making, in which the input from one becomes the output of another. Inspired by and named after the Krebs cycle, which describes the chemical reactions used by organisms that inhabit oxygenated environments, Oxman's version replaces the carbon compounds with the four modalities of human creativity.

The transitions from one domain to another generate intellectual energy, or CreATP. Science explains and predicts the world around us, converting information into knowledge; engineering applies scientific knowledge to the development of solutions for empirical problems, converting knowledge into utility; design produces solutions that maximize function and augment human experience, converting utility into behavior; and art questions human behavior and creates awareness of the world around us, converting behavior into new perceptions of information, presenting anew the data that initiated the cycle, in science.

The KCC's vertical axis leads from sky to earth, from the theoretical (or philosophical) to the applied (or economic). The north marks the climax of human exploration into the unknown; the south marks the products and outcomes this exploration produces. The horizontal axis leads from nature to culture, from understanding—describing and predicting phenomena within the physical world—to creating new ways of using and experiencing it.

fig. 13 Neri Oxman. The Krebs Cycle of Creativity II (Units). 2016

The Krebs Cycle of Creativity II replaces the domains of the first iteration with realms that explain, predict, change, and perceive the world, each of them identified by a unit associated with it. In order for design objects, structures, and interventions to be able to inform and influence the natural ecology, designers must be able to freely move between ways of seeing and making in the world and the units that build them—physical units (such as the elements of the periodic table), digital units (pixels), and biological units (genes).

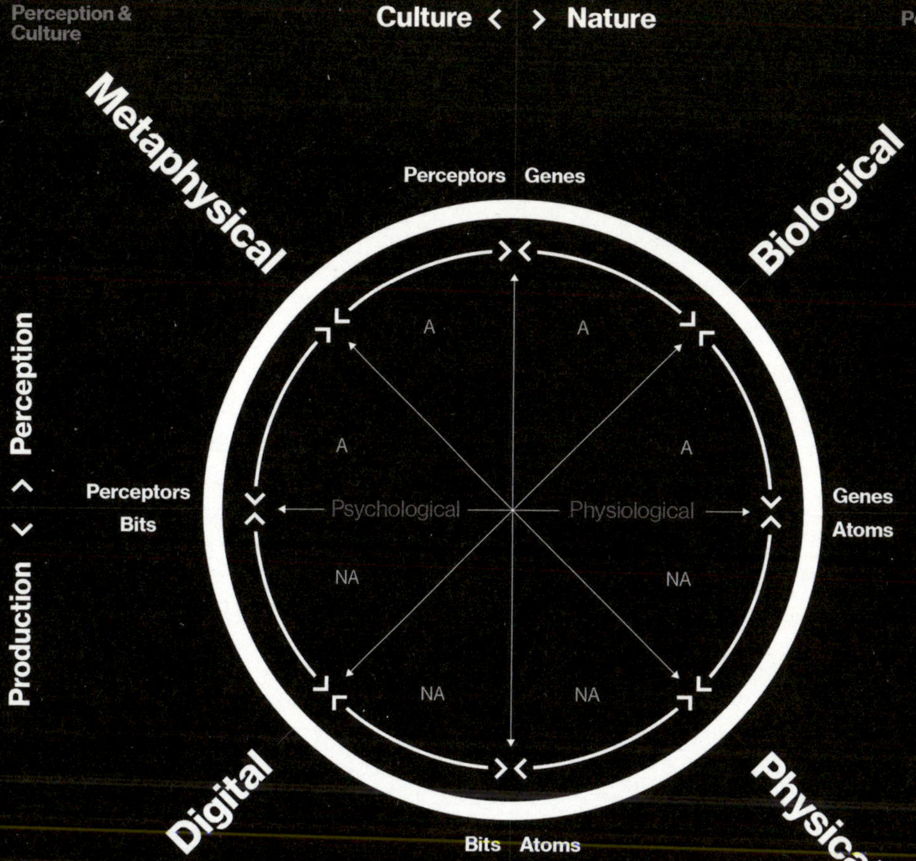

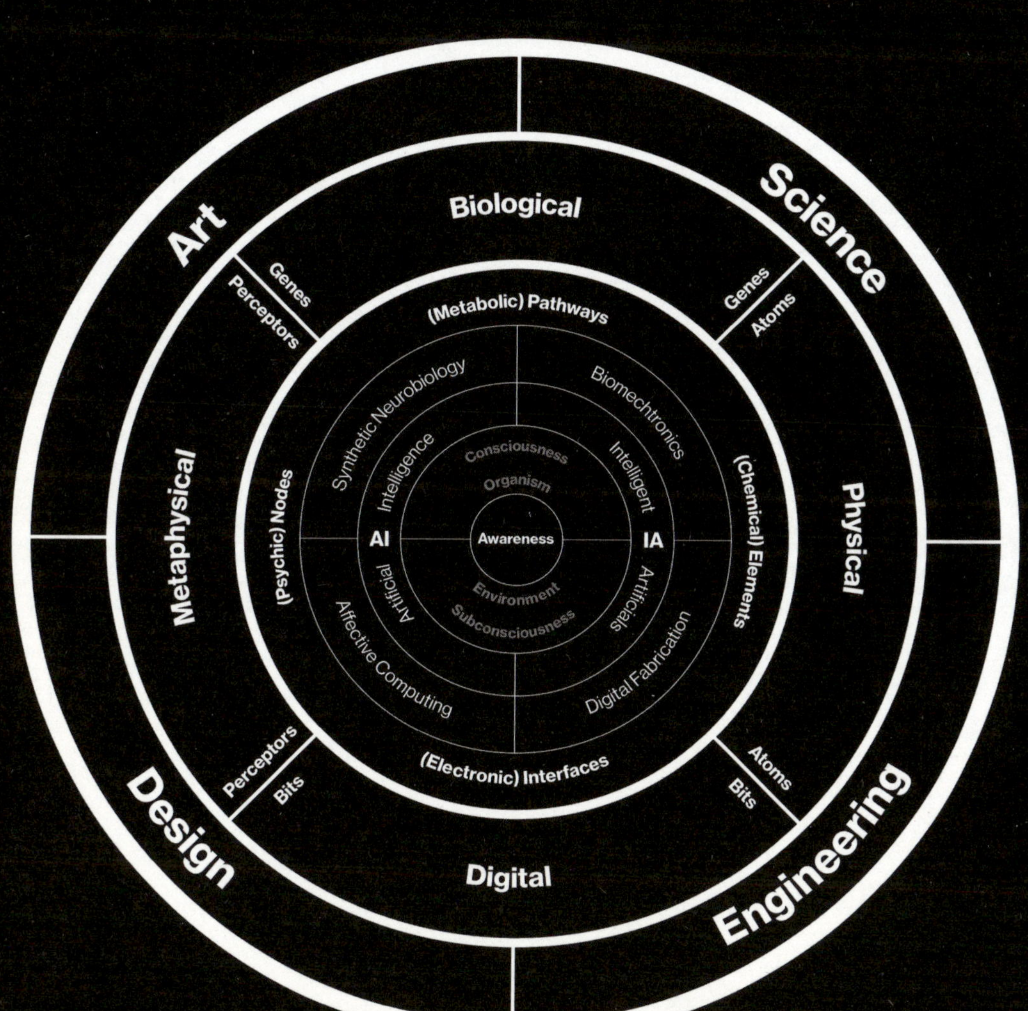

fig. 14 Neri Oxman. The Krebs Cycle of Creativity III (Domains + Units). 2016

The Krebs Cycle of Creativity III combines the first two iterations in a diagram of ideal interdisciplinary design practice, with unencumbered transitions among the realms and domains laid out in KCC I and KCC II. It can be interpreted from top to bottom as it relates to the organisms (top, expressed through metabolic pathways) and/or its environment (bottom, expressed through interfaces) and left to right as it relates to forms of Artificial Intelligence or Intelligent Artifacts.

KCC III was inspired by Walter Gropius's diagram of the Bauhaus curriculum, created in 1922, in which divergent trajectories are represented as a cycle, thus suggesting the school's mission of bringing together students from different disciplines in order to reform art, design, and society. The word *Bau*, in the center, is German for "building"; KCC III replaces that word with "awareness," suggesting that the future of design cannot address only the built environment and the objects designed for and deployed in it; instead, designers must combine the grown and the built, the natural and the artificial, the organism and its environment in novel ways that sustain, augment, and nourish our planet and its diverse inhabitants, in a bona fide Material Ecology.

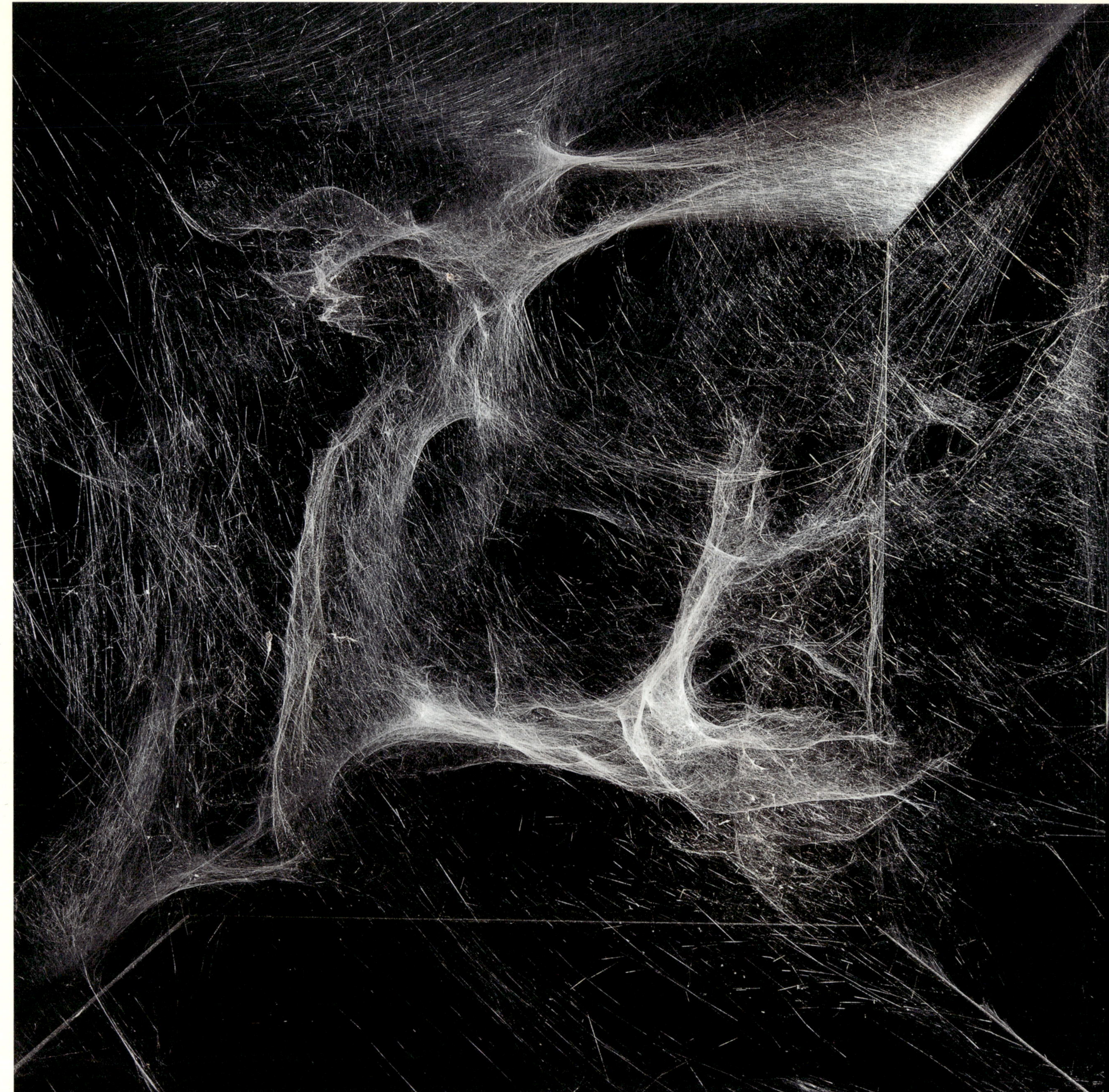

fig. 15 Tomás Saraceno. *Hybrid semi-social solitary musical instrument Arp87 built by: a couple* Cyrtophora citricola*—one month, one* Agelena labyrinthica*—two months, one* Cyrtophora moluccensis*—two weeks, and one* Tegenria domestica*—four months, turned 4 times 180 degrees on Z axis*. 2015.
Spider silk, carbon fiber sticks, and plexiglass, 37 ¼ × 37 ¼ × 37 ¼ in. (94.5 × 94.5 × 94.5 cm)

Oxman in fact took the compass a step further in the Krebs Cycle of Creativity (previous pages). In her scheme—which she calls a cartography but which in reality is a philosophy—she postulates not just adjacency and coincidence but entanglement. She outlined a metabolic biofeedback cycle that drives the interdependence of the areas of research necessary for applied creativity: "Knowledge can no longer be ascribed to, or produced within, disciplinary boundaries, but is entirely entangled," she wrote, citing the Krebs cycle—which normally describes the chemical reactions that the cells of oxygen-breathing organisms use to generate energy—as a model.[11] One of the great advantages of the Krebs cycle is that it could be expanded to include the whole MIT community, boarded at different stations by scientists and artists already working in all four spheres—artists like Tomás Saraceno, for instance, who spent time as a visiting artist at MIT's Center for Art, Science, and Technology (CAST) studying spiderwebs (**figs. 15, 16**), and Agnieszka Kurant, who has deepened her understanding of the future of labor and AI through contemplating the collective intelligence of termites (**fig. 17**), or synthetic biologists like Christina Agapakis, who moved from describing cellular organisms to imagining brand-new ones with new functions—and teaming up with designers as a matter of professional course. In a 2017 interview, Oxman offered a revised definition of Material Ecology that also emphasizes universal synthesis. She called it "a holistic approach to design which considers all environments—the built, the natural, and the biological—as one, postulates that any designed physical construct is—by definition—an integral part of our ecology. A practicing material ecologist will therefore engage multiple disciplines—computational design, digital fabrication, synthetic biology, the environment, and the material itself—as inseparable and harmonized dimensions of design."[12]

In 2019, Oxman began a sabbatical from MIT and in early 2020 launched a new practice in New York. The team includes some members of MMG and replicates its diversity, with its members representing six key disciplines: "Two chemists, two biologists, two product designers, two mechanical engineers, two scientists, like Noah's Ark, and architects galore!" There is also a wet lab "with creatures from all six kingdoms"—plants, animals, protists, fungi, archaebacteria, and eubacteria—"and robots."[13] In the Media Lab environment, Oxman and MMG conduct experiments under the gaze of interested and invested patrons. In the new enterprise, patrons will become clients asked to concede their interests to those of the entity that Oxman considers her primary client: nature.

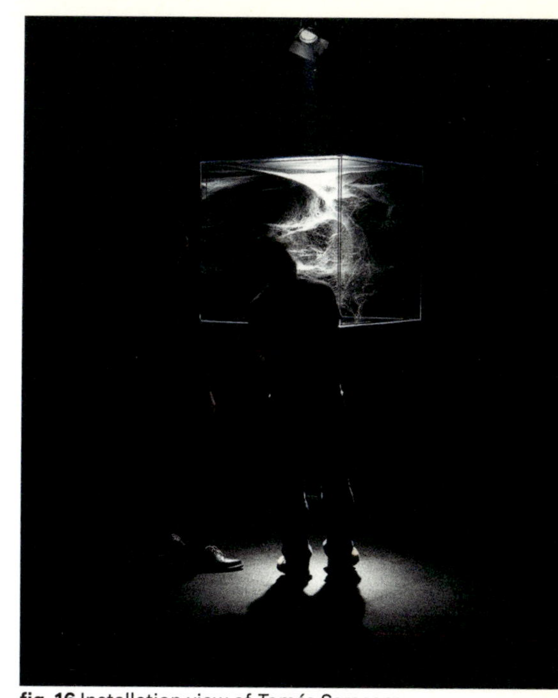

fig. 16 Installation view of *Tomás Saraceno: Hybrid solitary . . . Semi-social quintet . . . On Cosmic webs . . .*, Tanya Bonakdar Gallery, New York, March 26–May 2, 2015

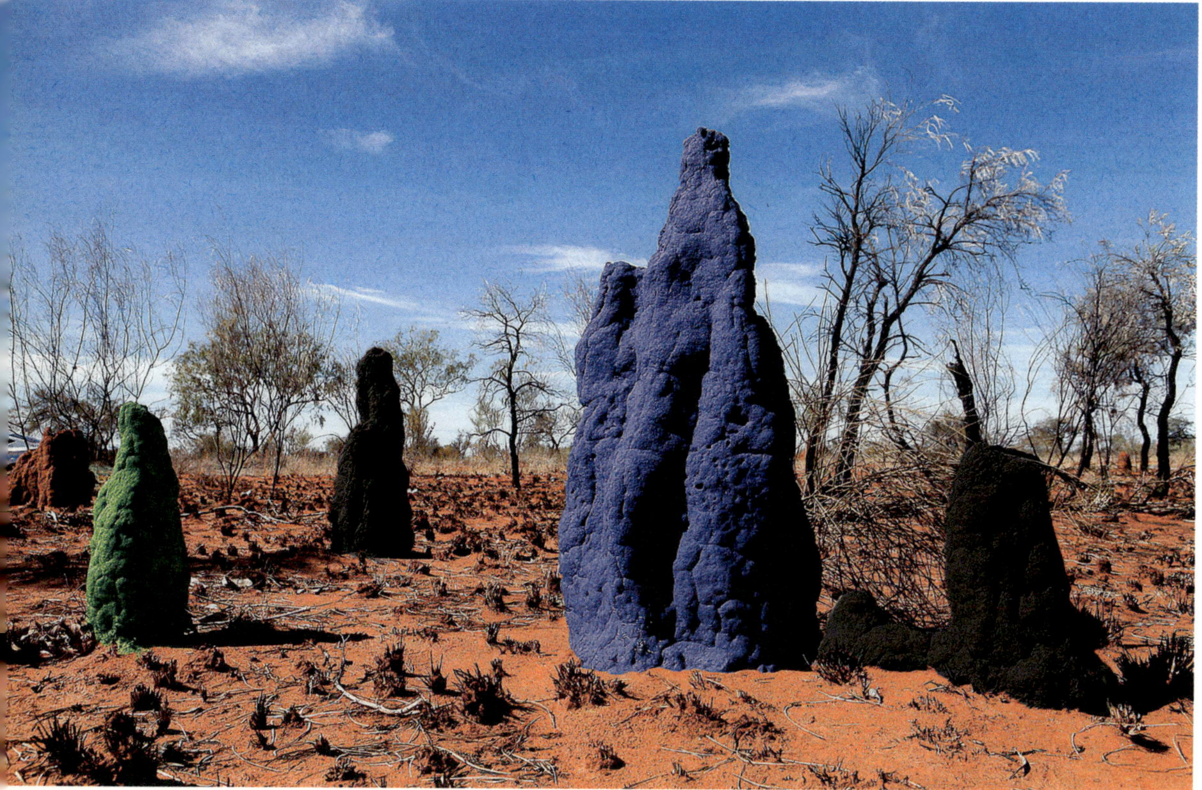

fig. 17 Agnieszka Kurant. *A.A.I.* 2014–present. Termite mounds built from colored sand, gold, and crystals, dimensions variable

[a]	[b]
affirmative	critical
problem solving	problem finding
design as process	design as medium
provides answers	asks questions
in the service of shareholders	in the service of society
for how the world is	for how the world could be
science fiction	social fiction
futures	alternative worlds
fictional functions	functional fictions
change the world to suit us	change the us to suit the world
narratives of production	narratives of consumption
anti-art	applied art
research for design	research through design
applications	implications
design for production	design for debate
fun	satire
concept design	conceptual design
consumer	citizen
makes us buy	makes us think
innovation	provocation
ergonomics	rhetoric
user-friendliness	ethics

fig. 18 Dunne & Raby (Anthony Dunne and Fiona Raby). "A/B Manifesto." 2009

Oxman has crystallized her practice's naturecentric philosophy in a series of principles to which human clients will be asked to commit. The principles have undergone several revisions and are still in flux. The most recent version is "Nine Commandments for a Material Ecology," a nevalogue, or statement in nine points. The commandments are constructive, a declaration of intent for a new design practice. The early versions were descriptive of Oxman's intellectual process and juxtaposed old and new, considering design before and after digital computation and the advent of a so-called Century of Biology at the turn of the millennium. In that, they sympathized with the growing pains of the designers and theoreticians Anthony Dunne and Fiona Raby in moving design away from the problem-solving paradigm and toward criticality and inquisitiveness, a path spelled out in their 2009 "A/B Manifesto" (**fig. 18**), twenty-one juxtapositions highlighting the difference between the past, present, and future of design in the postindustrial, pandigital era.[14] Here is Oxman's nevalogue, followed by selected quotes from previous versions of the principles, along with my own notes in brackets.

Nature as Client
"The natural environment at large constitutes the 'client' for every commissioned project, as well as its 'site' and material source."

Growth over Assembly
"Nature grows things. We will be able to create objects that will respond to their users, adapt to their environment, and even grow over time after they have been printed." [This principle conceptually moves design and production into the new age of biology, from the assembly line to the wet lab.]

Integration over Segregation
"The typical facade of a building . . . is made up of discrete parts fulfilling distinct functions. Stiff materials provide a protective shell, soft materials provide comfort and insulation, and—in buildings—transparent materials provide connection to the environment. In contrast, human skin utilizes more or less constant material constituents for both barrier functions (small pores, thick skin on our backs) and filtering functions (large pores, thin skin on our face). [In an ideal object,] barrier and filtering functions are integrated into a single material system that can at any point respond and adapt to its environment." [Oxman often describes her work as noncompositional, a whole in which the same material articulates dynamically the required functions.]

Non-Human-Centered Design
"The group considers all living creatures as equals." "The group aims to shift human-centric design to a design culture focused on conserving, improving and augmenting the natural environment though novel technological developments." [The past decade has seen an increasing preoccupation with interspecies design and questioning of anthropocentric values, and Oxman's new practice chooses this position as its baseline.]

Difference over Repetition
"Industrial products generated out of machines . . . consist of repeatable parts with identical properties." "Comprehending difference enables us to design repetitive systems—like bone tissue—that can vary their properties according to environmental constraints. As a consequence of this new approach we will be able to design behavior rather than form." [The form-giving design of yesteryear yields to the new concept of formation, in which a system adapts and performs—behaves.]

Decay over Disposal
"The Practice implements design workflows in which matter is synthesized by an ecosystem, implemented in human systems, and consumed by the same ecosystem upon obsolescence. . . . Designed decay is the process by which matter is programmed to rejoin an ecosystem's resource cycle and fuel new growth." [An embrace of circularity, this principle affirms the role of the architect within the ecosystem.]

Activist Design
"Any design commission becomes associated with a particular technology—invented or improved upon by the Lab—which embodies the value system associated with the group, and is directly linked to design and construction processes relevant to the commission." [Several enterprises, nonprofit and for-profit, are currently trying to integrate ethics and social and environmental responsibility into their practices, and to pass these values on to their customers—a brand of activism from within the system. Oxman's group will use technology as a vessel for change.]

System over Object
"The product—be it a product, a wearable device, or a building—is considered part of a *system* of interrelations between natural and designed environments including interactions between the entity and the human body as well as the entity and its environment."

Technology over Typology
"Moving away from the taxonomic classification commonly found in buildings and urban places, . . . topology—the way in which constituent parts are interrelated or arranged—is the driving force behind the design process, promoting condition-based programming as the approach for organizing spaces and making places."[15]

The Roots of Material Ecology: Critical Design, New Materiality, Performance-Based Design

This last principle in the "official" nevalogue represents a vow to approach every task at hand without being bogged down by past models, and yet Oxman's multifaceted projects can naturally be situated in various historical contexts. Critical Design, otherwise known—perhaps unfortunately—as Speculative Critical Design (SCD), accommodates the exploratory aspect of her work and the development of new prototypes, materials, and techniques in a speculative environment; these new design tools and technologies, she has said, "have the potential to redefine the way we make things, and *seeding them within speculative design contexts*."[16] Emerging around the turn of the millennium, Critical Design was centered on the Design Interactions course chaired at the Royal College of Art in London by Dunne between 2005 and 2015, at the apex of his and Raby's deep influence on the design world (**figs. 19–22**).[17] The products of Critical Design were, in most cases, objects—from fictional products to interfaces, interiors to buildings and infrastructures—designed in response to scenarios that imagined the possible future consequences of our present choices. They were produced by interdisciplinary teams that included experts on ethics, the life sciences, politics, philosophy, and the economy, among other fields.[18] Their objective was to instigate what the foresight expert Stuart Candy has defined as a "preferable future," in contrast with those that are "probable," "plausible," and "possible."[19]

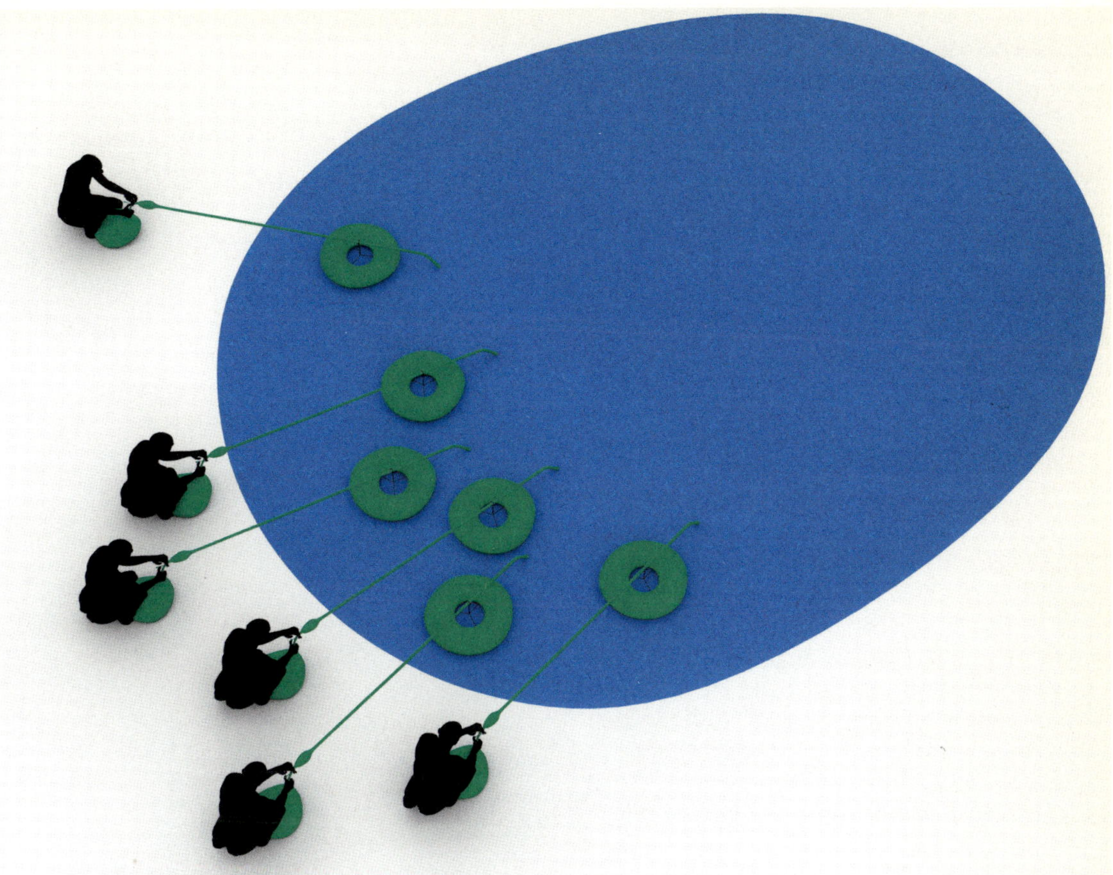

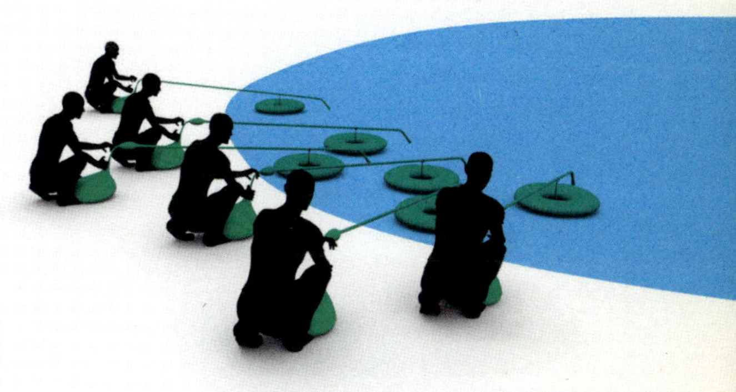

figs. 19, 20 Dunne & Raby (Anthony Dunne and Fiona Raby). Algae Digesters, from Designs for an Overpopulated Planet: Foragers. 2009. Video and prototypes. The Museum of Modern Art, New York. Gift of The Contemporary Arts Council of the Museum of Modern Art

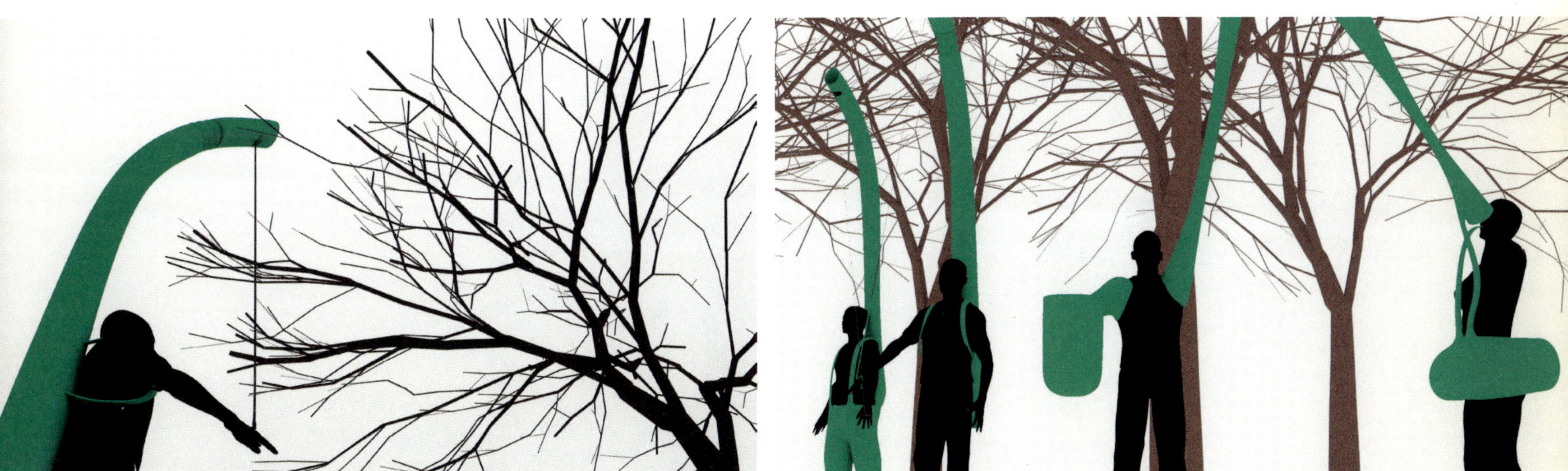

figs. 21, 22 Dunne & Raby (Anthony Dunne and Fiona Raby). Tree Processor/Digester, from Designs for an Overpopulated Planet: Foragers. 2009. Video and prototypes. The Museum of Modern Art, New York. Gift of The Contemporary Arts Council of the Museum of Modern Art

Much like Italian Radical Design of the 1960s and 1970s, Critical Design began at a moment of pivotal technological, sociopolitical, economic, and cultural change—in this case the ingraining of the digital era—and blossomed until it quickly fizzled into the academic mainstream. After the slow extinction of the RCA course, the Critical Design baton has passed to the Design Academy Eindhoven (already a long-time bona fide runner-up), to the Haute école d'art et de design (HEAD), in Geneva, and to the Aalto University, in Helsinki, among others. Despite its proliferation, much current SCD, sprinkled in design schools worldwide, has featured earnest but inconsequential exercises and clichéd storytelling. Overawed by urgent worldwide environmental and sociopolitical crises, and by Silicon Valley's gung-ho moonshots into the future, it has lost its bite. Further, it is now clear that there is little patience for fiction (at least in design practice), and no time to waste: the world needs to know that speculations, however far-fetched, can actually happen. The most significant heirs to the SCD approach then are the programs that start from foundations of engineering and science and build a bridge to design. Polytechnic institutions such as Carnegie Mellon University, Cornell University, and MIT model such approaches, and Oxman and MMG are exemplary. Creating "designs that have a profound connection with a specific ecosystem," they choose a preferable future and go for it with all the ammunition in their arsenal, making it first possible (as in an ambitious utopian plan for Tokyo in 2200, commissioned by the Mori Art Museum and the Mori Building Company [**fig. 23**]), then plausible, and hopefully in the end probable.[20]

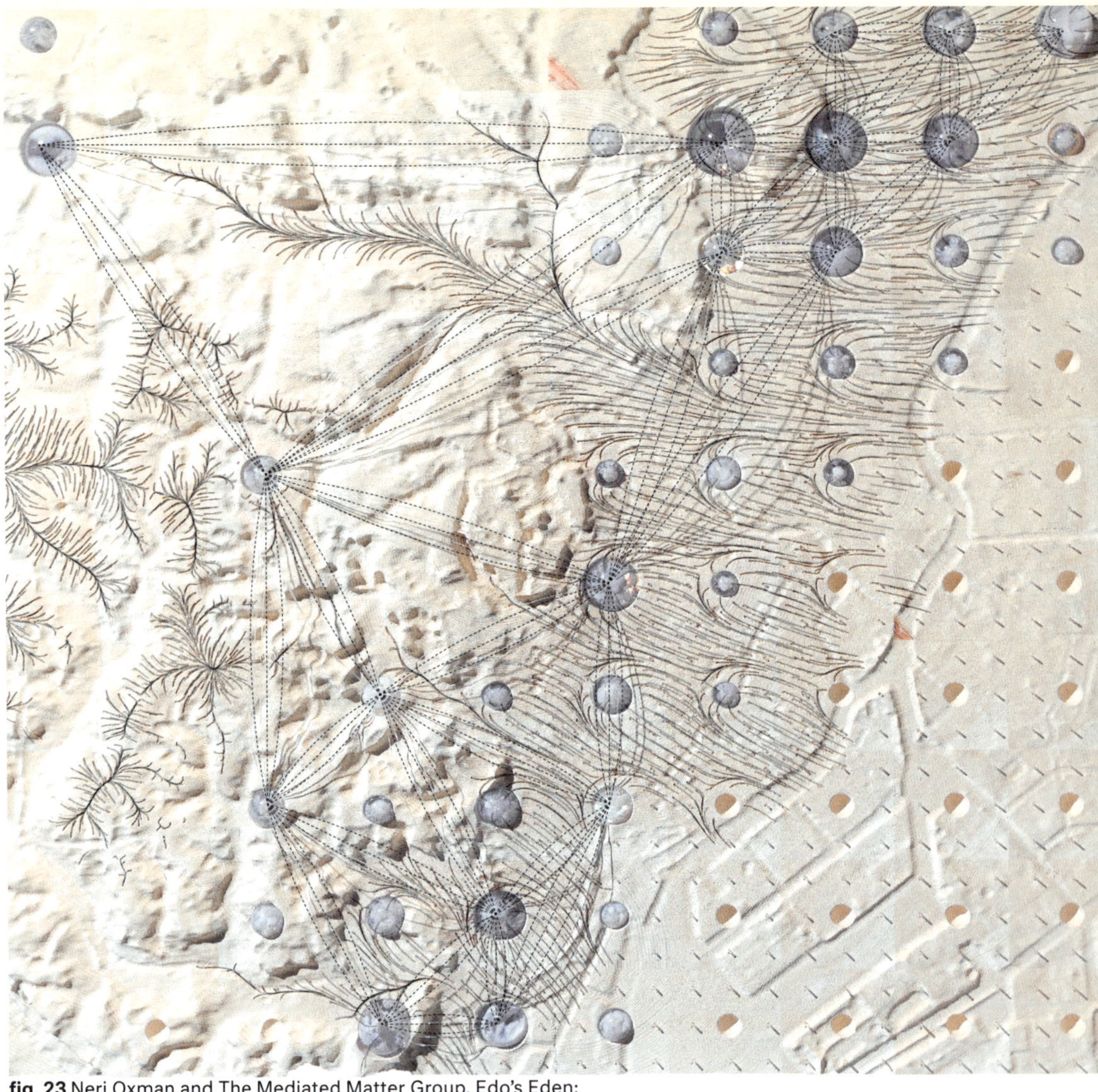

fig. 23 Neri Oxman and The Mediated Matter Group. Edo's Eden: Tokyo 2200. 2019. Wood (hinoki cypress from the Tono region of Japan), acrylic, and photopolymers, 39 ¼ × 39 ¼ × 51 in. (100 × 100 × 130 cm)

A single hinoki tree transformed into an urban-design model, for a future in which a new typology will grow out of the symbiotic relationship between nature- and human-made environments. The meganature is a megastructure with a controlled environment; the model is both time capsule and tree-seed bank.

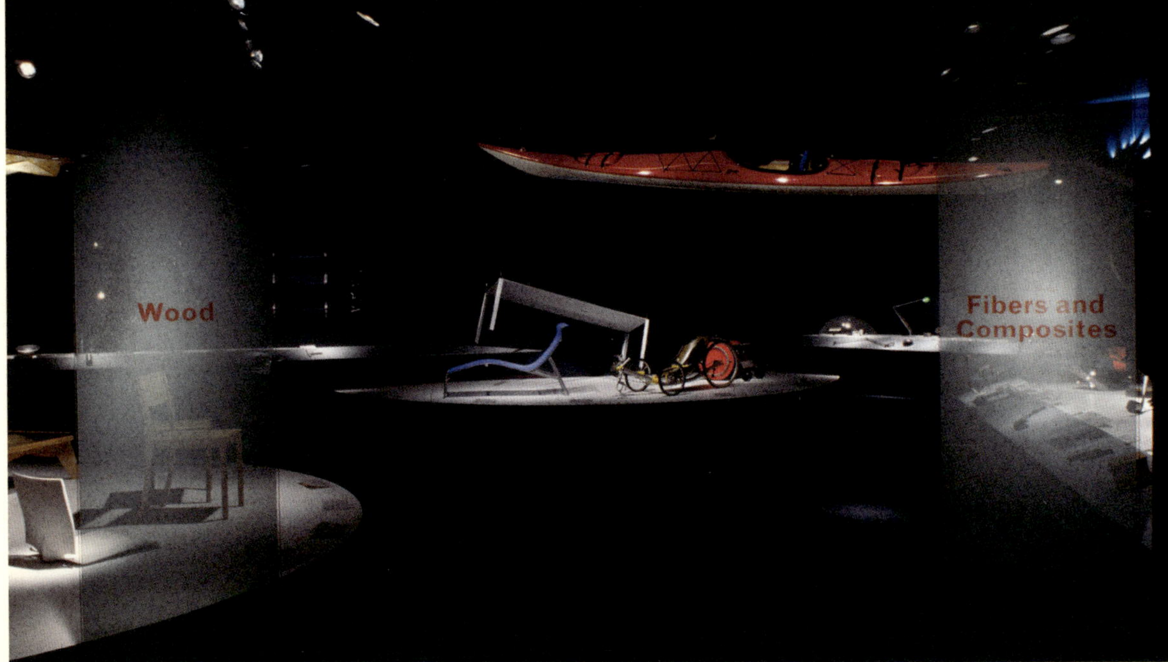

fig. 24 Installation view of *Mutant Materials in Contemporary Design*, The Museum of Modern Art, New York, May 25–August 22, 1995

Opposite, top:
fig. 25 Oron Catts, Ionat Zurr, and Guy Ben-Ary. The Pig Wings Project, from the Tissue Culture & Art Project. 2000–01. Developed at SymbioticA, The Art and Science Collaborative Research Laboratory, School of Anatomy and Human Biology, The University of Western Australia. Pig mesenchymal cells (bone-marrow stem cells) and PGA and P4HB biodegradable, bioabsorbable polymers, dimensions variable

Wings grown from the mesenchymal cells of pigs and shaped to mimic the wings of a bat, an imaginary angel, and a pterosaur.

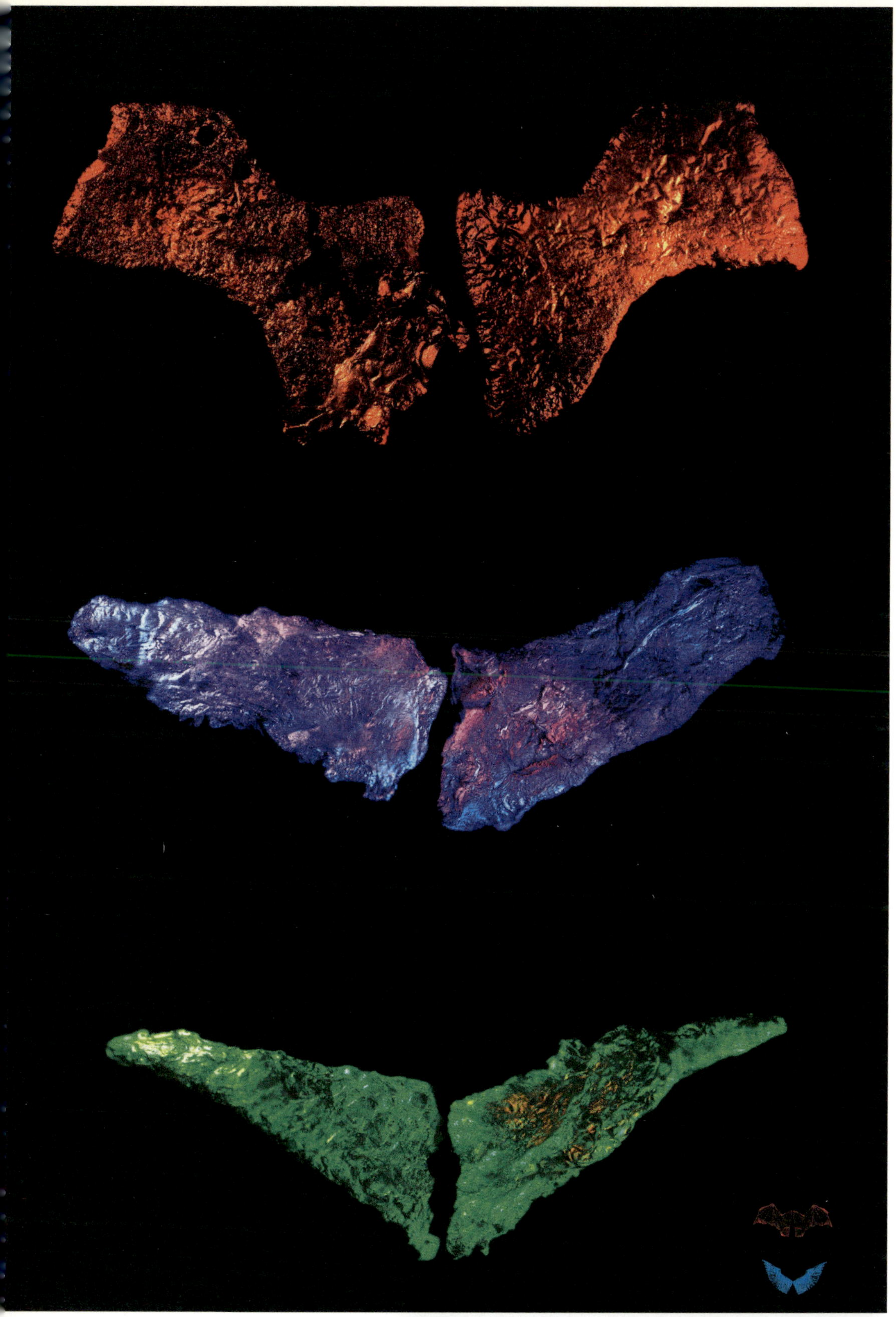

Another root of Oxman and MMG's work can be found in the design of materials. In the modern era, the evolution of the materials of architecture and design has provoked and supported extraordinary mutations in our built environment at all scales. Materials science, moreover, has had a significant impact on design culture at large, freeing not only the floors of buildings and the versatility of objects but also the imaginations of designers and architects—and giving each modern age its own distinct visual and spatial character. In 1995 I organized my first MoMA show, *Mutant Materials in Contemporary Design* (**fig. 24**). The show's thesis was that in a much welcome coup d'état, designers were able to take control of a new generation of materials and begin the design process with the design of the materials themselves, instead of receiving the offerings of engineering and chemical companies. Since 1995 designers' takeover of the material realm has evolved even further, from new resins and composites all the way to biopolymers and living tissues. Taking advantage of the curiosity and openness of synthetic biologists, chemists, physicists, and other scientists, designers like Oron Catts and Ionat Zurr (**fig. 25**), Suzanne Lee (**fig. 26**), Alexandra Daisy Ginsberg, Natsai Audrey Chieza, and Gionata Gatto, among others, have initiated a fertile exchange with science (with biology in particular) and ventured into territories new not only for design but for the whole world.[21] Oxman described this same evolution in her 2013 essay "Material Ecology," in the section "Designing (with) Nature: Towards a New Materiality": "Today, perhaps under the imperatives of growing recognition of the ecological failures of modern design, inspired by the growing presence of advanced fabrication methods, design culture is witnessing a *new materiality*. . . . Examples of the growing interest in the technological potential of innovative material usage and material innovation as a source of design generation are developments in biomaterials, mediated and responsive materials, as well as composite materials."[22] Indeed, Oxman and MMG have put such new materials to powerful use in their projects, freeing both their imaginations and their designs' potential.[23]

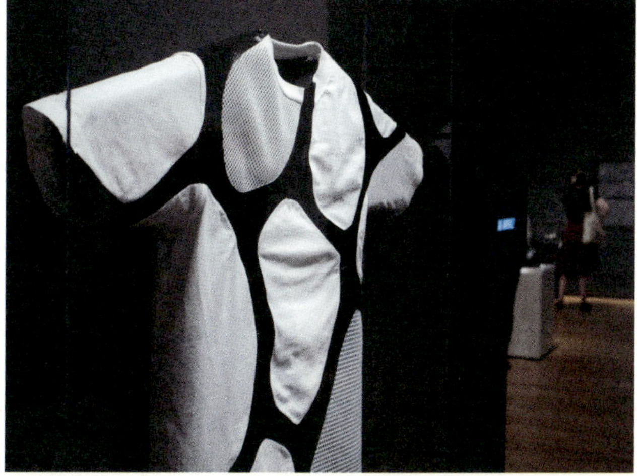

fig. 26 Suzanne Lee and Amy Congdon, Modern Meadow. Zoa. A new animal is born. 2017. Zoa biofabricated leather and cotton, 37 × 31 in. (94 × 78.7 cm). Installation view in *Items: Is Fashion Modern?*, The Museum of Modern Art, New York, October 1, 2017–January 28, 2018

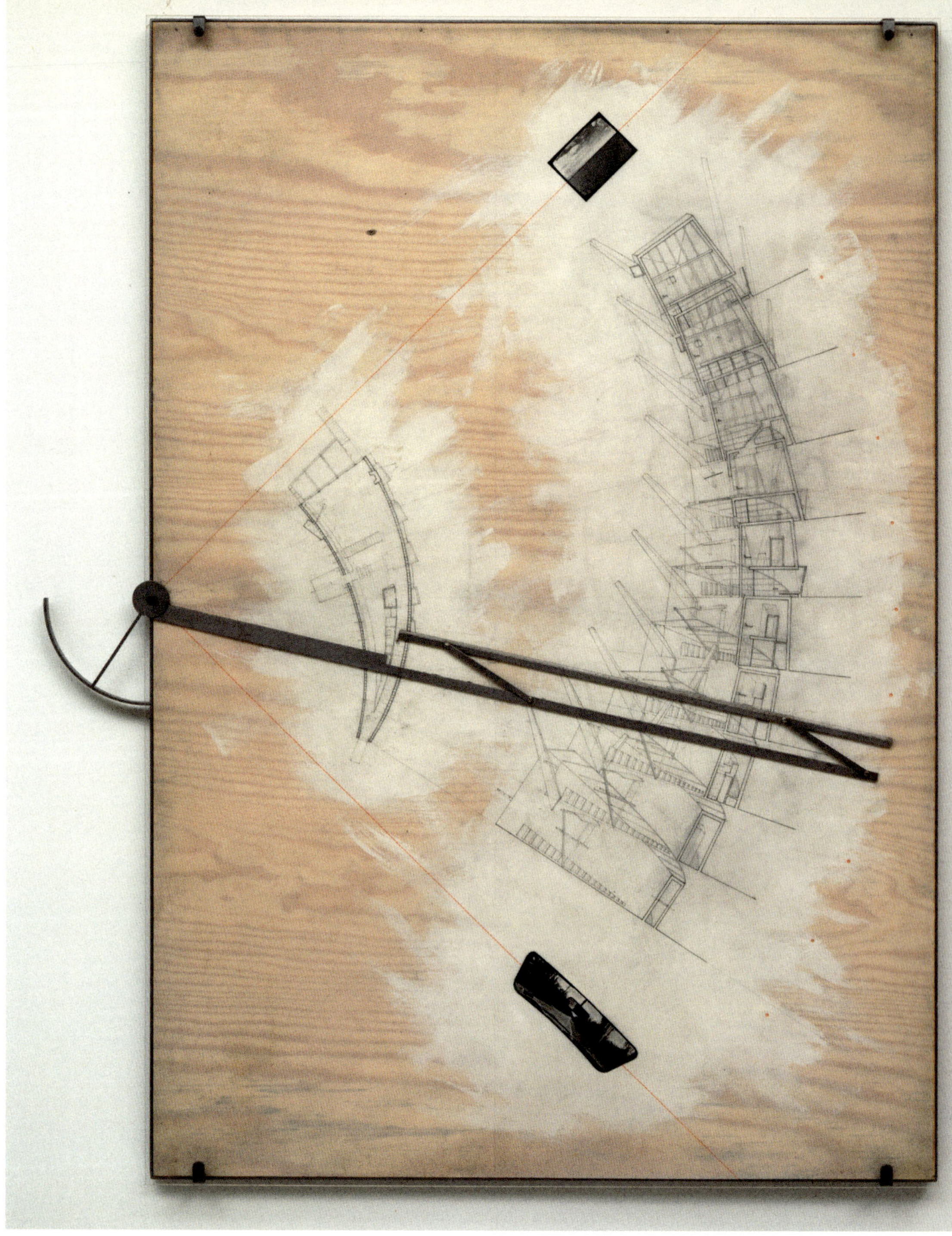

fig. 27 Diller + Scofidio (Elizabeth Diller and Ricardo Scofidio). Slow House Project, North Haven, New York. 1989. Plan of lower level and sections. Computer-generated print on frosted polymer sheet with graphite and colored ink, mounted on painted wood with metal, 47 ⅝ × 36 ½ × 1 ½ in. (121 × 92.7 × 3.8 cm). The Museum of Modern Art, New York. Marshall Cogan Purchase Fund and Jeffrey P. Klein Purchase Fund

A final historical thread in Oxman's work comes out of the exquisite synthesis of computation and inspiration that is performance-based design. Asked in 2019 to pick any object in MoMA's collection to discuss in a BBC podcast episode, Oxman chose the most flowing of all architectures—the Austrian architect Frederick Kiesler's legacy project Endless House (1950–60; pages 38–39, **fig. 1**; page 45, **fig. 13**).[24] As Oxman notes in the podcast, so many aspects of Kiesler's intellectual and professional life drew her to him—his spirituality and his nondenominational ease with art and artists among them. But more than anything, she admires the specificity of his architecture—how it adapts locally to changing conditions. The formal, if not conceptual, opposite of a modernist free plan, Kiesler's ideal space is elastic, biomorphic, even intestinal, attuning its curves, the thickness and pitch of the walls, the form and position of the windows, and the slope of the floor to the needs, habits, functions, and feelings of its dwellers. Among the rich pickings from the collection, Oxman was also tempted by Diller + Scofidio's Slow House (1988–90; **fig. 27**). "Conceived as a passage from physical entry to optical departure or, simply, a door to a window," the house is shaped like an acoustic horn opening up to the sea, moving from the smallest intake (the door) through the low ceilings of the sleeping quarters to the air and light of the kitchen and living room, which is open to the water.[25] The architects made sure that the ocean view, replicated in video projections, could be seen in the remote quarters near the entrance. Slow House is a poignant example of dynamic, conditional architecture, conceptual and yet realizable and livable—the video-era evolution of Kiesler's project. Endless House and Slow House both offer an apt introduction to Oxman's brand of performance-based architecture, which carries the conditional paradigm one step further, by integrating it in the technological sphere first, and the biological second.

As Michael Hensel wrote in his history of performance-based architecture, the approach first emerged in the 1960s, at the sunset of modernism, amid the Cold War and the space race, as rockets, space stations, biodomes, and bunkers were adopted as living paragons in both mainstream culture and by architects—the conceptions of Walter Pichler and Raimund Abraham among them, to mention two outstanding works in MoMA's collection (**figs. 28, 29**).[26] These highly technical dwellings guaranteed survival only by adhering to specific, scientifically established parameters and infrastructures; they were parts of ecosystems and constituted an ecosystem in their own right. At that same time, notions of cybernetics and automation made their way into architecture, casting buildings as complex machines that should be devised to support human interaction and functional definition. The great, early manifesto for this kind of approach was Cedric Price's deeply influential 1964 Fun Palace project (**fig. 30**). Inspired by and developed with the avant-garde theater director Joan Littlewood, relying on experts such as the cybernetician Gordon Pask and the psychiatrist Morris Carstairs, Fun Palace has best been described by the architecture scholar Stanley Mathews as "a unique synthesis of a wide range of contemporary

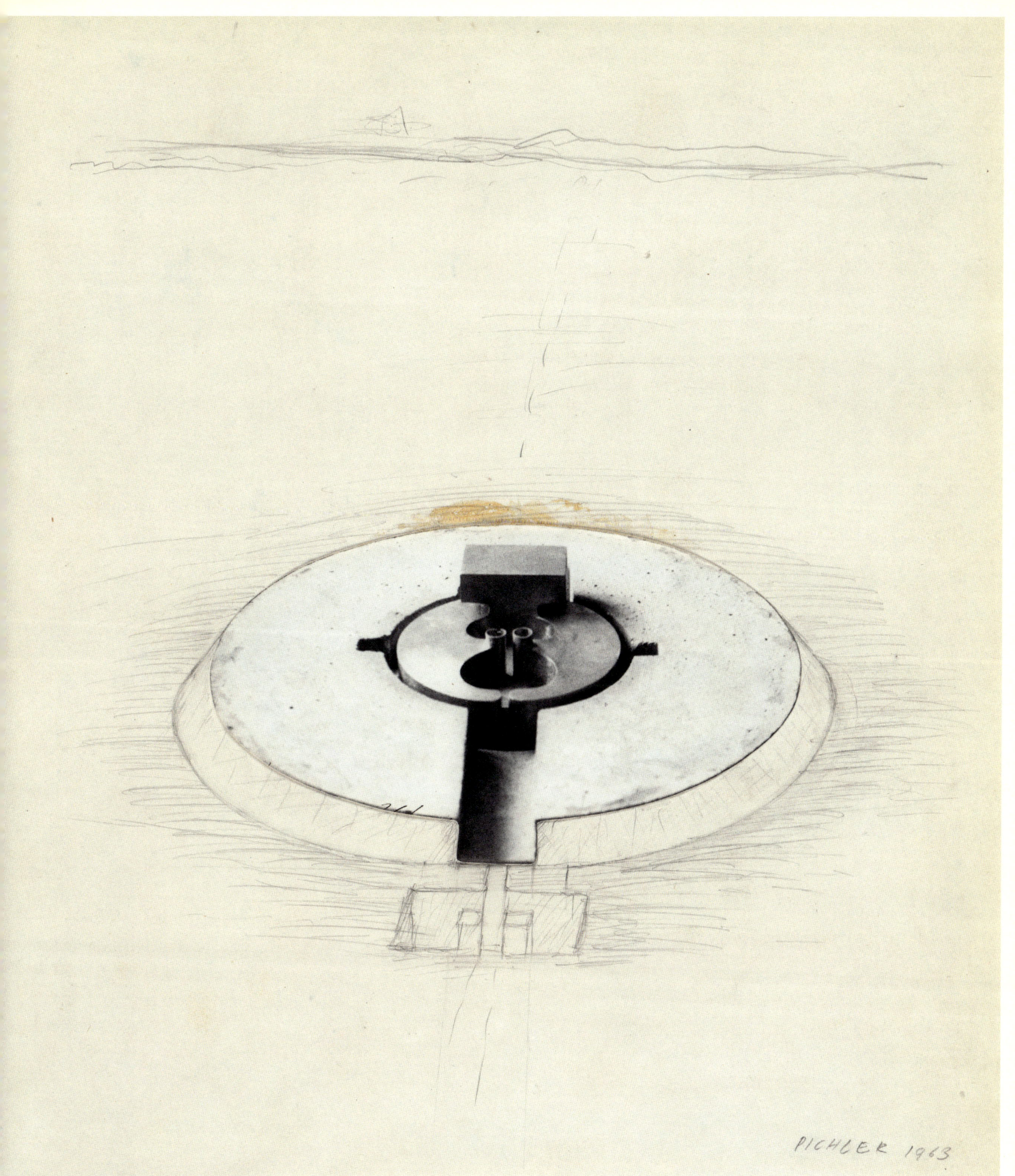

fig. 28 Walter Pichler. Entrance to an Underground City. 1963. Perspective. Photomontage and graphite, 13 ¾ × 11 ¾ in. (34.9 × 29.8 cm). The Museum of Modern Art, New York. Philip Johnson Fund

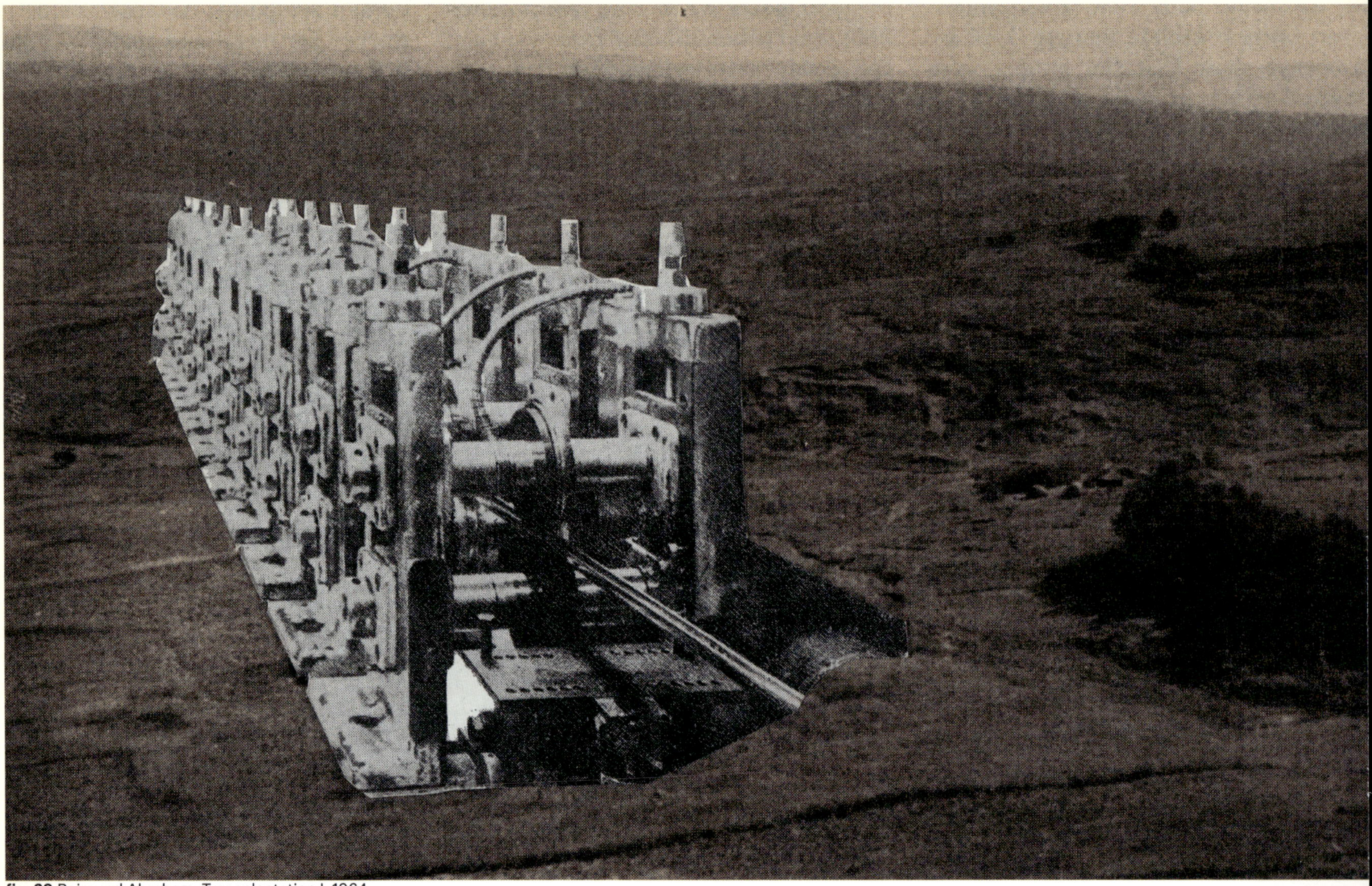

fig. 29 Raimund Abraham. Transplantation I. 1964. Perspective and plan. Collage, 4 ½ × 7 in. (11.4 × 17.8 cm). The Museum of Modern Art, New York. Philip Johnson Fund

fig. 30 Cedric Price. Fun Palace for Joan Littlewood Project, Stratford East, London. 1959–61. Perspective. Felt-tipped pen, ink, graphite, crayon, and ink stamp on tracing paper with tape, 6 ½ × 15 ⅞ in. (16.5 × 40.3 cm). The Museum of Modern Art, New York. Gift of The Howard Gilman Foundation

discourses and theories, such as the emerging sciences of cybernetics, information technology, and game theory, Situationism, and theater to produce a new kind of improvisational architecture to negotiate the constantly shifting cultural landscape of the post-war years."[27] Indeed, Fun Palace was designed to be activated at the whim of its dwellers, occupied, owned, and used in the way a design object would be. The machine could be inhabited, at last. Flexibility and adaptability were the key-words, and interfaces the new territory of architecture.

In August 1967, the American journal *Progressive Architecture* (**fig. 31**) dedicated an entire issue to the topic of performance design. Hensel notes that the issue focuses on "systems analysis, systems engineering and operations research," as well as "mathematical modelling towards optimization and efficiency." In other words, performance-based architecture is informed not only by the functional program and its formal synthesis but also contemplates a building as a complex system, shifting the focus "from what the building *is* to what it *does*."[28] In 1969 Reyner Banham echoed this call for integration in *The Architecture of the Well-Tempered Environment,* arguing that buildings should not be "divisible into two intellectually separate parts—*structures*, on the one hand, and on the other *mechanical services*."[29] Rather, he suggested, a successful approach to architecture consists in an organic integration and should investigate both external and internal factors, changes in use, changes in users' expectations, changes in the methods of servicing the users' needs and, most importantly for him, the mechanical environmental controls of the building.[30] Banham had in fact already begun arguing for such organic integration in 1965, in an essay for *Art in America* titled "A Home Is Not a House," for which the Canadian architect François Dallegret produced memorable drawings (**fig. 32**).

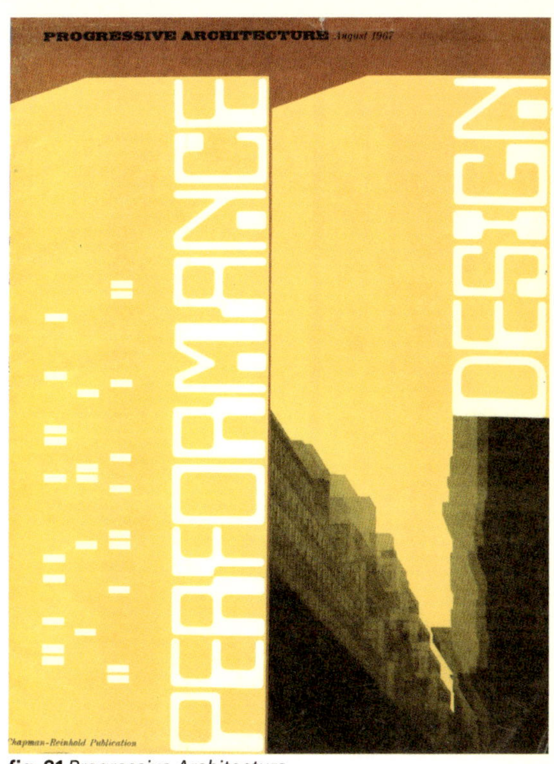

fig. 31 *Progressive Architecture*, "Performance Design" issue, August 1967. Cover designed by Richard C. Lewis

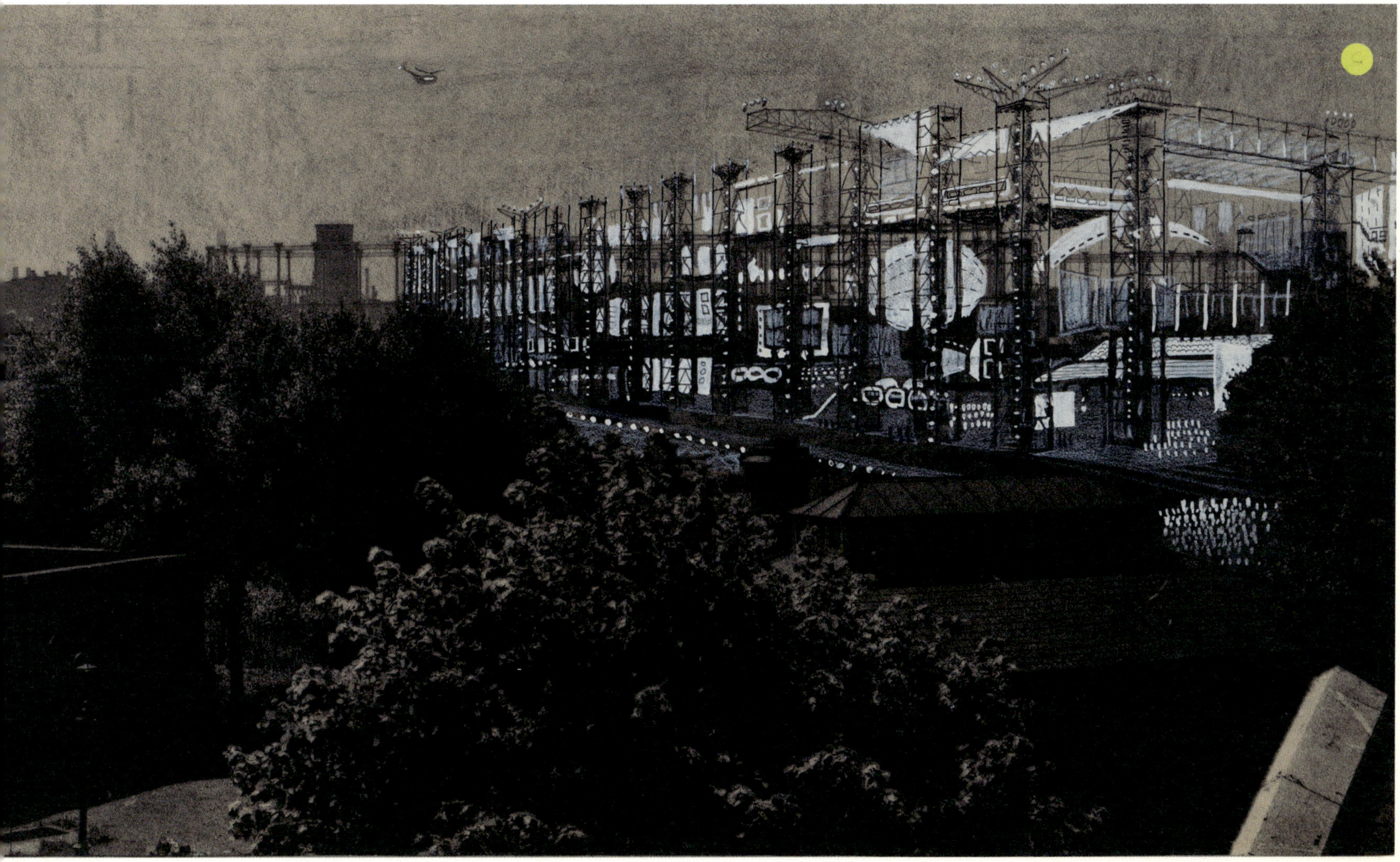

Nicholas Negroponte's *The Architecture Machine: Toward a More Human Environment* (1970; **fig. 33**) was among the first books in the United States dedicated to architectural performance.[31] Highlighting three key concepts—generation, evaluation, and adaptation—the book continued to explore the relationship between human and machine in architecture and design and looked at how emerging technologies facilitated human-machine partnerships. Negroponte's idea of the future was one in which architecture and design embraced these collaborations to the point at which it was not possible to distinguish each partner's contributions.[32] In the decades after the publication of *The Architecture Machine* and with the advancement of computation in architecture, many research projects sought to develop an integrated design system to facilitate performance-based architectural design. The concept was widely adopted toward the end of the twentieth century with the rise of sustainable and green architecture in the United States and, especially, in Europe as a response to environmental concerns.[33] The development of building performance–simulation technologies, which replicate aspects of building performance using a computer-based mathematical model, was also a key factor in this period.

Digital culture and the advent of digital fabrication shifted the focus of architecture's optimization from the mechanism to the algorithm—not only functionally, structurally, and systemically but also expressively. In the book and exhibition *Archaeology of the Digital*, Greg Lynn pointed to a moment in the late 1980s and early 1990s when architects outside of polytechnic schools anointed computation, and when early elements of it could be found in the work of established architects such as Peter Eisenman (**fig. 34**) and Frank Gehry. The introduction of Netscape in 1994 marked the beginning of a new era of "digital normal" for the world at large, during which new tools for creativity and processing led to the proliferation of exuberant new formal languages in all realms—art, literature, popular culture, and architecture and design. Theoreticians and practitioners ran wild with highly exploratory models, driven by the search for a new, "better" way of designing.[34]

fig. 32 François Dallegret. Anatomy of a Dwelling. 1965. Chinese ink on translucent film and gelatin on transparent acetate, 24 ¼ × 19 15/16 in. (61.1 × 50.6 cm). Frac Centre-Val de Loire collection

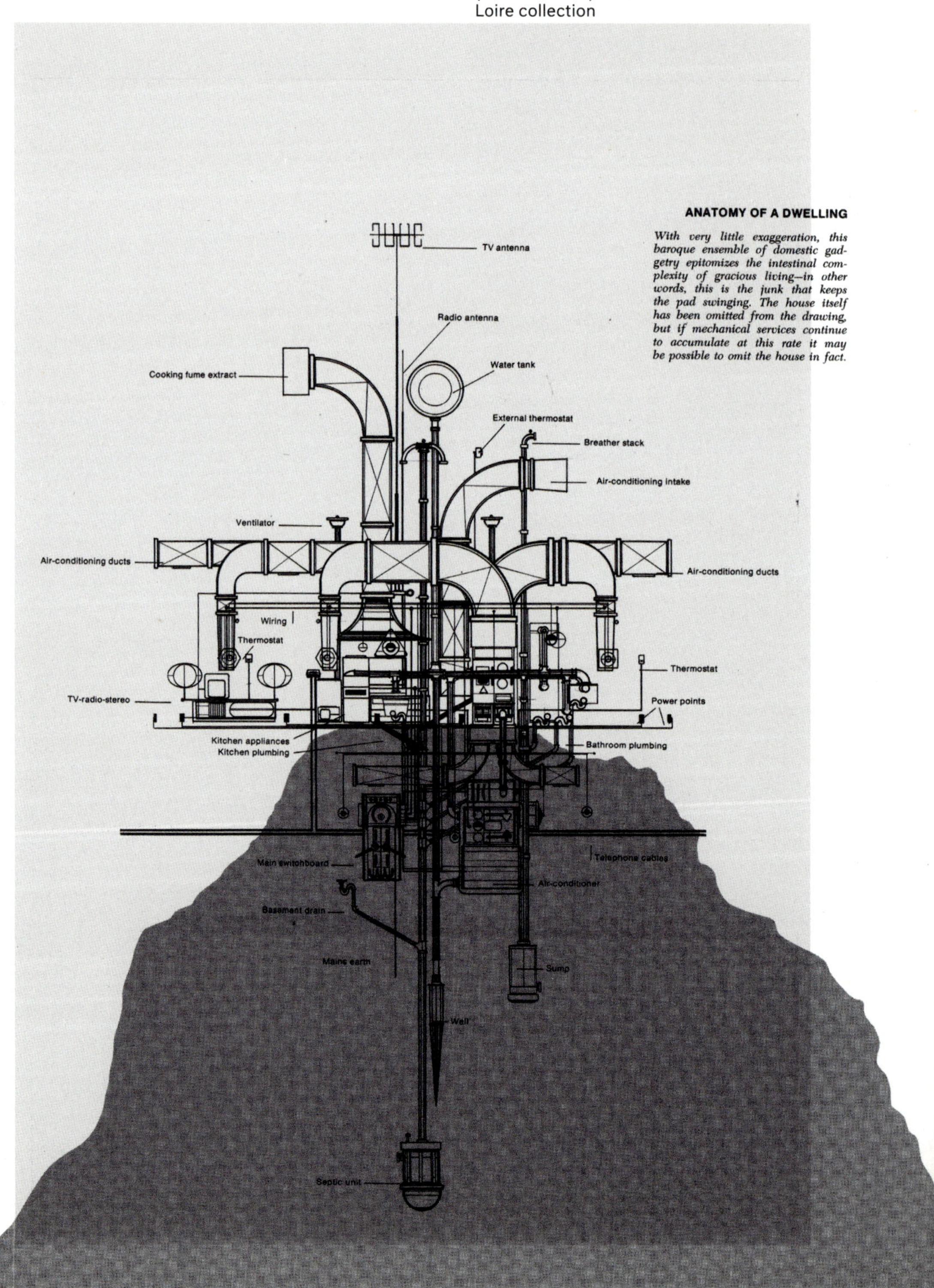

fig. 33 *The Architecture Machine*, by Nicholas Negroponte, 1970. Cover designed by Muriel Cooper

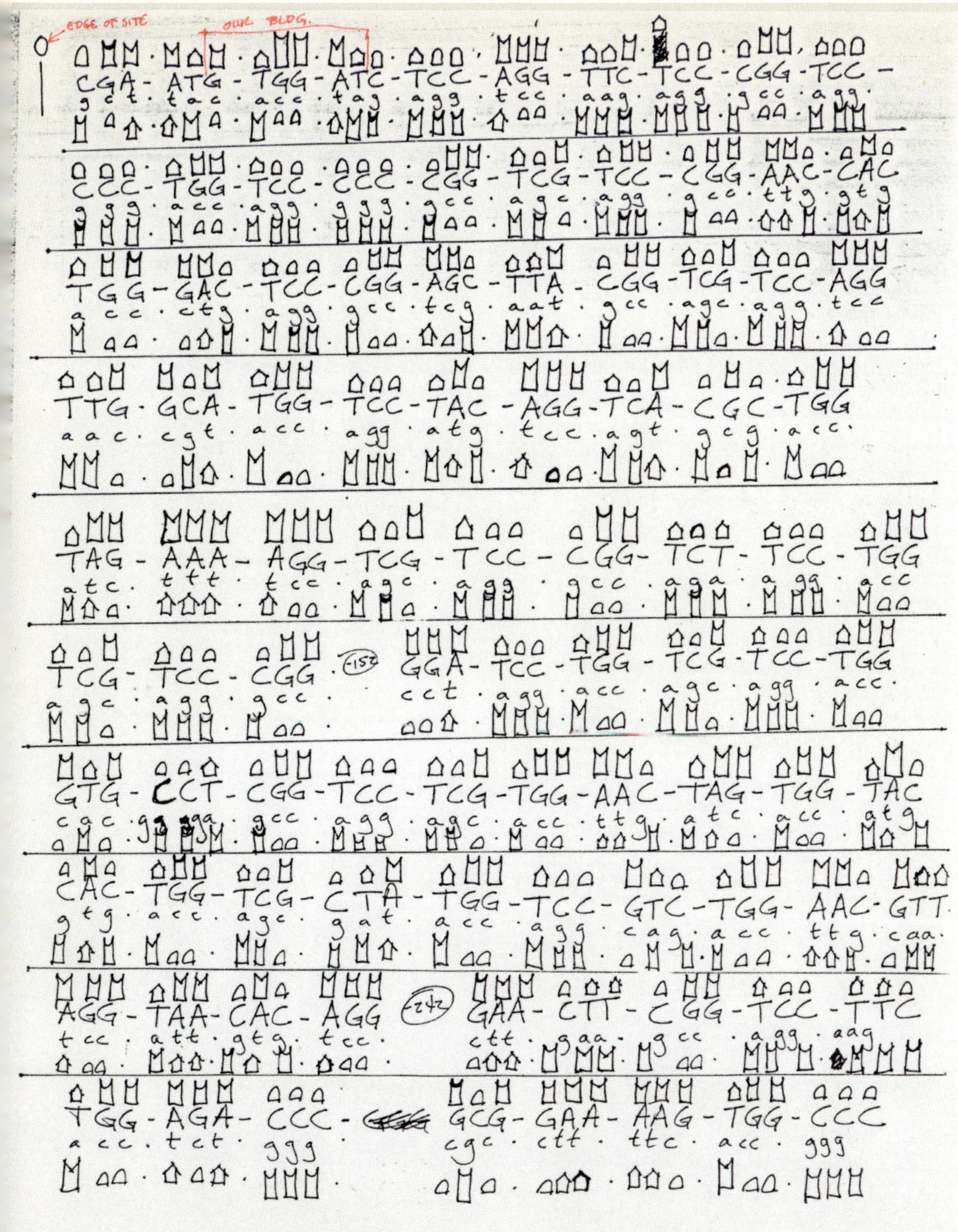

fig. 34 Eisenman/Robertson Architects (Peter Eisenman and Jaquelin T. Robertson). Schematic representation of a DNA sequence for Biozentrum, Biology Center for the J. W. Goethe University, Frankfurt am Main, Germany. 1987. Electrostatic print with ink notations, 11 × 8 ½ in. (28 × 21.5 cm)

The print was included in *Archeology of the Digital*, Canadian Centre for Architecture, Montreal, in 2013, and Yale School of Architecture, New Haven, Connecticut, 2014.

Generative architecture, in which the architect and the computer refine each other's program and design, has emerged over the past decades from this digital Big Bang. Oxman embraced this reciprocal attitude and evolved it into a new organicism, in which the algorithm is a starter for architecturally controlled growth, and the building is no longer a machine but an organism. Just as her description of the creative process had gone from geometric to metabolic, so the generative process in her architecture moved from the purely parametric to the organic. Rivka Oxman has argued that three components are key to performance-based design—the geometric model, the evaluative processes, and the interactivity of the designer.[35] In her daughter's work, however, the interactivity is not between architect and computer but rather between architect and nature, with the computer becoming a tool igniting an interdependency between the two and leading closer to the ideal way of designing and building: "Nature optimizes for a multiplicity of functions across scales: structural load, environmental performance, spatial constraint, and more . . . enabling multi-functionality that matches Nature."[36] This multiplicity leads, in turn, to specificity, the ability of the same design, even of the same material, to respond and adapt to specific conditions, which themselves might be in flux—an object handled by small or big hands, for instance, or a building facade at different times of day, seasonal temperatures, or levels of occupancy. As Kiesler did with his concrete sculptures, Oxman dreams of using one material, or at least one material system—whether silk, glass, or pectin—and taming it to behave differently in different parts of the object, an Endless House in a silkworm's cocoon.[37]

Just when the morphological intelligence of parametric and generative design had engendered a formal style that has become almost dated, Oxman and others introduced a new ingredient—biology—to performative and generative architecture in a way that has reinvigorated the field and diversified its production. Architects such as David Benjamin (**figs. 35, 36**), Skylar Tibbits (**fig. 37**), and Jenny Sabin (**figs. 38, 39**), for example, are each invested in such new expression. Not needing to control the whole process of form making, they let nature run its course, if under specific parameters, some distilling its processes, others integrating it directly in the architecture and infusing it with their own aesthetic along the way. Architects are not alone in embracing biology as the new North Star. Biology is in fact so prevalent in all fields that many intellectuals and opinion makers, including World Economic Forum founder Klaus Schwab, have named the current period the Fourth Industrial Revolution, an era in which every process is informed by organic considerations, in which culture has become wet, infused with an awareness of biology and ecosystems.[38] Oxman herself has noted this precise shift: "Unlike the Industrial Revolution, which was ecology-agnostic, this new approach tightly links objects of design to the natural environment."[39]

Thus the field of performance-based design has moved another notch, from the mechanism to the algorithm and now to the organism. If it seems too good to be true, and if parametric architecture cannot simply be considered sustainable when wet, as Christina Cogdell pointed out in *Toward a Living Architecture? Complexism and Biology in Generative Design* (2019), the effort is nevertheless evolutionary.[40] Despite the distance traveled from the pure volumes of modernism to the intricacy of parametrically enabled neo-organic design, the adaptability Oxman and her contemporaries pursue is not the opposite of the universality that modernist architects had sought. Rather, it is its essence distilled and freed from the homogenized forms of modernism, which were dictated not only by Corbusian dogma—the open plan especially—but also by twentieth-century building materials and techniques. Digital technology has enabled designers to get much closer to nature and its awe-inspiring pluralistic articulations. With that superpower, in the future some of modernism's antagonisms (homogeneity versus specificity, for instance, the former affordable and democratic, the latter pricey and elitist) will be rendered obsolete. New technologies will make specificity accessible. Computation and its expressions—such as performance-based architecture and the generative and parametric design that supports it—are the key to specificity, which is in turn the ability to accommodate different requirements and conditions within the same adaptable, algorithm-based design, perhaps even using the same material. One of the great promises of digital design and fabrication is the ad hoc, the customized, "just in time" production. No waste, no suffocating standardization, no need to differentiate anymore between prototypes and series. The result will be a truly universal kind of design, pluralism without sacrifice.

Opposite, top:
fig. 37 Skylar Tibbits, MIT Self-Assembly Lab, and Invena. Growing Islands. 2019. Digital images, drawings, and diagrams, dimensions variable

An exploration of the possibility of replenishing sandbars and protecting coastlines through the careful placement of objects able to harness the natural pattern of waves and tides.

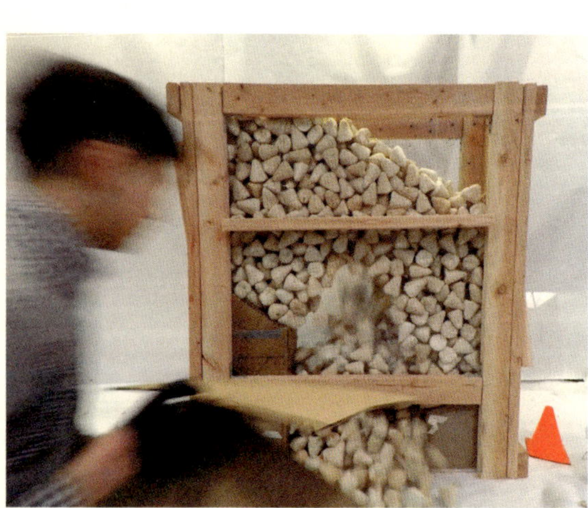

figs. 35, 36 David Benjamin of The Living. Jammed Bio-Welding. 2019. Corn stover and mycelium, dimensions variable

Investigation of the compression of biomaterials in formwork molds as a method of building. The materials grew into each other to form solid or semisolid structures.

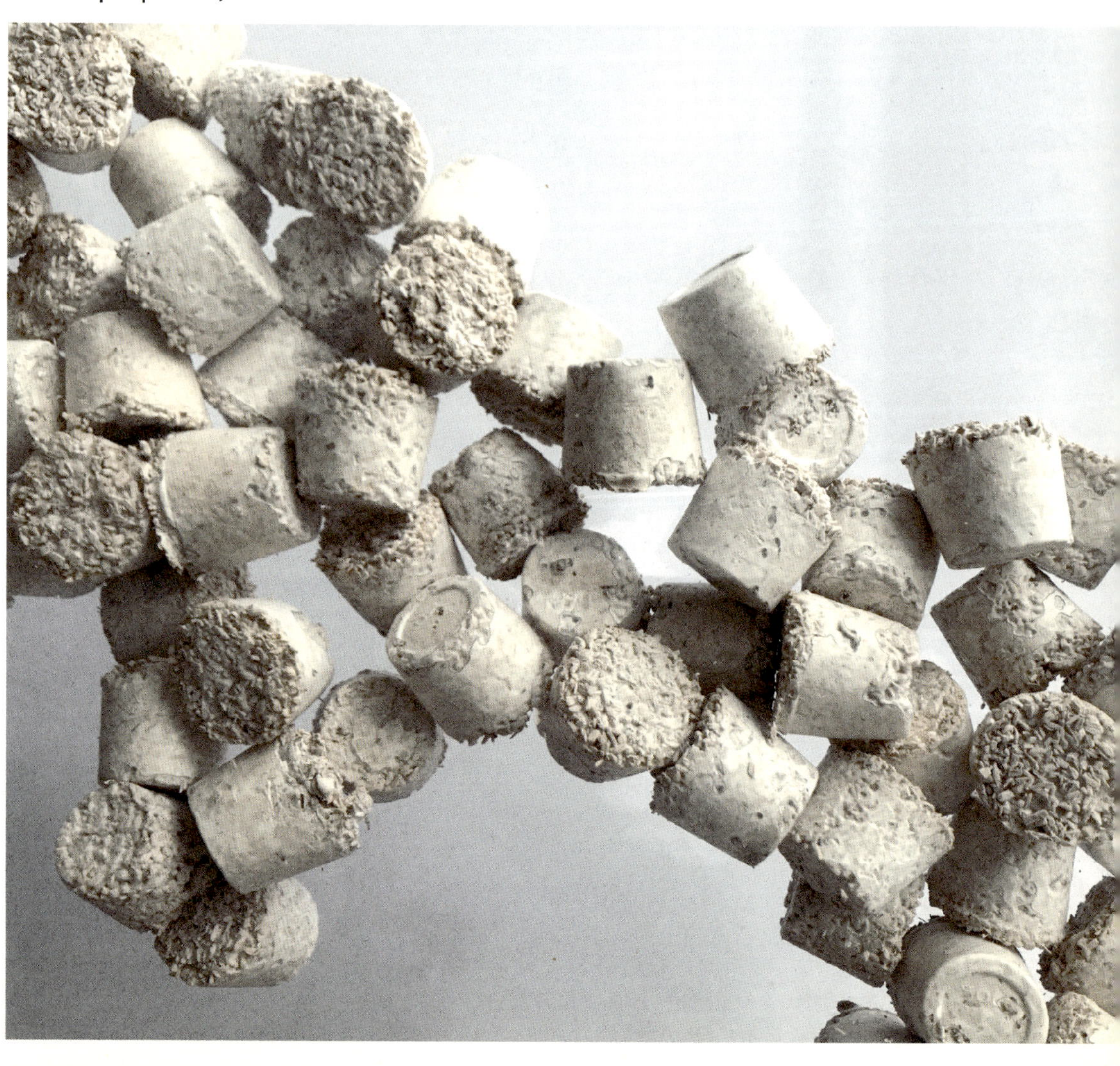

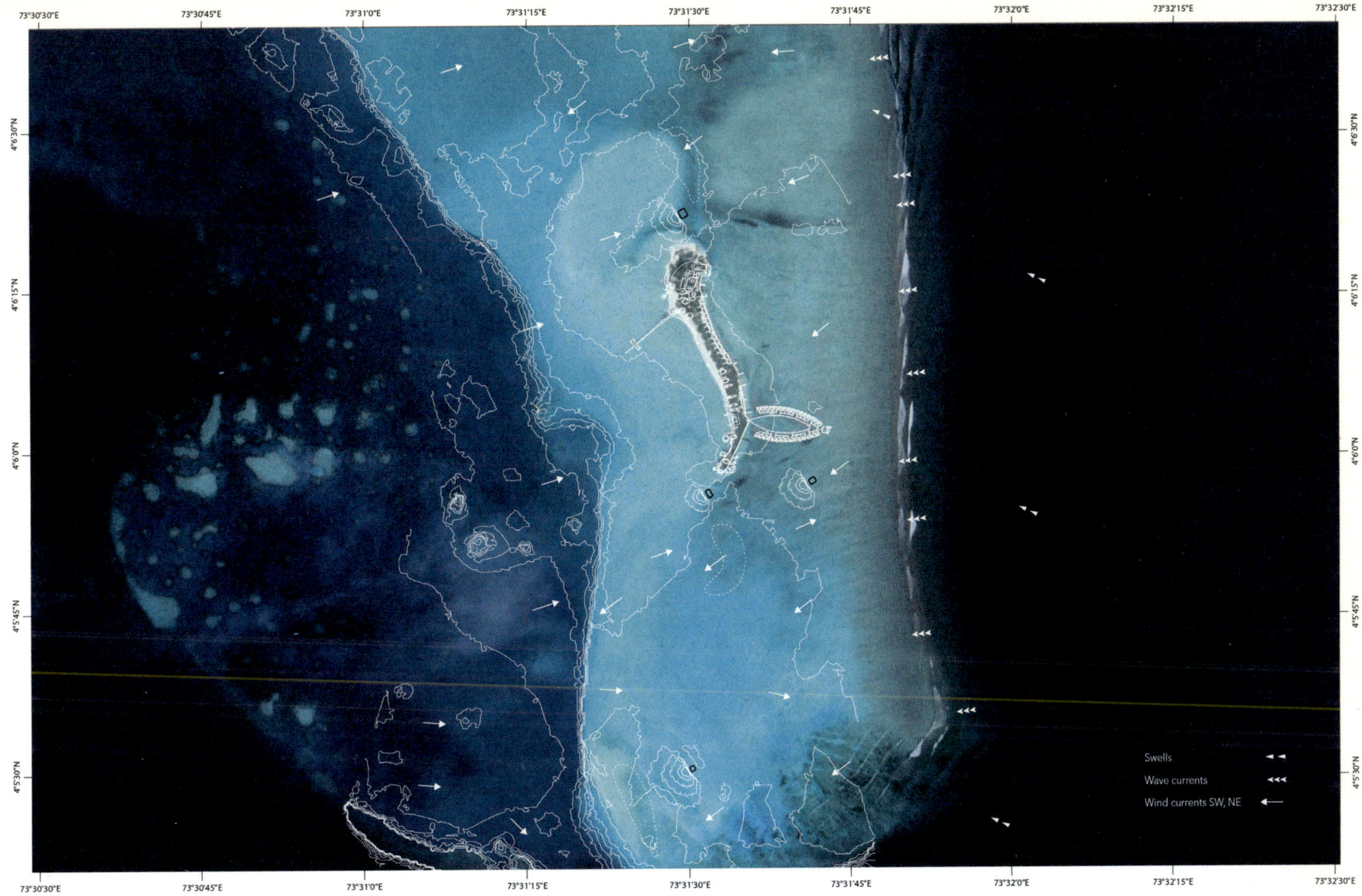

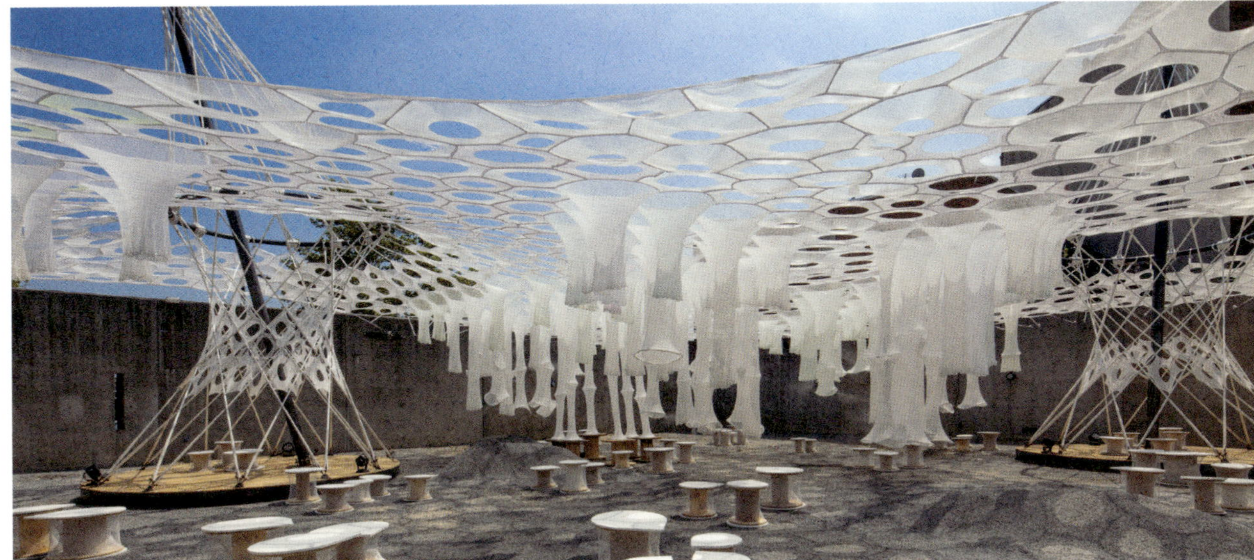

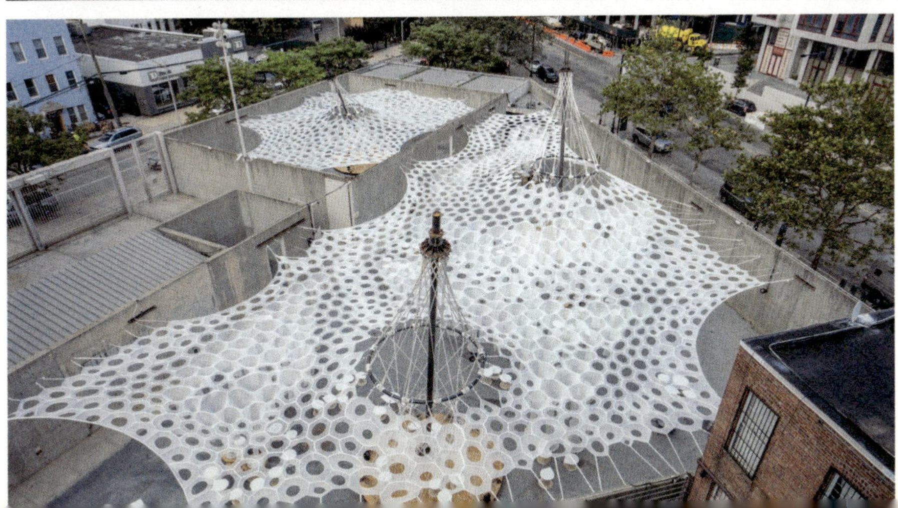

Figs. 38, 39 Jenny Sabin Studio. Lumen. Installation view in Young Architects Program, MoMA PS1, Long Island City, New York, June 29–September 4, 2017

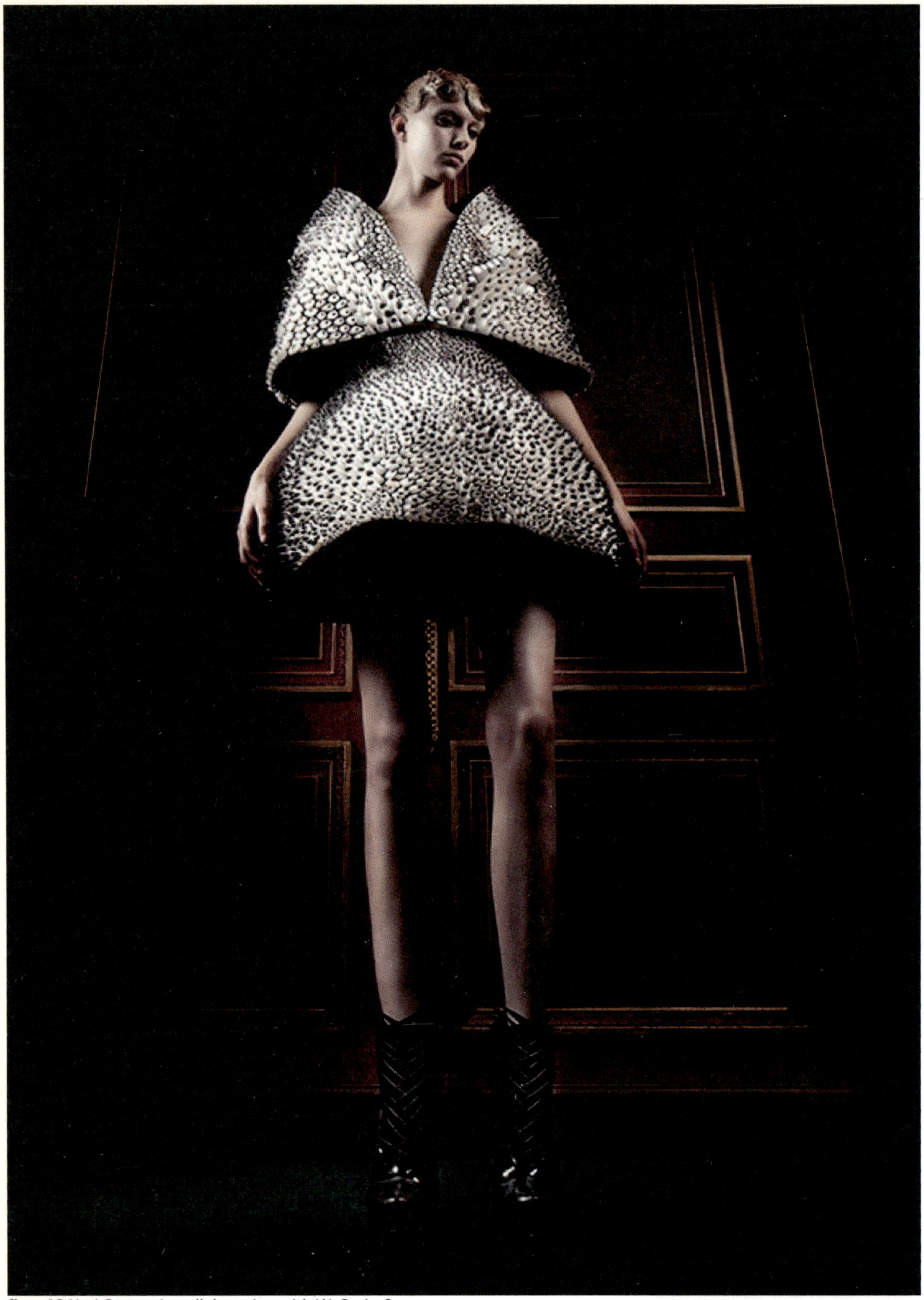

Material Ecology at MoMA

Material Ecology is an exhibition not only about the future of architecture and design but also about the future role of the designer as the initiator of a process that exists within systems and ecologies, rather than as the form-giver of an objec or the follower of functions. MoMA's curatorial team has selected seven projects some realized by Oxman independently, most of them with MMG, to highlight the impact that her attitude is having on the next generation of designers; of materials- computation-, and life-scientists; of engineers, clients, and—through her work and her public talks—citizens.

Every object in the *Material Ecology* exhibition, however finished looking and formally accomplished, is presented as a demonstration of new materials and new processes, as a door into a new way of designing and making. The curatorial team sought to highlight process over product in selecting works, privileging the most processual and speculative experiments over more immediately applicable ones, while also delineating the evolution in Oxman's work. This means we did not include, for instance, the cape and skirt ensemble Oxman designed for the Dutch couture designer Iris van Herpen in 2013 (**figs. 40, 41**), which is seamlessly 3D printed and made of thousands of quills that adapt to the body like a second skin. We did not include Wanderers (2014; **figs. 42, 43**), Oxman's "life-sustaining clothes for interplanetary voyages"—bacteria-infused, photosynthetic gear for space exploration— or Rottlace (2016; **figs. 44, 45**), the spectacular mask modeled on the understructure of Björk's face and worn by the artist on the outside

figs. 40 Neri Oxman in collaboration with W. Craig Carter and Iris van Herpen. Anthozoa. 2013. Photopolymers, dimensions variable. Presented at Paris Fashion Week, 2013

A skirt and cape produced with multimaterial 3D-printing technology by Stratasys, which enables both hard and soft materials to be incorporated into a single multifunctional material system.

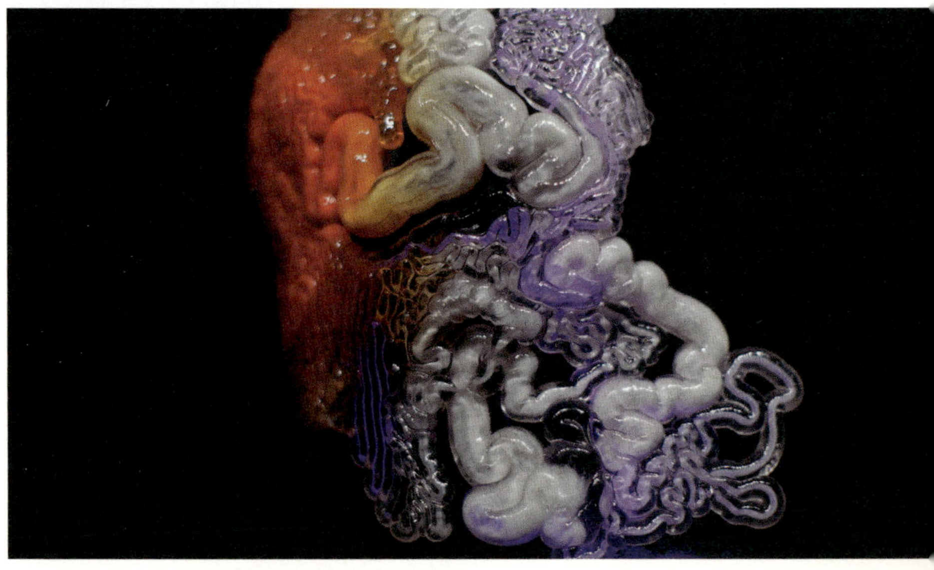

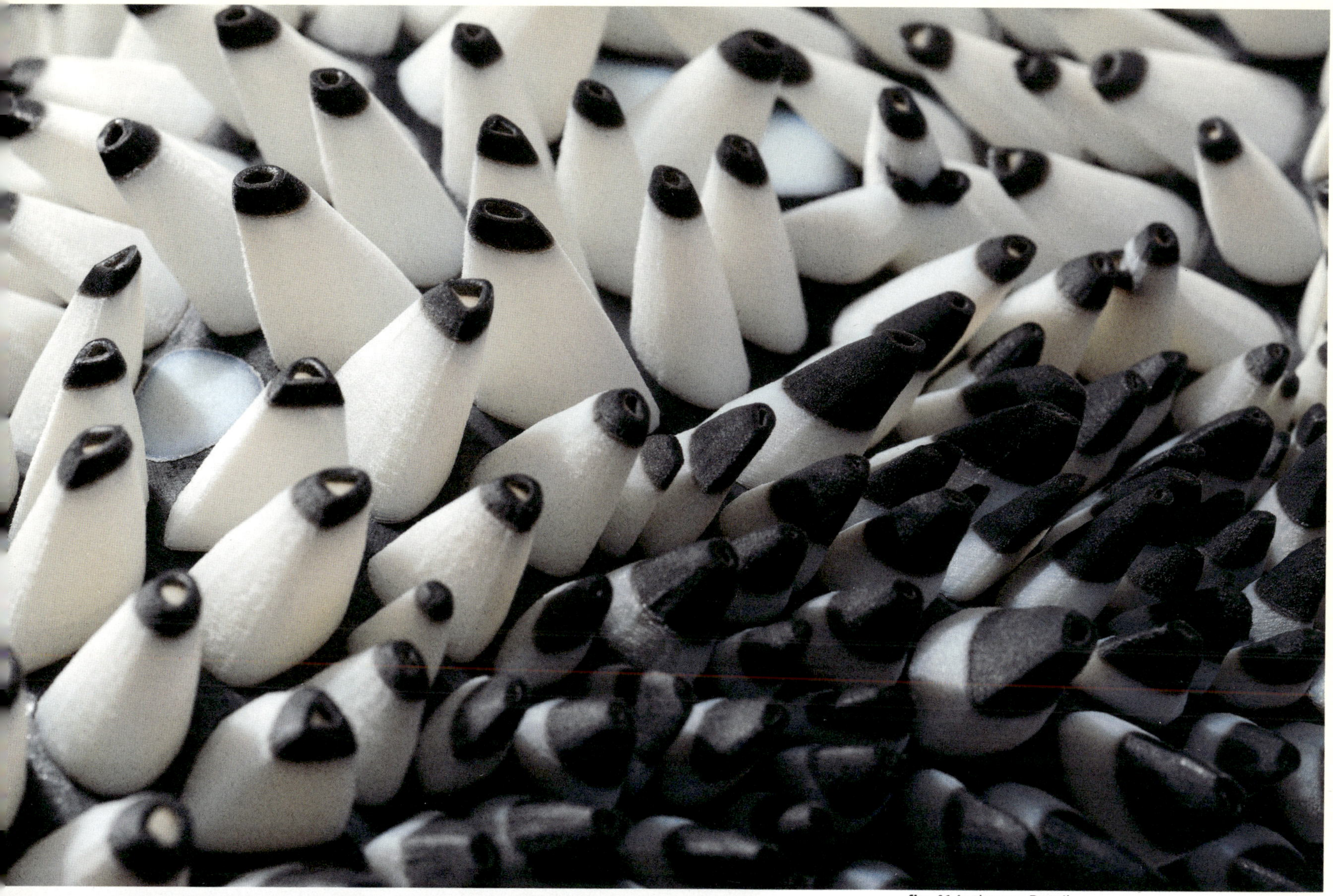

fig. 41 Anthozoa. Detail

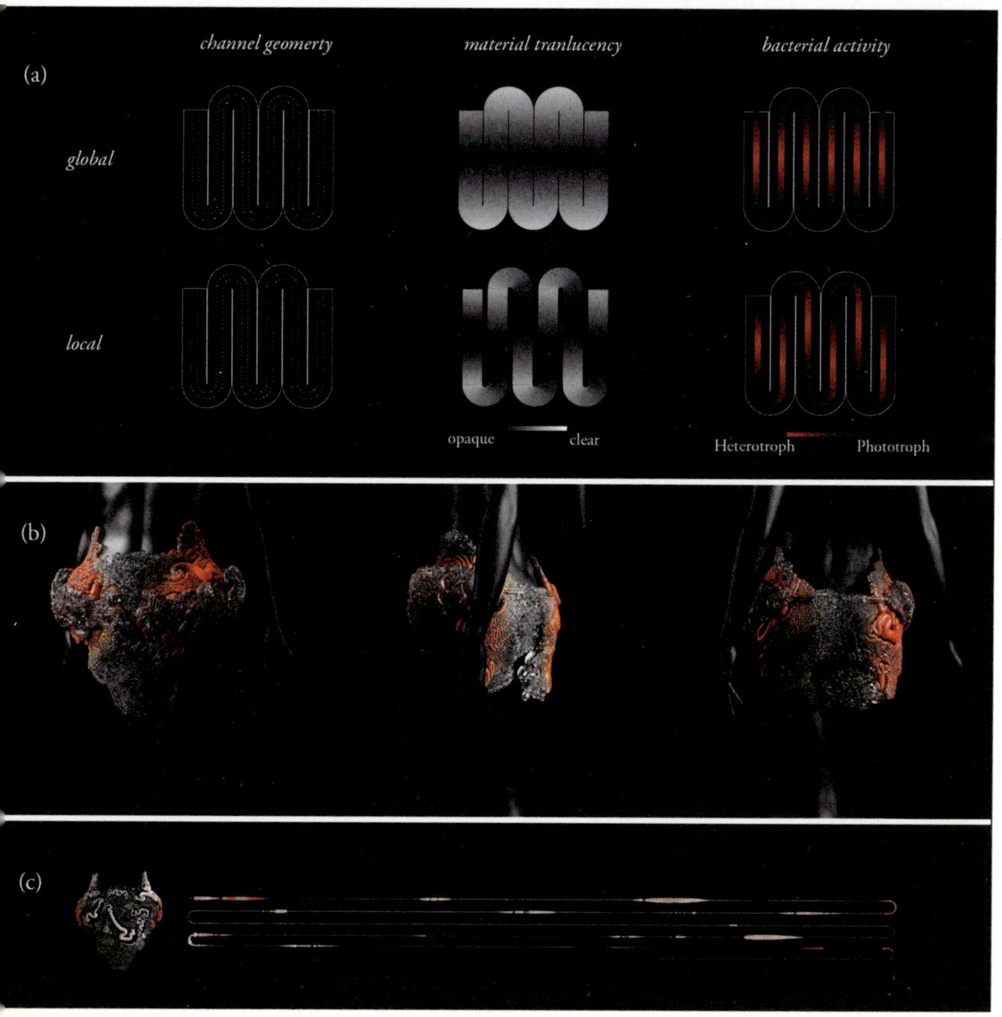

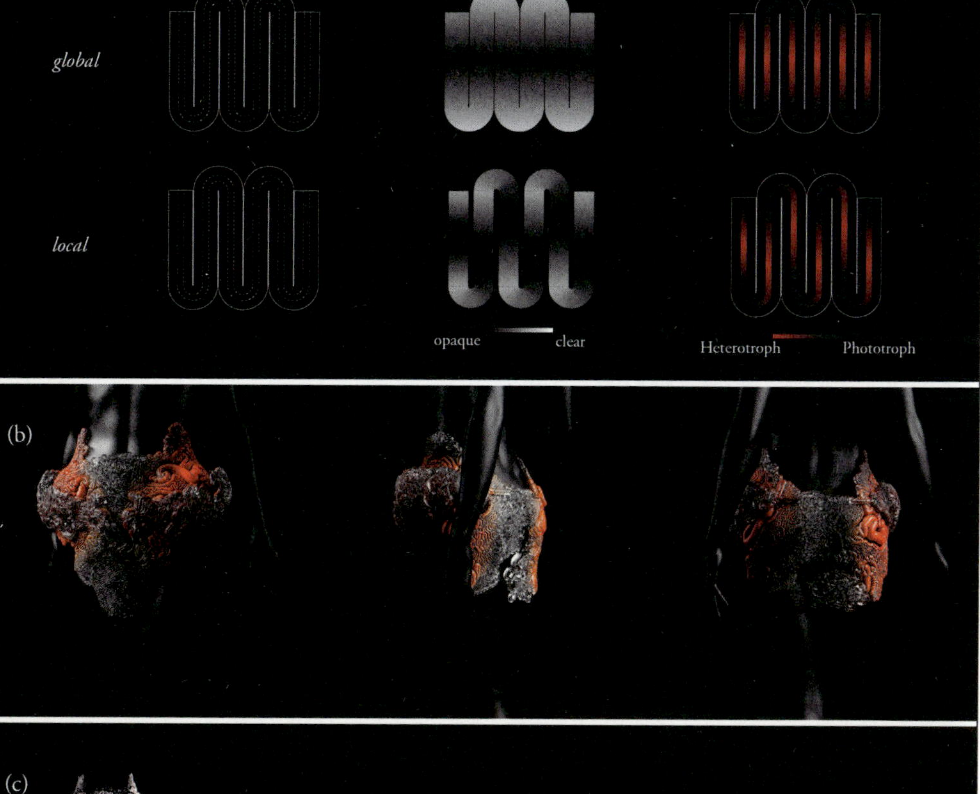

figs. 42, 43 Neri Oxman and The Mediated Matter Group, in collaboration with Stratasys Ltd. Mushtari, from The Sixth Element Collection. 2014. Photopolymers, 35 ½ × 15 ½ × 6 in. (90 × 40 × 15 cm)

This wearable organ system, a physiological augmentation for interplanetary travel, was designed using generative growth algorithms and produced on a Stratasys Objet500 Connex3 3D printer. The computational form generation processes mimic biological growth by generating recursive forms over many iterations of the algorithm.

The effect of material transparency on microbial activity (a); implementation of the global heterogeneous modeling approach on the wearable (b); visualization of the unfolded strand showing how the overall source-based approach influences local changes in opacity along the strand (c).

To stress the prototypical quality and meta-architectural value of the projects in the exhibition, we accompany them with documentation of their development, in the lab and outside, in video and in artifacts. Together with Oxman and with Irma Boom, who designed this book under the spell of Stewart Brand's immortal *Whole Earth Catalog* (**figs. 46–48**), we vowed to make the exhibition and the book "dirty" with process, to convey the feeling of the lab at MIT—which, albeit very clean, is ebullient with ideas, samples, and living beings, from humans to silkworms, ants, and cyanobacteria.[41]

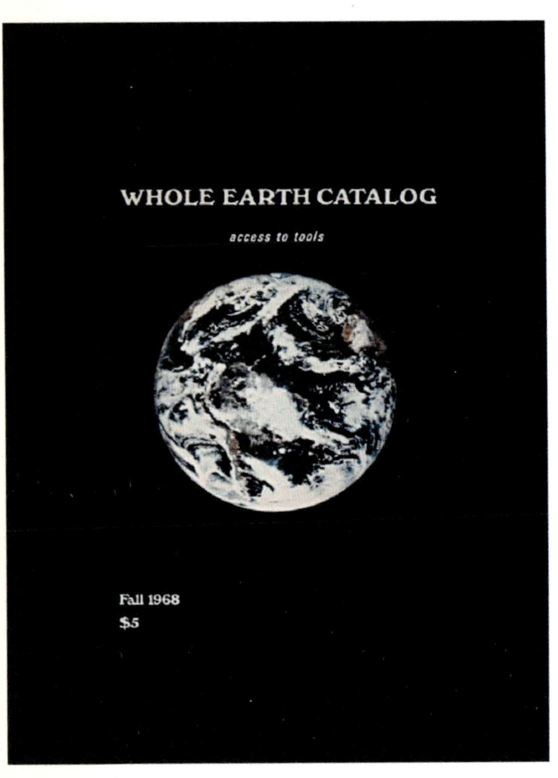

After an introduction to the early Materialecology oeuvre, this catalogue divides the selected works according to the family of techniques that generated them—between Extrusions and Infusions, a division left more loose in the gallery display.[42] If the 2007 Materialecology experiments demonstrated the ability to distill algorithms from natural behaviors in order to apply them to optimizing objects and buildings, the other works featured in the 2020 exhibition describe an evolution toward more complex entanglements with the biological. Extrusions are projects in which the object is obtained from a sometimes biologically augmented material that is extruded by a mechanical or animal nozzle, as in Silk Pavilion I and II (2013 and 2019; pages 96, 106), Aguahoja I and II (2018 and 2019; pages 74, 80), and Glass I and II (2015 and 2017; pages 116, 128). Glass, silk, chitosan, pectin, and cellulose are here excreted by three different printers—the first an innovative printer with a chamber that can function at molten-glass temperature, the second a "massively multiplayer" printer made of thousands of silkworms, and the third a rather traditional robotic arm.[43] Over this category and over the whole exhibition towers Silk Pavilion II, a site-specific installation designed for the MoMA gallery, which manifests the evolution of the project since it was first installed in the atrium of the MIT Media Lab building, in 2013. One of Oxman's and MMG's best-known projects, the pavilion is an arresting collaboration between humans and insects, powered by computation.[44]

In the Infusions group, a vessel is 3D printed and prepared to accommodate a biological soul in the form of organic elements humans share with the rest of nature—air, bacteria, pigments, and so on. The three series of death masks called Vespers begin with Lazarus (2016; page 144), inspired by Paul Kalanithi's unforgettable 2016 book *When Breath Becomes Air*, in which the young oncologist chronicles his own terminal cancer.[45] If Lazarus is proof of the concept of capturing a dying person's last breath through data-driven additive manufacturing processes, the three subsequent series culminate with the infusion of bacteria to color the masks, almost in a heat map of life fleeting. Another infusion project, Totems (2019; page 168), focuses on melanin ("the pigment of life," Oxman calls it) as a symbol of both nature's infinite ingenuity—its usefulness, adaptability, omnipresence, and specificity—and of humans' fatal proclivities toward racism and injustice. The project was initiated at Design Indaba in Cape Town in 2018 by the Nelson Mandela Foundation and presented in the 2019 Milan Triennale, *Broken Nature*.[46]

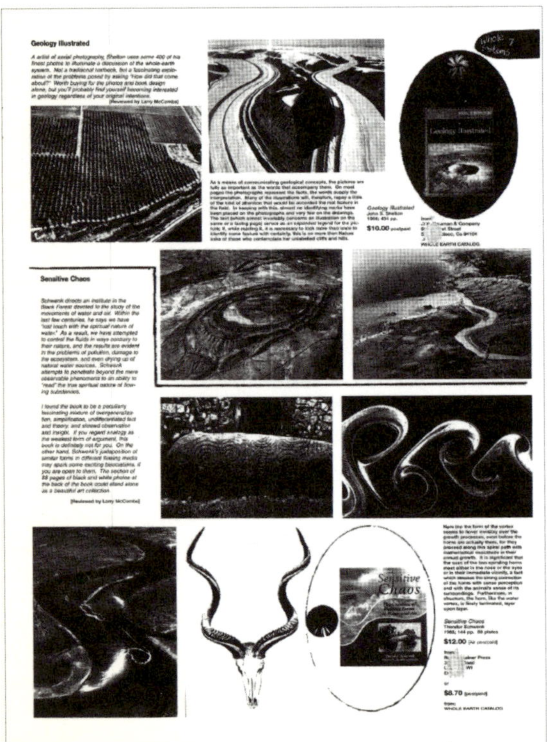
Geology Illustrated

Sensitive Chaos

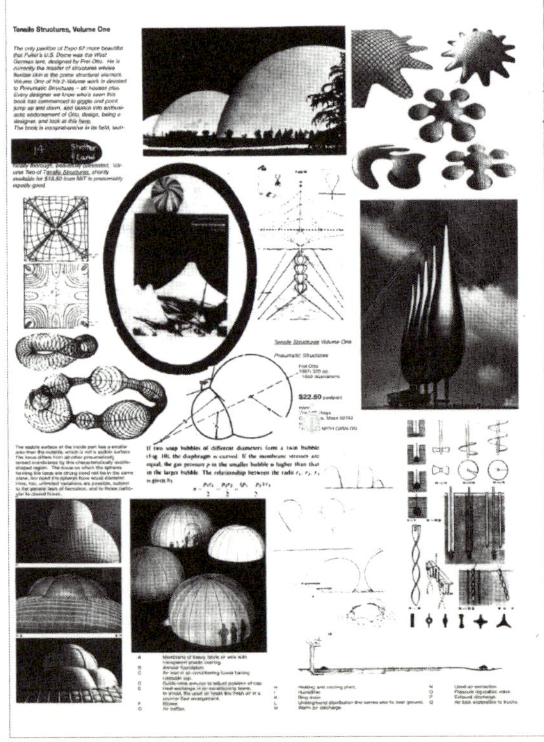
Tensile Structures, Volume One

figs. 46–48 Stewart Brand, ed., *Whole Earth Catalog*, Fall 1968. Cover, page 8, page 15

igs. 44, 45 Neri Oxman and The Mediated Matter
Group. Rottlace. 2016. Nano-enhanced elastomers,
8 ½ × 15 ½ × 8 ¾ in. (21.3 × 39.3 × 22.3 cm)

Below, top: Björk wearing Rottlace before performing
"Quicksand," Miraikan, National Museum of Emerging
Science and Innovation, Tokyo, 2016. The mask's tunable
physical properties recapitulate, augment, or control
acial form and movement.

Below, bottom: Conceptual rendering. Inspired by their
biological counterpart and conceived as muscle textile,
he masks are bundled, multimaterial objects, providing
ormal and structural integrity to the face and neck
while also allowing for movement.

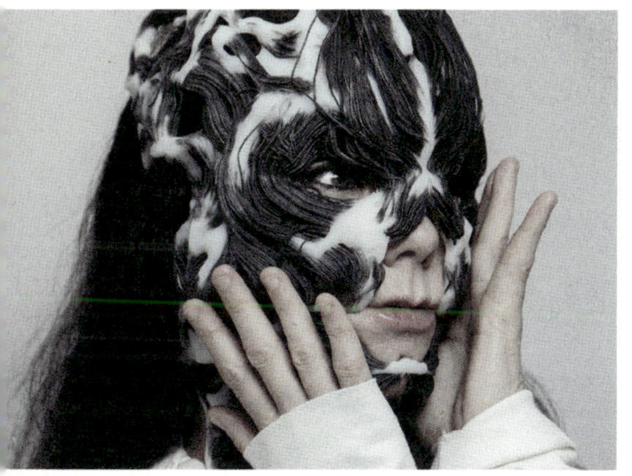

Design and Change

One of the major arguments and conclusions of *Design and the Elastic Mind* was that designers play an essential role in aiding and abetting change.[47] Designers take people by the hand and make change, if not imperceptible, at least acceptable, and even desirable. They have served in this role since the beginning of time, since the first spear-head, the first wheel, and the first brick. Along the way, architects and designers have built wonders and propelled progress, displaying the best aspects of human ingenuity and vision. In some cases, however, they have also offered us glimpses into the darkest corners of human nature. They have turned a blind eye to the exploitation of other human beings, participated in the depletion of resources and in the obliteration of other species, cemented social injustice, celebrated false value systems in order to please and enrich their masters—from tyrants to corporations—and subjugated their neighbors, turning them into citizen consumers. For centuries, they have justified their actions by shifting respon-sibility onto others, often claiming they were just work for hire.

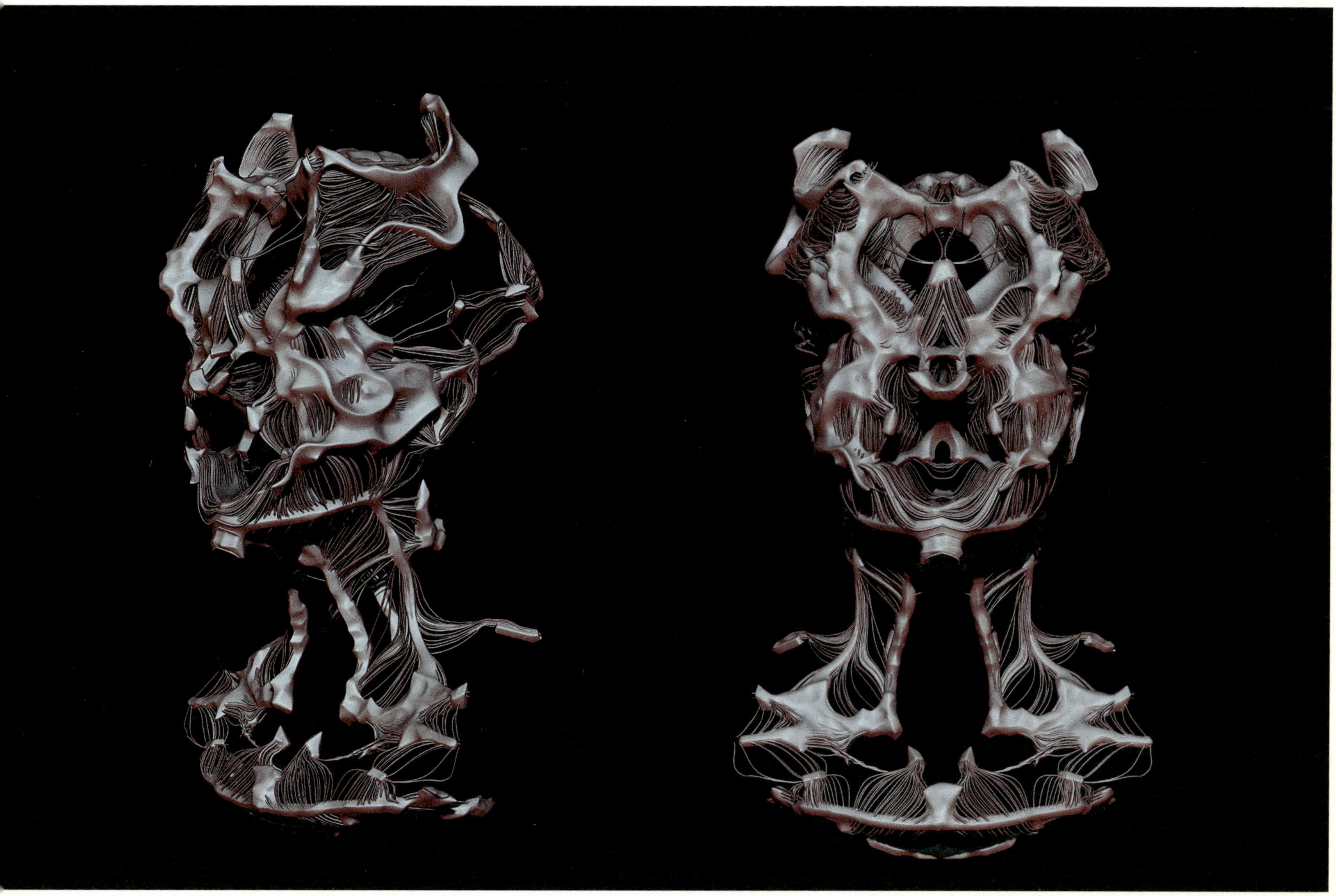

A positive consequence of our era's many crises, however, is that the ethics of all actions and transactions are now under strict and fundamental scrutiny. For many architects and designers, civic and human values have become a moral compass that guides the design process, even if one occasionally compromised by the need to maintain a practice and cover its overhead costs. One could argue that in design, ethics are sometimes more important than aesthetics and anyway intimately connected to it. It is a local articulation of a widespread movement toward justice, equality, and consideration of the needs and survival of others, in response to the antithetical and equally widespread exacerbation of inequality, conflict, and intolerance. "Without civic morality," Bertrand Russell wrote, "communities perish; without personal morality, their survival has no value."[48] The values of a morally sound system—the cultural patrimony and mission of an individual, a community, a corporation, a nation—informs behaviors and choices of all kinds. Today, many spheres of human activity and study, from governance to management and economics, have shifted from a dogmatic, pseudoscientific model to value-based systems, although, in truth, some grand pronouncements coming from corporations and financial institutions have left even the most credulous and optimistic among us puzzled by their well-sounding vows of social and environmental responsibility. Such statements, whether indications of real shifts or merely lip service, are nonetheless a sign of the establishment's need to reckon with public scrutiny and expectations.[49]

One of the most important developments of the past sixty years has been the shift to an approach that takes into account, or attempts to take into account, the polyhedric beliefs and interests of a pluralistic society, as well as the ecological needs of our planet, a shift that is systemic and fluid and embraces complexity. When cultural diversity matters as much as biodiversity, when we treat it as equally fundamental in order to build a better future for all, the joining of science, design, technology, and culture emerges as a promising and constructive approach.

Echoing this conviction, Oxman laid out in a nevalogue draft a principle for her work going forward: "Co-culture over single organism culture." This single tenet highlights and celebrates not only biodiversity but also a deeper entanglement and dependency among individuals and species. It recalls the biologist and evolution scholar Lynn Margulis's important concept of the holobiont, first defined in 1991 as the discrete ecological units formed by groups of biological entities depending on each other and combining their genome in a new, composite version.[50] "Better Together," a promising paper coauthored by Oxman, expands on this idea: "Communities dominate the microbial world; coexisting organisms cannot help but interact."[51] There is strength in unity, and, more precisely, robustness—"the ability to survive perturbation"—which "is an emergent property of microbial communities and is necessary for survival in the wildly fluctuating world."[52] The same can be said about all communities, and this is the most important lesson we can learn from architects and designers who, like Oxman, are looking for new ways forward.

Living in a crucial, complex, and deeply experimental moment, some architects and designers are doing their part by finding new centers of gravity in humans' relationships with other species, with nature, and with one another. They are no longer designing in a vacuum but rather thinking of how their work will be part of ecosystems; they are trying to influence and help manufacturers and developers to define their new powers and responsibilities in a changed context of environmental crisis and social mutations. Those who, like Oxman, engage nature directly, are writing new protocols for their practices. Some selected pairings from Oxman's evolving thoughts help us understand better not only her ideas and her design's evolution but also the essence of her relationship with nature. The "over" proposition that often links the two poles in her nevalogue, as a matter of fact, implies the designer's active role in leading the way, a philosophical choice that becomes a way of life and at the same time an affirmation of the central position of the human species. In Oxman's construct, the role of design is to initiate the reaction and provide interfaces of control. The designer's job, by extension, is "to promote and enable synergy between the built and the grown, the artificial and the natural. . . . The designer then becomes a mediator, a gardener, an alchemist operating across scales and domains to conduct rather than construct."[53] Indeed, it is just this attitude—toward responsive, cooperative, evolutionary change—that we need today.

Notes

1 "Panta chōrei kai ouden menei," attributed to Heraclitus, in Plato, *Cratylus*, trans. Harold N. Fowler (Cambridge, Mass.: Harvard University Press; London: William Heinemann, 1921), 402a. Translation amended by the author.

2 Neri Oxman, statement to the tenure committee, Massachusetts Institute of Technology, 2017. "We have created such frameworks that allow all or most degrees of freedom of the natural phenomenon to be embodied by the object. This advancement and salient feature presents the opportunity to develop material engineering, computational approaches, and digital fabrication techniques that allow us to introduce property gradients with high spatial resolution and multi-functionality across scales."

3 Neri Oxman, "Material Ecology," in Rivka Oxman and Robert Oxman, eds., *Theories of the Digital in Architecture* (London: Routledge, 2013), 322.

4 The series also marked the beginning of a fruitful collaboration with the U.S. 3D-printing company Stratasys, a partnership that is still very much alive today.

5 At the time of writing, the MIT Media Lab is undergoing a deep process of self-examination following the resignation of Joi Ito, its director, in 2019, in the wake of the revelation that the institution had received funds from Jeffrey Epstein, the convicted sex offender and financier, and invited him to tour its premises.

6 Heidi Legg, "Neri Oxman #99," *TheEditorial,* March 2017.

7 Oxman, "Design at the Intersection of Technology and Biology," presentation at TED2015: Truth and Dare" conference, Vancouver, Canada, March 2015.

8 Oxman, "Material Ecology," 321. See Gottfried Semper, *The Four Elements of Architecture and Other Writings* (1851; Cambridge: Cambridge University Press, 1989).

9 Rich Gold, *The Plenitude: Creativity, Innovation, and Making Stuff* (Cambridge, Mass.: MIT Press, 2007).

10 I contend that nothing will ever be accomplished until we also add design: STEAMD.

11 Oxman, "Age of Entanglement," *Journal of Design and Science*, no. 1 (January 2016). Oxman describes the genesis of the Krebs Cycle of Creativity and puts it in context with other events at MIT Media Lab and beyond.

12 Legg, "Neri Oxman #99."

13 Ibid.

14 Dunne & Raby, "A/B Manifesto," dunneandraby.co.uk/content/projects/476/0.

15 Three juxtapositions that don't appear here resonate as highly symptomatic of Oxman's new approach to design: "Scale over Size" (implying that the ecosystem reigns above all; physical dimension is less important than the impact it has on the system); "Process over Product" (investigating new materials and technologies, with objects produced as "muses" and demos); "Heterogeneity over Homogeneity" (an interest in biomaterials, which can be fine-tuned to behave differently in different parts of the designed organism, and to behave intelligently, varying their behaviors in time and space).

16 Legg, "Neri Oxman #99." In her tenure statement, she articulates this idea in better, although more technical, terms (see note 2). She goes on to say, "Thus, the ideas and principles behind *Material Ecology* inspired the need, and generated a scaffold, for innovation in computational form-generation and digital fabrication *increasing the dimensionality of the design space.* To address this need, my team and I are developing new approaches that enable design and production at Nature's scale to seamlessly vary the physical properties of materials at the resolution of a sperm cell, a blood cell, or a nerve cell. Stiffness, color, transparency, conductivity—even smell and taste—can be individually tuned for each 3D pixel (voxel) within a physical object. The generation of products is therefore no longer limited to assemblages of discrete parts with homogeneous properties. Rather—like organs—objects can be computationally 'grown' and 3D printed to form materially heterogeneous and multi-functional constructs." Oxman, statement to the tenure committee.

17 The term "critical design" was first used by Anthony Dunne in *Hertzian Tales: Electronic Products, Aesthetic Experience, and Critical Design* (London: RCA CRD Research Publications, 1999) and later in *Design Noir: The Secret Life of Electronic Objects* (Basel: Birkhäuser, 2001), written with Fiona Raby.

18 Dunne & Raby, "Critical Design FAQ," dunneandraby.co.uk/content/bydandr/13/0.

19 Stuart Candy, "The Future of Everyday Life: Politics and Design of Experimental Scenarios" (PhD diss., University of Hawai'i at Mānoa, 2010), 35. Futures studies as an academic discipline originated in the 1960s. Several authors have written about different classes of futures. In 1997 Weddell Bell explained that researchers should consider "what can or could be" (the possible), "what is likely to be" (the probable), and "what ought to be" (the preferable). Bell, *Foundations of Futures Studies: Human Science for a New Era*, vol. 1, *History, Purposes, and Knowledge* (New Brunswick, N.J.: Transaction, 1997), 73. Trevor Hancock and Clement Bezold created the "futures cone" in 1994, based on a categorization by Norman Henchey, who considered four different kinds of futures: possible futures (what may be); preferable futures (what should be); plausible futures (what could be); and probable futures (what will likely be). See Hancock and Bezold, "Possible Futures, Preferable Futures," *Healthcare Forum Journal* 37, no. 2 (1994): 23–29; and Henchey, "Making Sense of Futures Studies," *Alternatives* 7, no. 2 (1978): 24–28.

20 Oxman, statement to the tenure committee.

21 See William Myers, ed., *Bio Design: Nature Science Creativity* (New York: The Museum of Modern Art, 2012).

22 Oxman, "Material Ecology," 322.

23 Among the many innovative material technologies developed by Oxman and MMG is "high-resolution multi-material modeling and bitmap printing—which enables the design and digital fabrication of structures that can vary their mechanical and optical properties in high spatial and temporal resolutions (ones that often *transcend* the scale of the physical phenomena they are designed to embody)." Legg, "Neri Oxman #99."

24 "Neri Oxman and the Endless House," *The Way I See It* podcast, BBC Radio 3, bbc.co.uk/programmes/m0009bf5. The Endless House project preoccupied Kiesler for a decade, with the first model being built in 1950 and the very last iteration coming in 1960. A commission for MoMA's *Visionary Architecture* exhibition, the last and most important model envisioned the Endless House as a series of cavelike spaces. See also Pedro Gadanho and Phoebe Springstubb, "Endless House: Experimental Archetypes of Dwelling," *Architectural Review*, June 30, 2015.

25 Diller Scofidio + Renfro, "Slow House," dsrny.com/project/slow-house.

26 Michael Hensel, *Performance-Oriented Architecture: Rethinking Architectural Design and the Built Environment* (Chichester, U.K.: John Wiley & Sons, 2013), 23.

27 Stanley Mathews, "The Fun Palace: Cedric Price's Experiment in Architecture and Technology," *Technoetic Arts: A Journal of Speculative Research* 3, no. 2 (2005): 73–91.

28 David Leatherbarrow, "Architecture's Unscripted Performance," in Branko Kolarevic and Ali Malkawi, eds., *Performative Architecture: Beyond Instrumentality* (London: Routledge, 2005), 7.

29 Reyner Banham, *The Architecture of the Well-Tempered Environment* (London: Architectural Press; Chicago: University of Chicago Press, 1969), 11.

30 Ibid., 13.

31 Nicholas Negroponte, *The Architecture Machine: Toward a More Human Environment* (Cambridge, Mass.: MIT Press, 1970), 27. "Our interest," Negroponte wrote, "is simply to preface and to encourage a machine intelligence that simulates a design for the good life and will allow for a full set of self-improving methods. We are talking about a symbiosis that is a cohabitation of two intelligent species." Ibid., 7.

32 Tristan D'Estrée Sterk, "Building upon Negroponte: A Hybridized Model of Control Suitable for Responsive Architecture," in Wolfgang Dokonal and Urs Hirschberg, eds., *Digital Design: 21st eCAADe Conference Proceedings* (Graz, Austria: eCAADe, 2003), 407.

33 Xing Shi, "Performance-Based and Performance-Driven Architectural Design and Optimization," *Frontiers of Architecture and Civil Engineering in China* 4, no. 4 (December 2010): 512–18.

34 Greg Lynn, ed., *Archaeology of the Digital* (Berlin: Sternberg Press; Montreal: Canadian Centre for Architecture, 2013).

35 Rivka Oxman, "Performance-Based Design: Current Practices and Research Issues," *International Journal of Architectural Computing* 6, no. 1 (January 2008): 3–4.

36 Oxman, statement to the tenure committee.

37 Toward realizing this dream, Oxman and her team have recently developed Hybrid Living Materials (HLMs), which they describe as a "fabrication platform, which integrates computational design, additive manufacturing, and synthetic biology to achieve replicable fabrication and control of biohybrids." Rachel Soo Hoo Smith et al., "Hybrid Living Materials: Digital Design and Fabrication of 3D Multi-Material Structures with Programmable Biohybrid Surfaces," in *Advanced Functional Materials*, published ahead of print, December 18, 2019, onlinelibrary.wiley.com/doi/10.1002/adfm.201907401. The HLMs combine living and nonliving components to harness the biological capabilities of organisms within structural materials, thereby producing a new class of uniquely responsive and multifunctional products.

38 The term Fourth Industrial Revolution is commonly attributed to Schwab, who first used it in his book of that same name (London: Penguin Books, 2017). See also Schwab, "The Fourth Industrial Revolution," *Encyclopedia Britannica*, May 2018, in which he predicts that "the Fourth Industrial Revolution heralds a series of social, political, cultural, and economic upheavals that will unfold over the 21st century. Building on the widespread availability of digital technologies that were the result of the Third Industrial, or Digital, Revolution, the Fourth Industrial Revolution will be driven largely by the convergence of digital, biological, and physical innovations." Its theoretical framework, however, has been laid out throughout the end of the twentieth century and the beginning of the twenty-first by academics and entrepreneurs, such as Craig Venter and Daniel Cohen in "The Century of Biology," *New Perspectives Quarterly* 1, no. 4 (2004).

39 Oxman, statement to the tenure committee.

40 Christina Cogdell, *Toward a Living Architecture? Complexism and Biology in Generative Design* (Minneapolis: University of Minnesota Press, 2019).

41 Stewart Brand launched the *Whole Earth Catalog* at the height of the counterculture movement in San Francisco, inspired by R. Buckminster Fuller, DIY culture, and the belief that information, above everything, wants to be free. The first issue was published in the fall of 1968. Its now instantly recognizable covers featured some of the first images of Earth viewed from space, taken during the early space missions, along with the catalog's slogan and mission statement: "access to tools." Inside its pages, readers were offered Brand's vision of a new social order and were given the skills and tools necessary to embrace such a future. The catalog recommended everything from agricultural and mechanical equipment to books like Melford Spiro's *Kibbutz: Venture in Utopia* or Norbert Wiener's *Cybernetics, or Control and Communication in the Animal and the Machine*, to new digital technologies such as the Hewlett-Packard 9100A. Even though the publication had a short life span—it was published quarterly from 1968 through 1971, and then occasionally thereafter—it left a lasting impression, influencing everyone from Steve Jobs (who once described it as "Google in paperback form") and Steve Wozniak to *Wired* magazine founder Kevin Kelly.

42 The early works (Raycounting, Subterrain, Armour, and Cartesian Wax) and the Imaginary Beings series—poetically themed pretexts to fine-tune techniques, materials, and goals—represent moments of free experimentation that straddle the two categories.

43 This is an off-label use of a term from the video games world: a massively multiplayer online game is a game with thousands of players active at the same time on the same server.

44 Here, silkworms are treated the way an artisan would be treated by an Italian architect: counting on the artisans' skills and expertise, the architect would not insult them with too many details in the drawings. After all, silkworms are already highly accomplished builders.

45 Paul Kalanithi, *When Breath Becomes Air* (New York: Random House, 2016).

46 The XXII Triennale di Milano, titled *Broken Nature: Design Takes On Human Survival*, was organized by me, with Ala Tannir, Laura Maeran, and Erica Petrillo. The exhibition highlighted the concept of restorative design and studied the state of the threads that connect humans to their natural environments. Oxman's Totems was one of four projects commissioned for the show.

47 "One of design's most fundamental tasks is to help people deal with change. Designers stand between revolutions and everyday life. . . . [They] have the ability to grasp momentous changes in technology, science, and social mores and to convert them into objects and ideas that people can understand and use." Paola Antonelli, "Design and the Elastic Mind," in *Design and the Elastic Mind* (New York: The Museum of Modern Art, 2008), 14–15.

48 Bertrand Russell, *Authority and the Individual: The Reith Lectures* (1949; London: Routledge, 2009), 75.

49 See Jamie Dimon, the CEO of JPMorgan Chase, in Elizabeth Dilts, "Top U.S. CEOs Say Companies Should Put Social Responsibility above Profit," Reuters, August 2019.

50 Lynn Margulis and René Fester, eds., *Symbiosis as a Source of Evolutionary Innovation* (Cambridge, Mass.: MIT Press, 1991).

51 Stephanie G. Hays et al., "Better Together: Engineering and Application of Microbial Symbioses," *Current Opinion in Biotechnology* 36 (August 2015): 40. "Interactions include touching, using dedicated signals, horizontal gene transfer, 'competitive or cooperative' scenarios where microbes compete for a provide resources, and alteration of environment to influence the growth of neighbors."

52 Ibid.

53 Oxman, e-mail to the author, July 7, 2019.

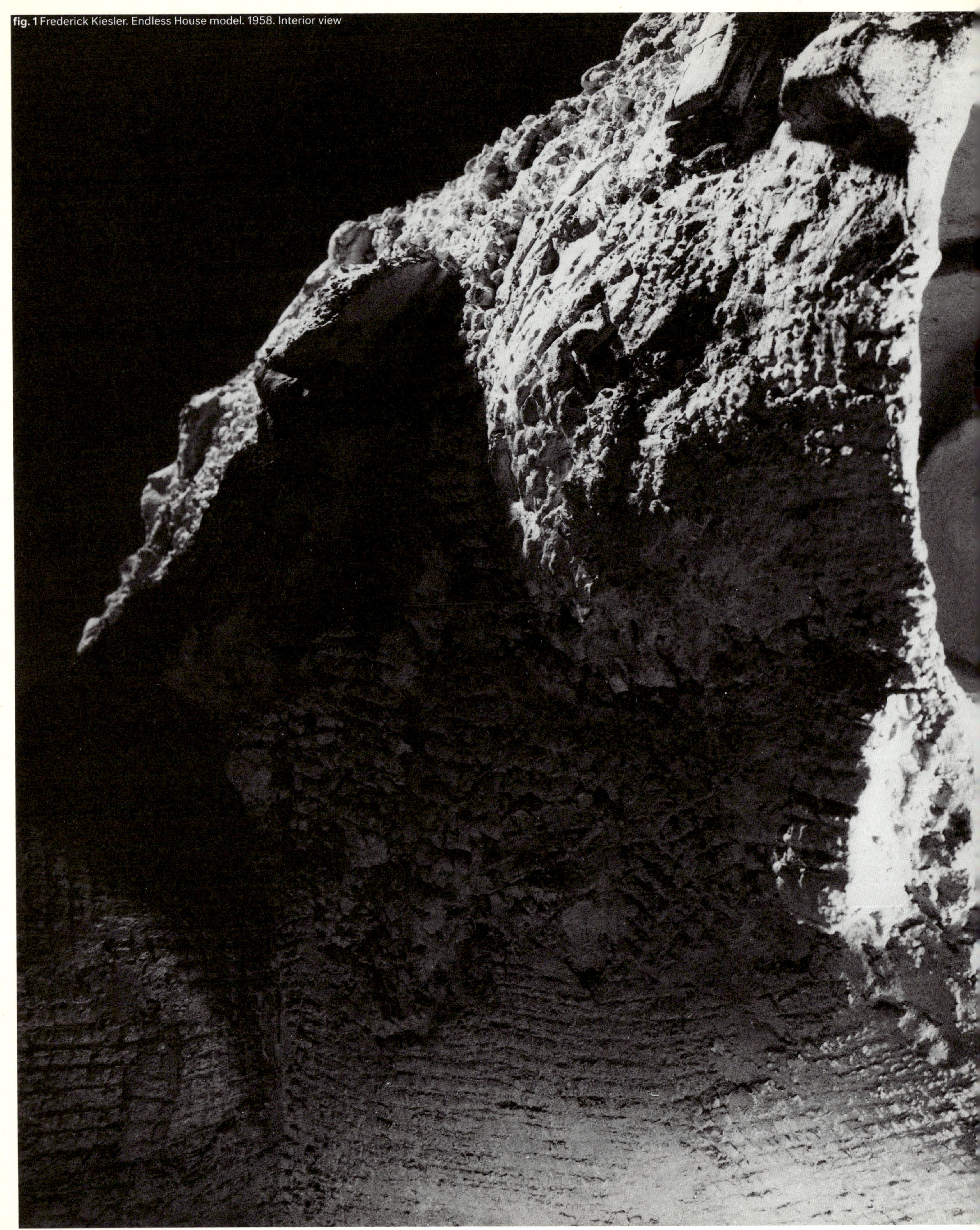

fig. 1 Frederick Kiesler. Endless House model. 1958. Interior view

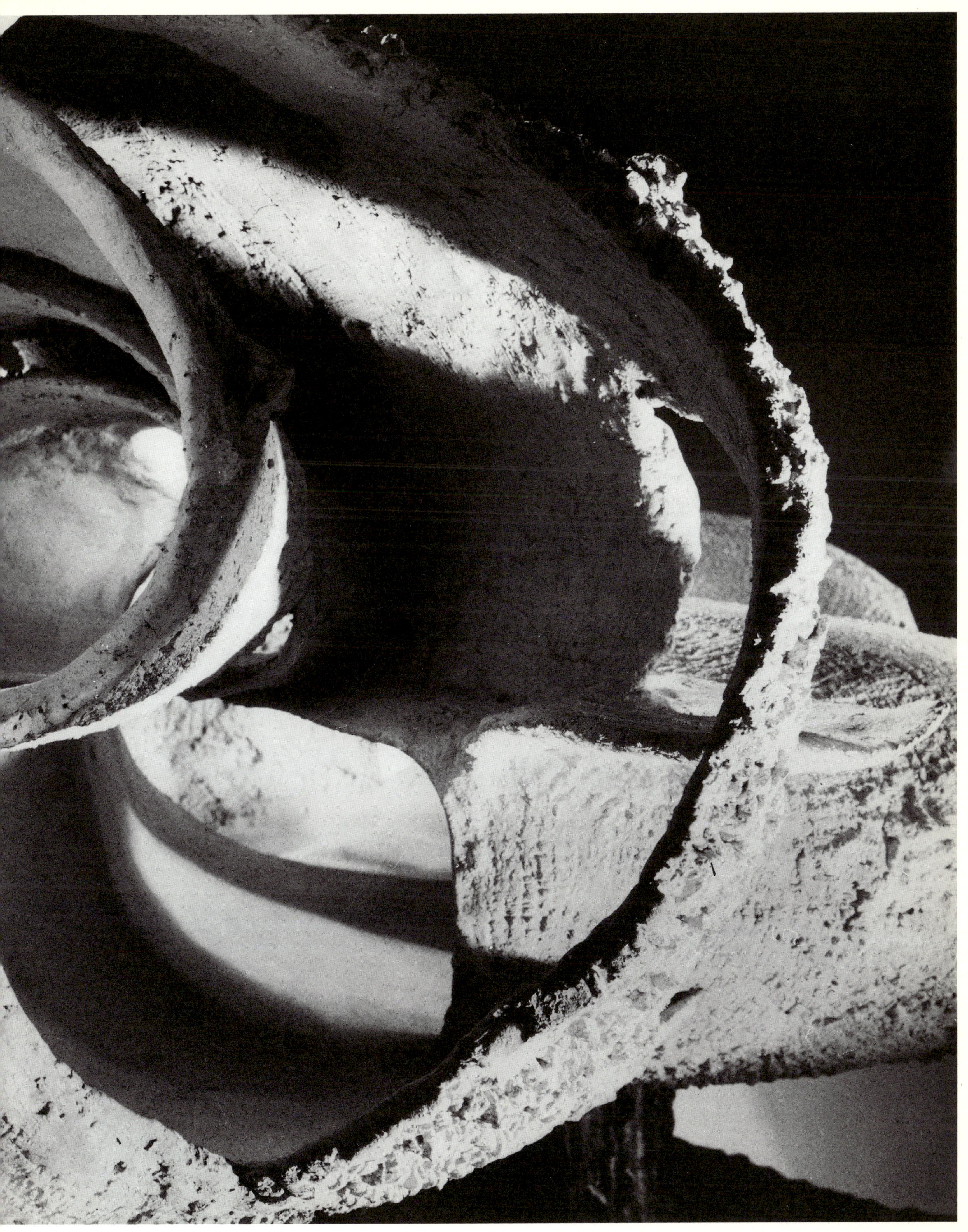

LIMBS OF NATURE

HADAS A. STEINER

OUR WORK IS NOT A PHILOSOPHY AND NOT A SYSTEM FOR ACQUIRING COGNITION OF NATURE, IT IS A LIMB OF NATURE.
—EL LISSITZKY, "NASCI," 1924[1]

Neri Oxman has taken an insinuation found in ancient scripture as a professional challenge.[2] The earth, as biblical commentary tells us, was directed to yield a fruit-bearing tree that was physically continuous from root to crop. Instead, nature delivered a variety fragmented into trunk, branch, leaf, and fruit.[3] Thus did a variable material condition become the norm, rather than the intended cohesive ideal. Oxman aspires to correct this primal insubordination and fabricate nature as it could have been. An interest in parables, though, does not obscure the contemporary spirit of a practice that requires the latest tools. The provocation offered by Oxman's work, moreover, is particularly timely, directed as it is at the exhausted dichotomy of organic growth and machine assembly, which constrains architects from envisaging better environments.

While the techniques that sustain Oxman's work are recent, its particular fusion of biological, technological, and design expertise has a robust history. That historical foundation is itself built on the classical alignment of architecture with the natural world. Once Leon Battista Alberti demonstrated in *De re aedificatoria* (*On the Art of Building*, 1485), the first book to be dedicated to the art of architecture, that the principles of harmonious unity demonstrated in nature were reproducible in the built world, the association between them has been taken for granted (**fig. 2**).[4] What was culturally designated by the term "nature," however, mutated over time. When, for example, biological disciplines emerged, early in the nineteenth century, as independent from the larger scope of natural history, the focus of previous centuries on anatomy and classification yielded to evolutionary reasoning in both science and design (**fig. 3**).

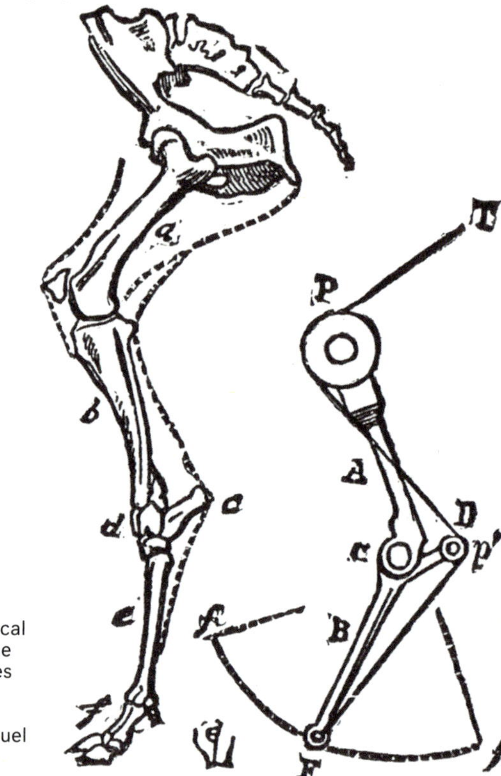

fig. 3 A mechanical application of the action of muscles and tendons, illustrated by Eugène-Emmanuel Viollet-le-Duc, 1854–68

As industrialization progressed, however, technology vied with nature as a factor in architectural design. By the end of the nineteenth century, the theory that architecture developed in an evolutionary manner had become increasingly correlated with its expression as an industrial technology; by the turn of the twentieth, industrial forms could even be construed to be the result of an organic process—the functional outcome of technological optimization. Thus a wide ideological range of modern architects, from Louis Sullivan to Ludwig Mies van der Rohe, laced their technological writing with biological terminology without an eyebrow being raised among their contemporaries (**fig. 4**).[5]

Despite the embrace of biological thinking by early modernists, detractors of the style dismissed modernism as the product of mere mechanical logic over the subsequent decades.[6] Yet a renewed interest in the integration of biology, technology, and design is now emerging in response to societal and disciplinary transformations. Design practices that utilize

fig. 2 Karl Blossfeldt. *Adiantum pedatum*. 1898–1926

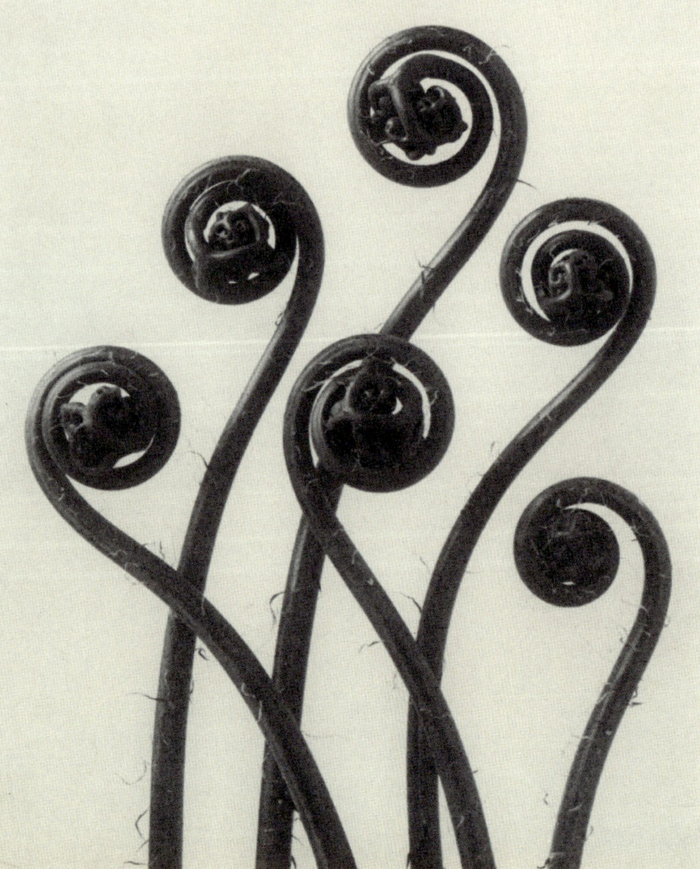

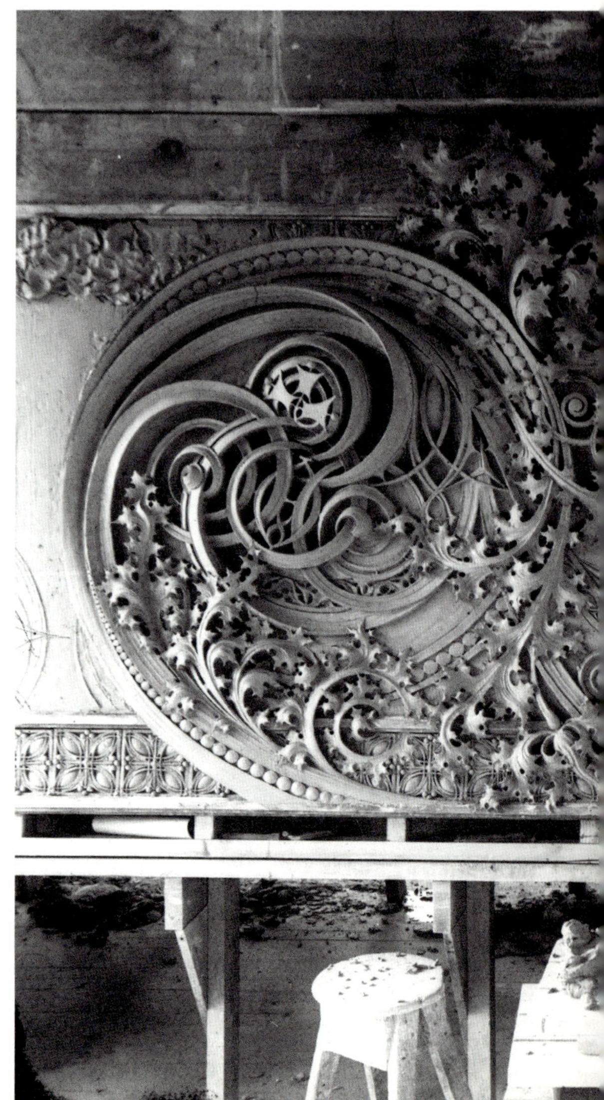

computation in ways that mimic natural routes of growth have proliferated, as have the software platforms that support the generation of biomorphic forms (**fig. 5**); practices focused on conservation have been emulating organic systems as an antidote to environmental degradation. The flow of information from science to design in these cases, however, is almost always one-sided and analogical: design modeled on nature as found. Genetic algorithms and green buildings thus reinforce the metaphorical ways in which architecture has deployed nature.[7]

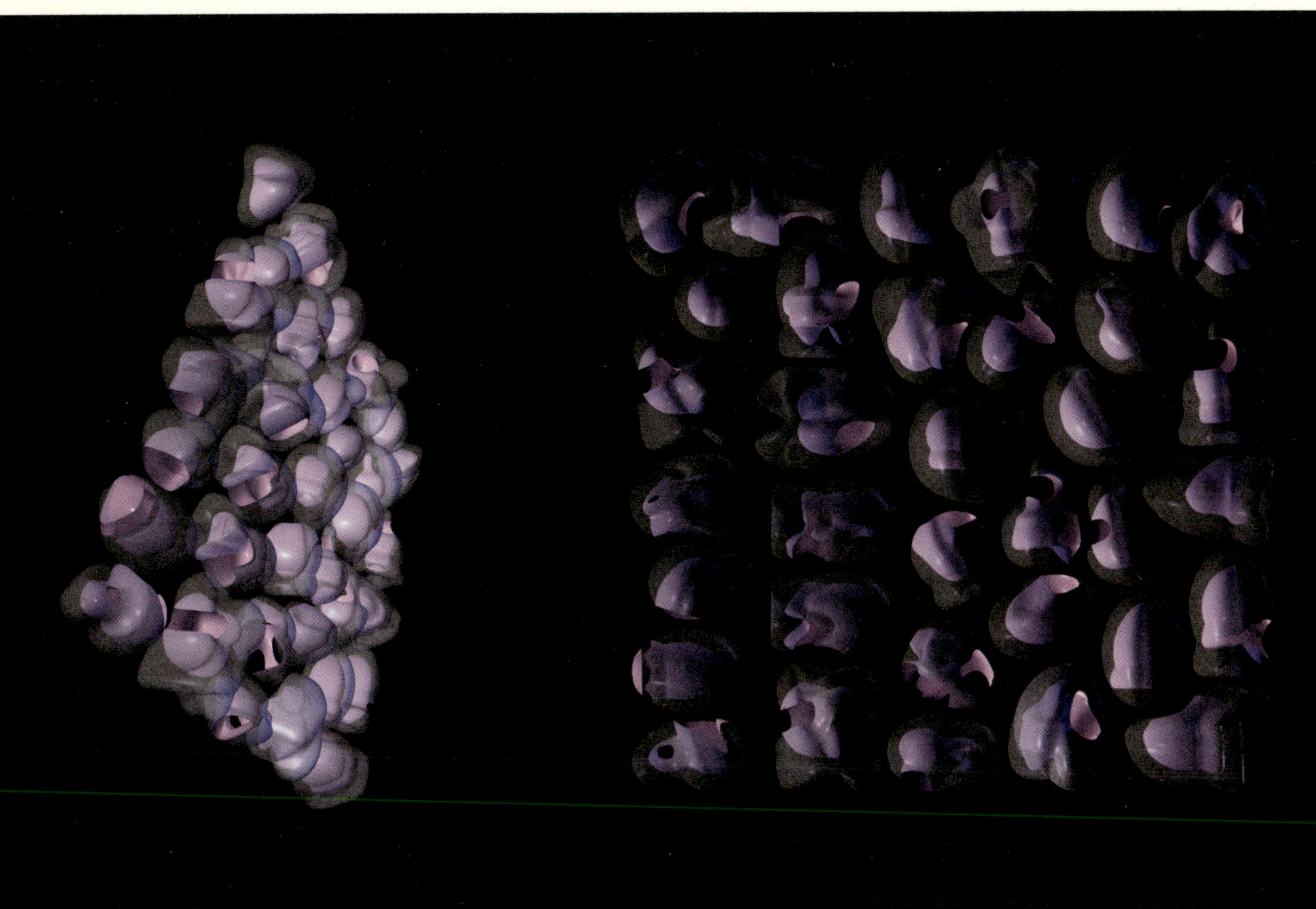

ig. 5 Greg Lynn. Embryological House. 997–2001. Rendering

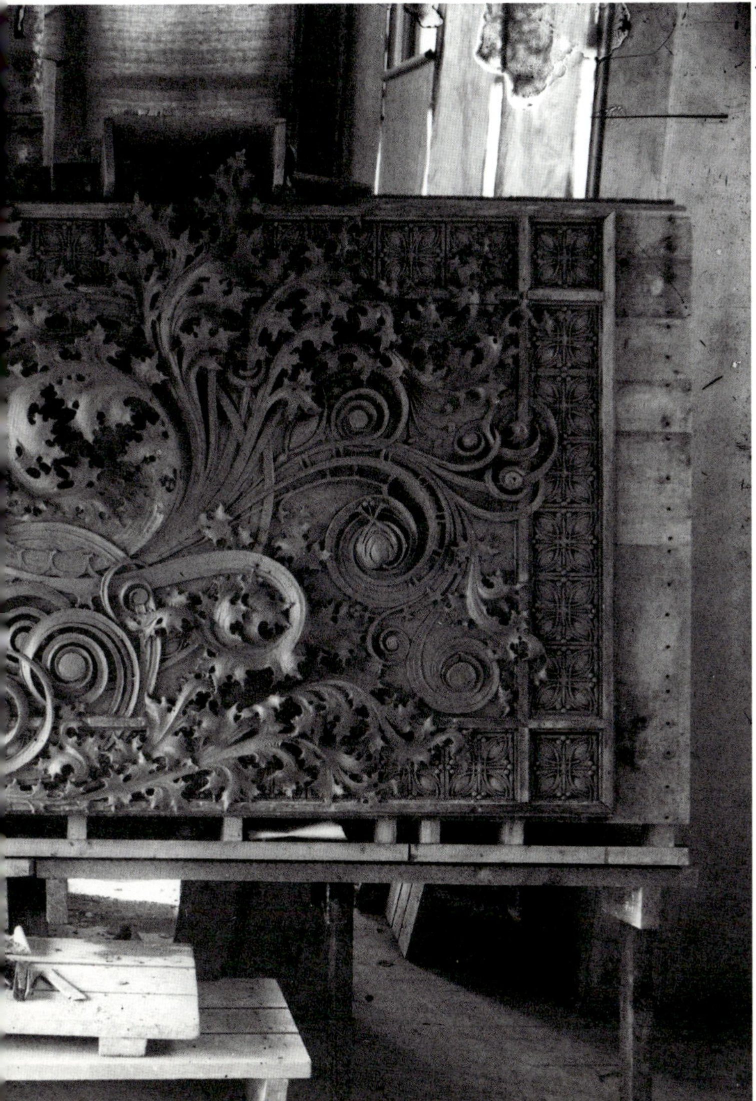

fig. 4 Louis Sullivan. Carson, Pirie, Scott and Company Building, Chicago. 1899. Terracotta detail

Oxman, by contrast, isolates moments in which design is a catalyst for technological development, so that design augments—rather than replicates—the natural state of things. This objective revives the aspirations of the avant-garde designers who labored at the architectural margins well before the development of the tools that enabled such a reciprocal relationship. R. Buckminster Fuller, who referred to himself as a design scientist, is perhaps the best-known champion of the conviction that designers should build a better nature. Using an interdisciplinary approach, Fuller devised products at scales ascending from the body to the astronomical, which addressed social afflictions from domestic hygiene to environmental depletion. Analogical thinking of any kind, he believed, would only ensure continuation of the inadequate status quo.[8] Human sustainability necessitated that found conditions be improved through an expansion of material and technological palettes—with the result of any chemical reaction defined as a natural substance: nature was a property of behavior rather than of genesis. As such, Fuller embraced the construction potentials of synthetics, including steel and plastics. While an obvious disparity separates manufactured compounds from the biopolymers in

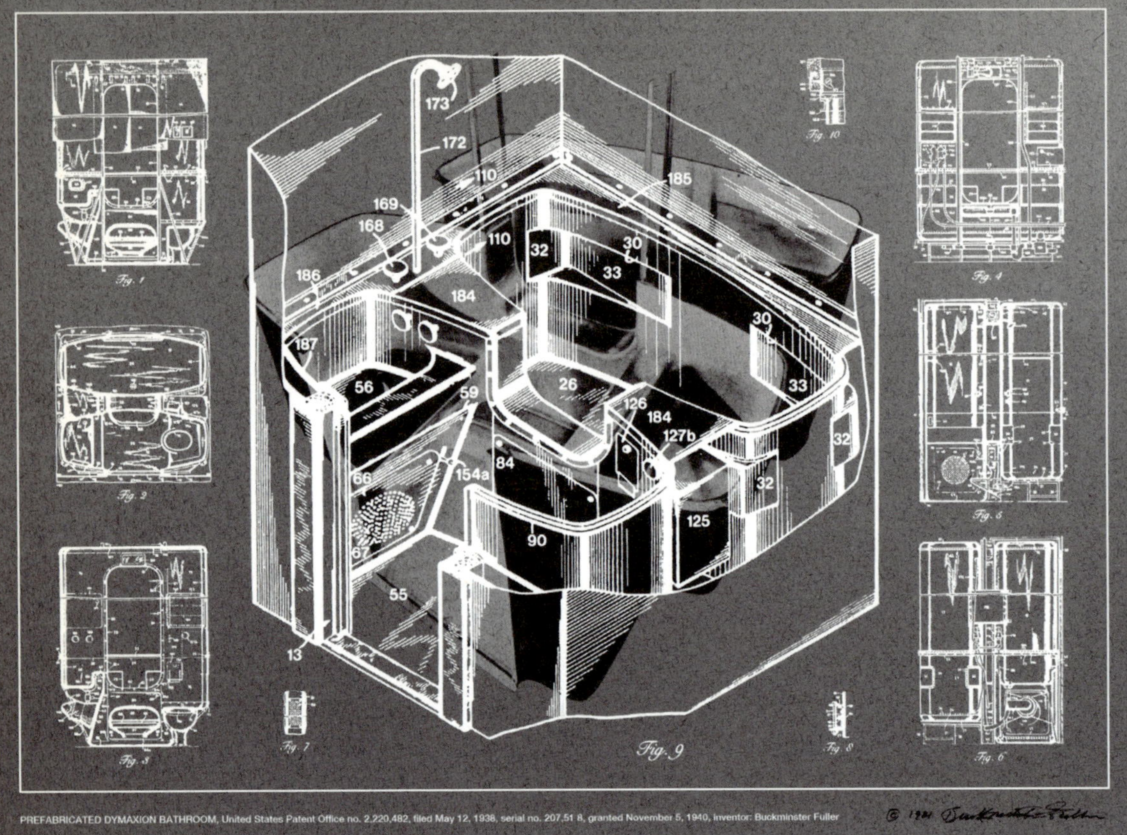

fig. 6 R. Buckminster Fuller. Prefabricated Dymaxion Bathroom. 1936

crustaceans, Fuller and Oxman share an interest in the social benefits made possible by the continuous properties of new materials: Fuller, for example, promoted the plastic version of his seamless Dymaxion Bathroom (1936) as a design critical to public health, as it eliminated the bacteria-harboring crevices where diseases such as tuberculosis festered (**fig. 6**).

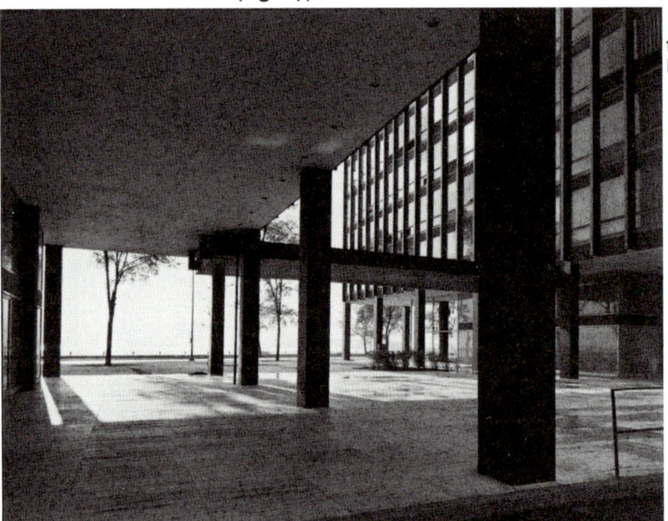

fig. 7 Ludwig Mies van der Rohe. Lake Shore Drive Apartments, Chicago. 1951

Oxman, in a similar vein, has devised seamless outfits and helmets out of acrylic polymers that respond to the physiology of the body. Each of the Imaginary Beings (2012; page 134), a collaboration with Craig Carter based on the fantastical creatures of Jorge Luis Borges's book of 1957, intensifies the capabilities of the human body through an application of material technology. In 2013, with the Dutch fashion designer Iris van Herpen, Oxman developed a series of printed ensembles that, like human skin, would take on distinctive properties through modulation rather than the accretion of elements. This experimental approach to materiality distances Oxman, as it did Fuller, from the reiteration of the industrial module that was the hallmark of mainstream architectural modernism by the 1950s (**fig. 7**). While Fuller's geodesic domes and tensegrity-based constructions—innovative structural typologies that responded to the behavior of contemporary materials under the forces of tension and compression—were assemblages of factory-produced components, the elements in these integrated systems functioned as a responsive continuum (**fig. 8**); true to Fuller's beliefs about material behavior, the distribution of force undercut the conventional logic in which physical composition varies based on function. This logic can also be defied by biomaterials, as it is in Oxman's Aguahoja I (2018; page 74), which uses crustacean-derived biopolymers that oscillate in structural performance. Structural dynamism far outweighs formal concerns in these designs.

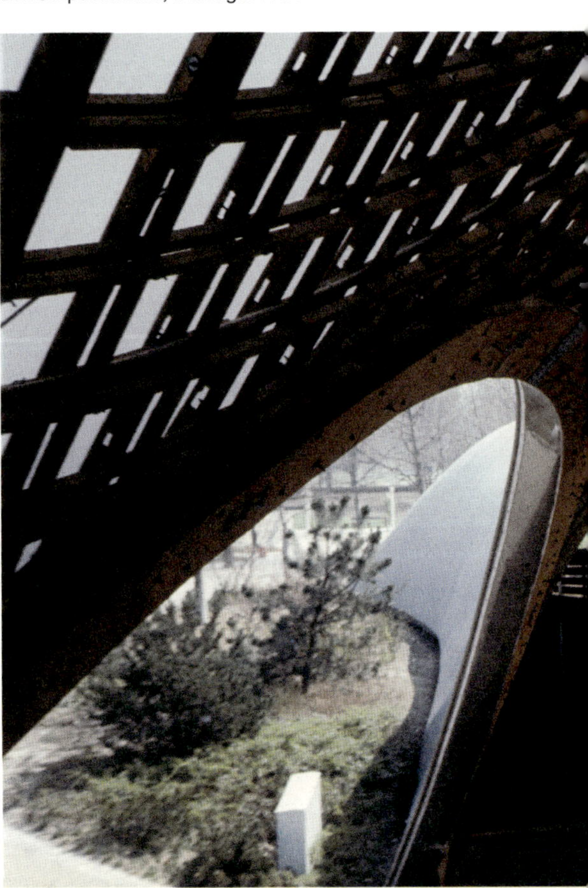

Although the possibilities of compositional continuity went mostly ignored by Fuller's peers, the implications of his dynamic lightweight structures were finally appreciated in the 1960s and '70s by neo-avant-garde architects such as Frei Otto, who was innovative in his use of tensile and membrane structures to produce organic forms, and Cedric Price, who employed tensegrity in his 1962 collaboration with the structural engineer Frank Newby for a bird habitat in the London Zoo (**figs. 9, 10**).[9] When the responsiveness of new materials was combined with the adaptive attributes of new digital technologies, as it was in the work of the avant-garde group Archigram, architecture was even more definitely articulated as a consumer product to be placed at the service of lifestyle (**fig. 11**). The reemergence of the inclination for structural dynamism in the 1960s, then, came with a political shift from the architectural to the human scale: if the advent of industrial technology had shaped the early modernist agenda to shelter the homeless masses after World War I, the possibilities of emergent digital technologies were now encouraging the pursuit of architecture responsive to daily routine, lifespan, the weather. Schemes for such reflexive architectures advanced the idea that new technologies would bring about a consumer-oriented way of life.

ig. 8 Buckminster Fuller with ensegrity model, 1979

Housing was the first program addressed collectively by modernists. In this, they had been responding to an urgent collective need, but the home also served as a test case for the principle of *Existenzminimum*, or the minimum essentials required for living. Since the basic architectural unit of the house formed the cladding of the elemental social unit—the family, the couple, the loner—the expression of an individual nucleus could be multiplied as needed to create urban conditions. When the idea of the dwelling was called upon again in the 1960s and '70s, it was in order to investigate the values of personal freedom that had come to replace postwar ideals of austerity. Architecture at the personal scale became a site of radical investigation. Marshall McLuhan's notion of clothing as the most immediate layer of enclosure, for example, led designers to develop wearable housing schemes that wrapped architecture close to the body.[10] The Suitaloon (1966) by Michael Webb, for example, replaced the paradigm of the house as machine with a proposal for a garment that could be inflated into a habitable space whenever and wherever the wearer wanted (**fig. 12**). Wearable architecture was animated by technology to adapt to the changing conditions, needs, and aspirations of the body.

ig. 9 Carlfried Mutschler and Frei Otto. Iultihalle, Mannheim, Germany. 1975. iterior view

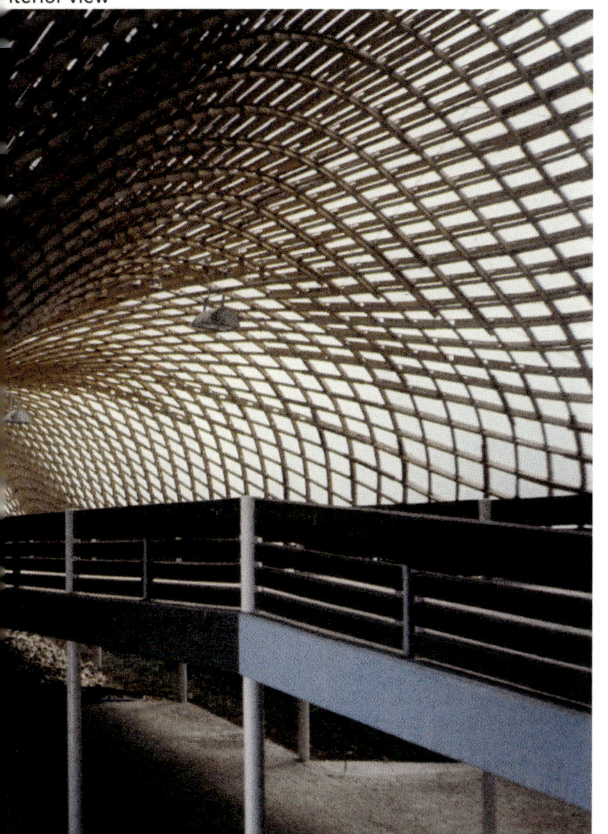

fig. 10 Cedric Price and Frank Newby. Snowdon Aviary, London Zoo. 1964

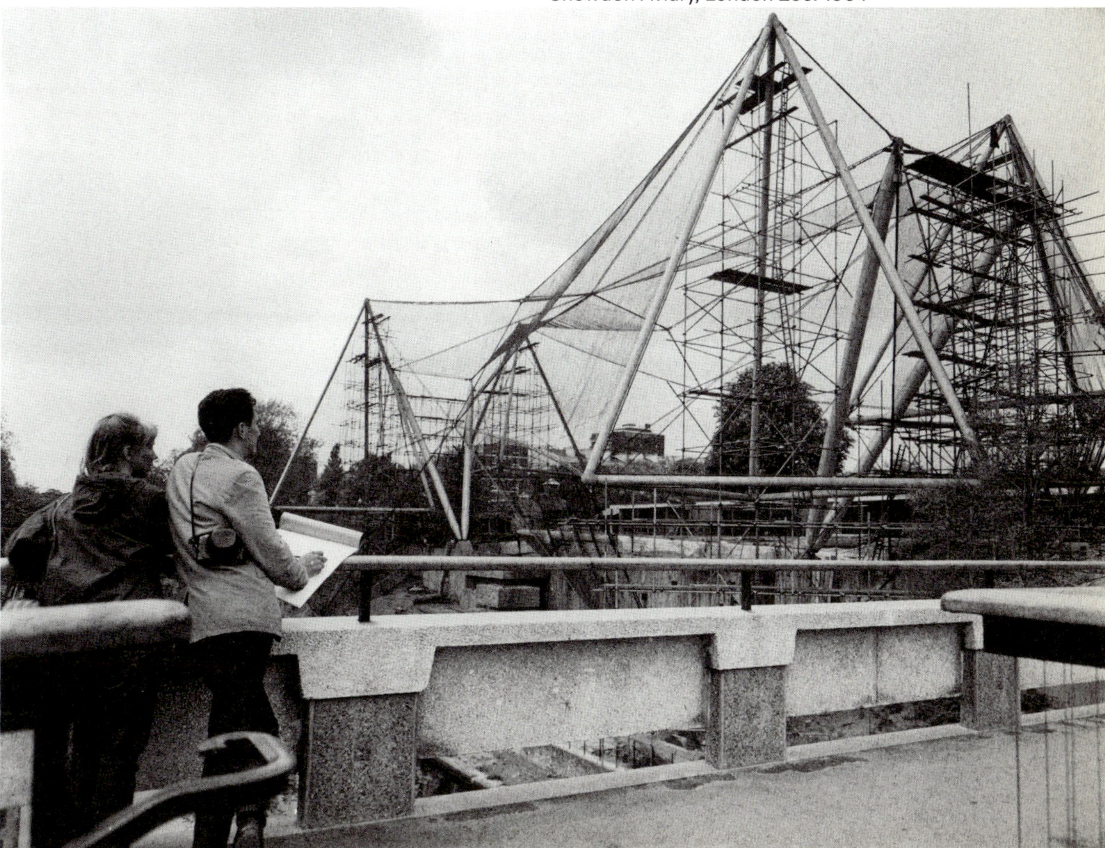

Another critical demonstration of evolving domestic architecture endowed with the responsibility of cultivating society's physical and mental well-being was Frederick Kiesler's ongoing Endless House (1947–60; **figs. 1, 13**).[11] The project was the direct extension of Kiesler's belief that the fundamental interconnectedness of all domains of life should be enhanced by architecture. Endless House, with its continuous, fluid design, was intended to displace the conception of architecture as a discrete series of angular and alienating boxes. The primacy of form that had held sway in architectural production would be reduced to but a momentary capture, or what El Lissitzky expressively called "a stopping-place on the road of becoming."[12]

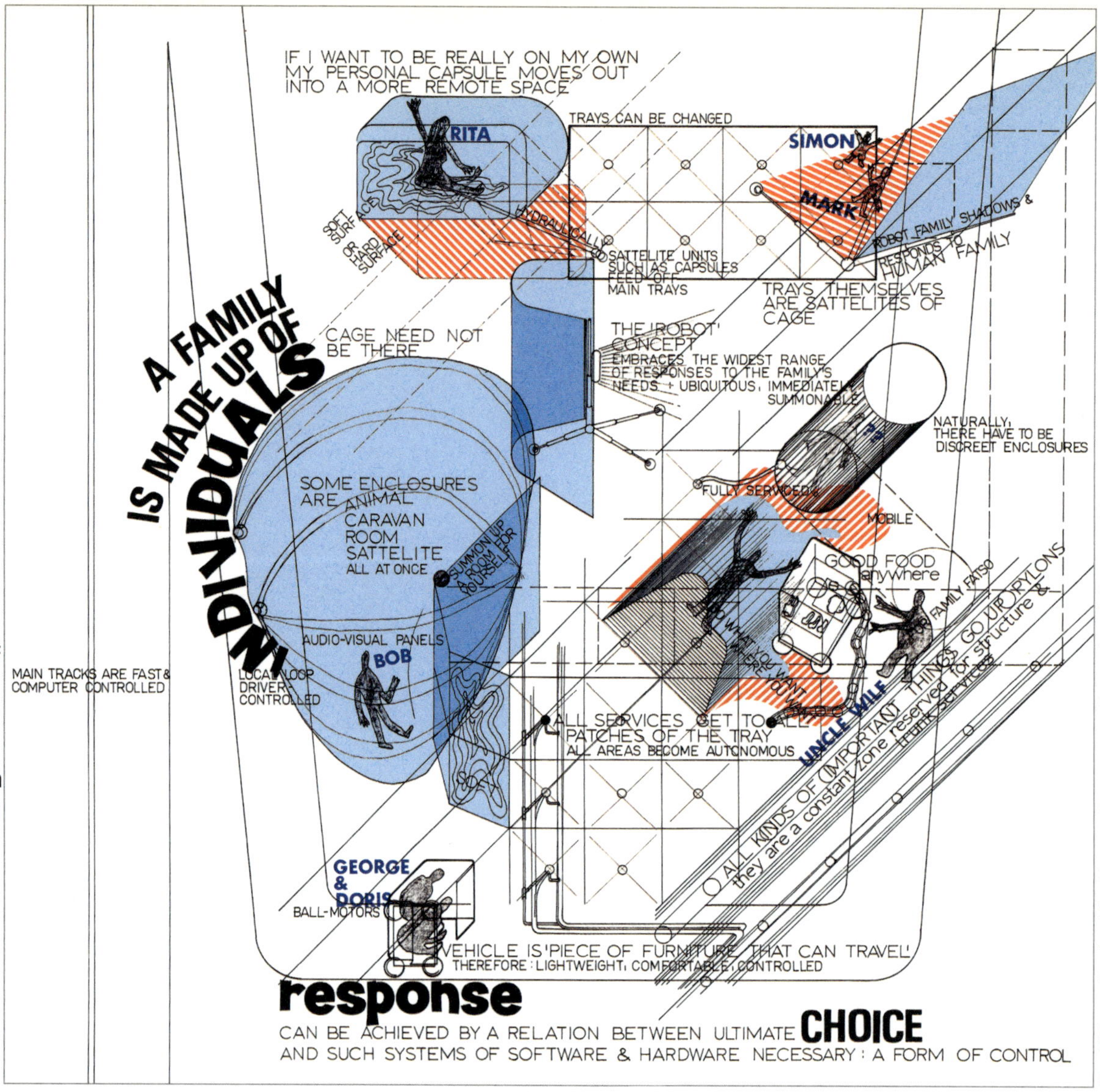

fig. 11 Peter Cook of Archigram. Responses, Control and Choice. 1966

Kiesler's interactive philosophy took its cue from a theory known as biotechnics, which had been influentially propagated among avant-garde practitioners such as László Moholy-Nagy and Mies by the writings of the botanist Raoul Francé. Biotechnics maintained that the interconnectivity of nature, technology, and design would ensure a better *Lebenslehre*, or way of living well.[13] But Kiesler, like Fuller, wanted design not only to enable life but also to *shape* it along ecological lines. Organisms would structure their environments and be fashioned by them in return, a relationship more like the current understanding of the interplay between genes and the environment that has destabilized the earlier duality of organismal and molecular biology. If the environment is understood to have an impact on the development of living things, then design can be implemented as an intentional tactic in the organization of nature. Structure, as Oxman has demonstrated in the Silk Pavilion (2013 and 2019; pages 96, 106), can thus integrate users with all aspects of life, along lines that Kiesler could only have conceived.

Experimental design practices in the biotechnical spirit—in which design is an engine of nature—have the potential for impact across environmental scales, especially now that technology has caught up with speculation. Within the optimism of this tale, however, there remains a valuable historical caution about faith in the technologically driven *Lebenslehre*. Avant-garde and commercial forces have been intertwined since the mass production of the first synthetic materials; the production of steel and concrete may have allowed for the spatial conceptions of modernism, replete with social implications, but it also produced the economic juggernaut of the modern construction industry and its environmental impact. Now that technological development is so visibly bound up with the aesthetics of consumerism, the packaging of architecture and design in terms of lifestyle is a practice that must be open to ethical consideration: what role, after all, is cutting-edge architecture required to play in the improvement of social conditions? From the ecology of shrimp farming required to produce the biopolymers used in Oxman's architectural applications to her utilization of silkworm labor in the fabrication of enclosures, the relationship of nature to design will always exact its own costs.

fig. 12 Michael Webb of Archigram. Dave and Pat join their Suitaloons by pressing together their kneepads. 1966

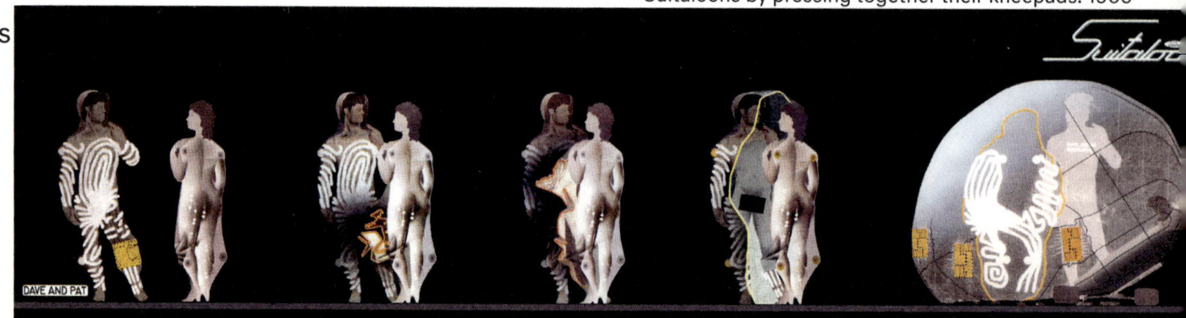

:aptions

ig. 1 Frederick Kiesler. Endless House model. 958. Interior view. Architecture & Design Study Center, The Museum of Modern Art, New York

ig. 2 Karl Blossfeldt. *Adiantum pedatum*. 898–1926. Gelatin silver print, 11 11/16 × 9 3/8 in. 29.7 × 23.8 cm). The Museum of Modern Art, New York. Thomas Walther Collection. Gift of James Thrall Soby, by exchange

ig. 3 Illustration of a mechanical application of the action of muscles and tendons, in *Dictionnaire raisonné de l'architecture française du XIe au XVIe siècle*, by Eugène-Emmanuel Viollet-le-Duc, 1854–68. Vol. 4, fig. 48

ig. 4 Louis Sullivan. Carson, Pirie, Scott and Company Building, Chicago. 1899. Terracotta detail. Ryerson and Burnham Archives, The Art Institute of Chicago

ig. 5 Greg Lynn. Embryological House. 997–2001. Rendering. Canadian Centre for Architecture, Montreal. Gift of Greg Lynn

ig. 6 R. Buckminster Fuller. Prefabricated Dymaxion Bathroom. 1936. Silk-printed llustration for patent, 30 × 40 in. 76.2 × 101.6 cm)

ig. 7 Ludwig Mies van der Rohe. Lake Shore Drive Apartments, Chicago. 1951. Photograph by Julius Shulman, 1963. Getty Research Institute, Los Angeles

fig. 8 Fuller with tensegrity model, 1979. Estate of R. Buckminster Fuller

fig. 9 Carlfried Mutschler and Frei Otto. Multihalle, Mannheim, Germany. 1975. Interior view. Südwestdeutsches Archiv für Architektur und Ingenieurbau (saai), Karlsruhe, Germany

fig. 10 Cedric Price and Frank Newby. Snowdon Aviary, London Zoo. 1964

fig. 11 Peter Cook of Archigram. Responses, Control and Choice. 1966. Photomechanical print on board, from original ink drawing, 18 5/8 × 15 11/16 in. (47.4 × 39.8 cm). Archigram Archives

fig. 12 Michael Webb of Archigram. Dave and Pat join their Suitaloons by pressing together their kneepads. 1966. Ink, airbrush, and overlays on tracing paper with digital additions, 31 × 8 1/4 in. (78.7 × 21 cm). Archigram Archives

fig. 13 Frederick Kiesler. Endless House Project, Plan. 1951. Marker and color pencil on tracing paper, 14 × 17 1/2 in. (35.6 × 44.5 cm). The Museum of Modern Art, New York. Purchase

Notes

1 El Lissitzky, "Nasci" (1924)," in Sophie Lissitzky-Küppers, *El Lissitzky: Life, Letters, Texts* (Greenwich, Conn.: New York Graphic Society, 1968), 351.

2 Neri Oxman, "Design at the Intersection of Biology and Technology," presentation at "TED2015: Truth and Dare" conference, Vancouver, Canada, March 2015.

3 Rashi, Genesis 1:11. The Earth was cursed for this insubordination.

4 On Alberti's position, see Caroline van Eck, *Organicism in Nineteenth-Century Architecture: An Enquiry into Its Theoretical and Philosophical Background* (Amsterdam: Architectura & Natura, 1994), 20–21.

5 See Detlef Mertins, "Where Architecture Meets Biology," interview by Joke Brouwer and Arjen Mulder, in Brouwer and Mulder, eds., *Interact or Die!* (Rotterdam: V2 Publishing, 2007), 110–31.

6 This assessment of modernism was launched by postmodern critics such as Robert Venturi and Charles Jencks. See for example, Venturi, *Complexity and Contradiction in Architecture* (New York: The Museum of Modern Art, 1966); and Jencks, *The Language of Post-Modern Architecture* (New York: Rizzoli, 1977).

7 The subject of biological analogy is thoroughly covered by Philip Steadman in *The Evolution of Designs: Biological Analogy in Architecture and the Applied Arts* (Cambridge: Cambridge University Press, 1979).

8 Fuller often critiqued the works of modernism as using form by way of analogy, rather than actually using technology to make real advances to the human condition.

9 Hadas A. Steiner, "Birds of a Feather: Habit, Habituate, Habitat, Habitivity," in Charissa N. Terranova and Meredith Tromble, eds., *The Routledge Companion to Biology in Art and Architecture* (New York: Routledge, 2017), 71–89.

10 Marshall McLuhan, "Clothing," in *Understanding Media* (1964; Cambridge, Mass.: MIT Press, 1994), 119.

11 In 2019, Oxman chose Endless House as the subject of a podcast produced by MoMA and BBC Radio 3 on objects in MoMA's collection. "Neri Oxman and the Endless House," *The Way I See It* podcast, BBC Radio 3, bbc.co.uk/programmes/m0009bf5.

12 Lissitzky, "Nasci," 347.

13 See Mertins, "Where Architecture Meets Biology." Architects in general came to know of biotechnics through the work of Raoul Francé. It was persuasively deployed as a theory by the planner Patrick Geddes and the critic Lewis Mumford, among others. See Oliver A. I. Botar and Isabel Wünsche, eds., *Biocentrism and Modernism* (Farnham, U.K.: Ashgate, 2011).

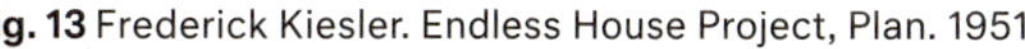

ig. 13 Frederick Kiesler. Endless House Project, Plan. 1951

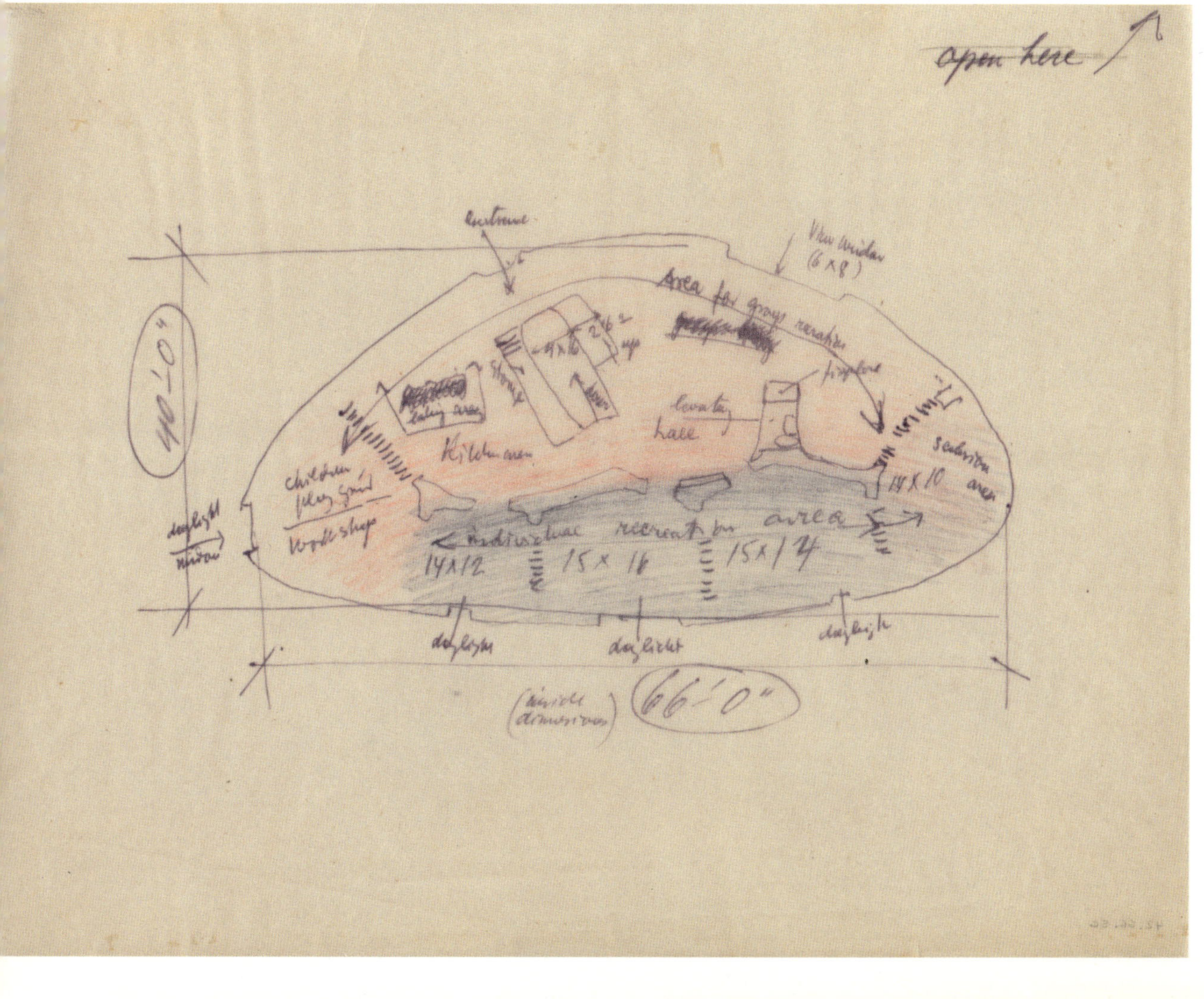

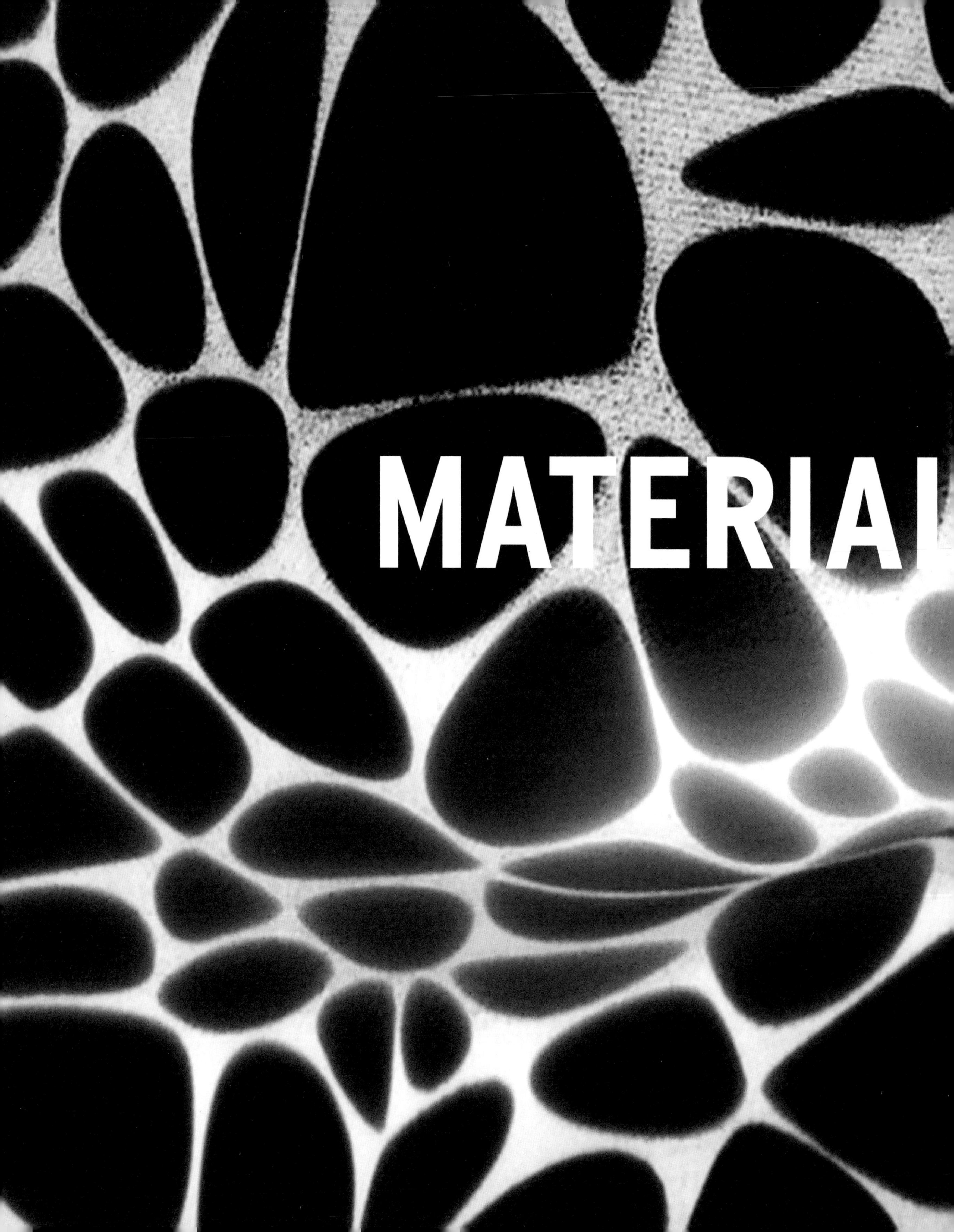

MATERIAL

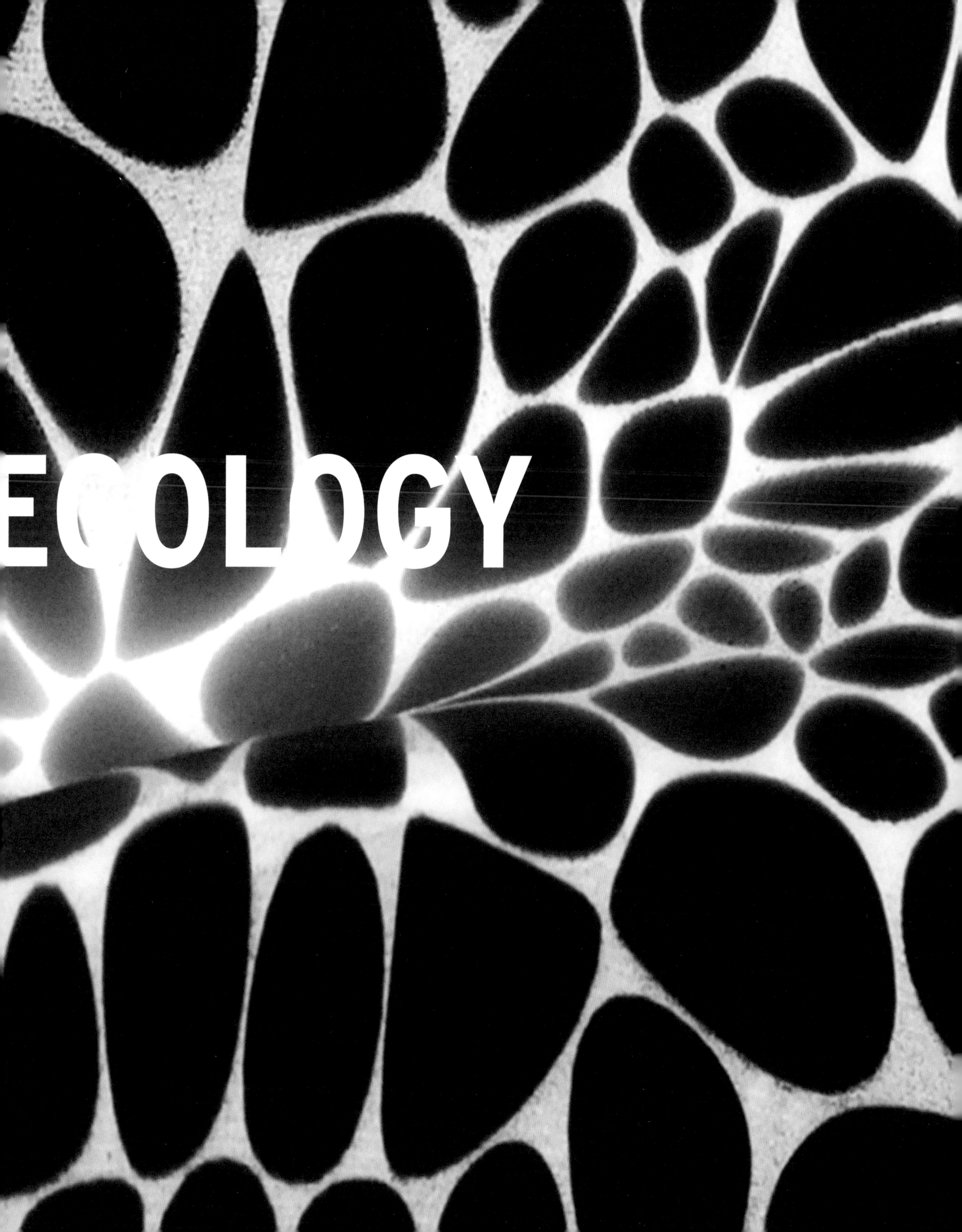

ECOLOGY

ARMOUR / METAMESH

Neri Oxman
Armour. 2007
Photopolymers
17 ½ × 7 ½ × 5 in.
(44.5 × 19.1 × 12.7 cm)
Produced by Stratasys Ltd.
Collaborators and contributors: ARRK Product Development Group Ltd., Tangible Express

Neri Oxman and
The Mediated Matter Group
MetaMesh. 2015
Computational model
An MIT Media Lab project
Research team: Jorge Duro-Royo, Katia Zolotovsky, Laia Mogas-Soldevila, Swati Varshney, Mary C. Boyce, Christine Ortiz, Neri Oxman
Collaborators and contributors: Institute for Collaborative Biotechnologies, MIT Institute for Soldier Nanotechnologies, National Security Science and Engineering Faculty Fellowship Program, Stratasys Ltd.

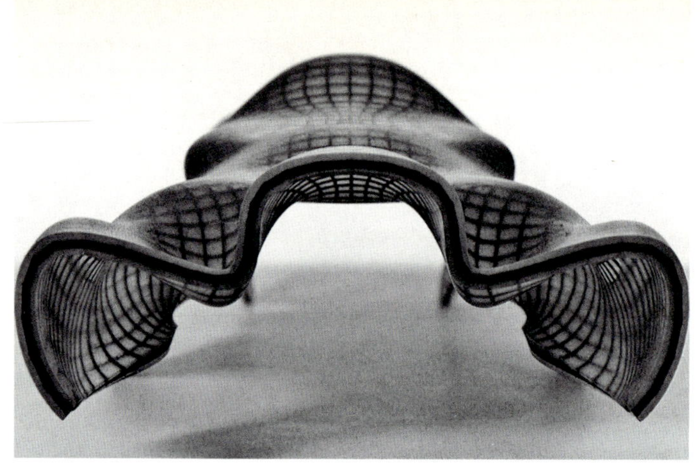

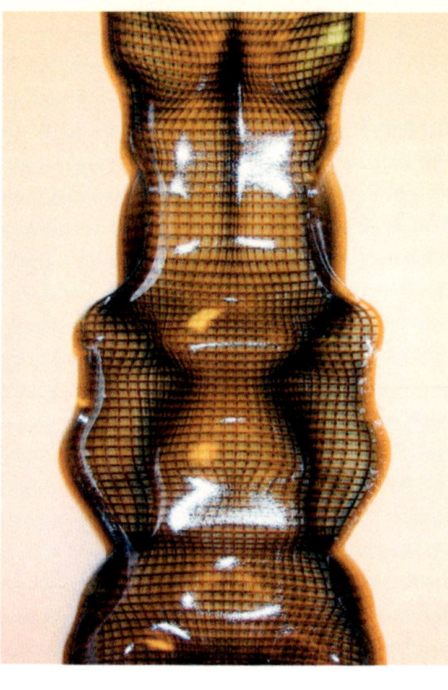

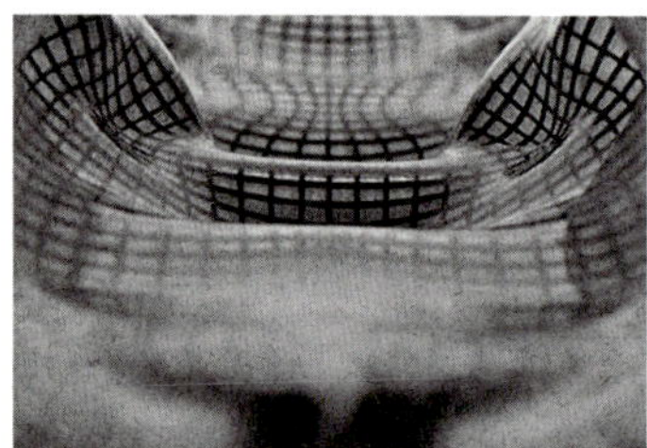

The structures of organisms in the biological world perform a multiplicity of functions at different scales, simultaneously managing structural load, environmental pressures, and spatial constraints: a living tree is the equivalent of any engineering masterpiece. In the designed world, however, human-made materials and the technologies necessary to create them have not matched the refinement and sophistication of natural ones. Bricks are dumber than cells, and synthetic fibers have yet to fire electrical signals into the textiles they inhabit.

With the design approach we call Material Ecology, we created a set of principles and technologies that promote and enable the production of smart objects that respond to their environment and function more like bodies than manufactured entities, accommodating multiple functions rather than just one. With such technologies, designers can create relationships between artificial structures and their environments at a resolution that will match and eventually outperform nature.

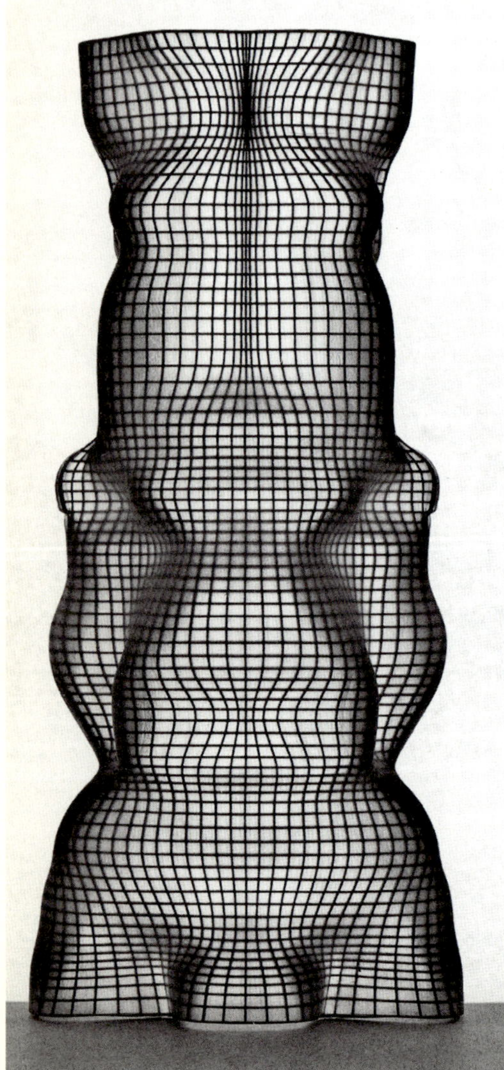

Materialecology is the name of a series of prototypes in which individual technologies explored how a design object might relate to its environment. Using multifunctional materials, high spatial resolution in manufacturing, and computational algorithms, we experimented with ways to create environmentally informed and responsive objects and buildings. Armour (pronounced ar-MOOR) was precursor of these technologies—a rethinking of modern construction techniques.

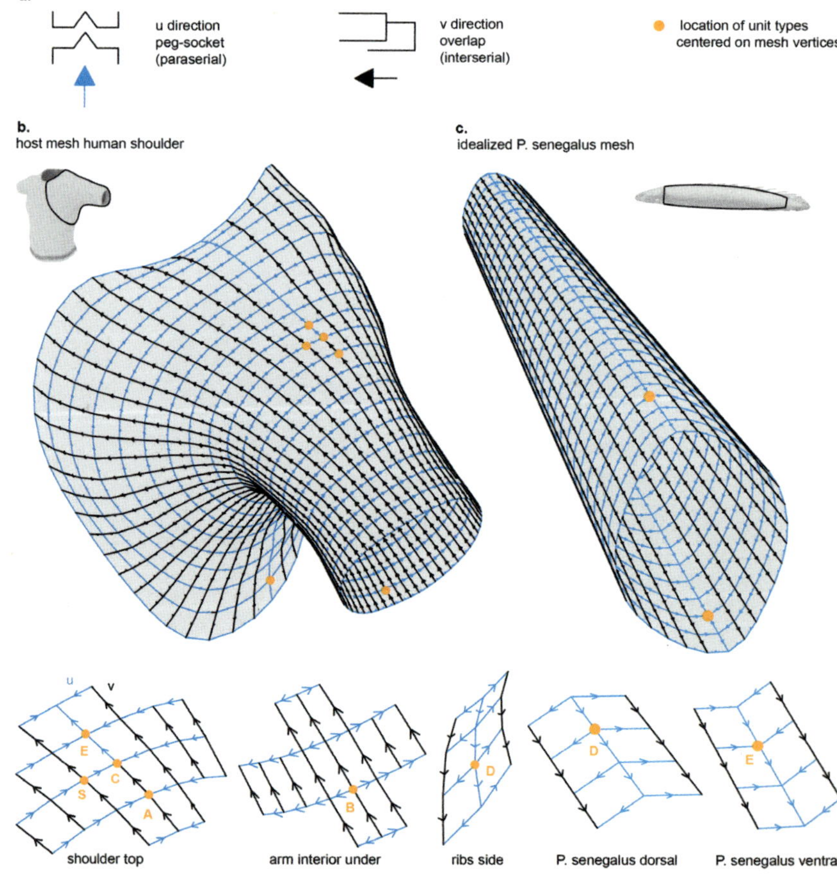

For MetaMesh, individual units were analyzed according to the predicted mobility of the areas they would cover. Directionality model analyzing mesh for a human shoulder (left) and for the ancient armored fish *P. senegalus* (right).

nlike nature's strategies for material distribution, rchitectural design and construction have primarily een based on strategies of material assembly and roperty assignment. The I-beam—one of the quintes-ential structural components of the modern move-nent in architecture and a hallmark of the Industrial evolution—originated in 1850s, with a single piece of teel rolled into a beam with an *i*-shaped cross section. s horizontal elements (known as flanges) resist bending, nd the vertical element (the web) resists shear forces. ecause I-beams are not as effective in resisting torsion, ney tend to be used as vertical structures in, for example, urtain-wall facades. Armour is a small-scale prototype f a beam designed to act as skin and structure at the ame time. Stiff structural components are embedded a soft skin, creating a composite that is able to carry ertical, horizontal, and rotational loads. The beam's ectional profile and structural thickness can be varied epending on the anticipated load. The black elements epresent anticipated trajectories of structural forces, nd the translucent skin represents the curtain wall or skin) containing them.

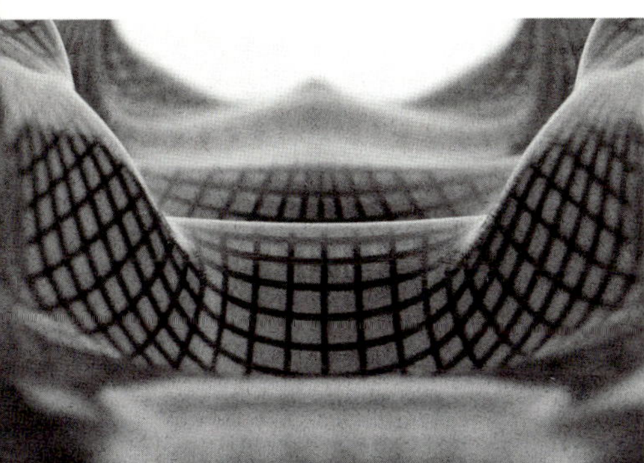

Photographs of different views of the Armour prototype. Its design was generated with a method similar to the one used for Raycounting (page 52).

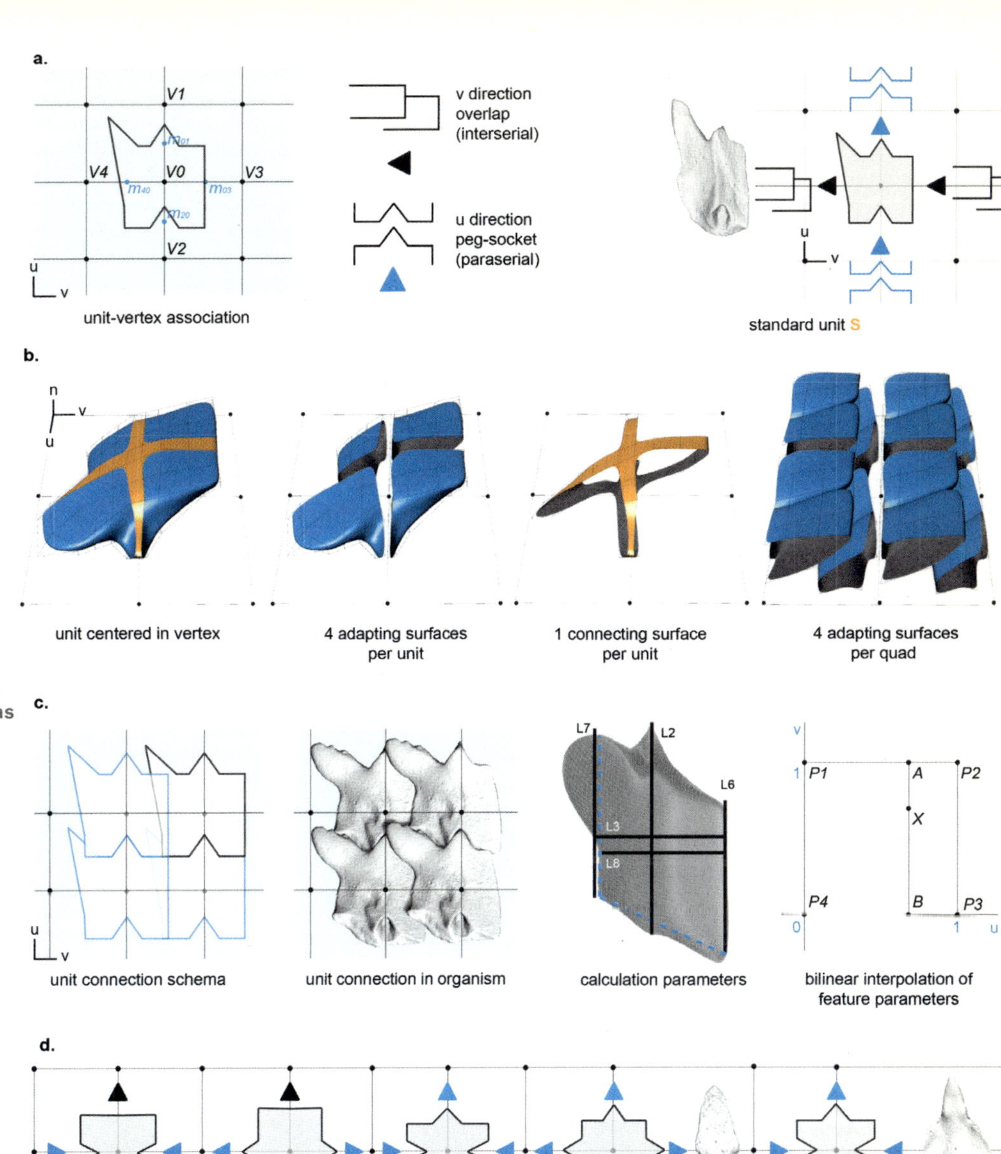

Standard armor units were defined in terms of how they connect to each other. Centering a scale unit over a mesh vertex (a), constructing a unit from connecting surfaces, and contact surfaces (b), calculating how multiple units connect (c), specialized unit types (d).

he armor's properties were optimized to suit various bodies nd body parts. *P. senegalus*'s armor (represented by a torus and ptimized for swimming) provides maximum protection near the ead and maximum flexibility near the tail (a); human shoulder rmor provides maximum protection above the shoulder and aximum flexibility under the arm (b).

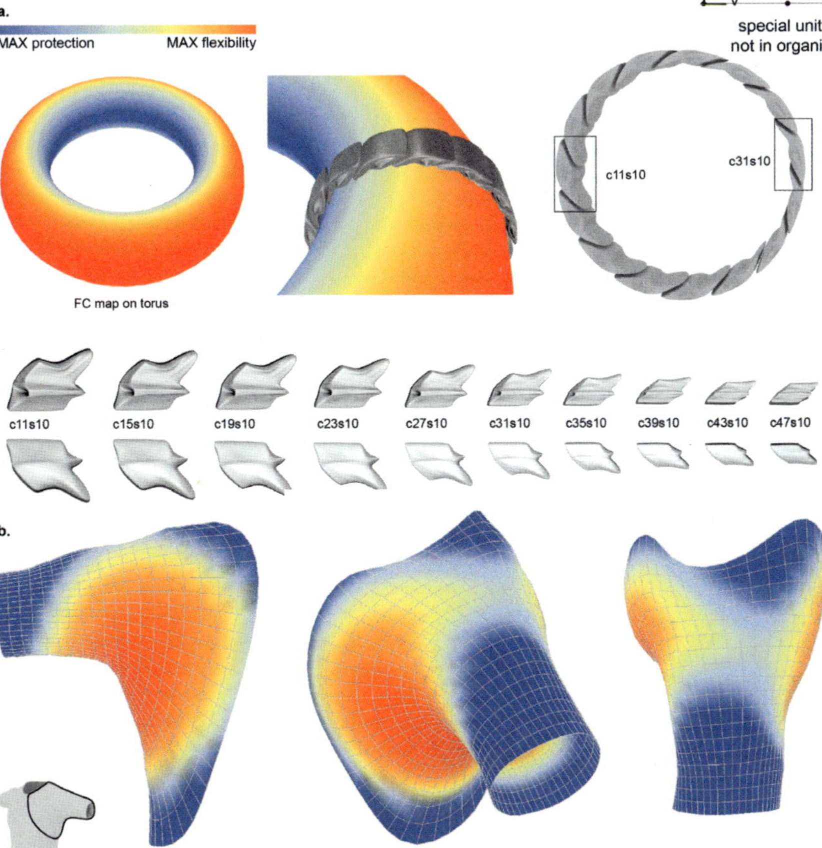

Using the research behind Armour, we developed processes that move beyond the questioning of existing construction. MetaMesh, based on a similar design method, is a functional exoskeleton created by adapting the configuration of fish scales, with a combination of protective rigid components and a flexible underlying skin that enables sophisticated articulation. As was true in other Materialecology projects, MetaMesh demonstrates the heterogeneity and differentiation of material properties in a structural skin, with shear stress and surface pressure distributed over the object in components of varying thickness. The 3D-printing technology developed for the project can print parts and assemblies from multiple materials in a single build as well as create composite materials based on preset combinations of mechanical properties.

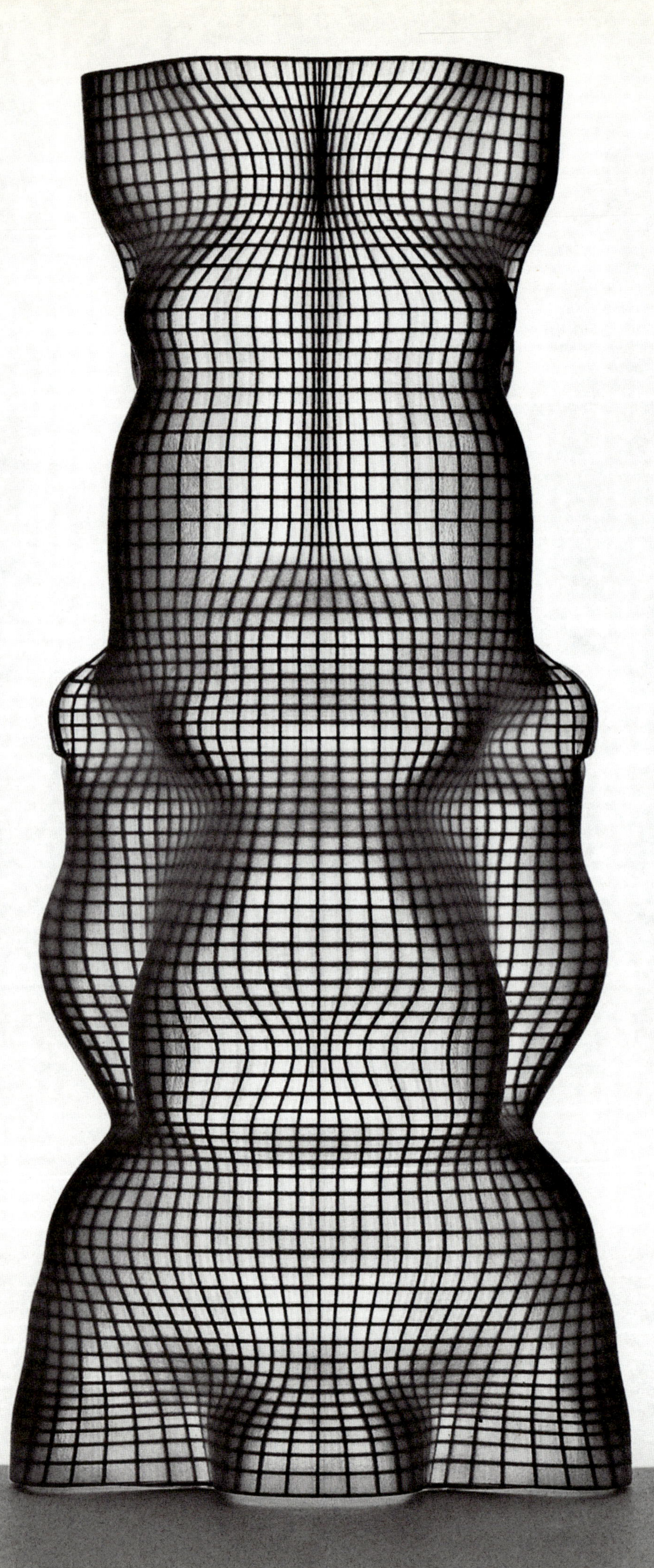

Materialecology: Armour

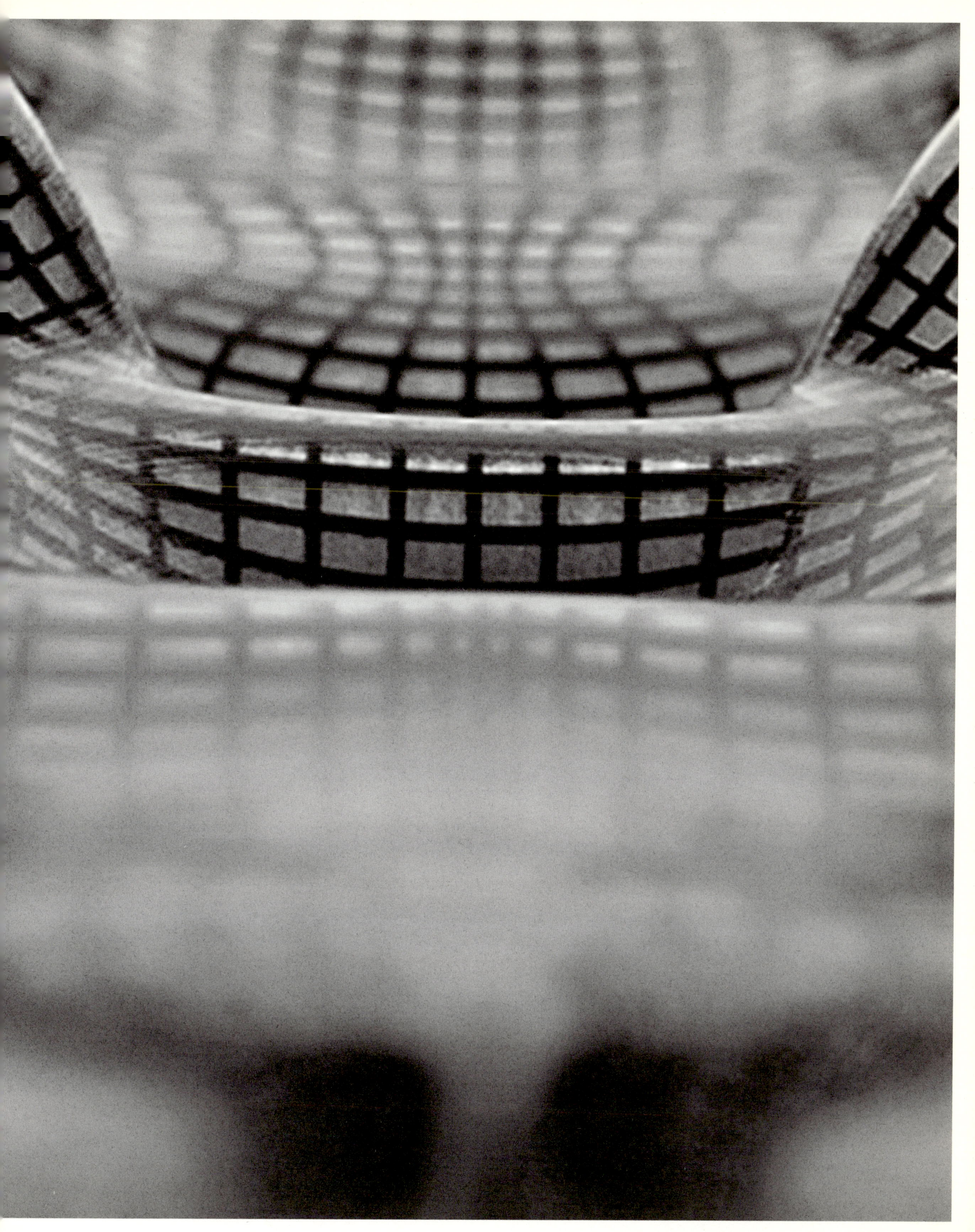

RAYCOUNTING

Neri Oxman
Raycounting. 2007
Silk-coated nylon and acrylic-based polymer
17 × 11 × 10 in. (43.2 × 27.9 × 25.4 cm) and
19 3⁄16 × 10 × 6 in. (48.7 × 25.4 × 15.2 cm)
Collaborators and contributors:
Tangible Express, AARK Product Development Group Ltd.

In Raycounting, we registered the intensity and orientation of natural light to compute the form of a 3D-printed construction. The project was inspired by nineteenth-century photo-sculpture, in which three-dimensional replicas were created by projecting photographs of objects taken from different angles onto sheets of wood and then carving and assembling them. For our version we created an algorithm to calculate the intensity, position, and direction of one or more light sources and then assigned curvature values to points in space. The models explore the relation between geometry and light performance from the perspective of computational geometry; the results are sunshades perfectly suited to their environmental conditions.

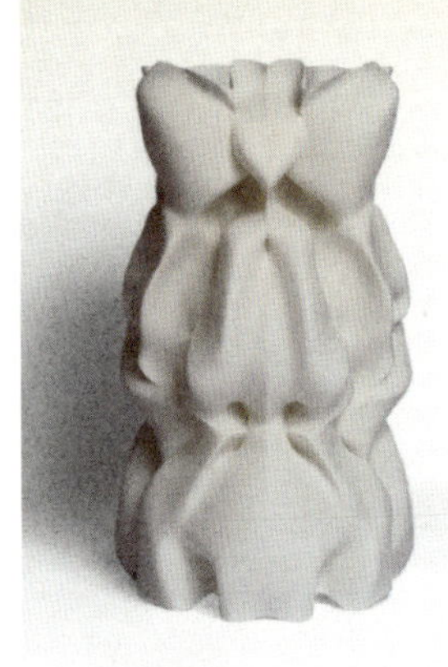

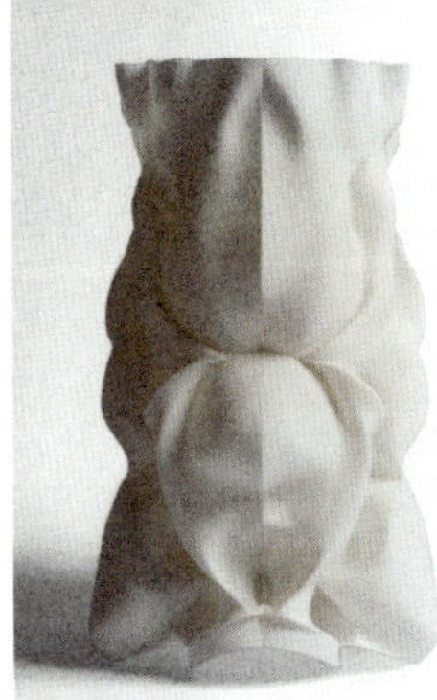

Structure 3D printed with silk-coated nylon.

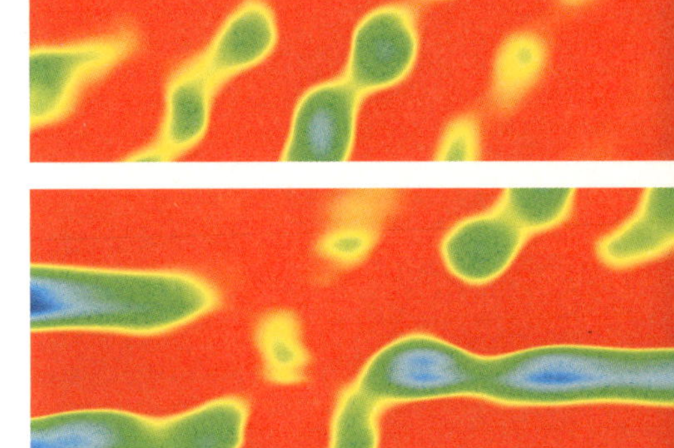

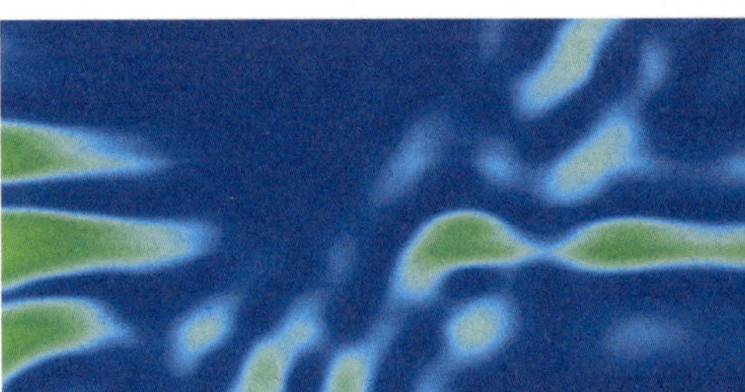

Visual analysis of a surface's draft angle (curvature in relation to a viewing point).

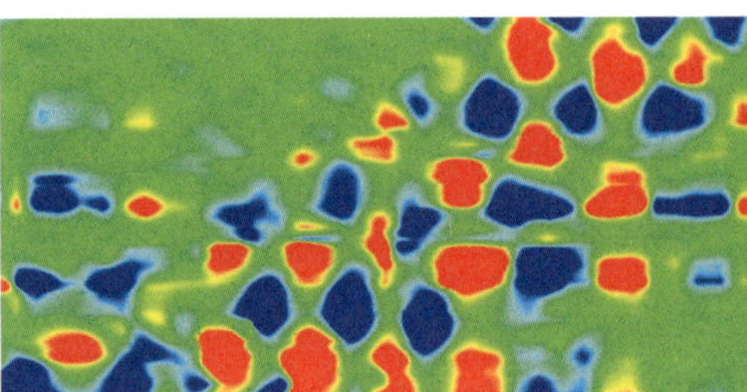

Visual analysis of a surface's curvature. Different colors correspond to bowl (or saddlelike) curvature.

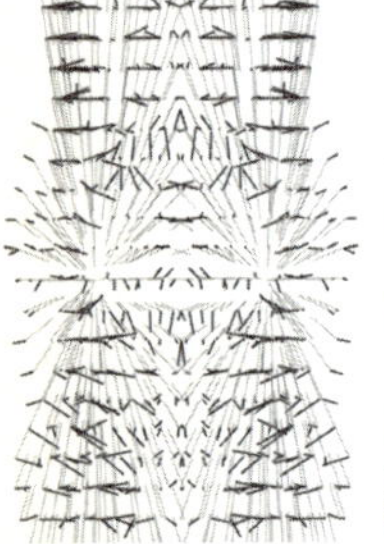
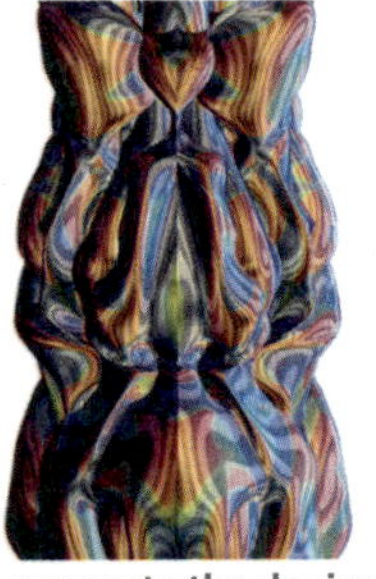
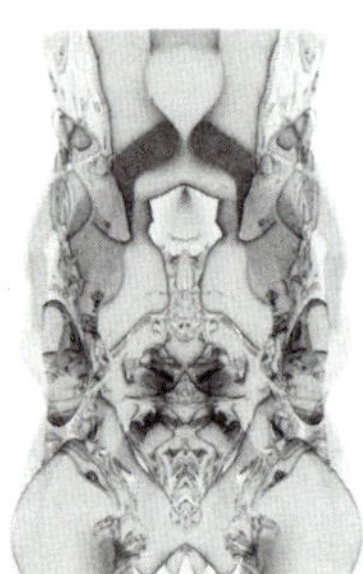
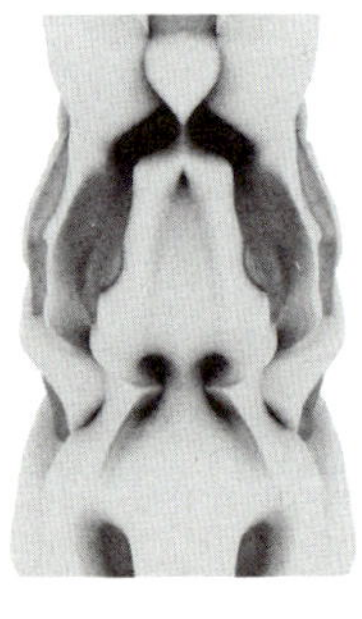

The method used to generate the design.

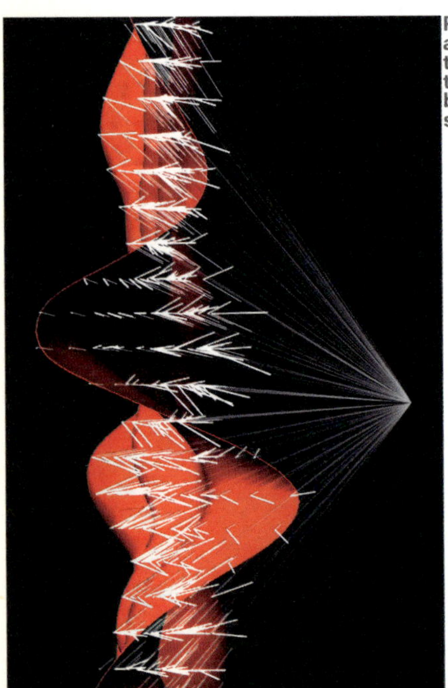

Rendering of a phase of the draft angle analysis tool generating the design of the object. The computation took into account the angles between the surface and a light source (coming from the right).

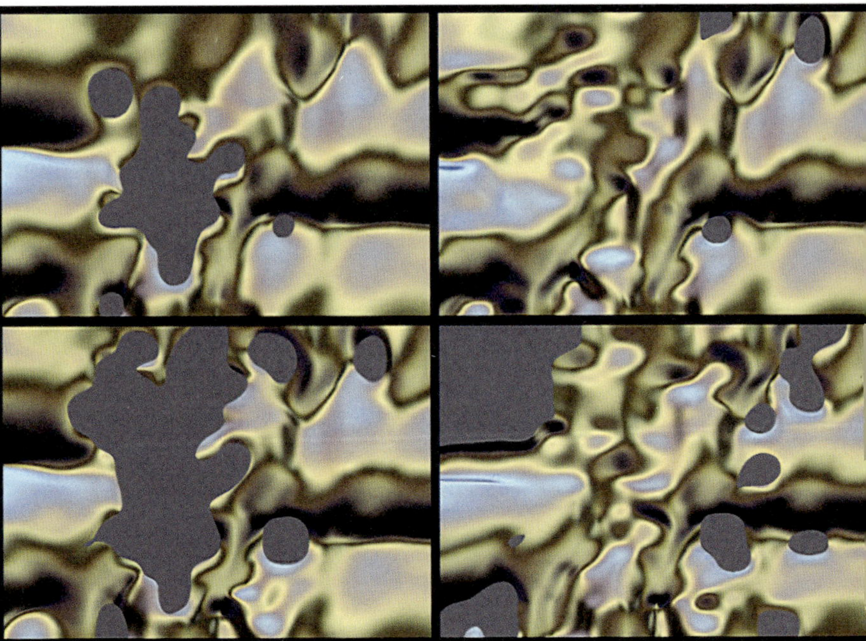

Results of four iterations of the draft angle method. Holes in the surface of the object appeared when the angle between the light source and the surface approached a minimal threshold value.

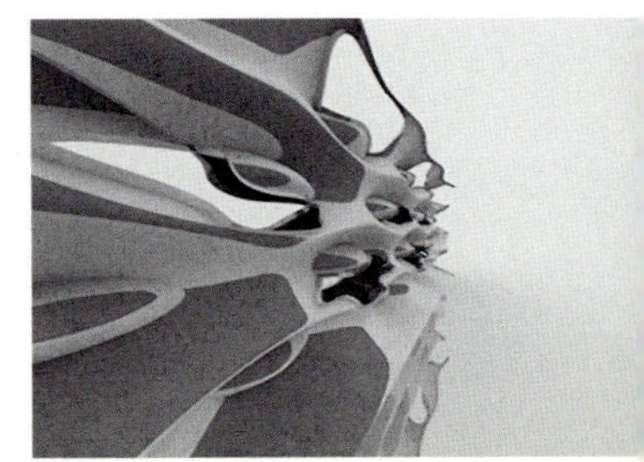

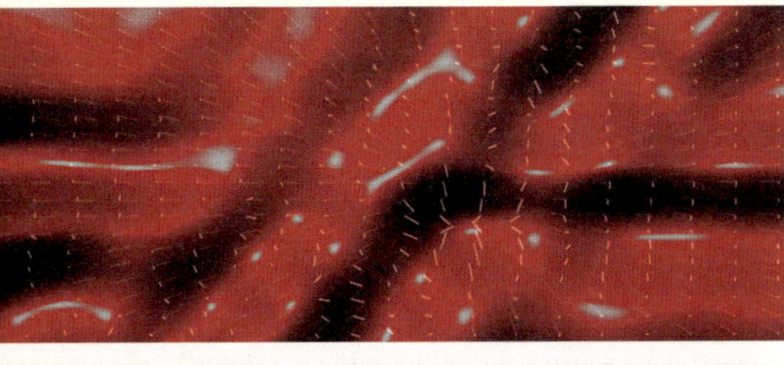

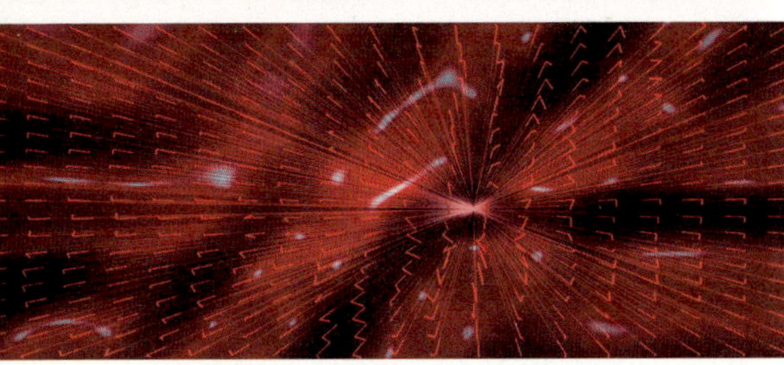

Phases of the draft angle method used to generate Raycounting's design: vector reparameterization and computation of light-source angles relative to the surface (top and center); final user-generated surface (bottom).

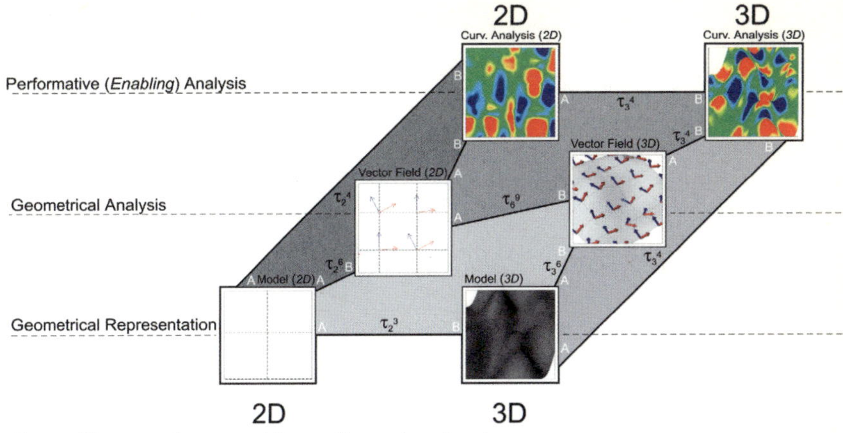

Above: Diagram of two- and three-dimensional reciprocal transformations, which increase in complexity from geometrical representation (bottom) to analysis (middle) to performative analysis (top).

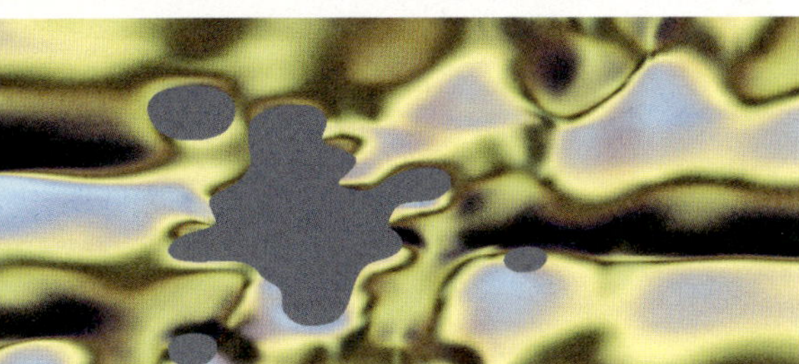

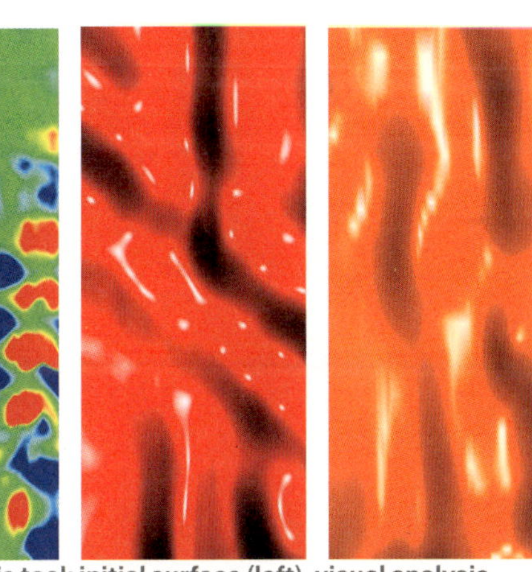

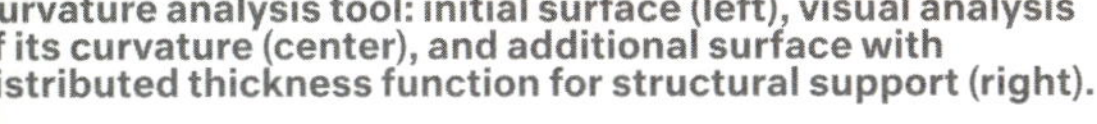

Curvature analysis tool: initial surface (left), visual analysis of its curvature (center), and additional surface with distributed thickness function for structural support (right).

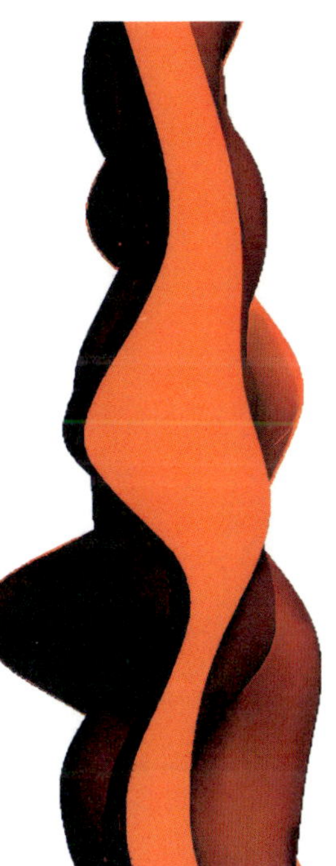

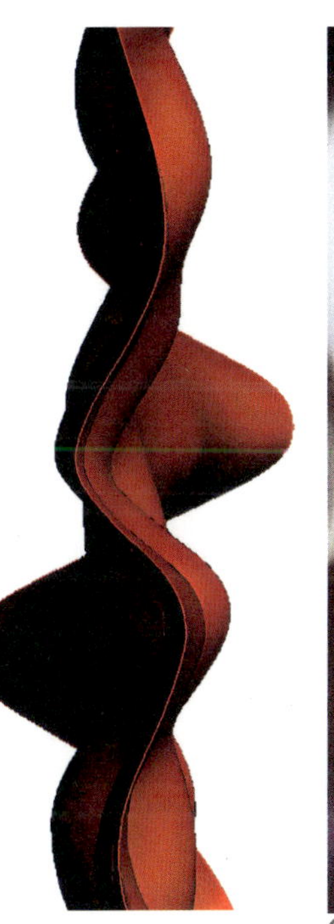

Renderings of a surface with and without thickness.

Customized light-shading constructions were generated by registering the intensity and orientation of light rays in a given environment.

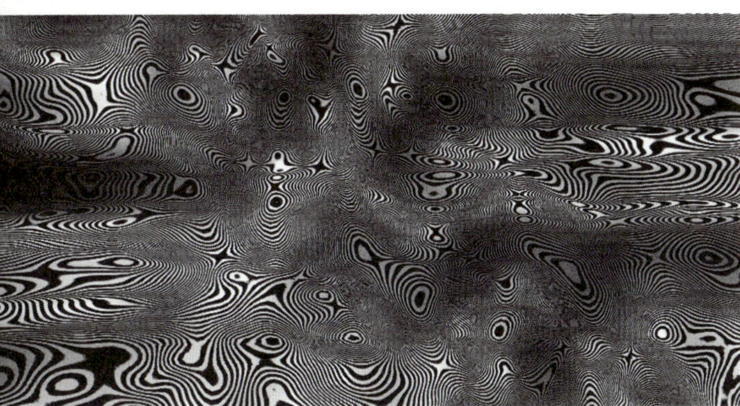

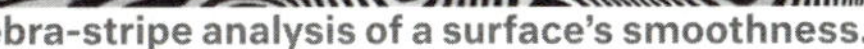

Zebra-stripe analysis of a surface's smoothness.

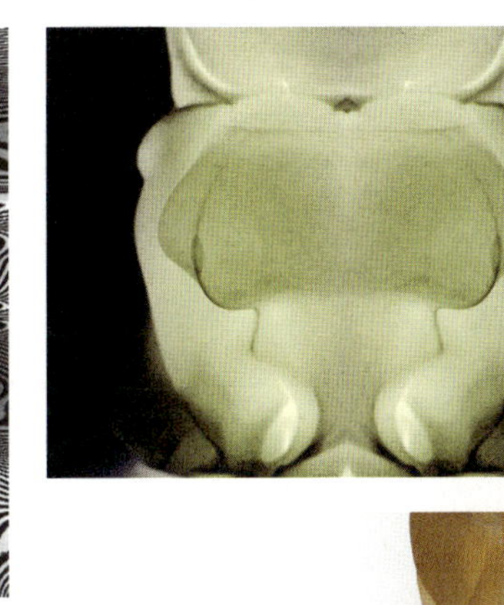

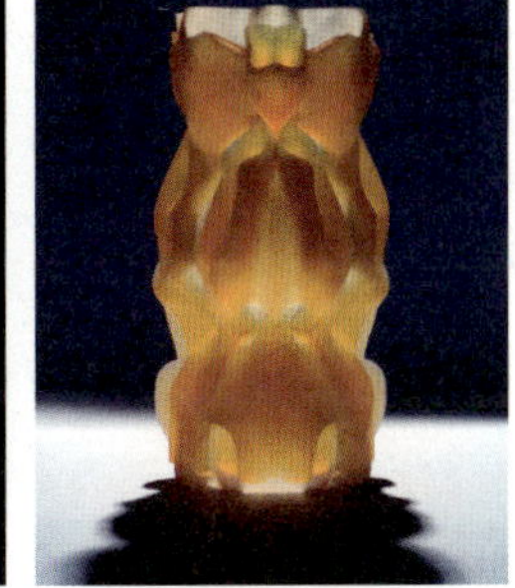

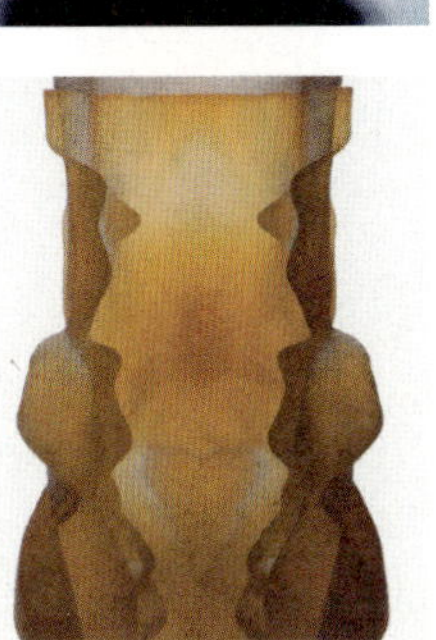

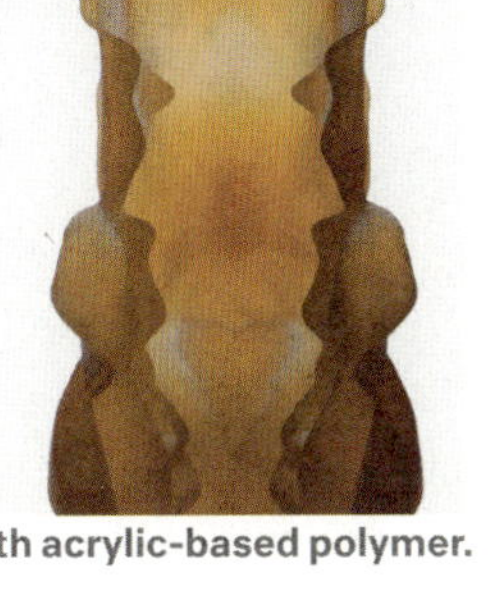

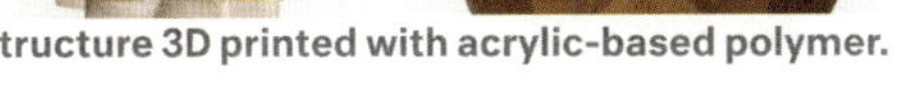

Structure 3D printed with acrylic-based polymer.

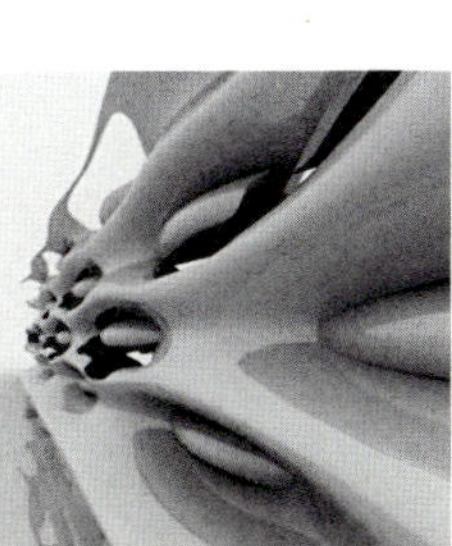

Project using the Raycounting method at Wiesner Student Art Gallery, MIT. The design corresponded to specific light-condition parameters and structural requirements.

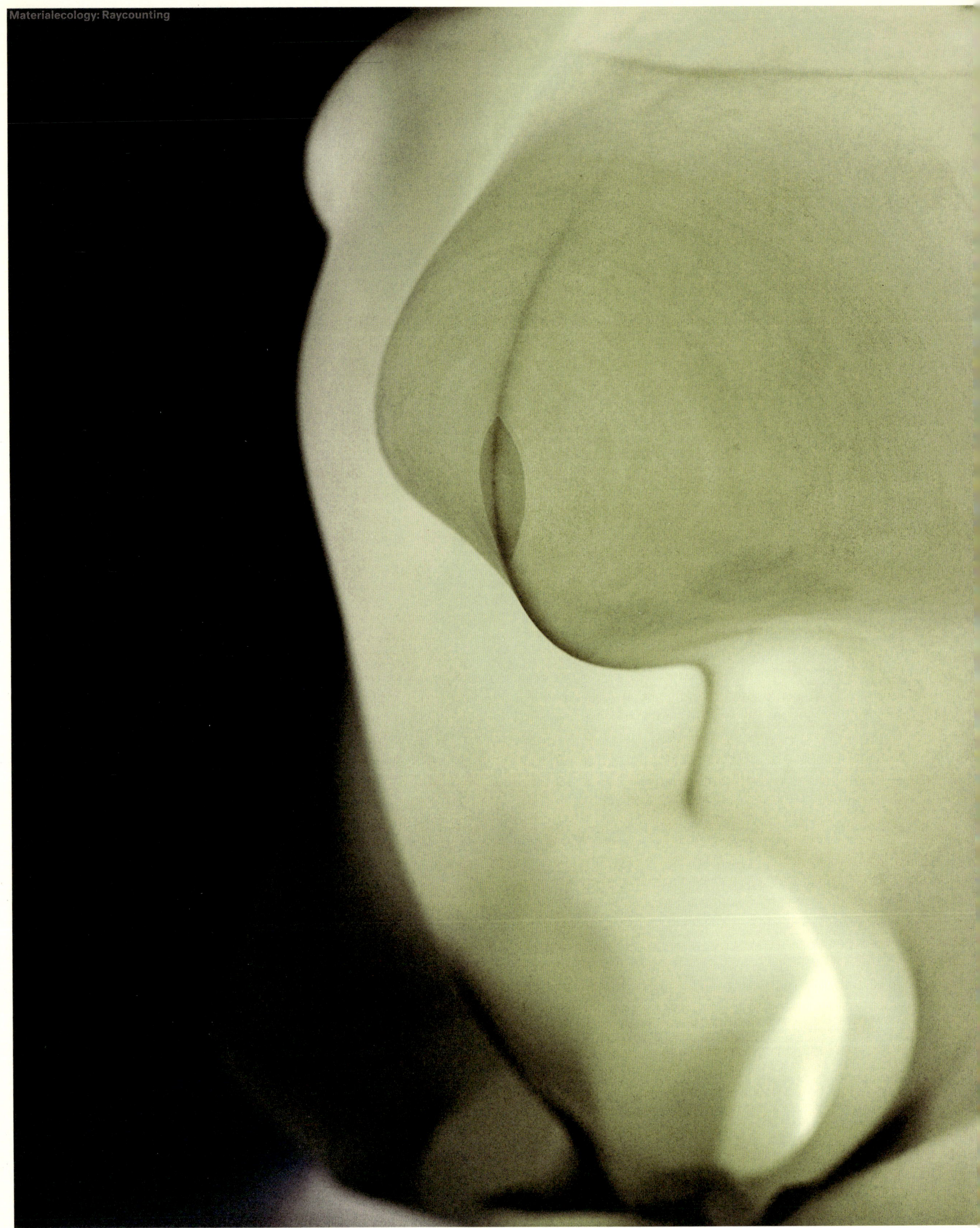

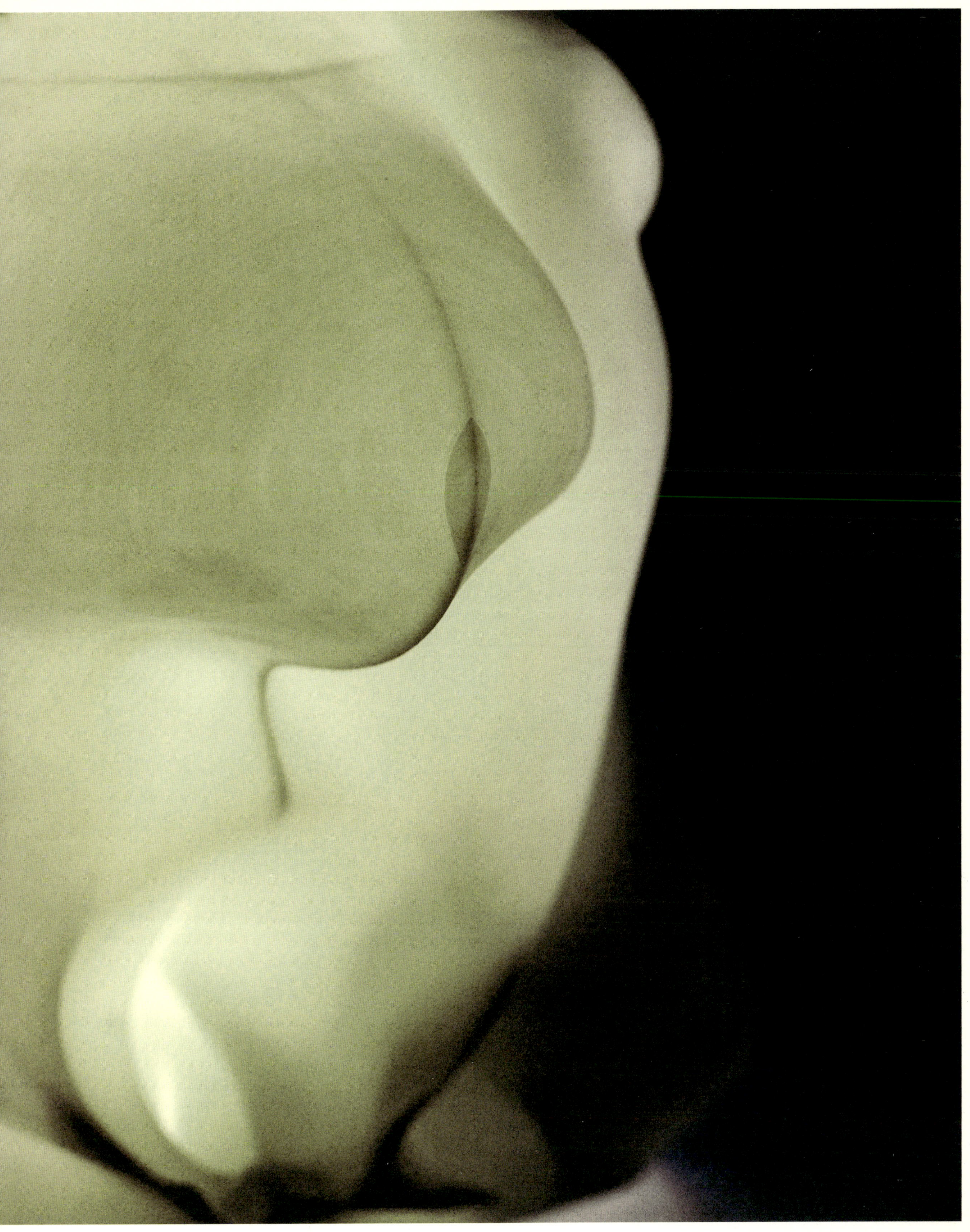

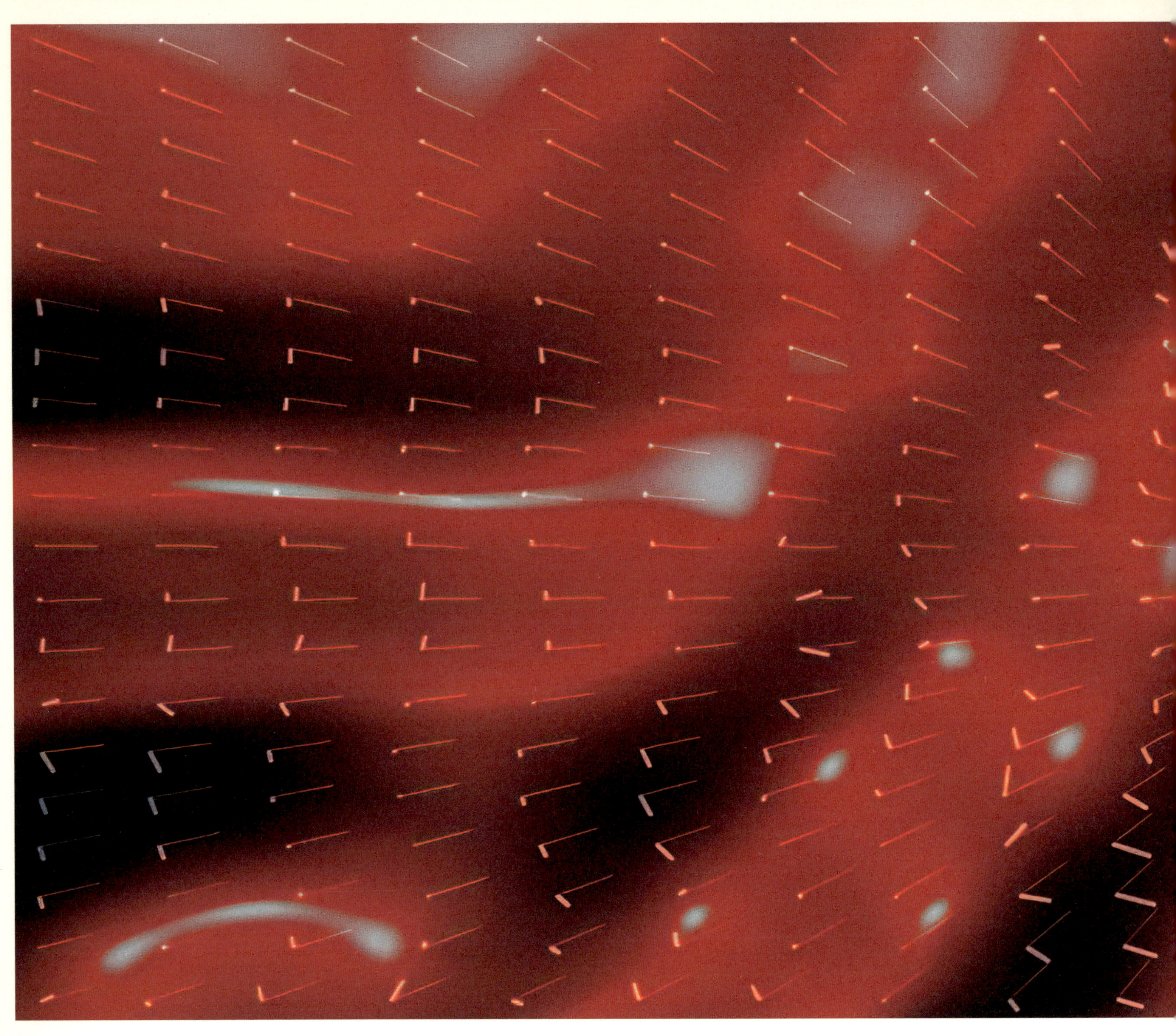

Materialecology: Raycounting
Phase of the draft angle method used to generate Raycounting's design.

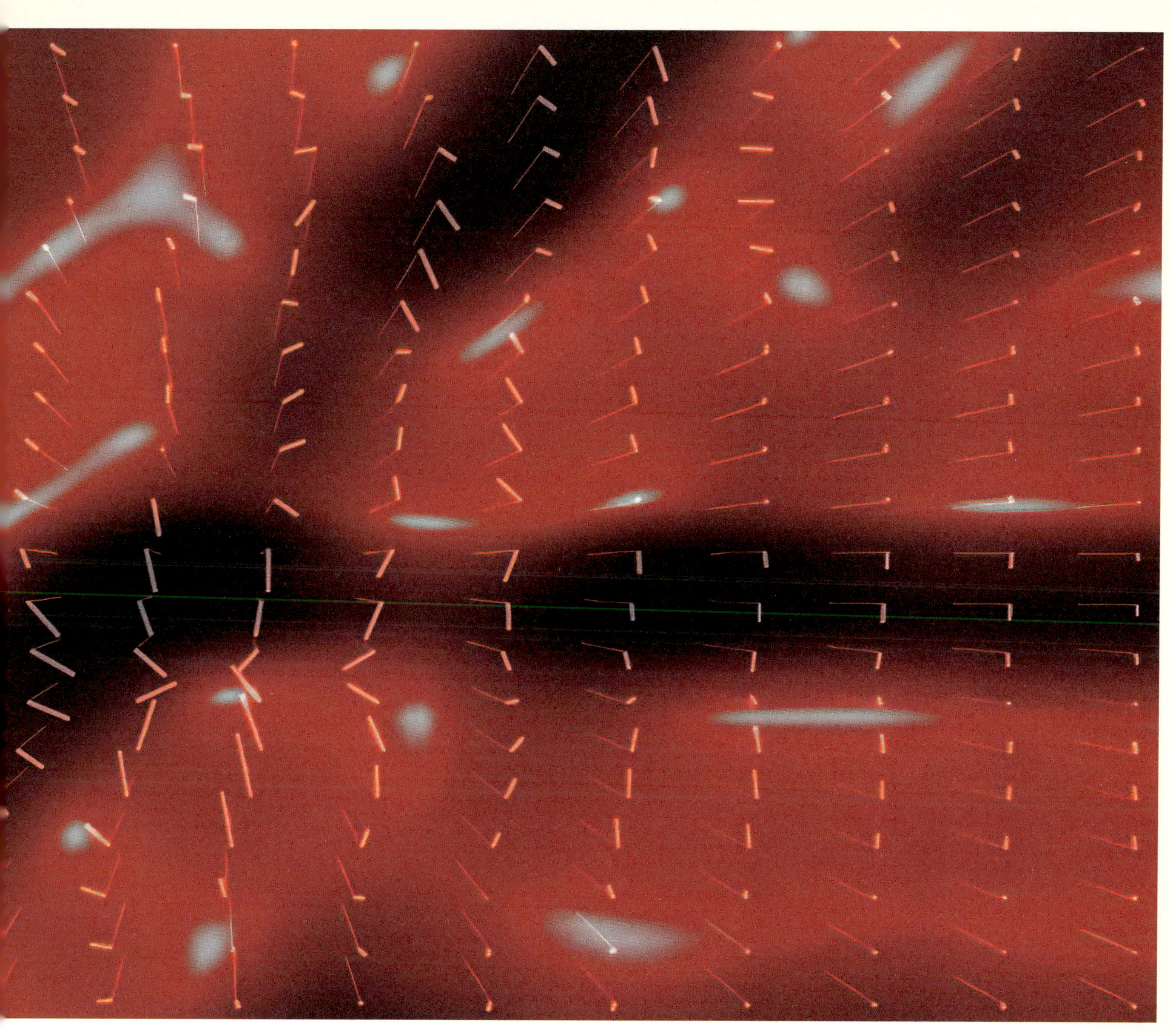

CARTESIAN WAX

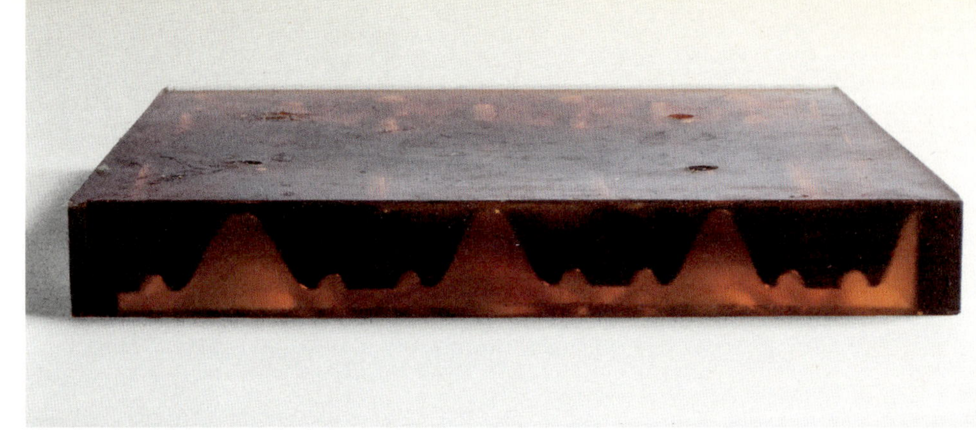
Tile showing thicker areas in darker resin.

Neri Oxman
Cartesian Wax. 2007
Rigid polyurethane casting resin composite and machinable wax
20 × 20 × 2 ¼ in. (50.8 × 50.8 × 5.7 cm)
Collaborators and contributors: Mikey Siegel, MIT Center for Bits and Atoms

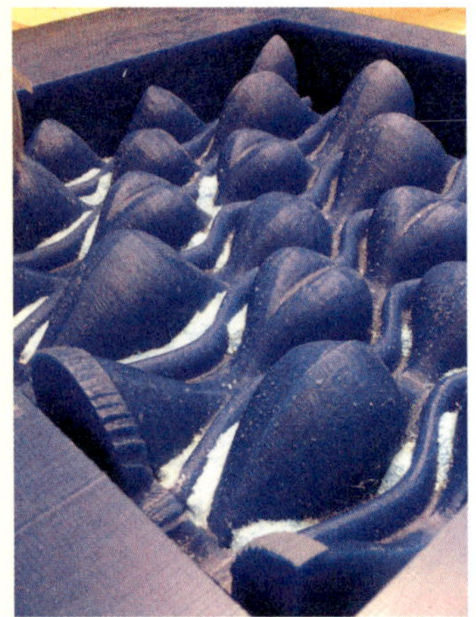
The 3D-milled machinable-wax mold in which the tiles were cast.

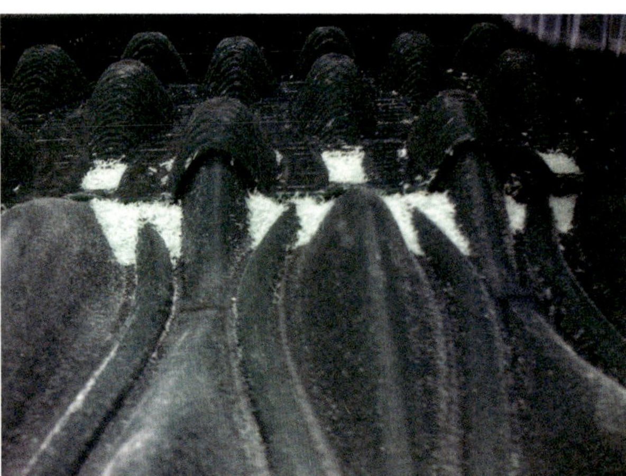
Each tile was cast in high temperature-curing resin.

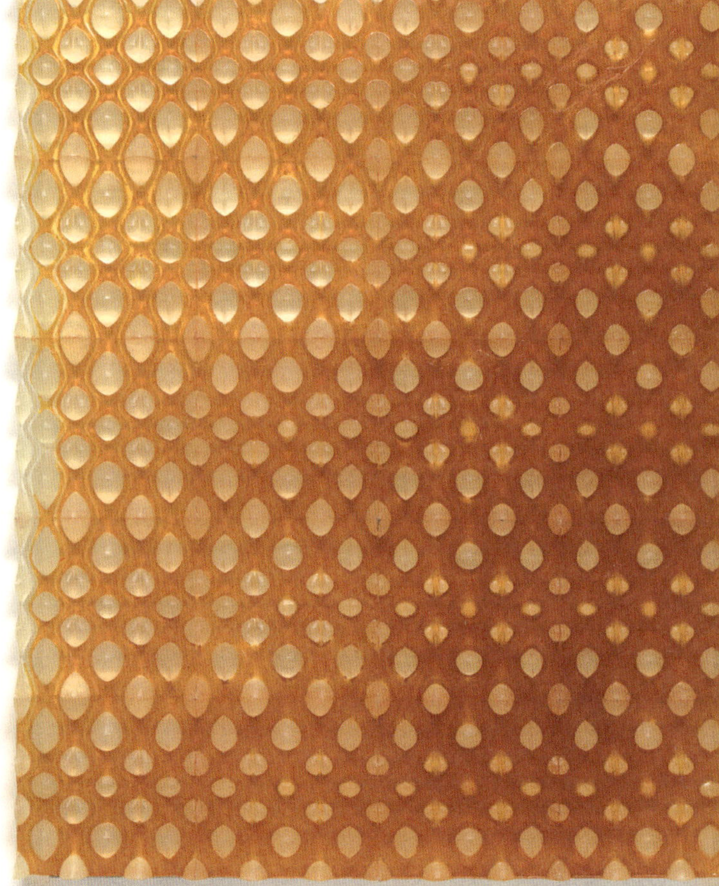
Twenty tiles of multiple resin types arranged along a continuum, from dark, stiff, and opaque to light, soft, and translucent.

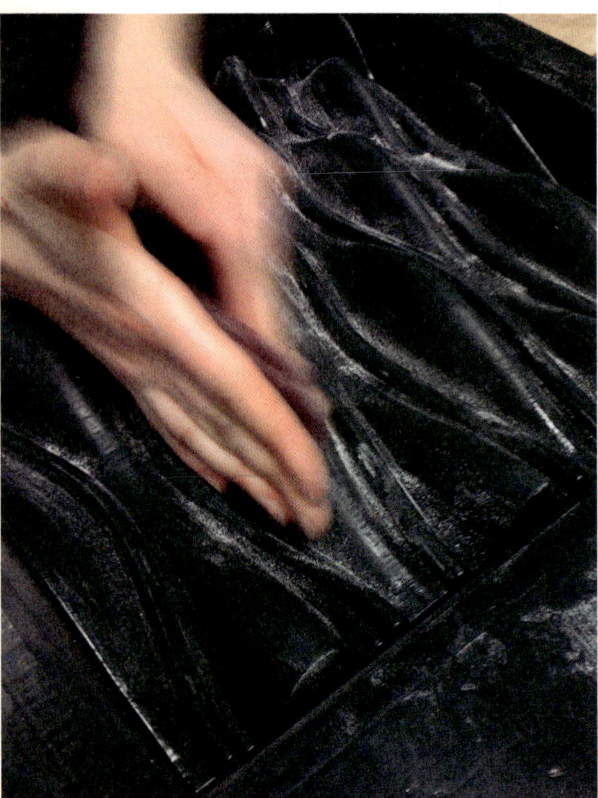
The mold was sanded before the resin was poured in.

Cartesian Wax is a wall-like surface made up of tiles from a single 3D-milled semiadjustable wax mold. We cast each tile in resin and cured it at a high temperature, in a process that increasingly deformed the original mold as each tile was made. We then produced further geometrical and physical-property differences in the tiles by changing the temperature across the mold, so that each resulting tile, although similar in shape, varied in material properties, such as stiffness and opacity, thus reflecting and responding to the conditions that formed it. We used these variations to correct and, in certain areas, augment structural performance—for example, using stiffer composites to ensure structural support in areas of tile that were made thinner because of the mold's deformation.

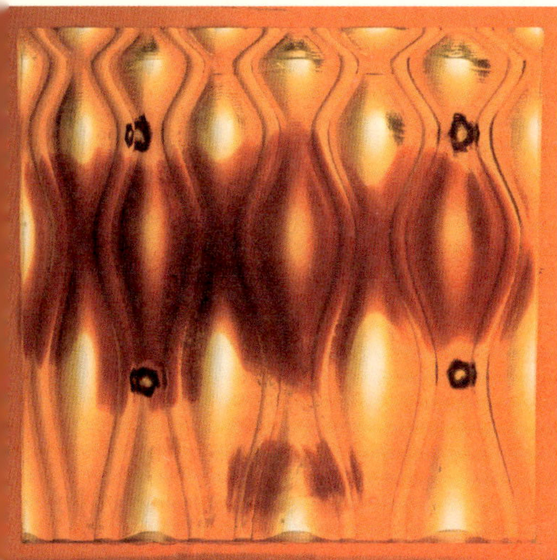

bove and below:
sometric views of a single tile.

With twenty of these tiles assembled in a sequence of rigid and flexible types, we created a continuous surface fabricated in response to physical conditions such as light transmission, heat flux, stored energy modulation, and structural support—thicker where it would be structurally required to support itself, darker where there would be more sunlight, to provide shade. These processes experiment with light- and heat-sensitive construction techniques specific to any given environment, resulting in structures designed and built to fit organically with their surroundings.

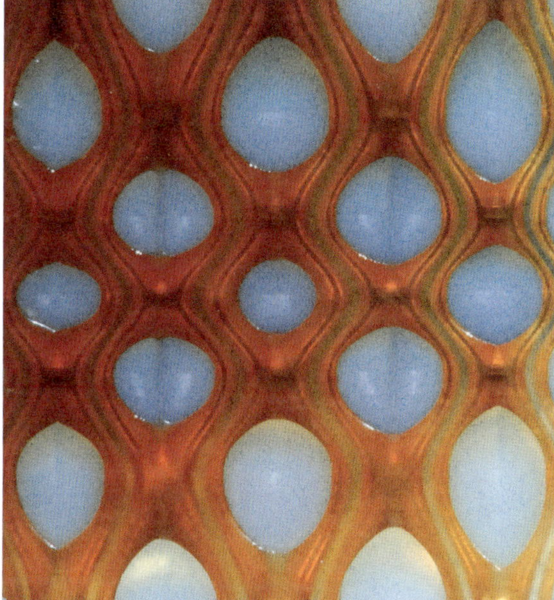

Detail of a single tile attached to the mold.

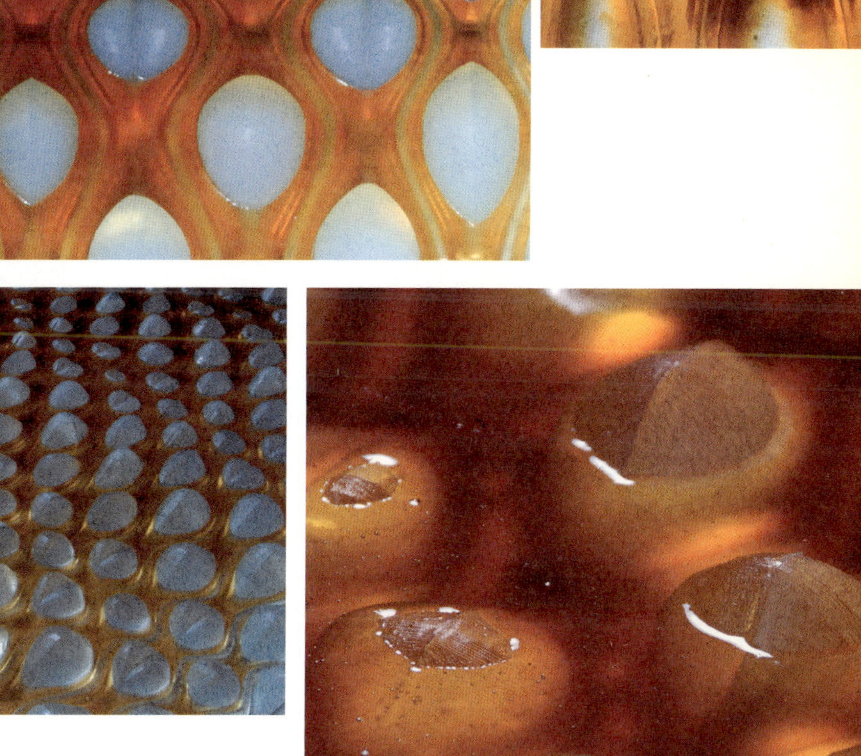

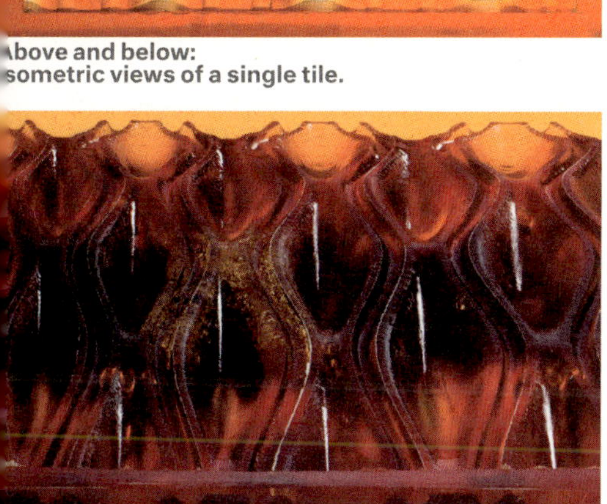

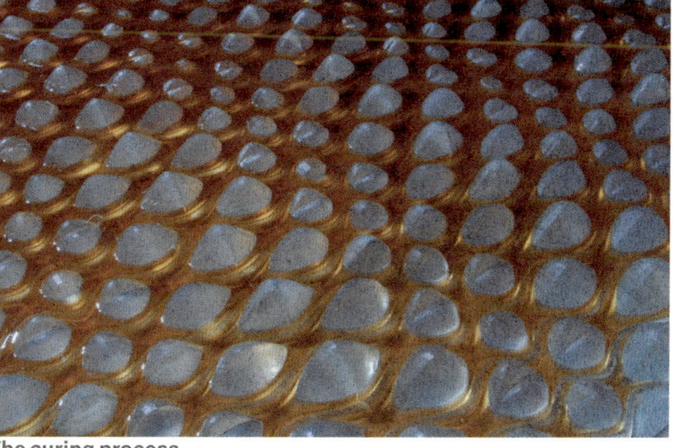

The curing process.

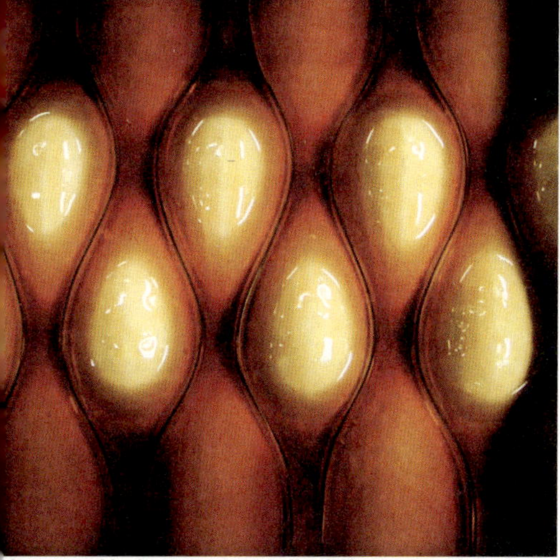

bove and below: A materially differentiated nvironmental screen, CNC milled and molded. lodulating the size, thickness, density, and overall rganization of the tiles produced different nechanical and optical effects.

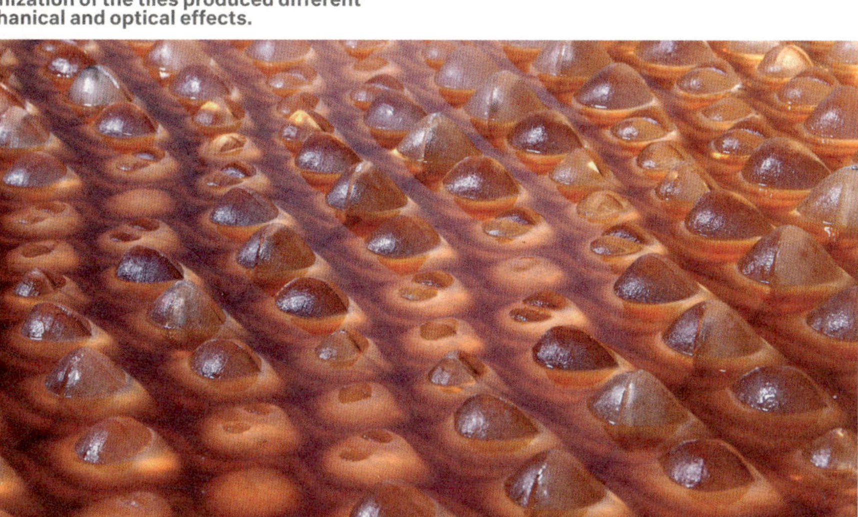

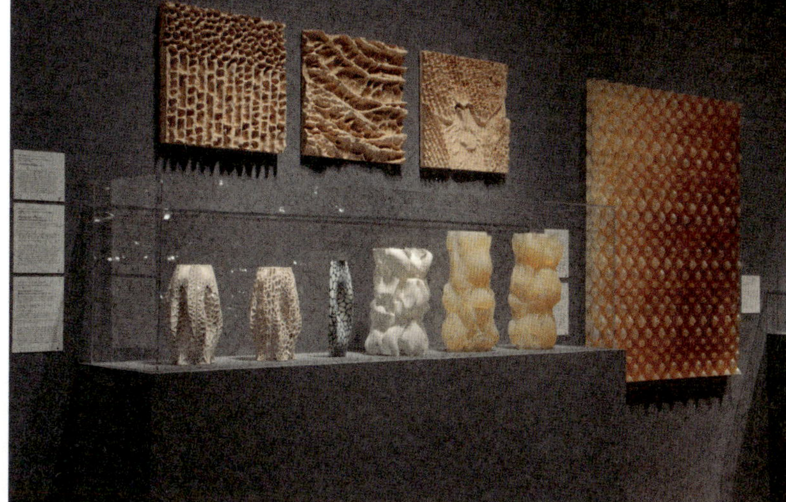

Installation view of *Design and the Elastic Mind*, The Museum of Modern Art, New York, 2008.

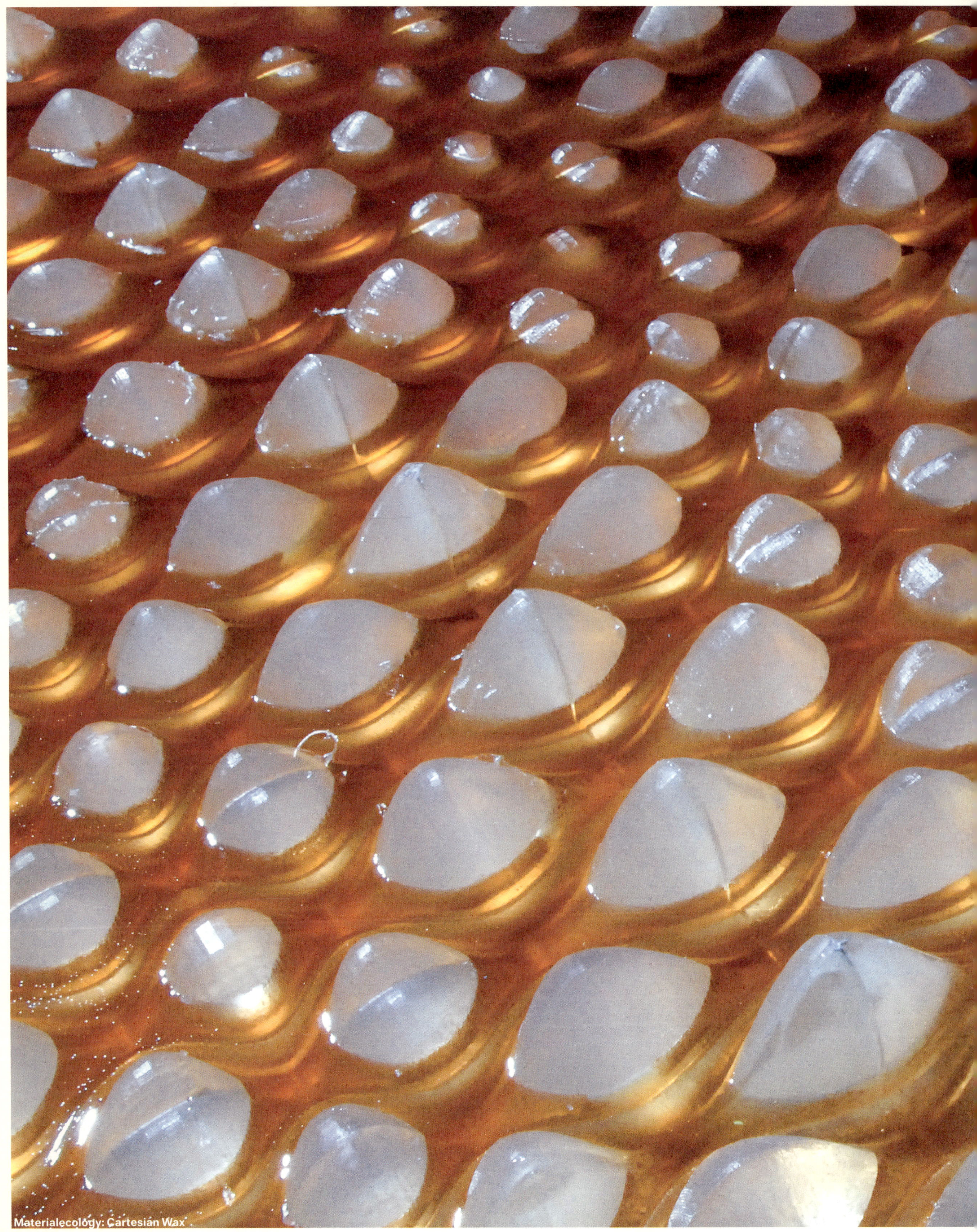

Materialecology: Cartesian Wax

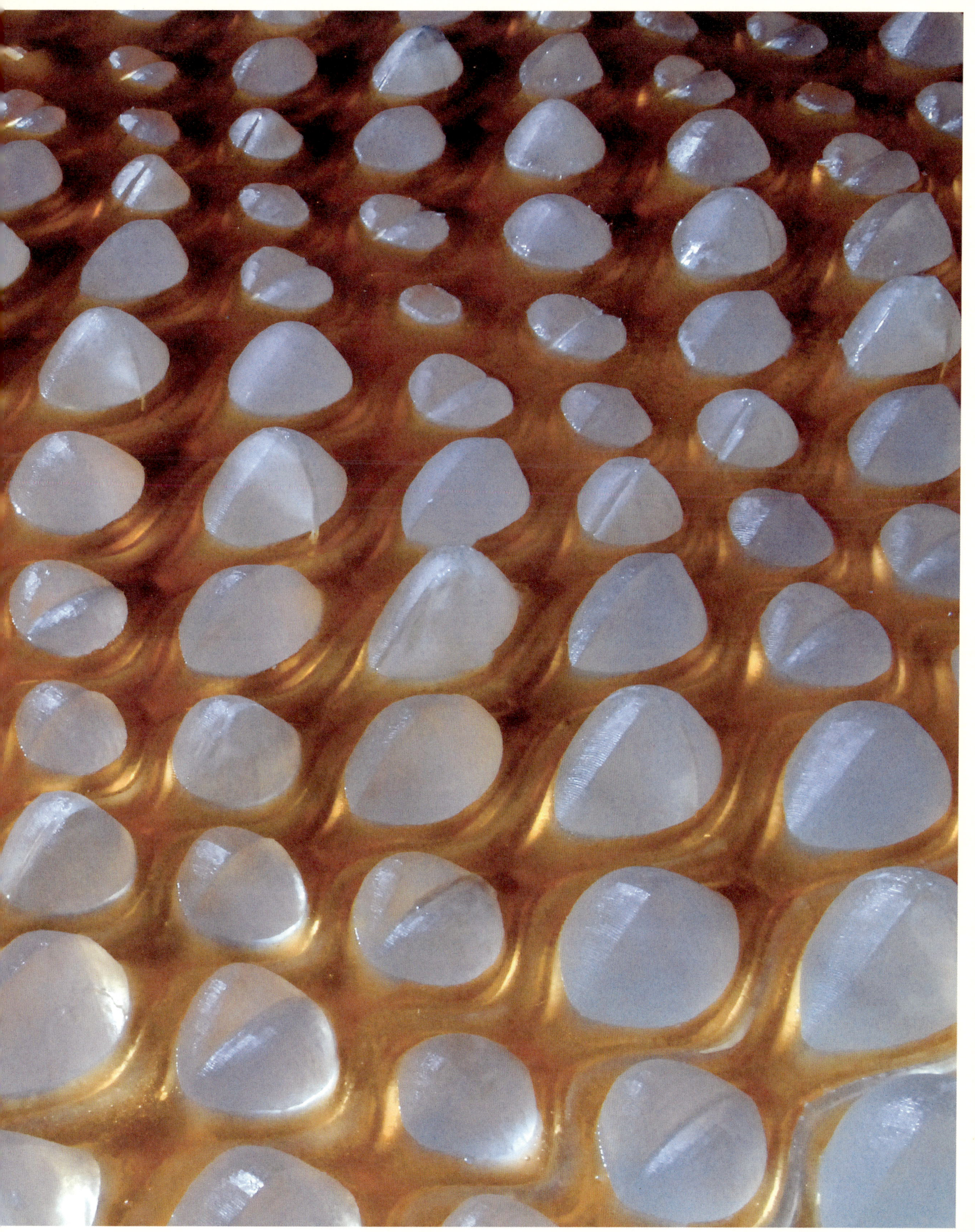

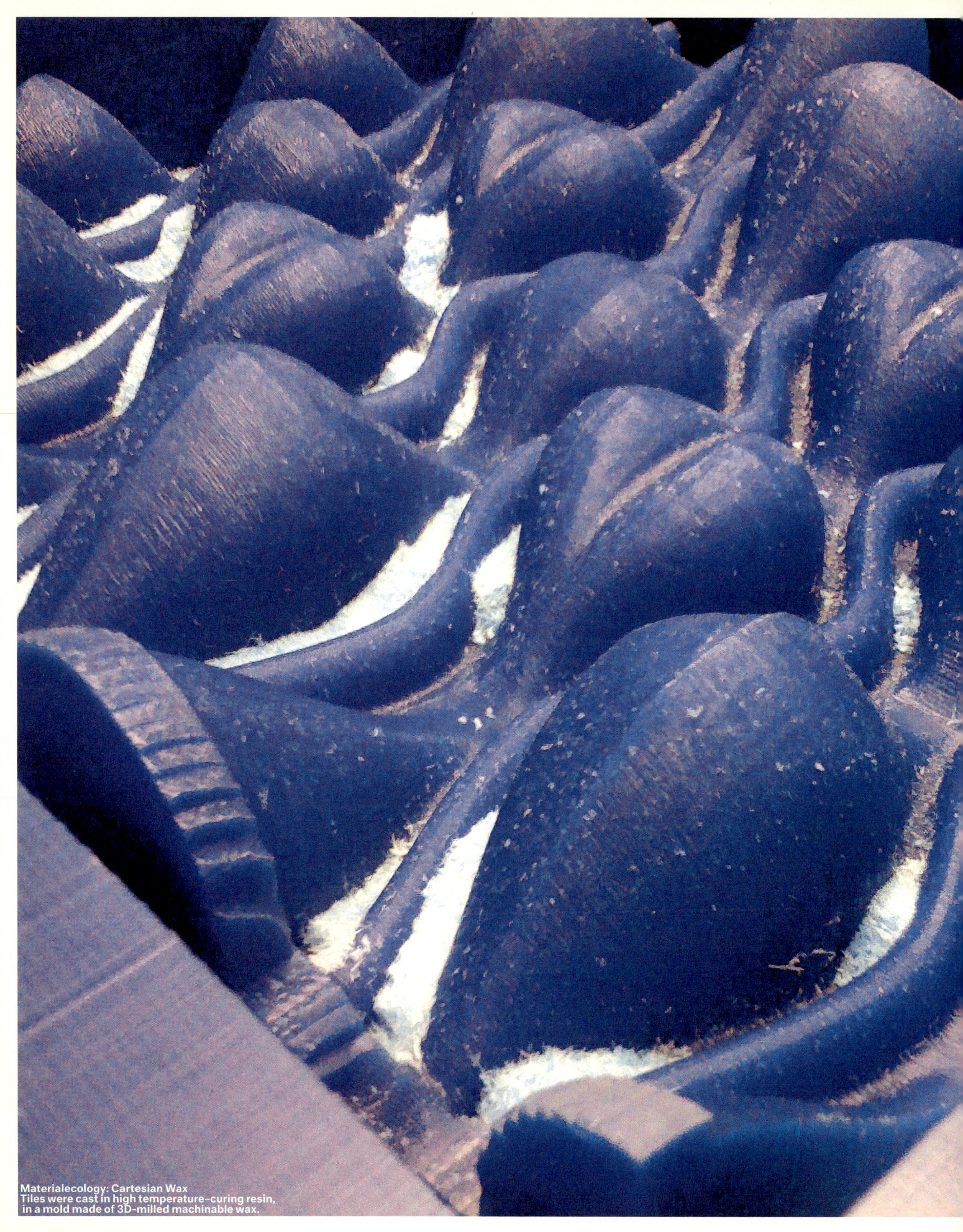

Materialecology: Cartesian Wax
Tiles were cast in high temperature–curing resin, in a mold made of 3D-milled machinable wax.

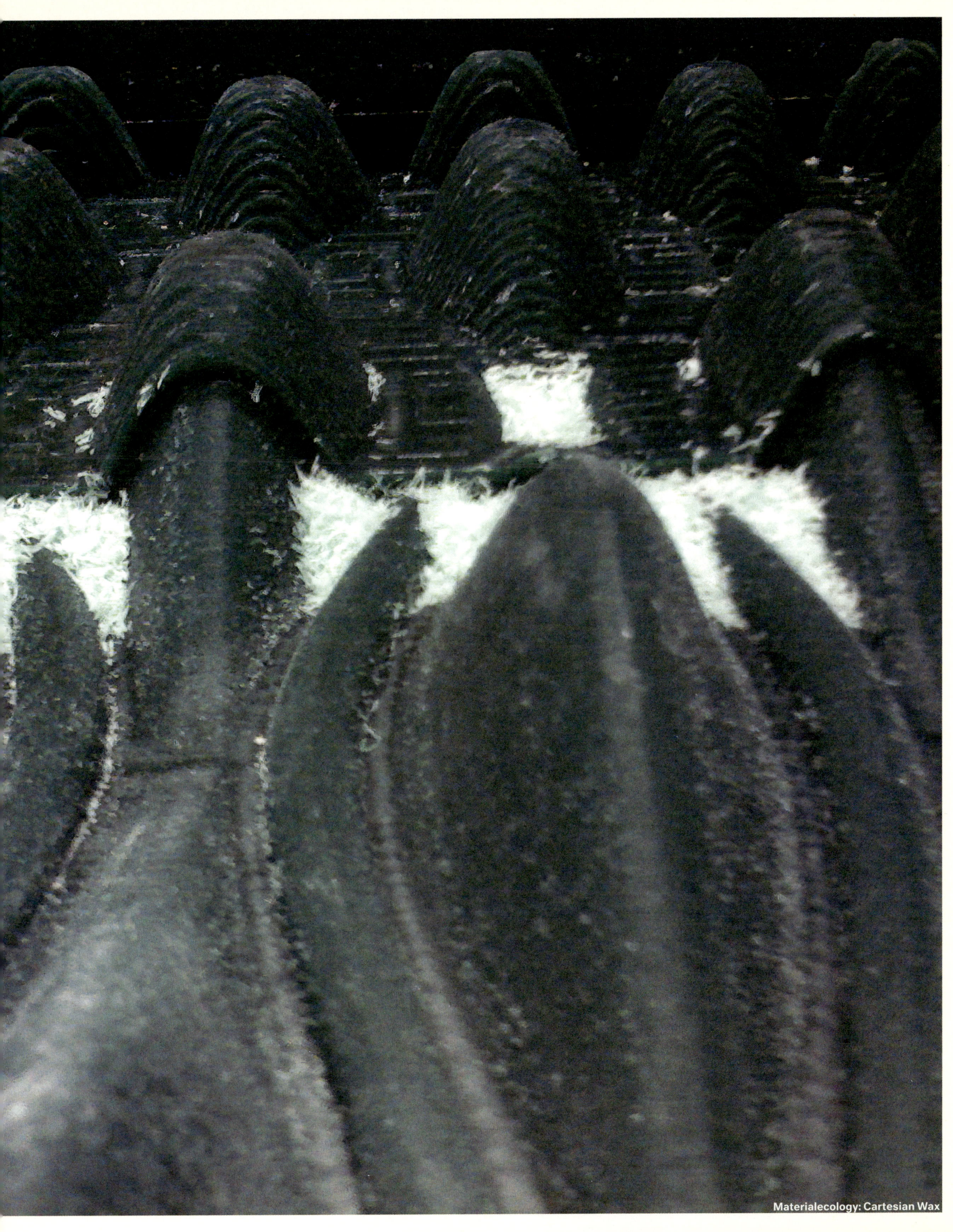

Materialecology: Cartesian Wax

MONOCOQUE / BEAST

Neri Oxman
Monocoque. 2007
Photopolymers and starch and gypsum powder with sandstone finish
3 ¼ × 14 × 5 in. (8.3 × 35.6 × 12.7 cm),
13 × 7 ¼ × 8 in. (33 × 18.4 × 20.3 cm),
13 × 7 ¼ × 8 in. (33 × 18.4 × 20.3 cm)
Produced by Stratasys Ltd. and Z-Corp.

Neri Oxman in collaboration with W. Craig Carter
Beast. 2008
Photopolymers
25 ⅜ × 12 ⅝ × 12 ⅞ in.
(6 × 32.1 × 32.8 cm)
Produced by Stratasys Ltd.

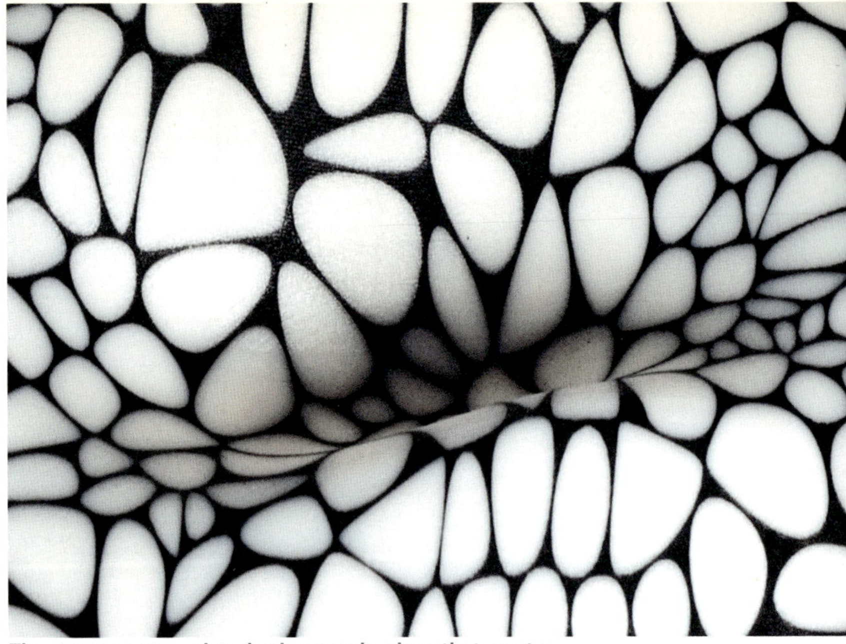

The structure was printed using a technology that creates parts and assemblies from multiple materials in a single build, producing composite materials with predefined combinations of mechanical properties.

Unlike designs that distinguish between internal structural frameworks and nonbearing skin elements, the objects in Monocoque, from the French word for "single shell," support their own load with a structural skin. We designed them in a Voronoi tessellation pattern, with veinlike elements that distribute shear stress and pressure over their surfaces. Monocoque's objects were among the first to be produced in high spatial resolution using a multimaterial printing technology that enables the production of structures whose properties vary within a single part.

The densities of Monocoque's structural skin's Voronoi pattern correspond to multiscalar loading conditions. Shear stress and surface pressure are distributed over the object through the allocation and relative thickness of veinlike elements built into the skin.

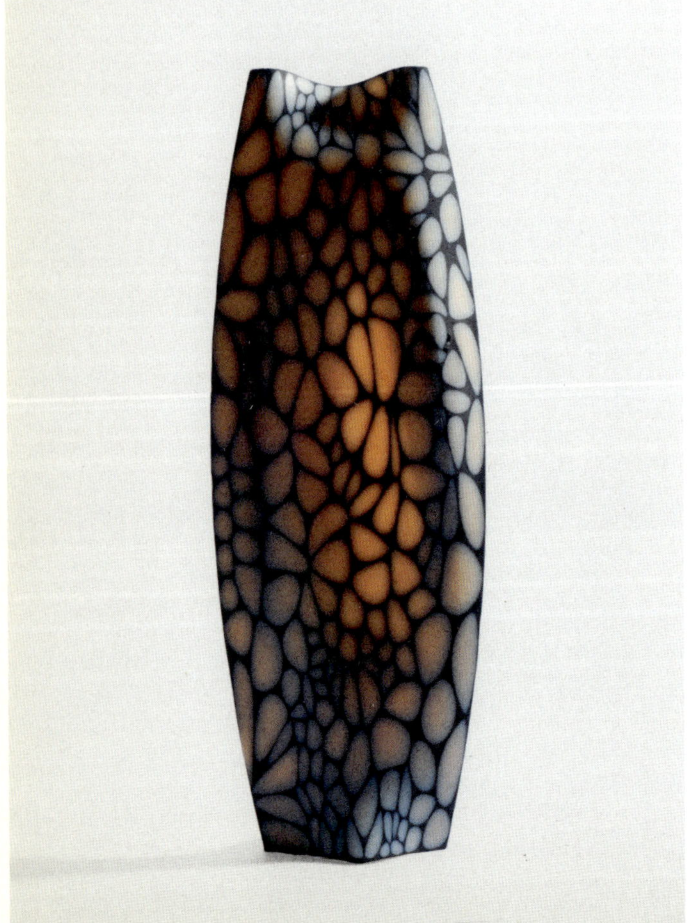

On the object's surface, smaller polygons were allocated to regions of higher stress to increase the surface area connecting the elements. Larger polygons were allocated to regions of lower stress.

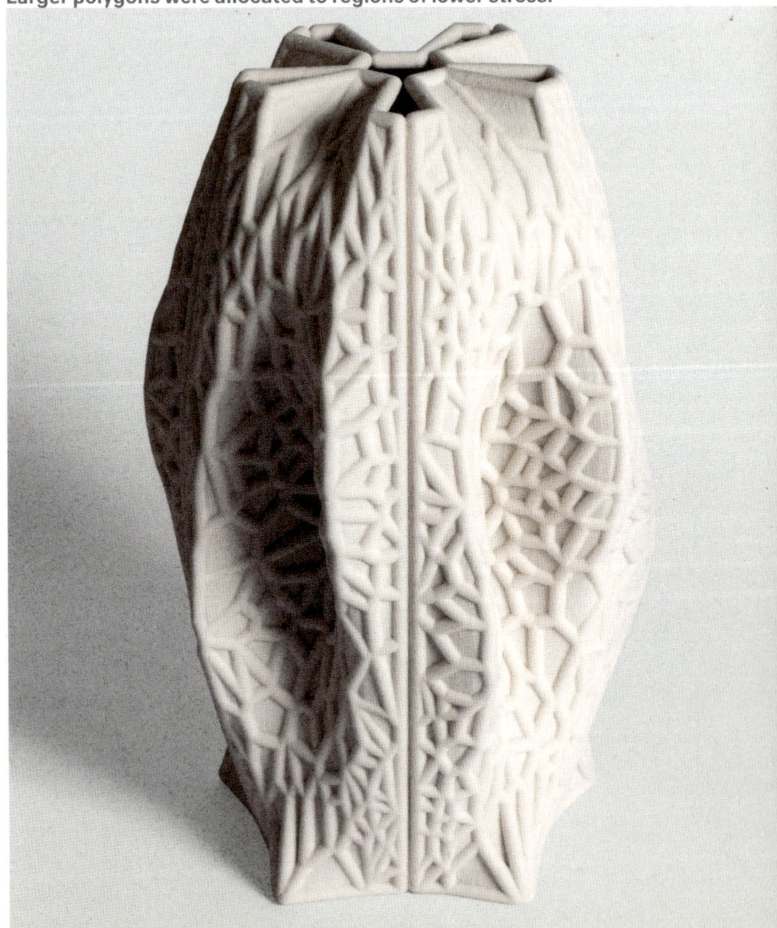

he structural load is supported by the object's external skin. ;ontrary to traditional building design, in which skin and tructure are made from different materials, the material ı Monocoque could serve as both.

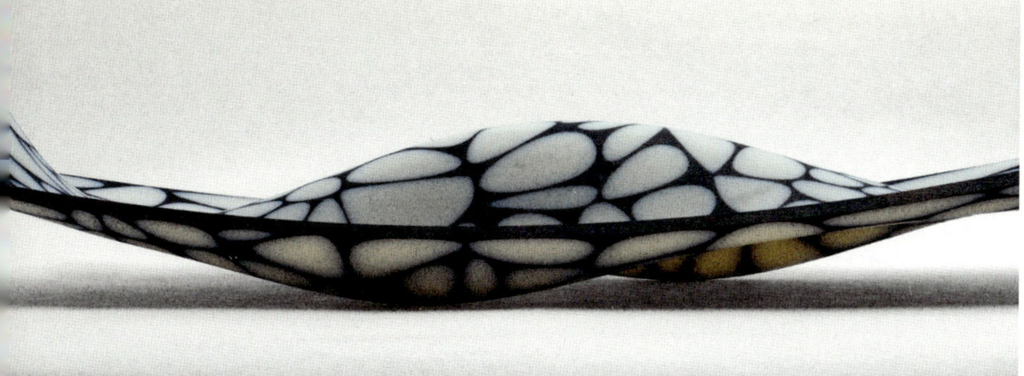

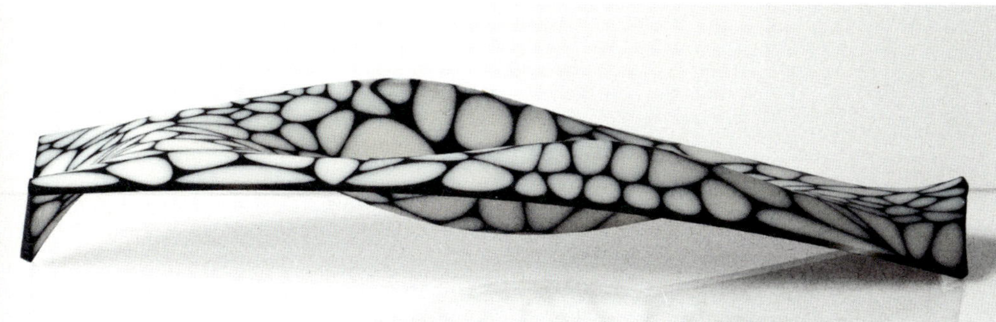

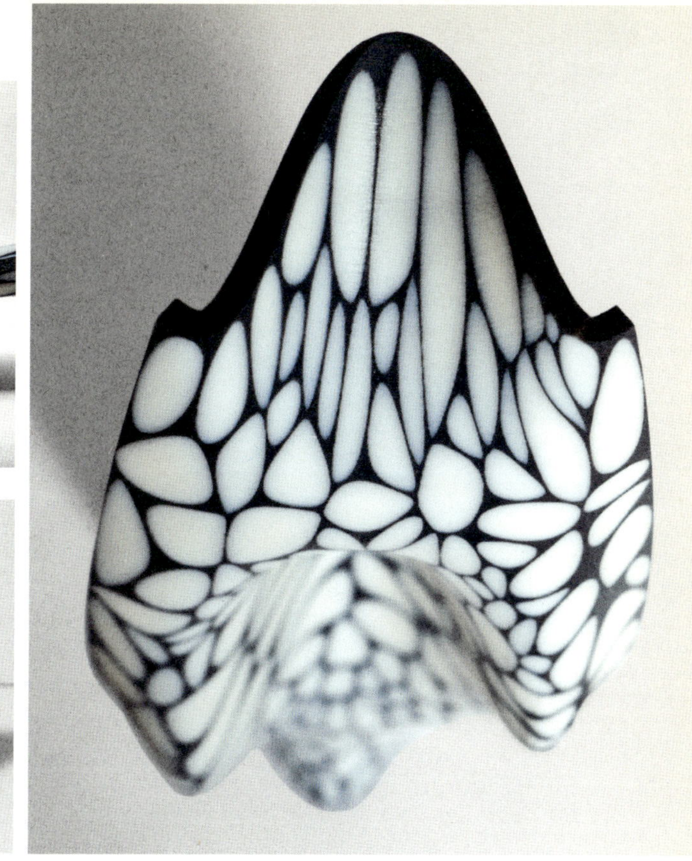

Monocoque's tessellation was determined by performance criteria that included ıe amount of mechanical load. The cell size and density vary based on force vectors ıat emulated this load's magnitude and direction.

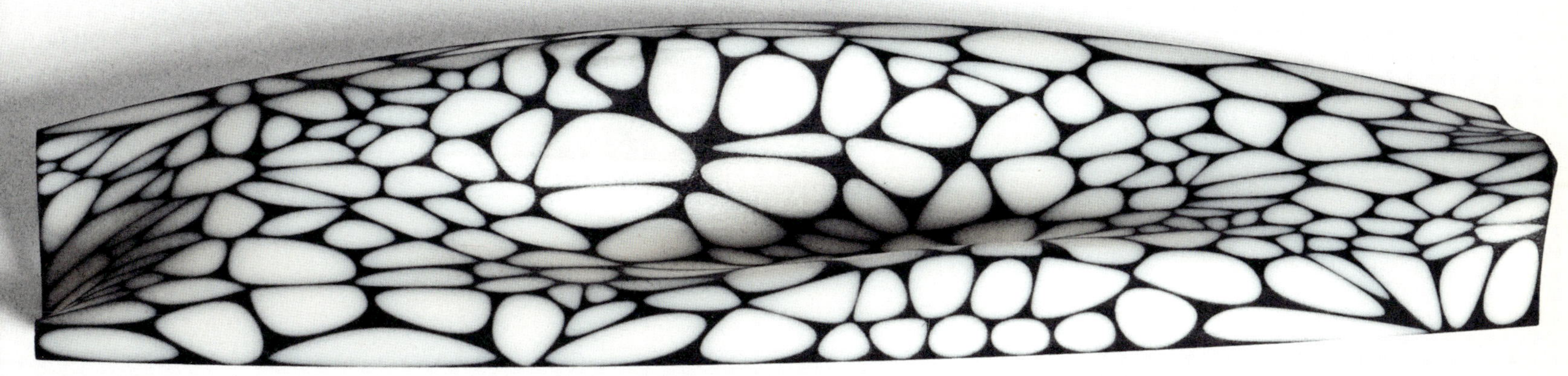

rchitectural design is often inspired by the textures and patterns of nature, but ı the synthetic world, tiling patterns are defined by geometrical considerations, ehavioral constraints, and material choices. In Monocoque's material-based pproach, each tile or group of tiles represents different mechanical and environ-ıental properties that were an integral part of the form-generation process.

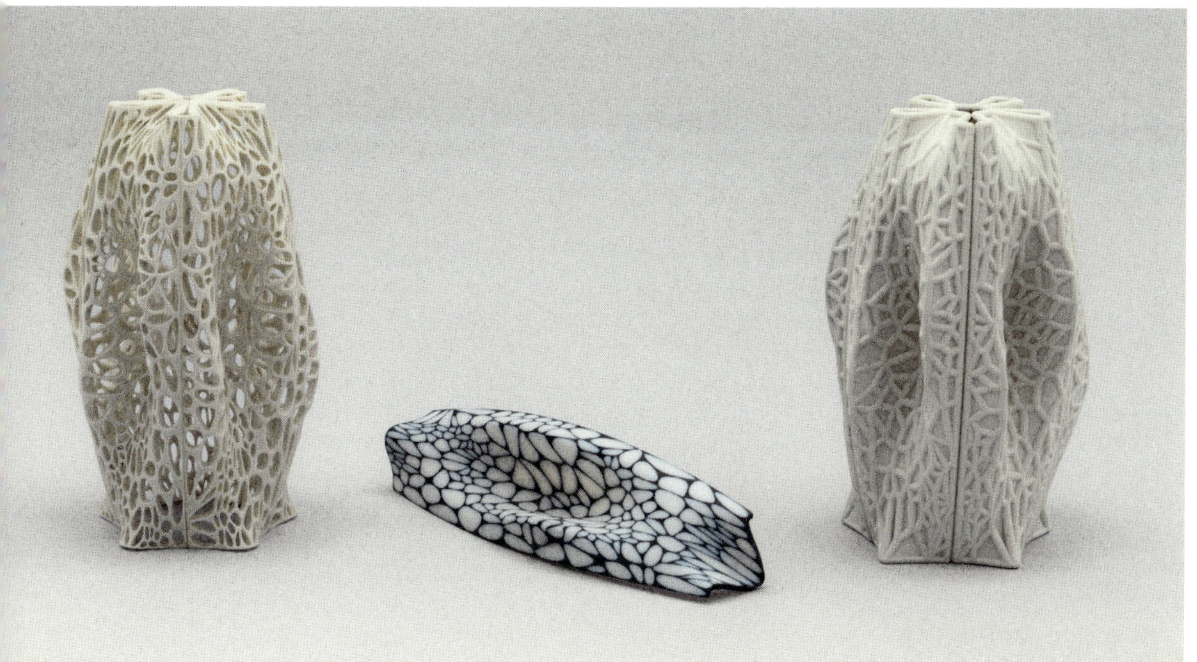

Monocoque is made up of three prototypes made with composite materials that represent three strategies for the design and fabrication of single-shell structures.

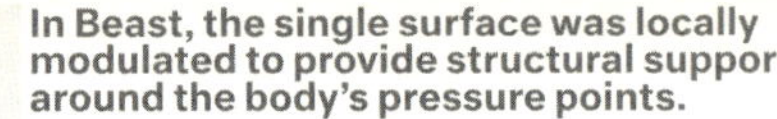

In Beast, the single surface was locally modulated to provide structural support around the body's pressure points.

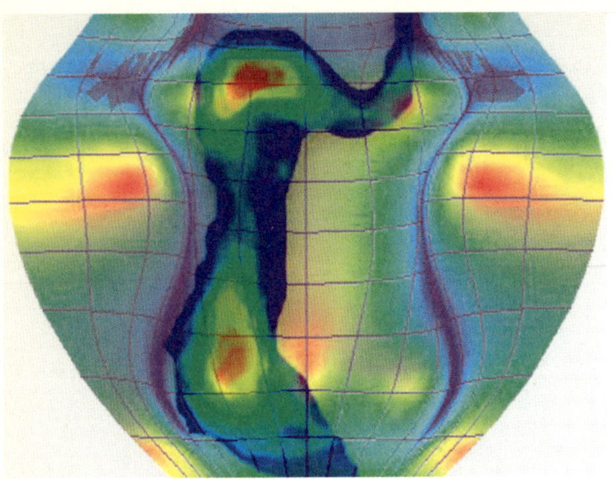

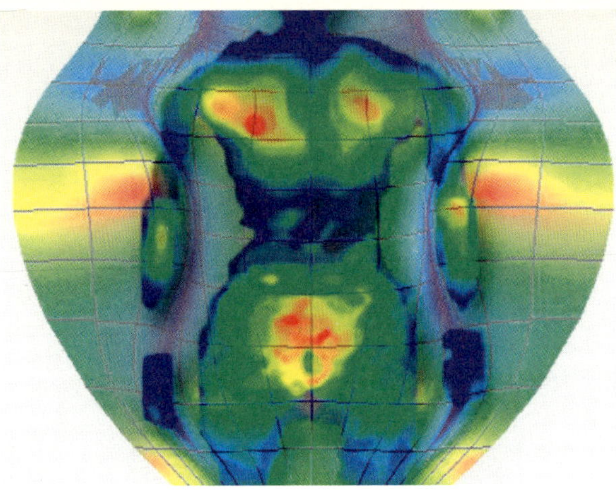

Monocoque is part of a series called Beasts. The continuous surface of the series's main project, Beast—a prototype for a chaise longue—acts as both structure and skin and is locally modulated to fit a human body. Beast's structural, environmental, and corporeal performance was achieved by adapting its thickness, pattern density, stiffness, flexibility, and translucency to load, curvature, and skin-pressure areas.

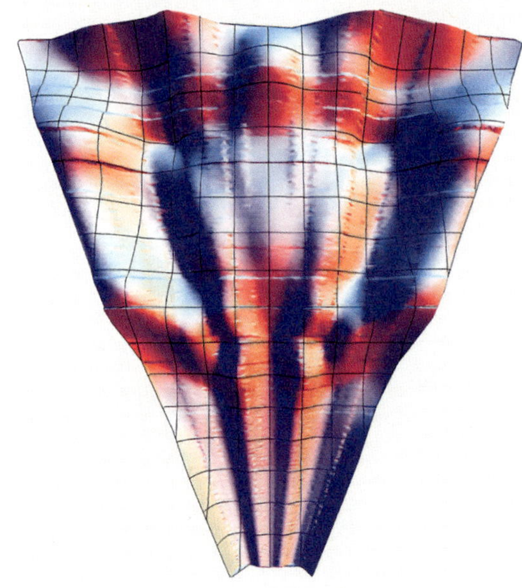

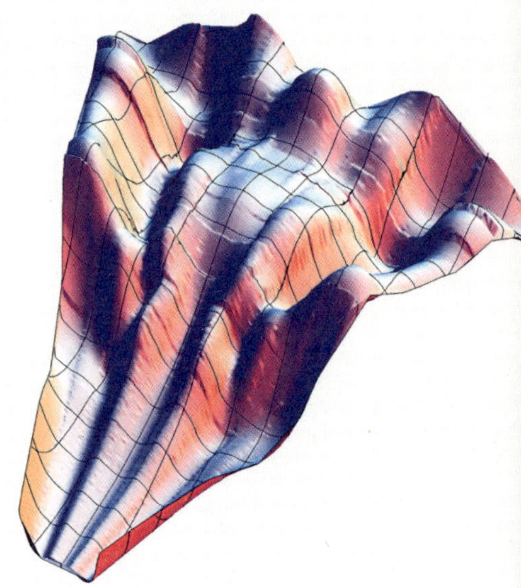

Point cloud (a set of data points in space) determining the density of the distribution of cell-like components on Beast's surface area. Because stiffness and tactility depend on local tile size, the object's geometry was intrinsically coupled with material behavior.

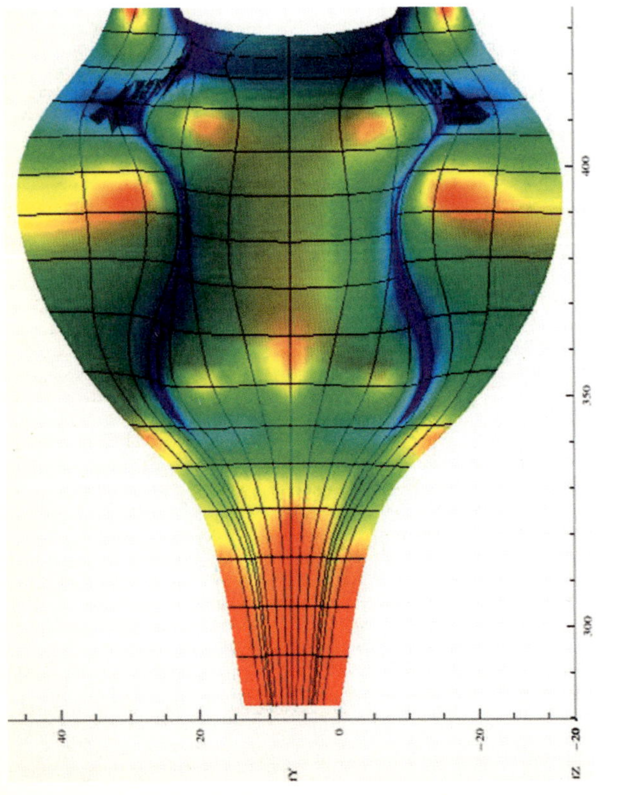

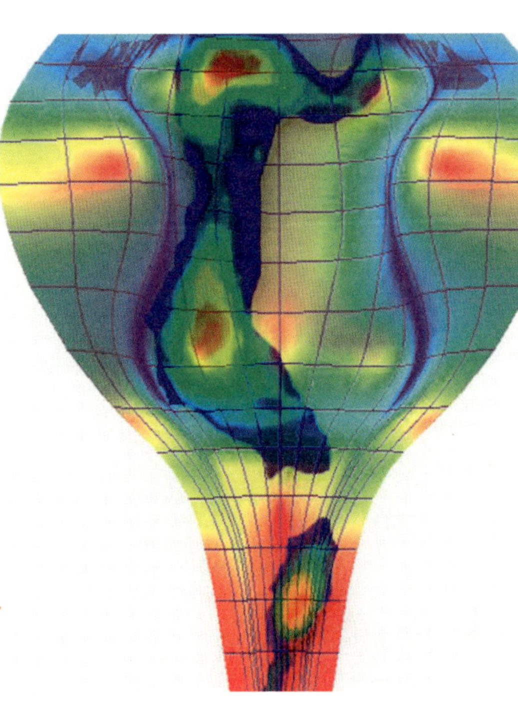

Curvature on the chaise's surface calculated by the distribution of tiles. Regions of higher curvature required smaller tiles, and regions of lower curvature were covered with larger tiles.

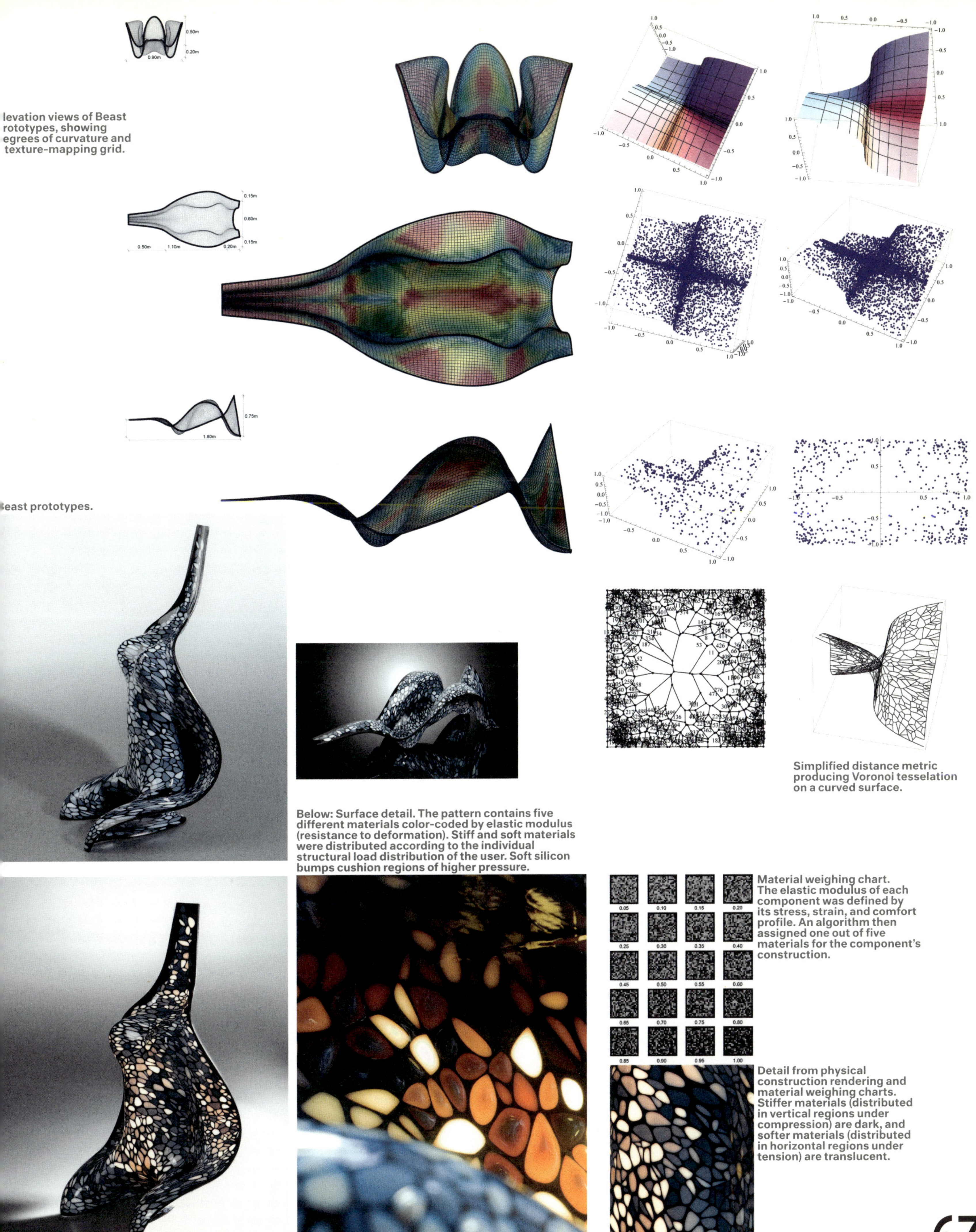

levation views of Beast rototypes, showing egrees of curvature and texture-mapping grid.

east prototypes.

Simplified distance metric producing Voronoi tesselation on a curved surface.

Below: Surface detail. The pattern contains five different materials color-coded by elastic modulus (resistance to deformation). Stiff and soft materials were distributed according to the individual structural load distribution of the user. Soft silicon bumps cushion regions of higher pressure.

Material weighing chart. The elastic modulus of each component was defined by its stress, strain, and comfort profile. An algorithm then assigned one out of five materials for the component's construction.

Detail from physical construction rendering and material weighing charts. Stiffer materials (distributed in vertical regions under compression) are dark, and softer materials (distributed in horizontal regions under tension) are translucent.

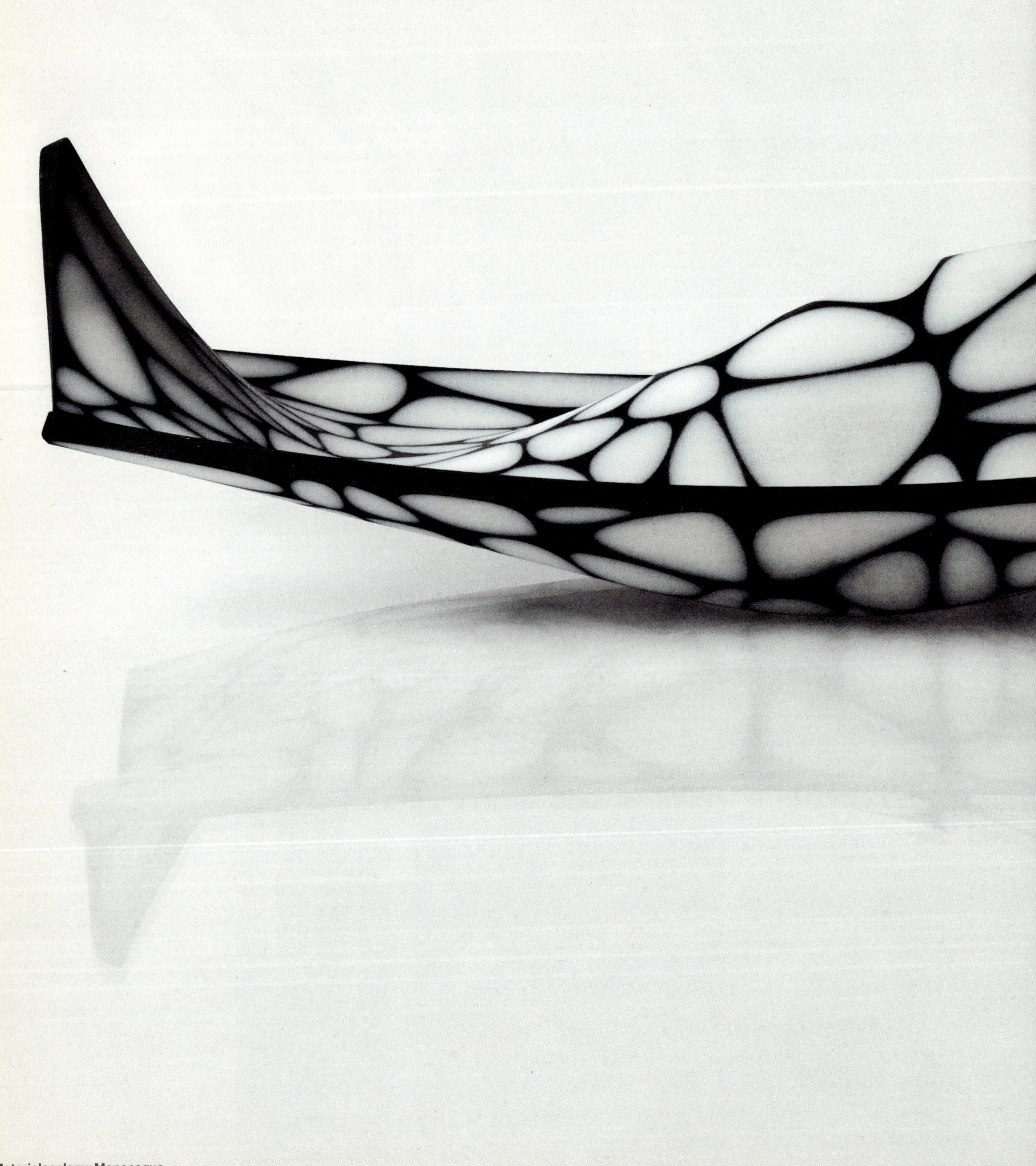

Materialecology: Monocoque

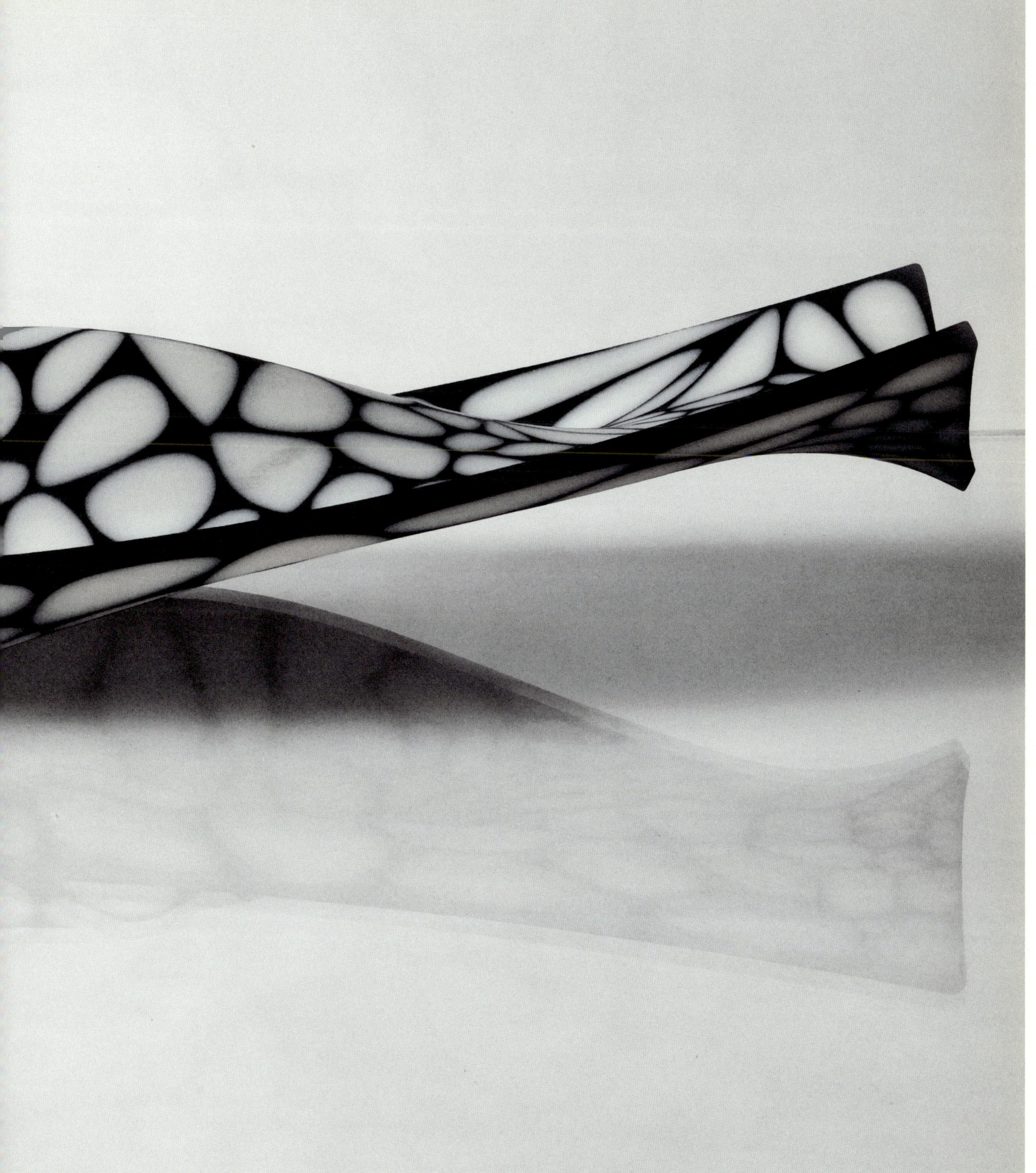

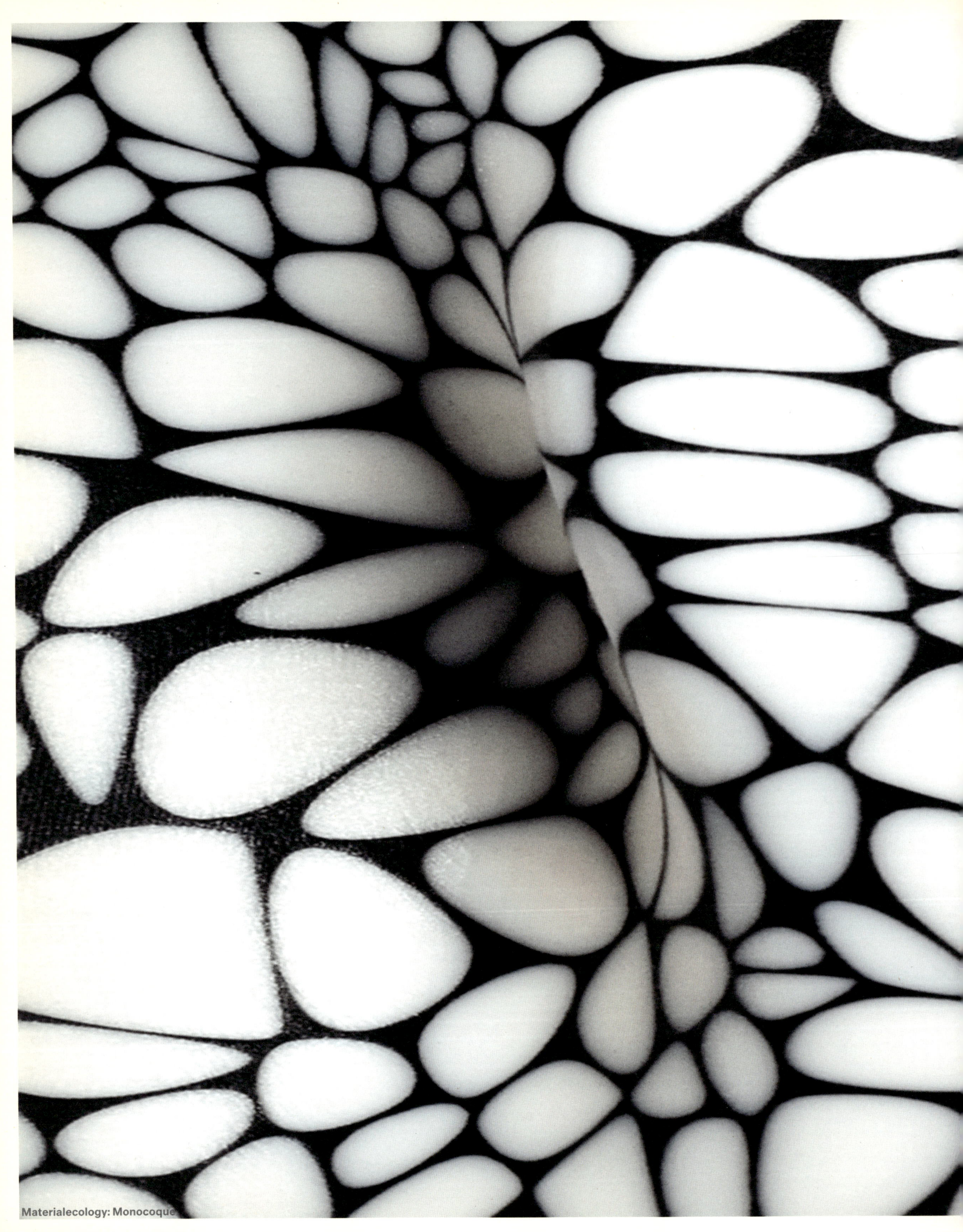

Materialecology: Monocoque

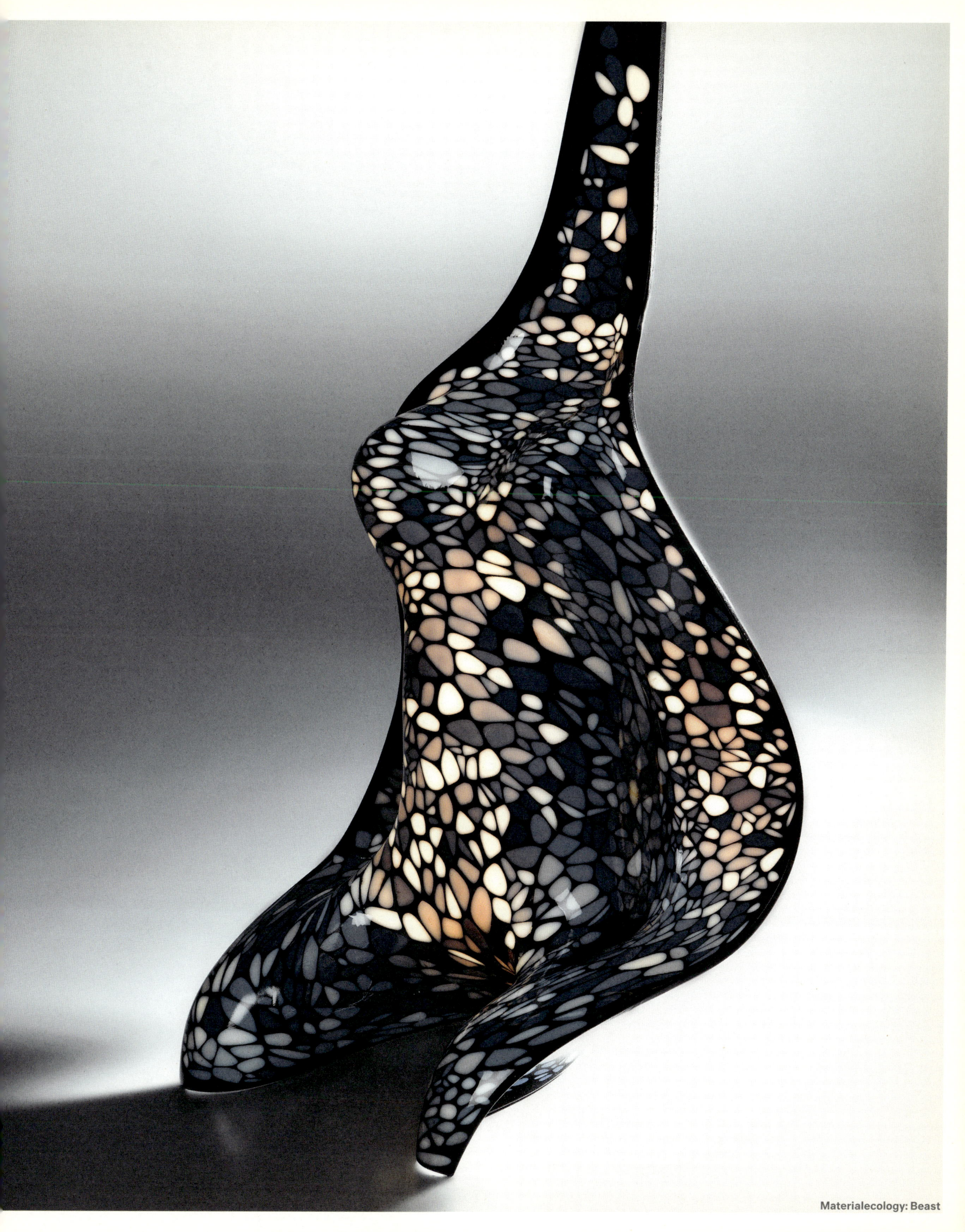

Materialecology: Beast

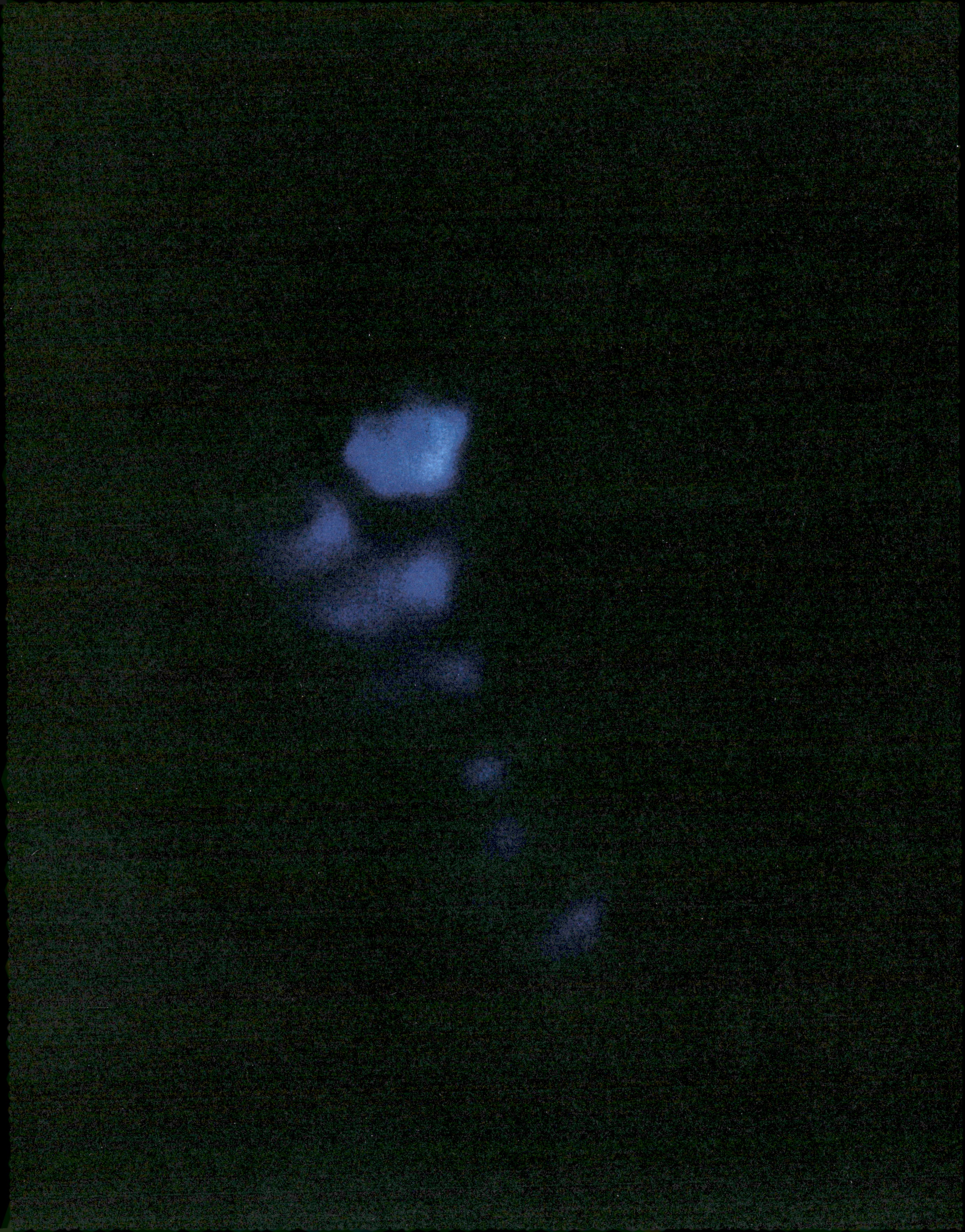

EXTRUSIONS

AGUAHOJA I

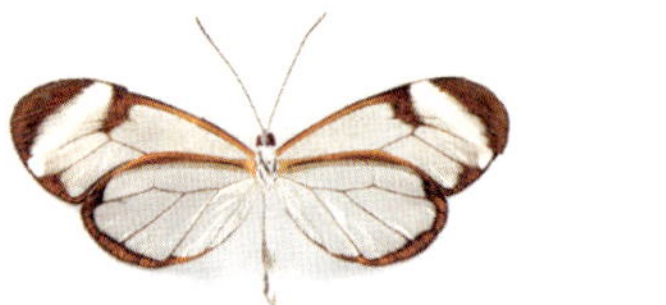

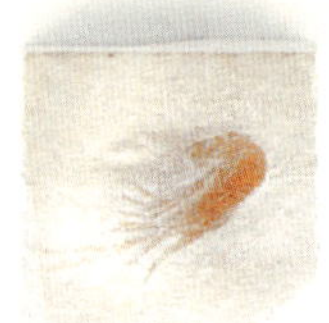

Neri Oxman and
The Mediated Matter Group
Aguahoja I. 2018
Wood-pulp cellulose, apple pectin, calcium carbonate, acetic acid, vegetable glycerin, and chitosan (85% deacetylated)
39 ⅜ in. × 39 ⅜ in. × 16 ft. 6 ⅜ in.
(100 × 100 × 504 cm)
Pavilion produced by Stratasys Ltd.
An MIT Media Lab project
Research team: Jorge Duro-Royo, Laia Mogas-Soldevila, Daniel Lizardo, Joshua Van Zak, Yen-Ju (Tim) Tai, Andrea Ling, Christoph Bader, Nic Lee, Barrak Darweesh, Sunanda Sharma, James C. Weaver, Neri Oxman
Undergraduate researchers: Matthew Bradford, Loewen Cavill, Emily Ryeom, Aury Hay, Yi Gong, Brian Huang, Joseph Faraguna
Collaborators and contributors: Shaymus Hudson, Tzu-Chieh Tang, Tim Lu and the Lu Lab, MIT Research Laboratory of Electronics
Acknowledgments: TBA-21 Academy (Thyssen-Bornemisza Art Contemporary), GETTYLAB, Robert Wood Johnson Foundation, Autodesk BUILDSpace

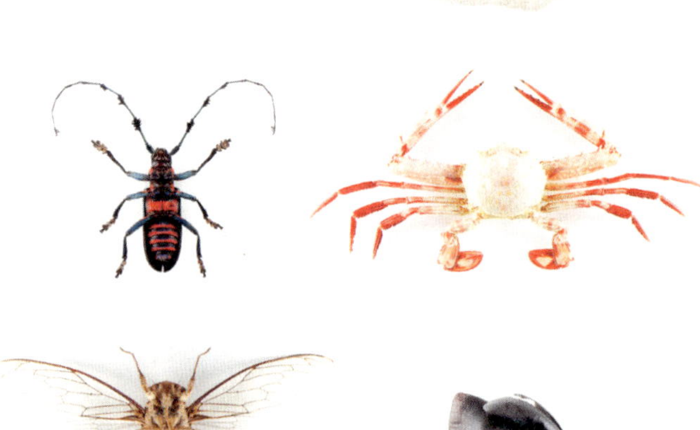

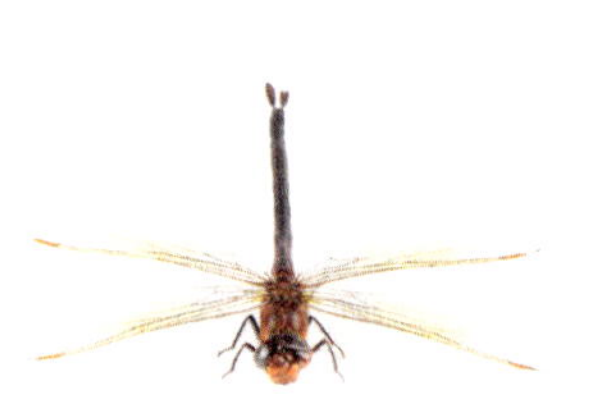

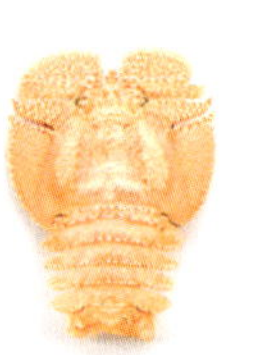

Chitin, the second-most abundant biopolymer on earth—forming crustacean shells, insect exoskeletons, and the soft tissue of fungi, among other things—is amenable to small chemical modifications that produce water-soluble chitosan.

More than 300 million tons of plastic are produced every year. Less than one tenth of this material is recycled, with vast quantities being transported to landfills or circulating indefinitely in ocean currents. The cycle of overproduction, obsolescence, and waste accumulation is not limited to plastics but permeates all industrial material production. By contrast ecosystems produce and consume material in a perpetual cycle in which

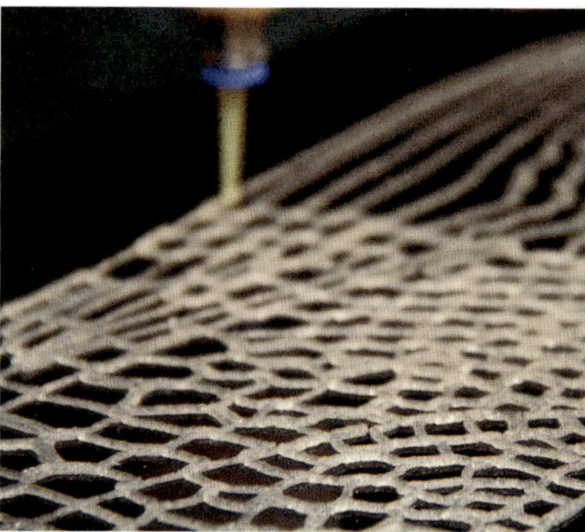

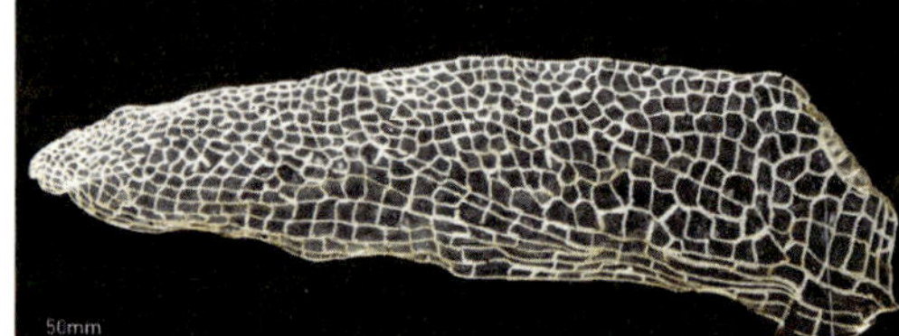

Early prototype with a low concentration of chitosan.

Up to sixty percent of the adult human body is water; about seventy-one percent of the Earth's surface is covered by water; and about ninety-six percent of all the water on the planet is in its oceans. Water shapes who we are and what we become: the presence or absence of water affects how biological organisms grow and develop, and also how they decompose.

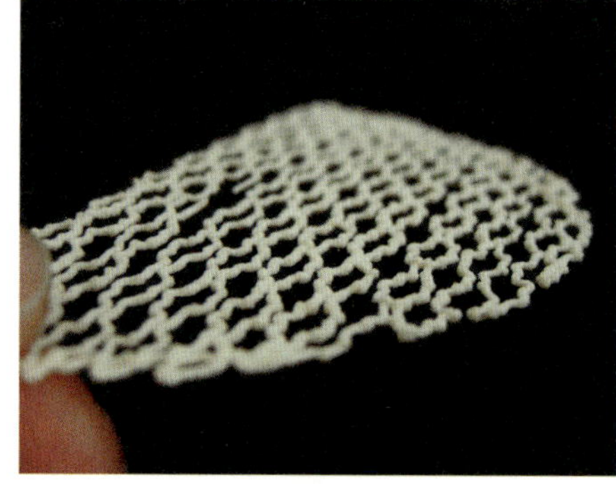

Above: Early prototype made of cellulose and chitosan.

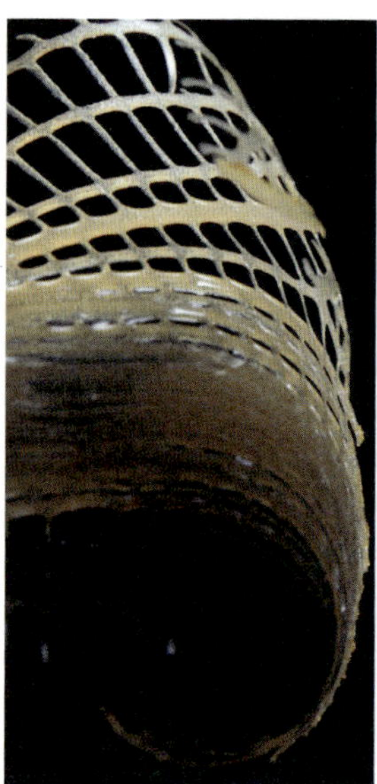

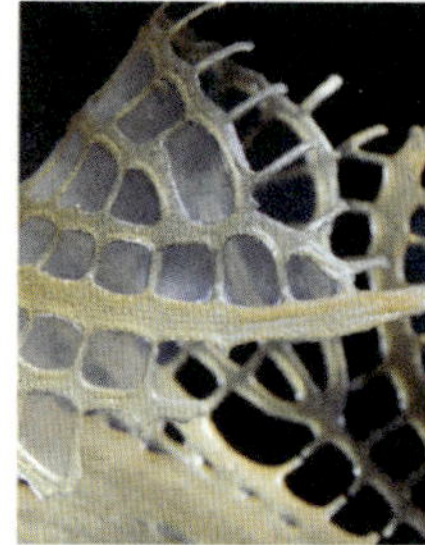

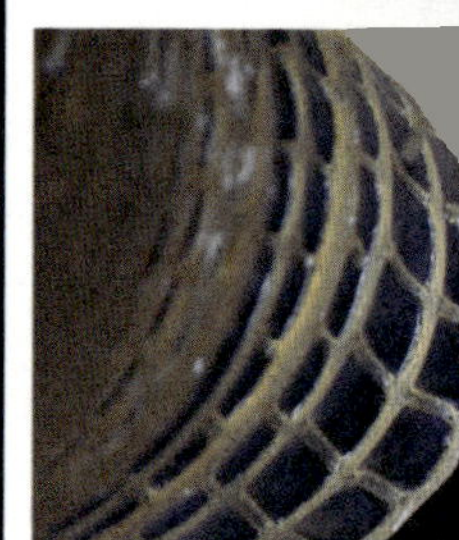

Rigid veins contained a higher concentration of chitosan, in contrast to the transparent cells.

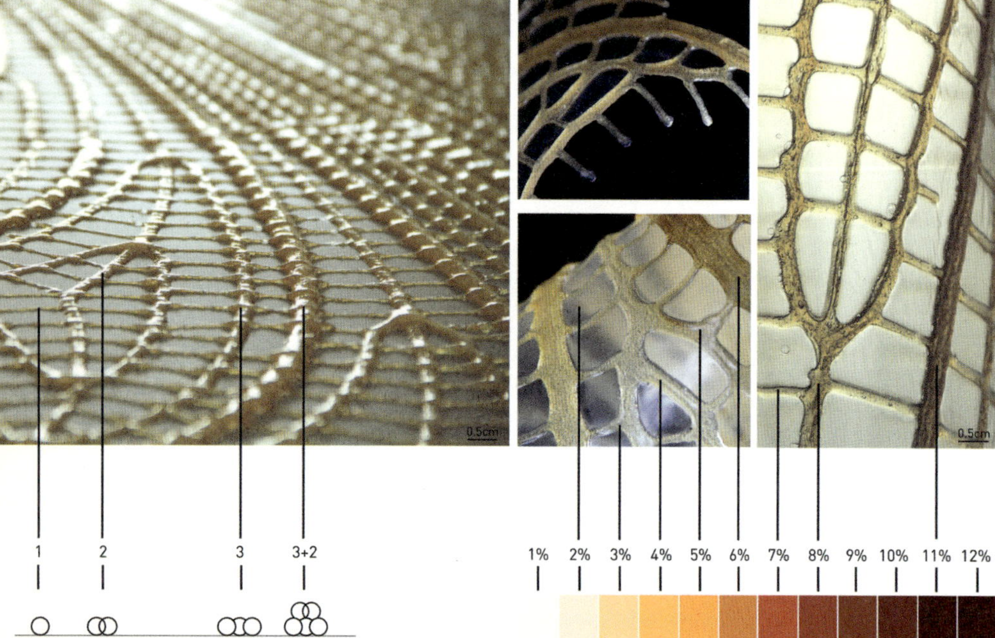

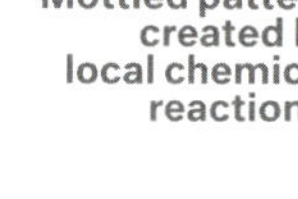

A hierarchical distribution of materials controlled the chitosan composite's mechanical properties and geometry. Greater concentrations of chitosan caused increased rigidity and deformities.

Right: Mottled pattern created by local chemical reactions.

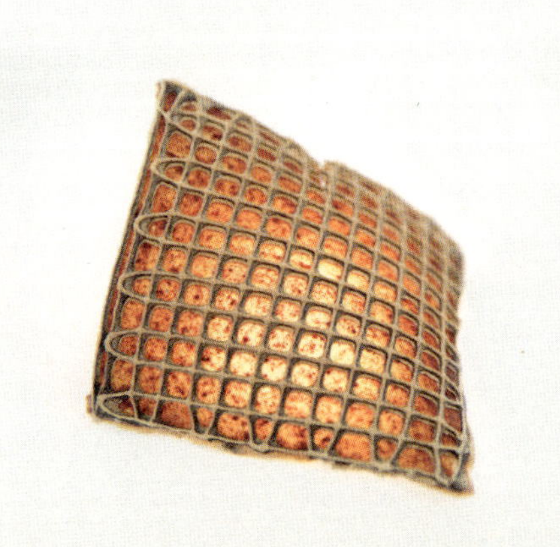

vaste is nonexistent. Matter is grown n response to environment; it lives and unctions; and upon its death it is consumed to fuel new growth. From a elatively small palette of molecular components, natural systems construct an extensive array of functional materials with no synthetic parallels. Chitin, for instance, nanifests in the form of thin, transparent dragonfly wings, as well as in the soft tissue of fungi. Cellulose makes up more than half of plant matter. These materials, and the iving systems they grow in, outperform human engineering not only in their diversity of functions but also in their esilience, sustainability, and adaptability. To leverage the power of natural resource cycles and the matter they produce, we designed digital-fabrication systems for he additive manufacturing of natural biopolymers. Our Water-Based Digital Fabrication platform converts cellulose, chitosan (deacetylated chitin), pectin, and other abundant biopolymers into high-performance sustainable materials that can be printed into objects for applications spanning scales and disciplines.

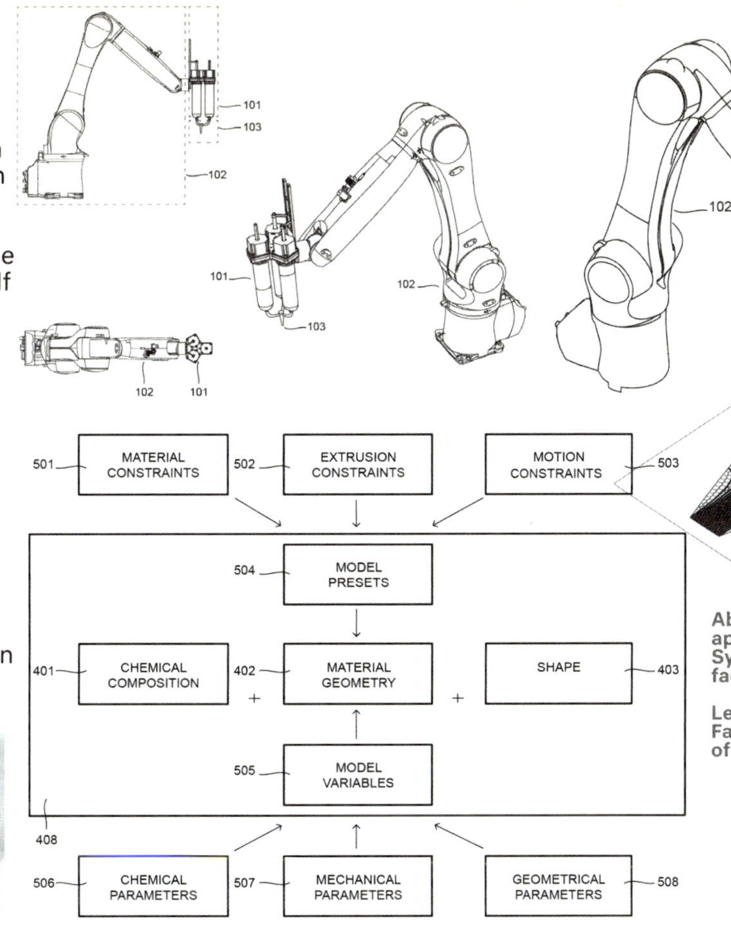

Above: Drawings submitted with the patent application for the Water-Based Digital Fabrication System, showing the platform for additive manufacturing of biopolymer composites across scales.

Left: Workflow diagram of the Water-Based Digital Fabrication System, demonstrating the translation of design parameters into fabrication information.

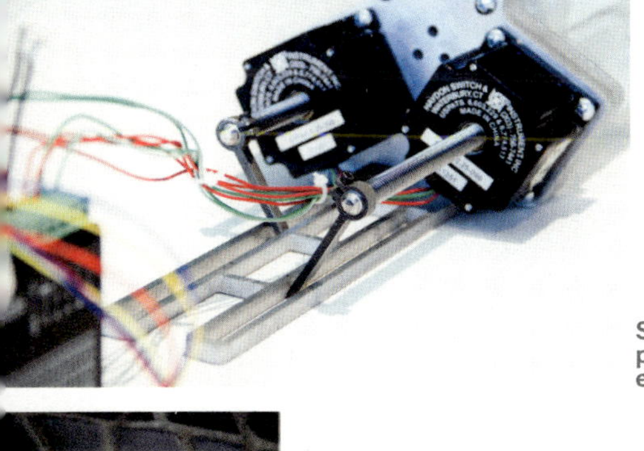

Section of the pneumatic extrusion system.

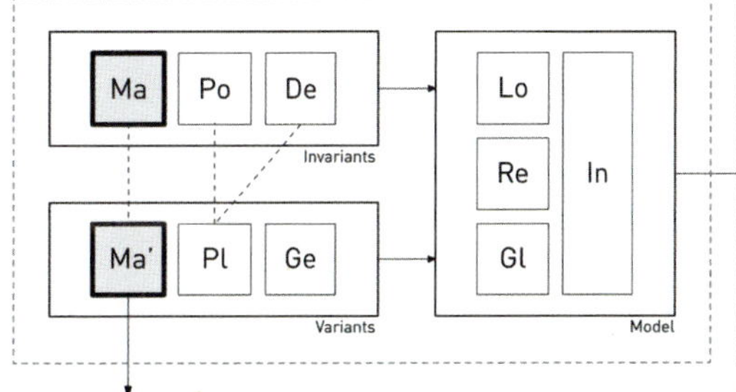

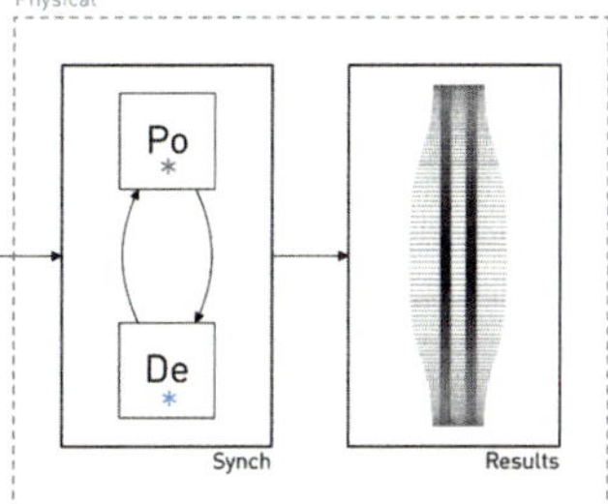

Computational workflow of the biopolymer-printing process. Experiments with heat sensing during fabrication continually informed the actions of a robotic extruder.

INPUT INVARIANTS AND CALCULATED VARIANTS

Ma	materials database (C-cellulose, I-chitosan, S-starch, P-pectin)
Po	positoning platform
De	deposition platform
Ma'	material composites data
Pl	Po and De synchronized
Ge	geometrical designs

MODEL HIERARCHY AND MODEL OUTPUT

Lo	local level of resolution (material distribution and concentration)
Re	regional level of resolution (extrusion geometry types)
Gl	global level of resolution (overall resulting shape)
In	fabrication instructions output that synchronizes Po and De

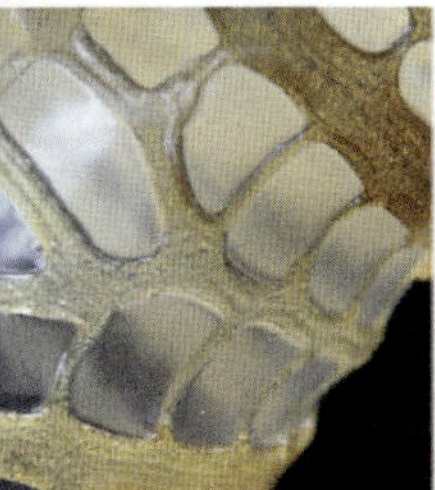

Different concentrations of chitosan lent different mechanical and optical properties.

Below: Biopolymer composites responded to humidity by expanding and becoming pliant.

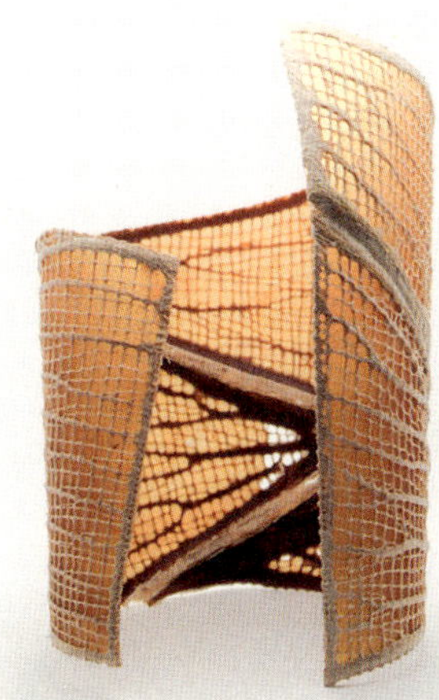

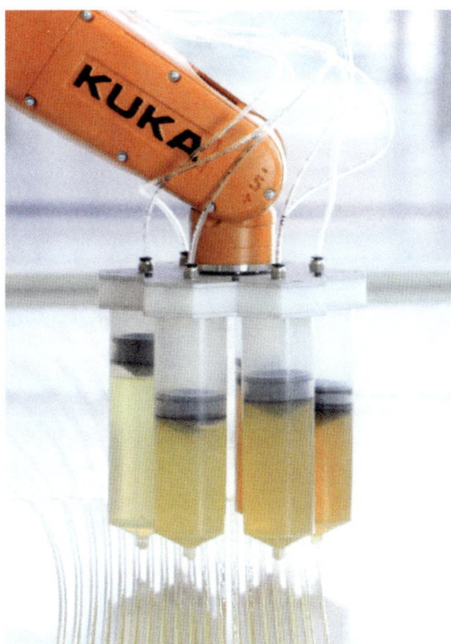

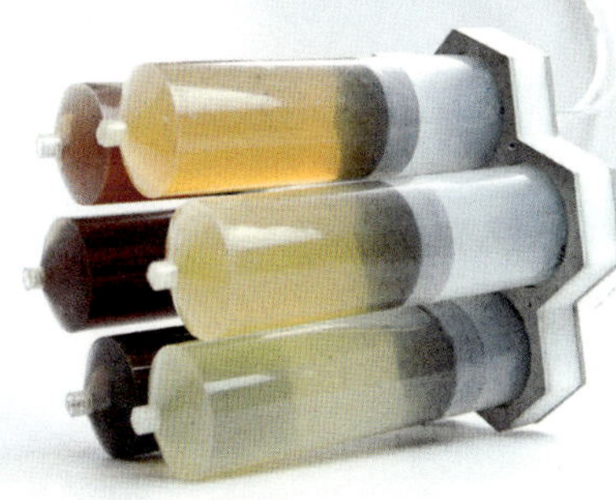

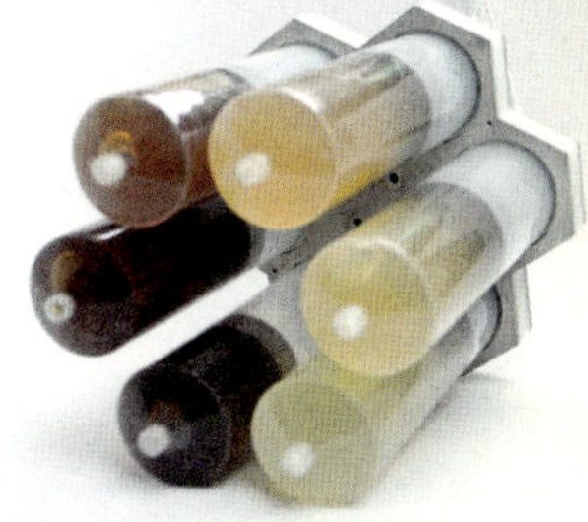

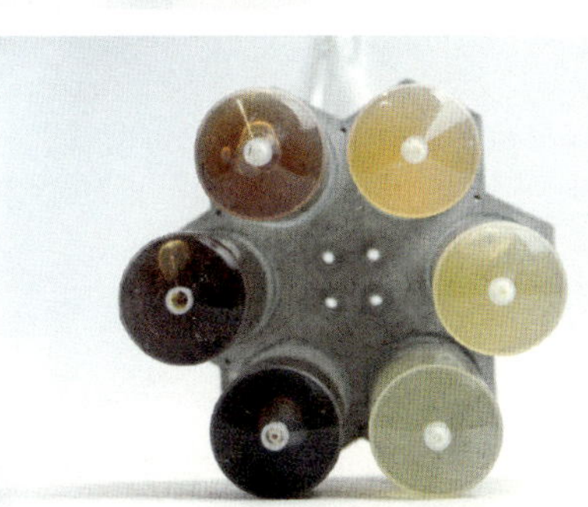

Multichamber extruder holding six different concentrations of chitosan for additive manufacturing. The system was affixed to an industrial robot for precision additive manufacturing.

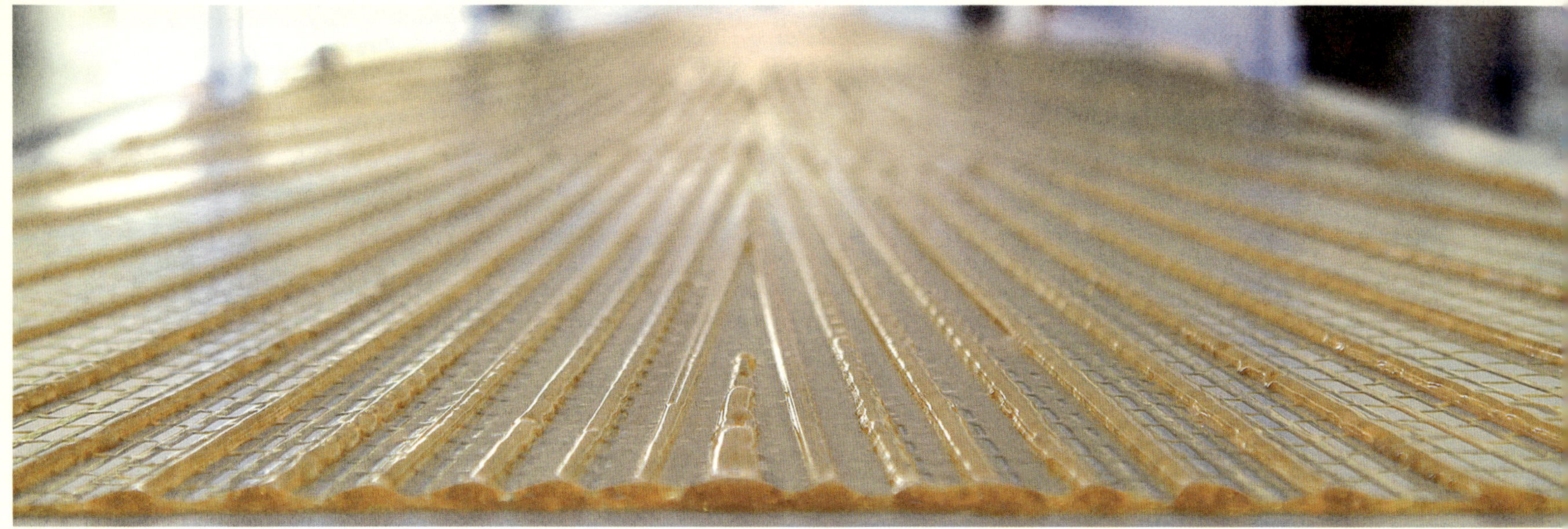

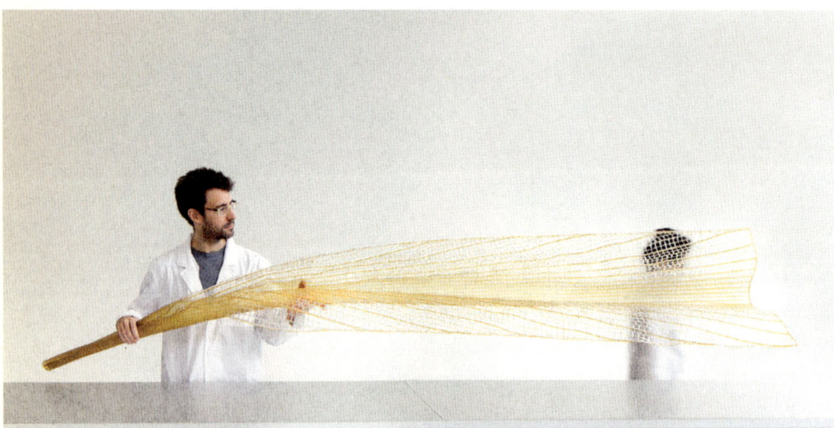

Above: The mechanical properties and geometry of a chitosan-composite object were fine-tuned by controlling the evaporation of water.

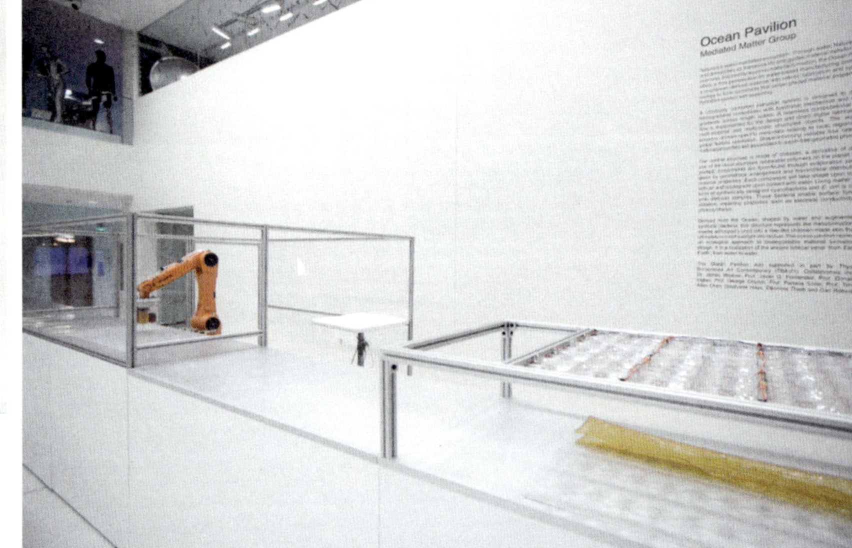

Robotic fabrication system for generating large-scale objects made of chitosan. A multichamber pneumatic extrusion system allowed for the multimaterial printing of a biocompatible object.

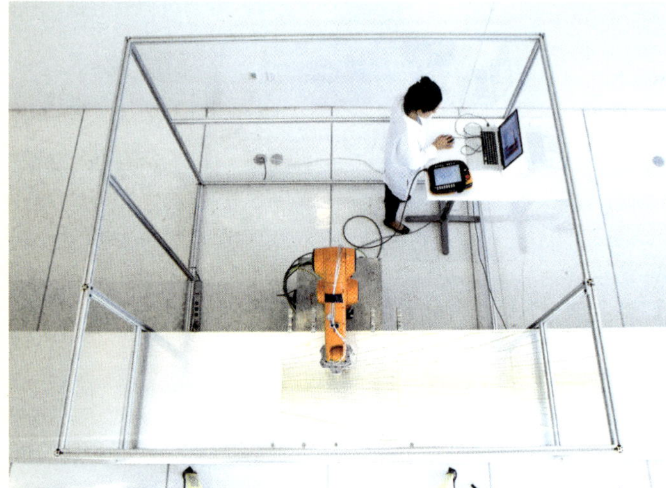

Fabrication of a composite.

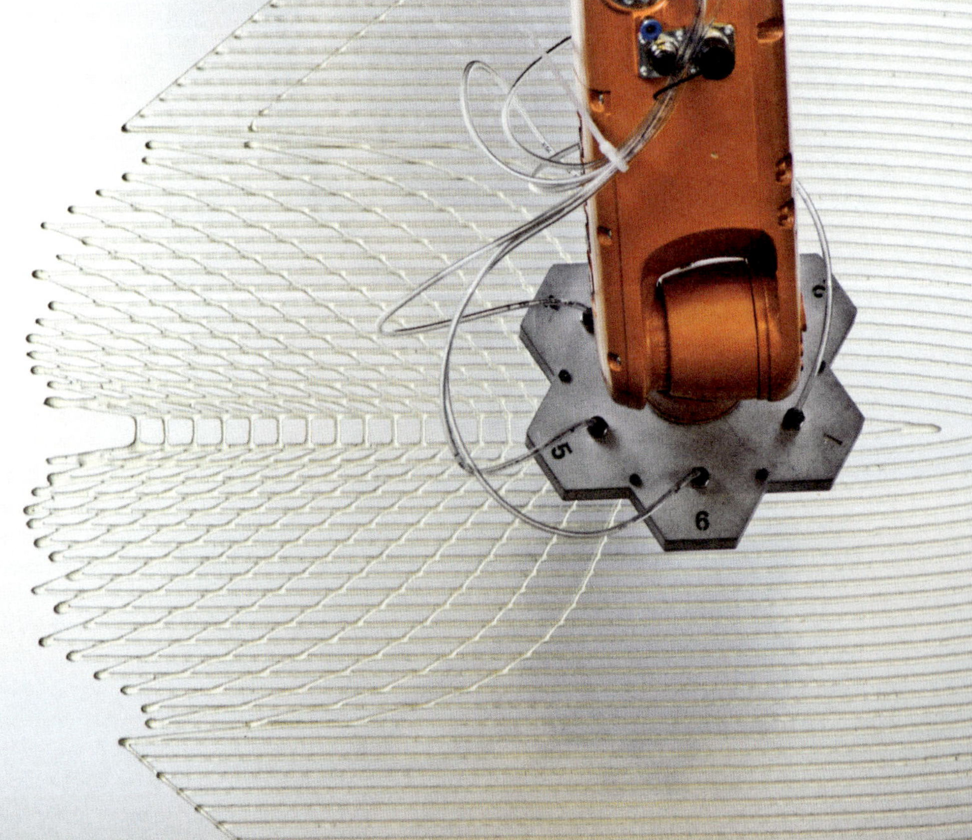

With Aguahoja we propose both a material alternative to plastic and a water-based approach to design. The work is derived from organic matter—biopolymers such as cellulose, chitosan, pectin, and calcium carbonate, which become printable once they are mixed with water. The objects are printed by a robot and shaped through hydration, using a technology that integrates material formation, digital fabrication, and physical behavior at a variety of scales that approach—and often match—that of the biological world. As a result, designers will be able to alter and control the properties of the substances they work with, which in turn become aspects of computational design and digital manufacturing; they will be able to design an object's decay as well as its shape.

ɜoth versions of Aguahoja consist of library of material experiments and collection of hardware, software, and vetware tools and technologies. These ulminated in a pavilion that demon-trates the architectural potential of the naterials. The biocompatible objects vere made from some of the most bundant biopolymers on our planet—he same materials found in trees, rustaceans, bones, and apples, among thers. The structures were designed as grown: their form, which we induced on he basis of previous tests and experi-nents, was guided by the process of heir formation, and their construction equired little to no assembly.

ɅguahojA embodies our design approach o material formation and decay by lesign—from water to water.

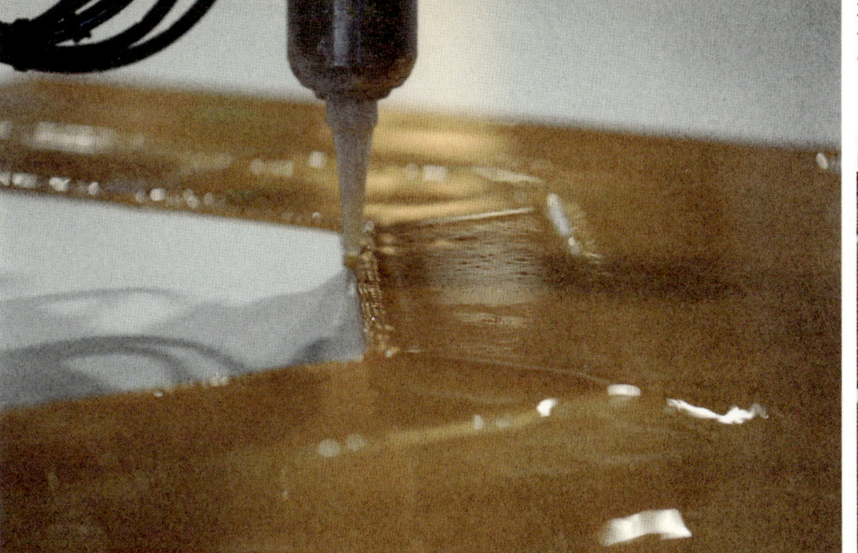

A membrane of pectin, which forms the flexible connective tissue of most vegetation, was extruded to stabilize a biopolymer composite.

Pavilion detail, with fore- and backlighting of its biopolymer skin.

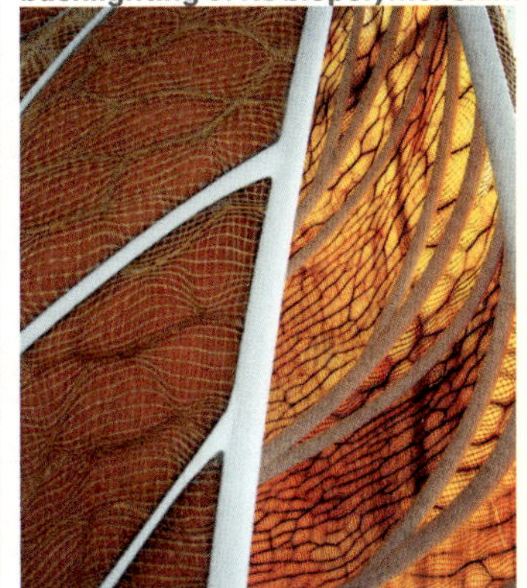

ayered composite of pectin skin, vhich stabilized rigid paths of hitosan-cellulose.

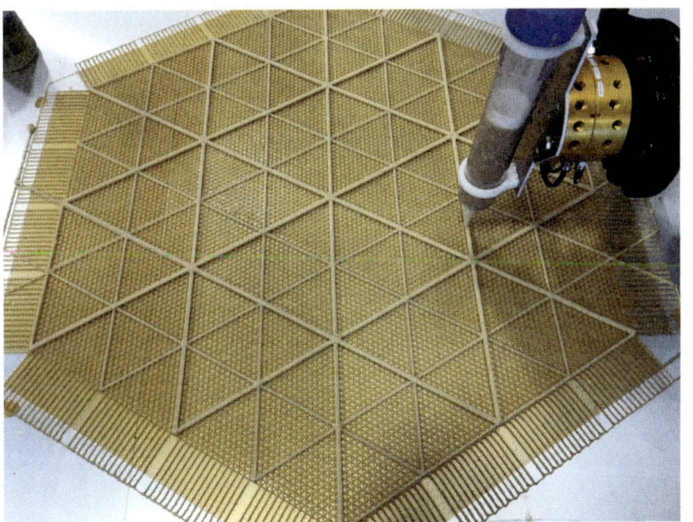

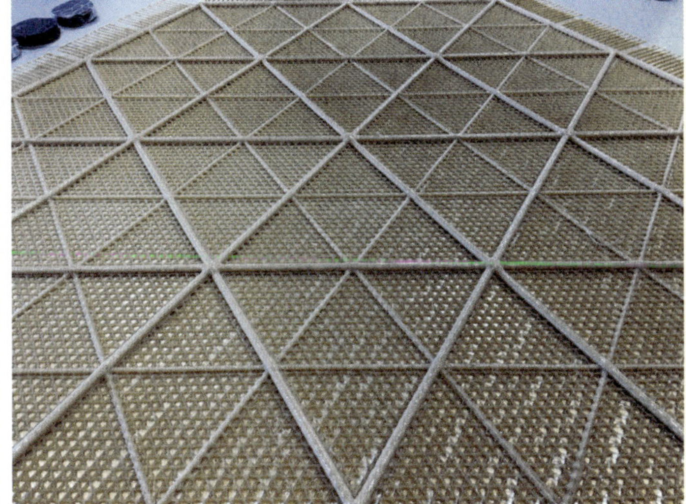

Map of material distribution in the Aguahoja I avilion. Different degradation rates and mechanical roperties—of pectin content and surface pattern in articular—governed the structure's decomposition in lobal and regional hierarchies of structural behavior.

NANO-SCALE

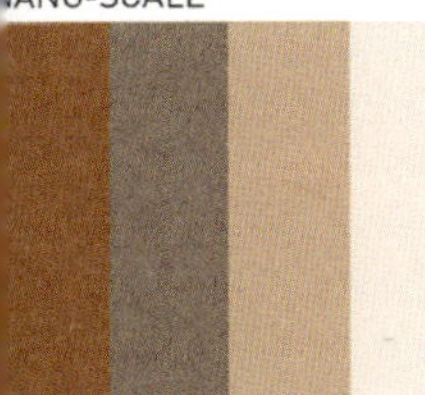

Max to Min acidity ranges orresponding to tonal palette nd oxydation levels

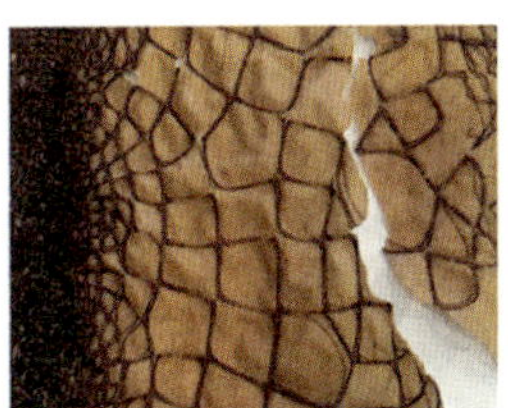

Higher hydrophilicity for blend-based humidity-induced flexibility and decay

MESO-SCALE

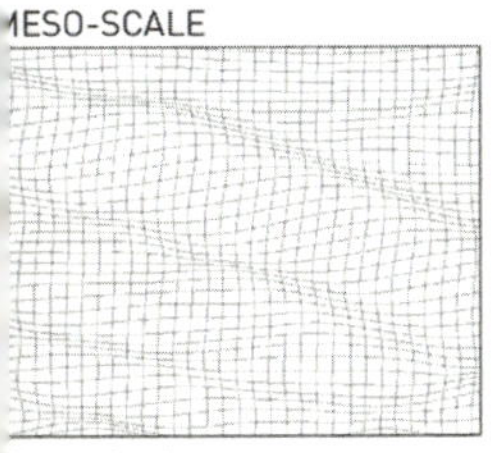

Hierarchy 1 Simple

WET ARRANGEMENT
1x0.7 DRY DIMENSIONS (mm)
FINAL DRY SHAPE

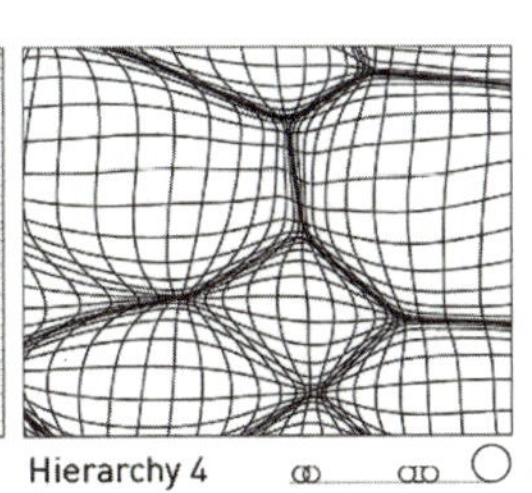

Hierarchy 4 Aggregate

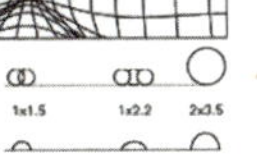

MACRO-SCALE

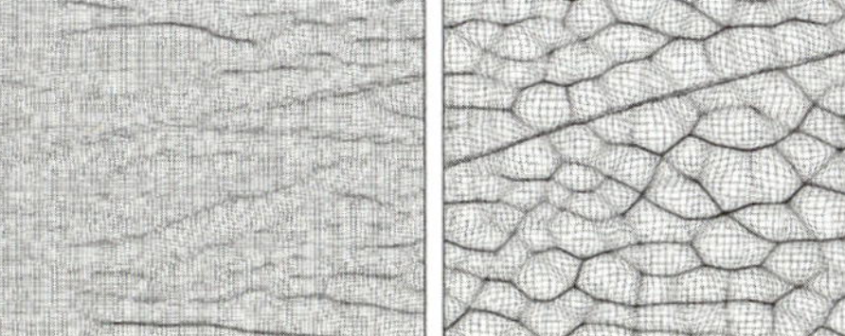

High density patterning or shading, rigidity, and ermanence

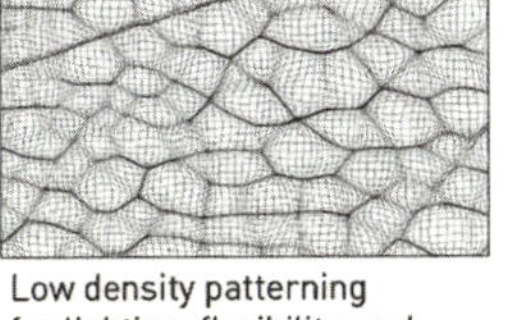

Low density patterning for lighting, flexibility, and dissotiation

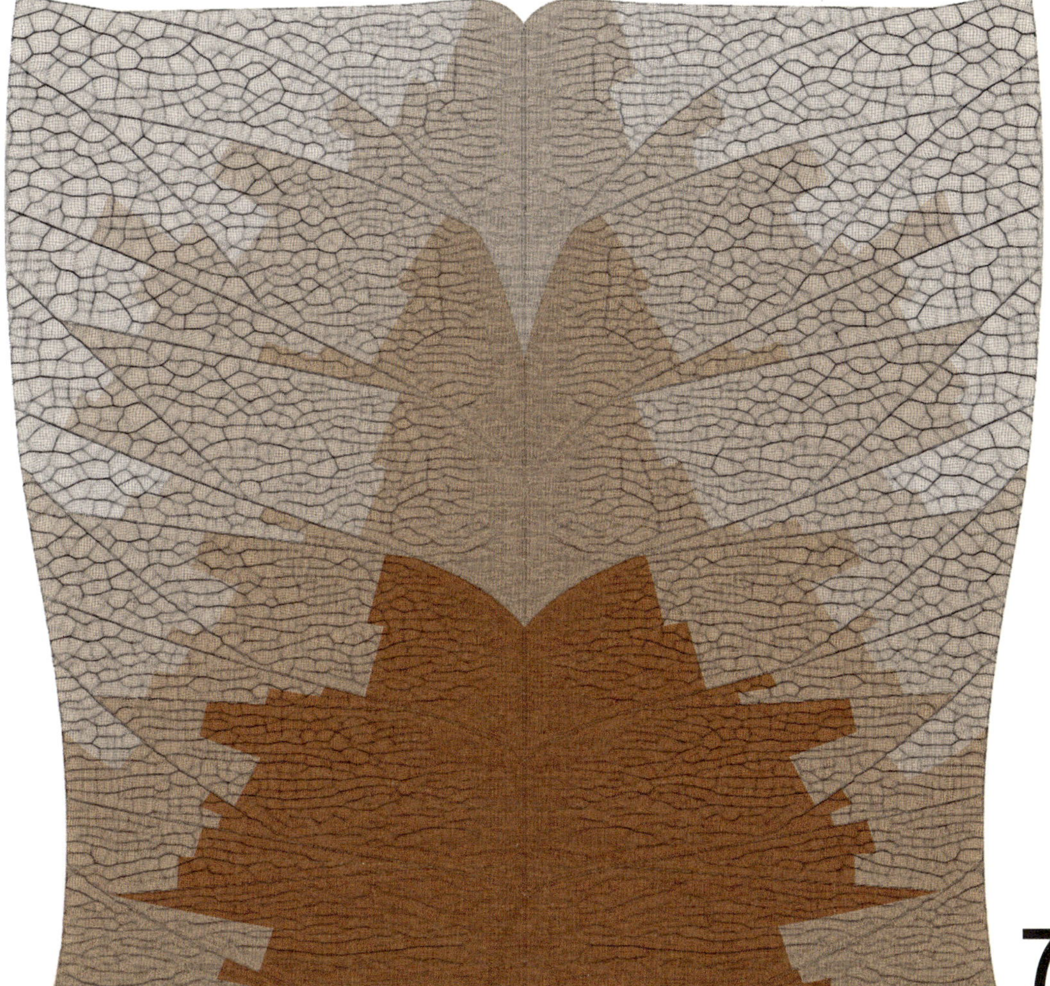

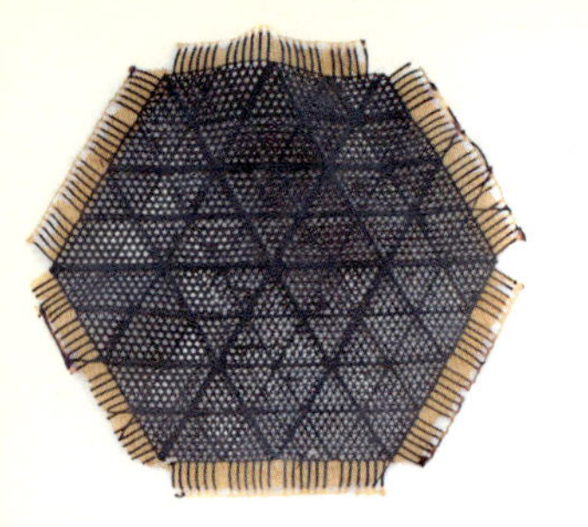

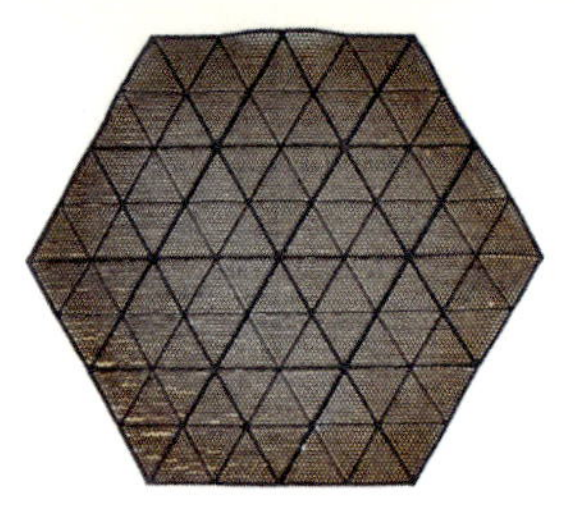

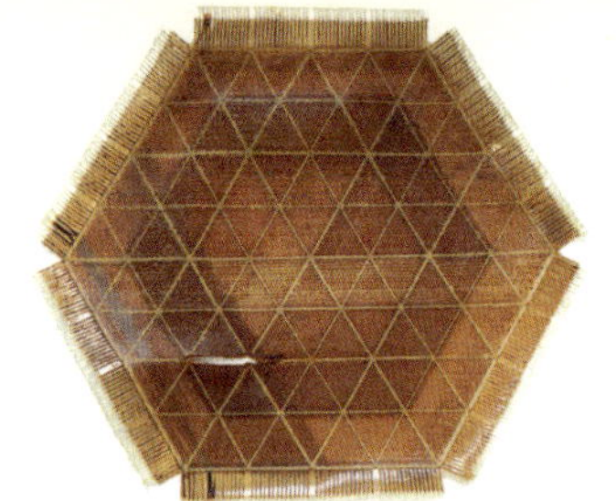

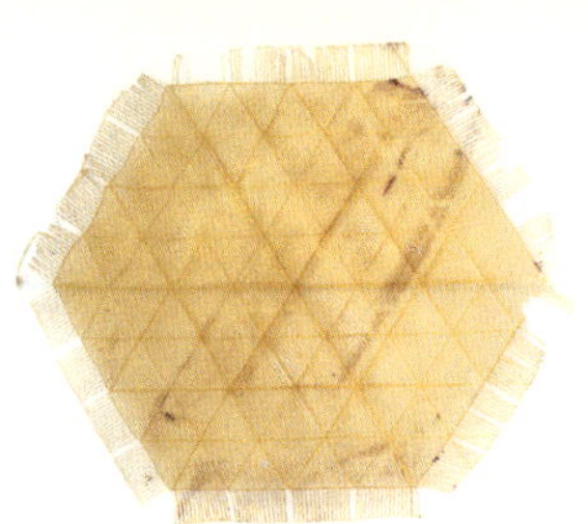

Four artifacts demonstrating the dramatically different appearances and mechanical properties elicited by minimal variations in geometry and chemistry.

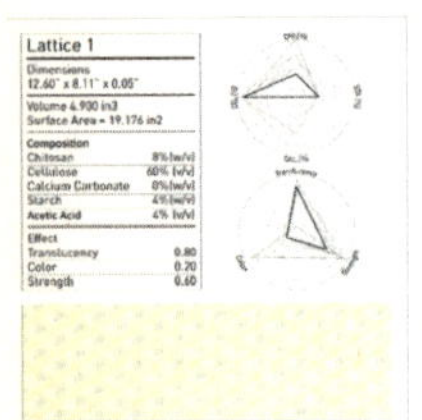

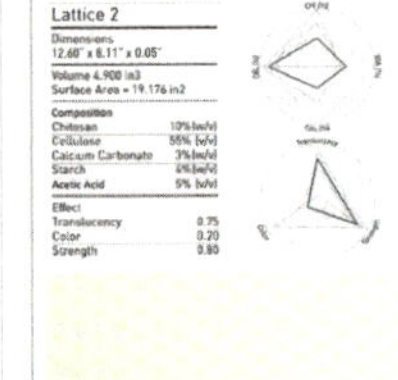

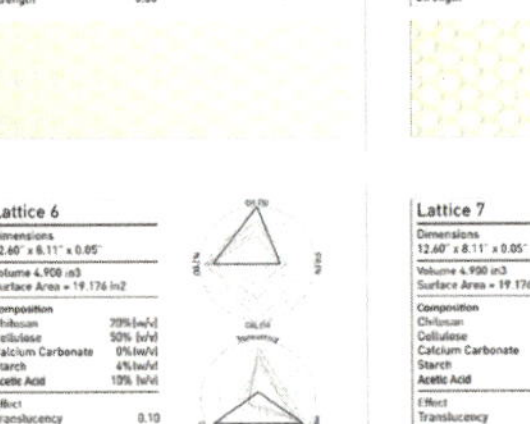

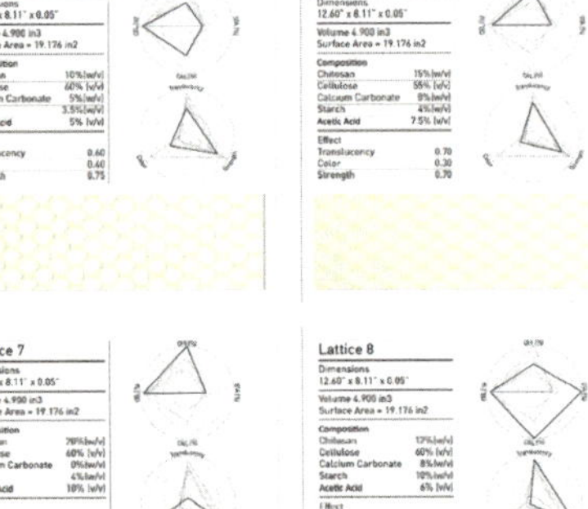

Biopolymer compositions with corresponding mechanical and optical properties, demonstrating a parametric chemistry approach to Fabrication Information Modeling (FIM).

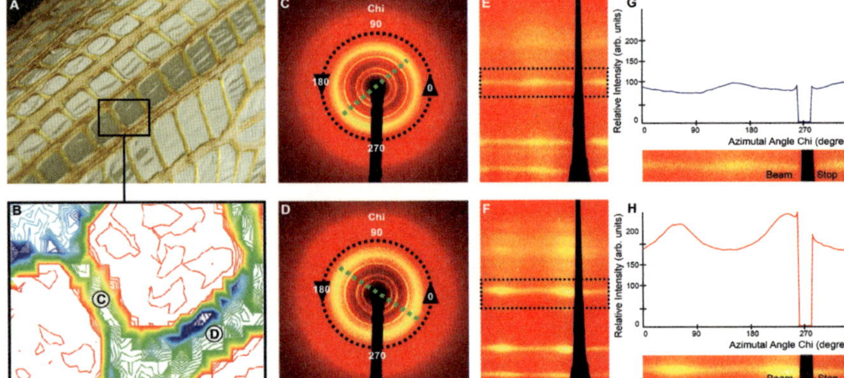

Microscale analysis of polymer alignment in printed chitosan structures. The molecular forces induced macroscale changes in shape and anisotropy.

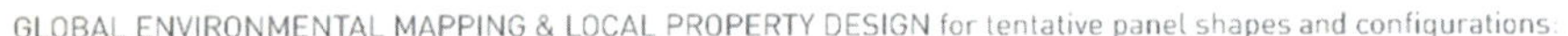

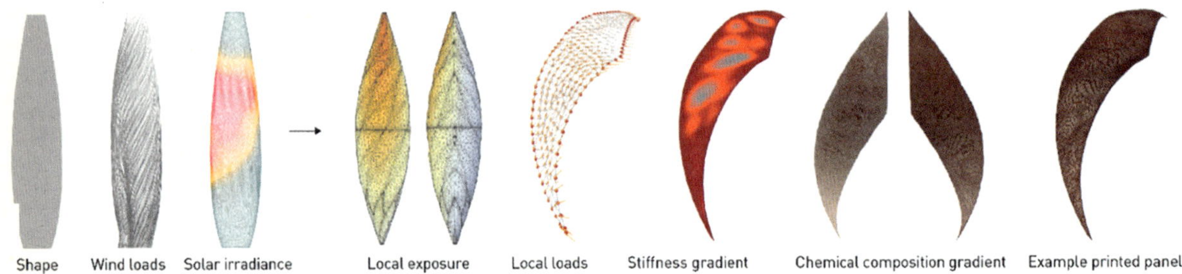

Computational data from structural and daylight simulations guided the geometry and material composition of an early version of the pavilion.

INITIAL SHAPE, PANEL ASSEMBLY & PATTERN EXPLORATION:

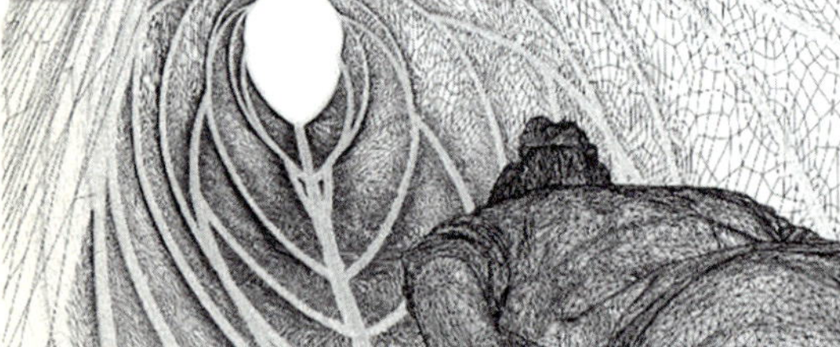

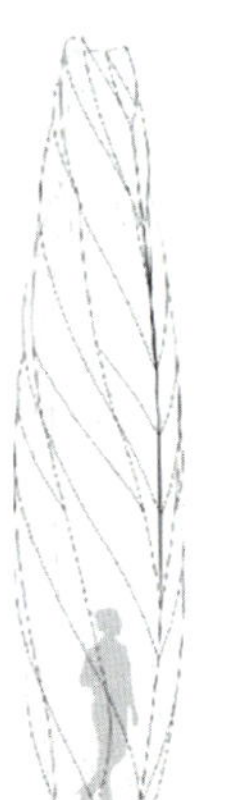

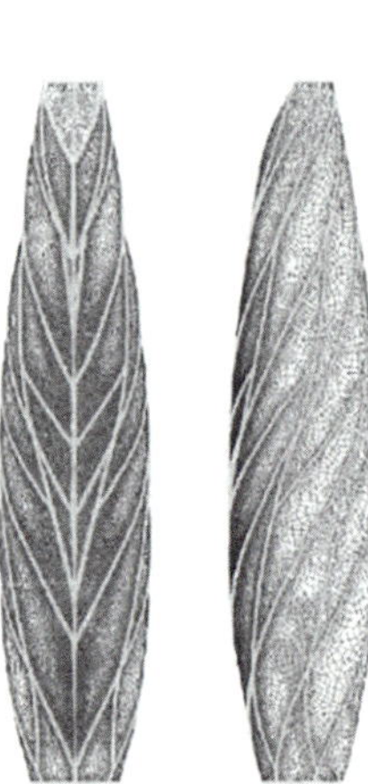

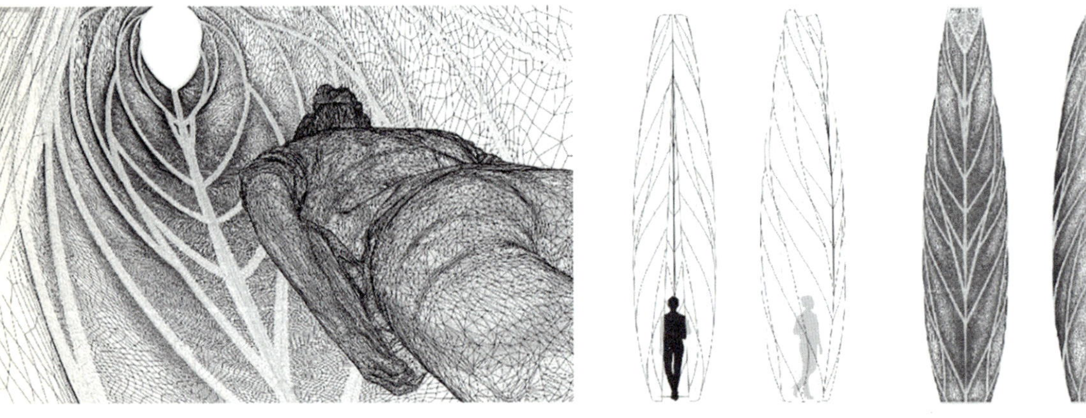

Workflow diagram illustrating feedback and templating during design and fabrication.

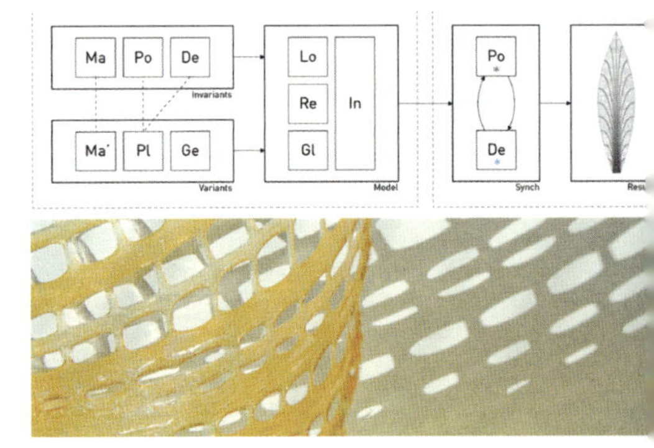

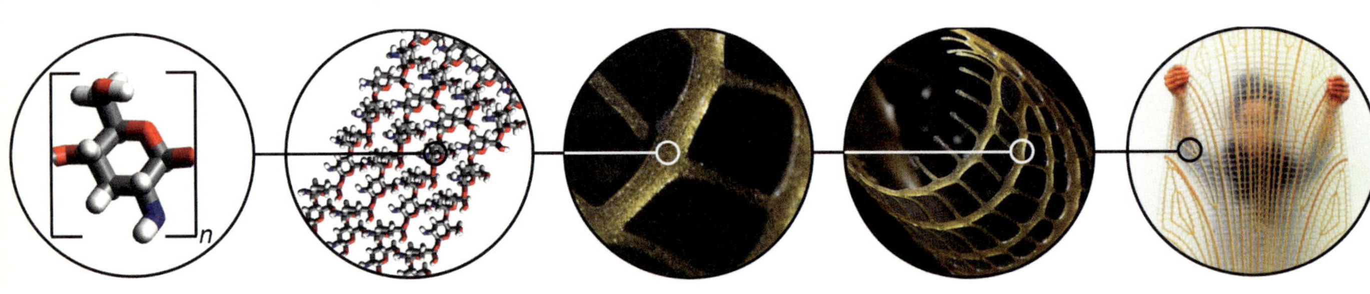

Microscale/local, small/regional, and large/global hierarchical geometries in a chitosan-based object.

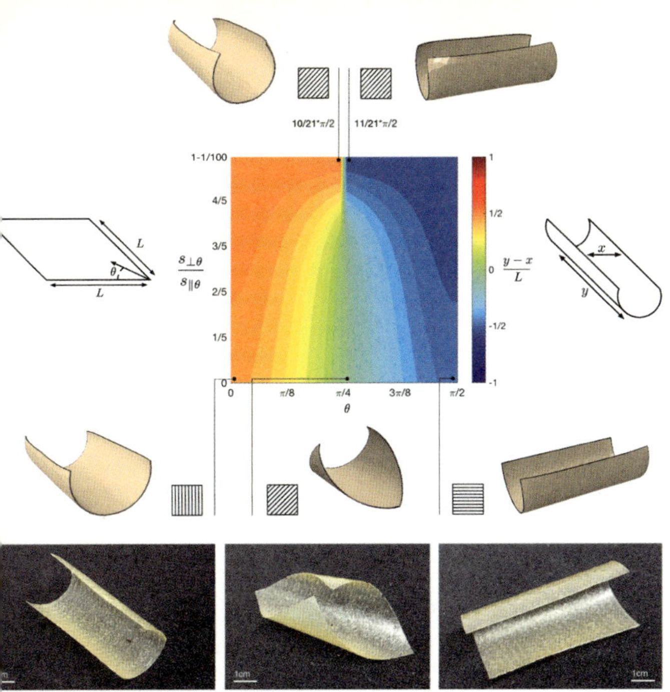

icroscale alignment of extruded
opolymers. Shape-changing behaviors
ould be tuned in a form of 2.5D printing—a
ew process that produced highly detailed
imensional" sheets akin to bas-reliefs.

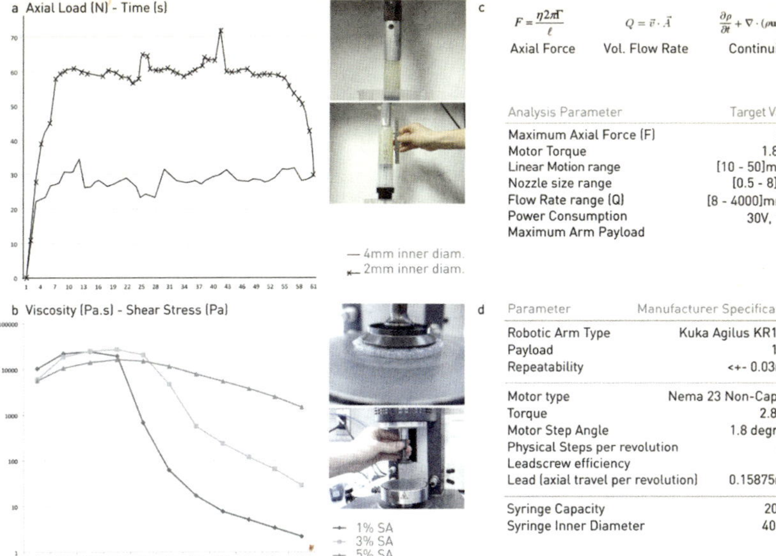

$$Q = \vec{v} \cdot \vec{A} \qquad \frac{\partial \rho}{\partial t} + \nabla \cdot (\rho u) = 0$$

Analysis Parameter	Target Value
Maximum Axial Force (F)	75N
Motor Torque	1.8Nm
Linear Motion range	(10 - 50)mm/s
Nozzle size range	(0.5 - 8)mm
Flow Rate range (Q)	(8 - 4000)mm³/s
Power Consumption	30V, 1.6A
Maximum Arm Payload	4kg

Parameter	Manufacturer Specification
Robotic Arm Type	Kuka Agilus KR1100
Payload	10kg
Repeatability	<+- 0.03mm
Motor type	Nema 23 Non-Captive
Torque	2.8Nm
Motor Step Angle	1.8 degrees
Physical Steps per revolution	200
Leadscrew efficiency	0.5
Lead (axial travel per revolution)	0.15875mm
Syringe Capacity	200cc
Syringe Inner Diameter	40mm

Tensile and compression analyses of 3D-printed biopolymer structures.

STRUCTURAL FORCES - GLOBAL SUPPORT | EXTRUDED MEMBER'S THICKNESS VARIATION | GLOBAL FIM LOGIC

max rigidity / max flexibility

Values of nozzle speed are assigned along deposition paths, and are related to material distribution needs.

Speed-based thickness variation provides global gradient of mechanical performance.

EXTRINSIC ENVIRONMENTAL FORCES - REGIONAL CHEMISTRY SOLUTION | MATERIAL BLEND COMPOSITION VARIATION | REGIONAL FIM LOGIC

max rad & light / min

Chitosan-Cellulose-Starch-CalciumCO material blends are designed related to heat and light control needs.

Chemistry- & Time-based molecular reactions provide tonal changes for regional reflectivity needs.

INTRINSIC HYDRATION FORCES - LOCAL GEOMETRY SOLUTION | LINE PATTERN DENISTY VARIATION | LOCAL FIM LOGIC

open cell / closed cell

Density-based patterning design responds to distribution of closed and open cells for controlled degradation.

Hydration-based swelling affects cell size differently, and provides local changes in structural decay.

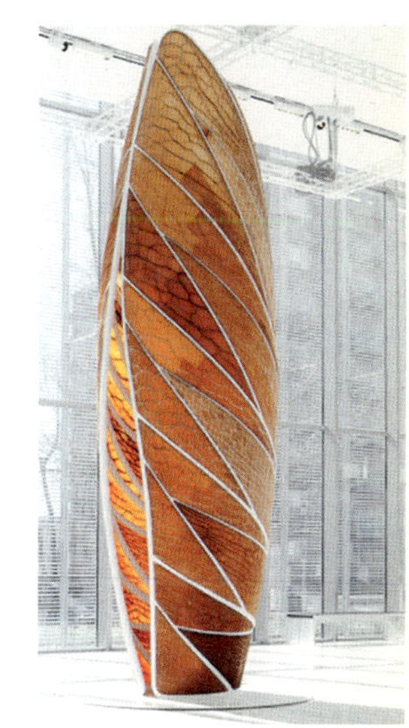

The pavilion was fabricated from pectin, chitosan, and cellulose, with the material distributed according to environmental parameters. The structure dynamically responded to the environment and changed its characteristics when exposed to heat and humidity over the course of days, months, and years.

Ashby chart plotting the density and strength of biopolymers in a printed object.

a) Material properties chart (M.F. Ashby, L.J. Gibson et al. 1995)

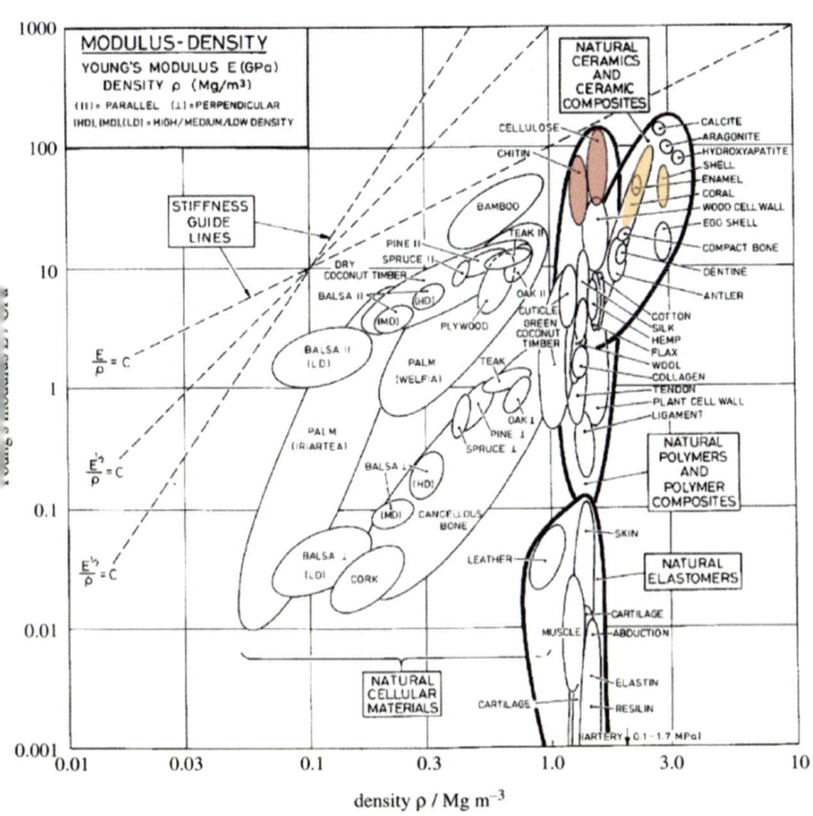

b) Natural origin of our materials of interest:

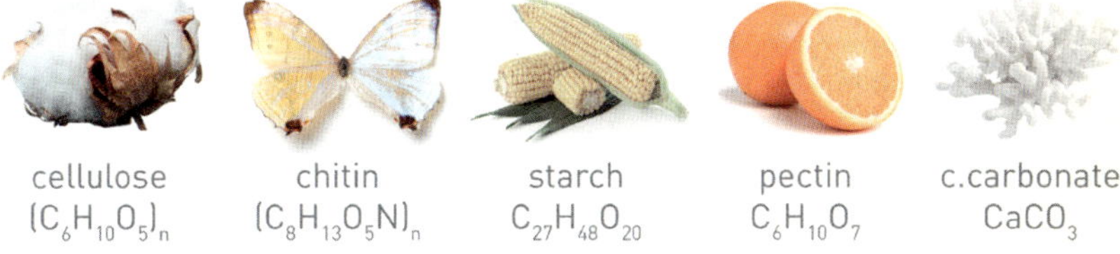

cellulose $(C_6H_{10}O_5)_n$

chitin $(C_8H_{13}O_5N)_n$

starch $C_{27}H_{48}O_{20}$

pectin $C_6H_{10}O_7$

c.carbonate $CaCO_3$

c) LARGE-SCALE STRUCTURES design requirements

Design parameters:

- fully-tuned optical and mechanical behavior
- future support of biological augmentation
- overall lightweight
- full biodegradability

Digital fabrication constraints:

- constant physical-to-virtual feedback
- nano-to-macro material behavior control
- integration of trans-disciplinary data sets

AGUAHOJA II

Neri Oxman and
The Mediated Matter Group
Aguahoja II. 2019
Wood-pulp cellulose, apple pectin, chitosan (85% deacetylated), vegetable glycerin, beet, turmeric, squid-ink melanin; pavilion: cotton canvas blend, steel brackets, beams, and aluminum pins
51 3/16 in. × 51 3/16 in. × 17 ft. 4 11/16 in.
(130 × 130 × 530 cm)
Pavilion produced by Stratasys Ltd.
An MIT Media Lab project
Research team: Nic Lee, Joshua Van Zak, Joseph H. Kennedy Jr., Ramon Weber, Christoph Bader, João Costa, Sunanda Sharma, James C. Weaver, Jorge Duro-Royo, Neri Oxman
Undergraduate researchers: Joseph Faraguna, Danielle Grey-Stewart, Amelia Wong, Ava Iranmanesh, Pam Silver
Collaborators and contributors: Paula Aguilera; Jeremy Flower; Jonathan Williams; Stratasys Direct Manufacturing; Wyss Institute and Department of Systems Biology, Harvard University; San Francisco Museum of Modern Art; GETTYLAB; Skolkovo Institute of Science and Technology; Quest AI; NOE LLC; MIT Research Laboratory of Electronics; Robert Wood Johnson Foundation

Cosponsored by the Esquel Group, for Esquel's Integral Exhibition Hall, Guilin, China, following its debut in *Nature: Cooper Hewitt Design Triennial*, Cooper Hewitt, Smithsonian Design Museum, New York, 2019

Above left:
Optical microscope image of a pectin membrane.

Above right:
Optical microscope image of a high-concentration pectin membrane.

Left:
Optical microscope image of moss nodules in a pectin membrane.

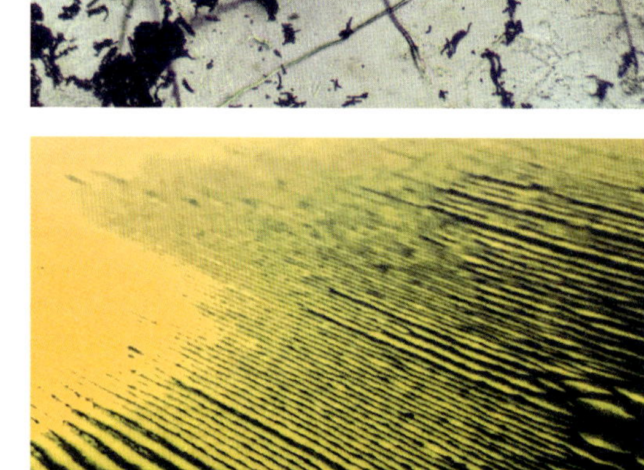

Composite materials dried over a longer period, giving them time to fuse and blend, thus enabling the creation of fine gradients.

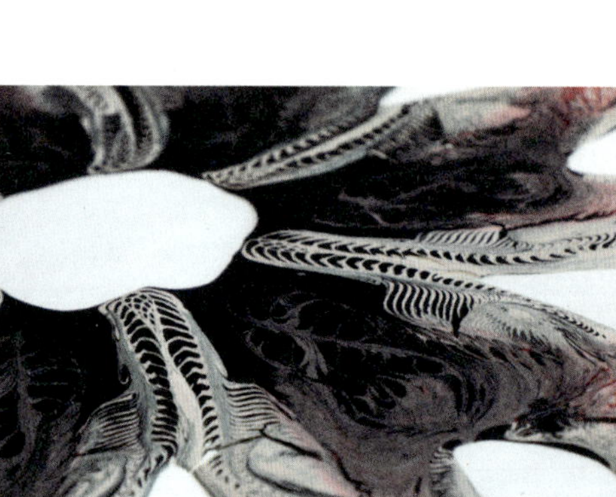

Prototype biocomposite containing blue spirulina, beet, and calcium carbonate in a pectin membrane.

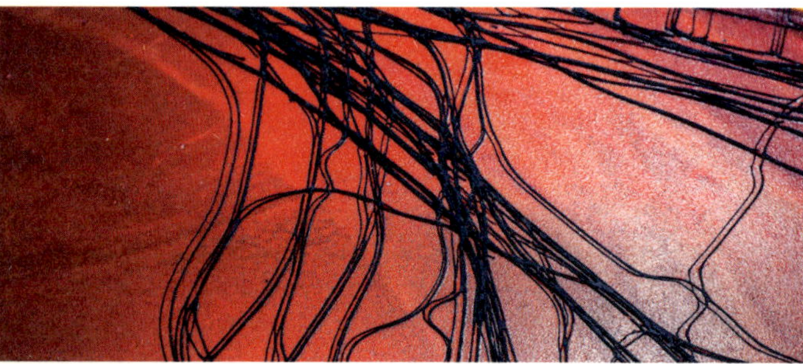

Rigid reinforcement patterns were printed directly into a flexible skin as it dried. The patterns followed a generative path, responding to environmental light and intrinsic structural data.

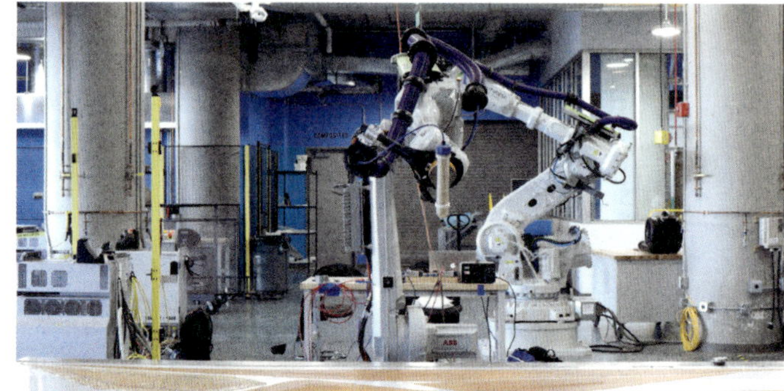

Large-scale robotic fabrication system used for fabrication of the pavilion.

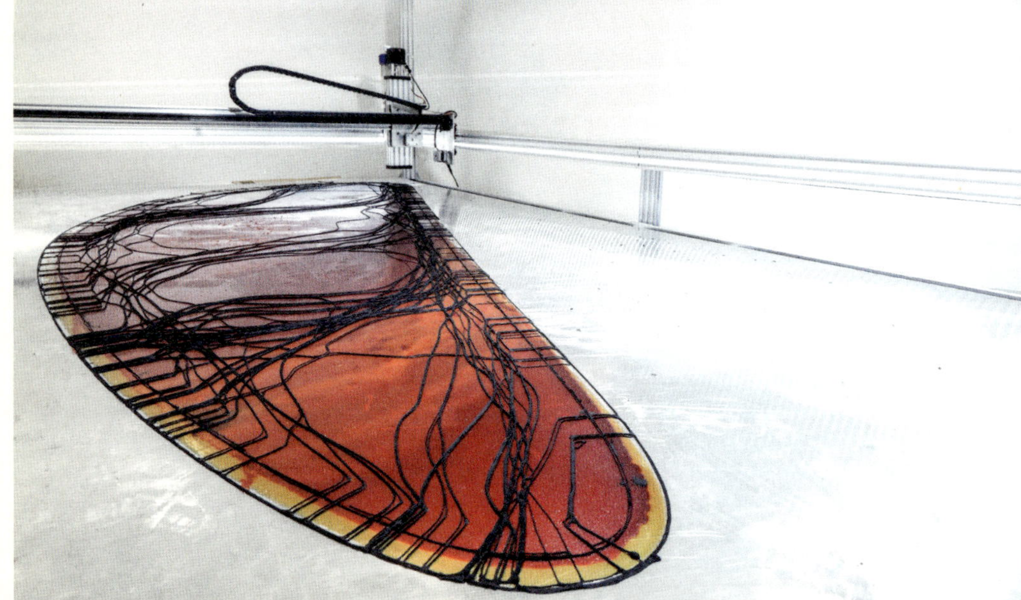

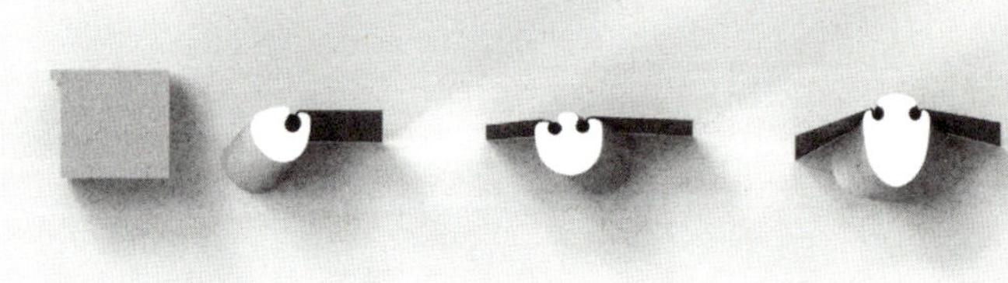

Cross section of scaffold sections into which the skin was friction fit.

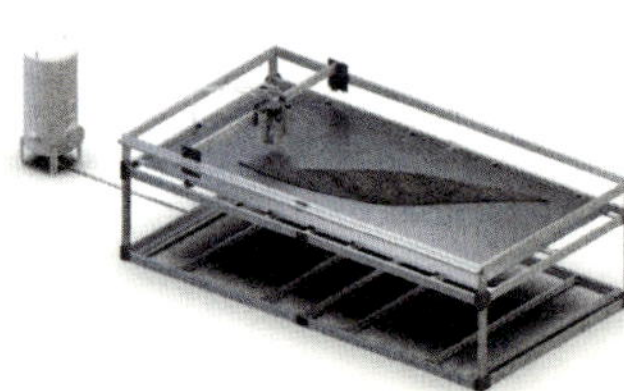

Rendering of the pneumatic extruder and positioning system.

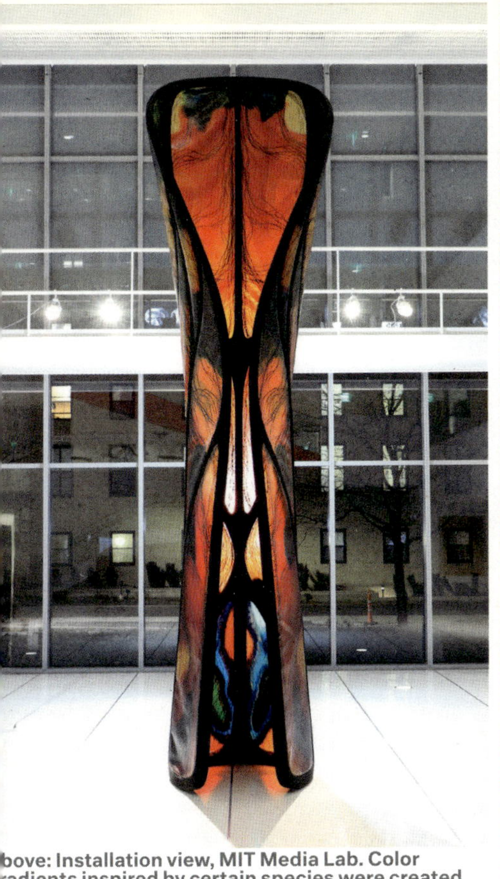

bove: Installation view, MIT Media Lab. Color radients inspired by certain species were created om the functional gradation of natural dyes. The ellow, red, and black of the lower region draw on the posematic patterns that certain fungi and venomus animals use to repel mammals; the blue and ellow gradients of the upper region attract key polliator species including ladybugs and honeybees.

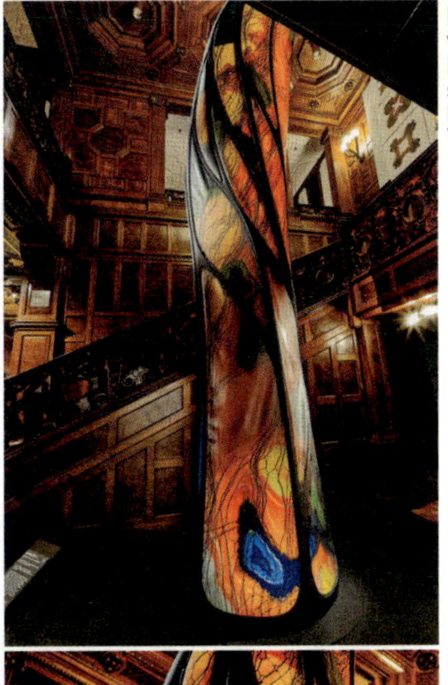

Installation views, Cooper Hewitt, Smithsonian Design Museum, New York, 2019.

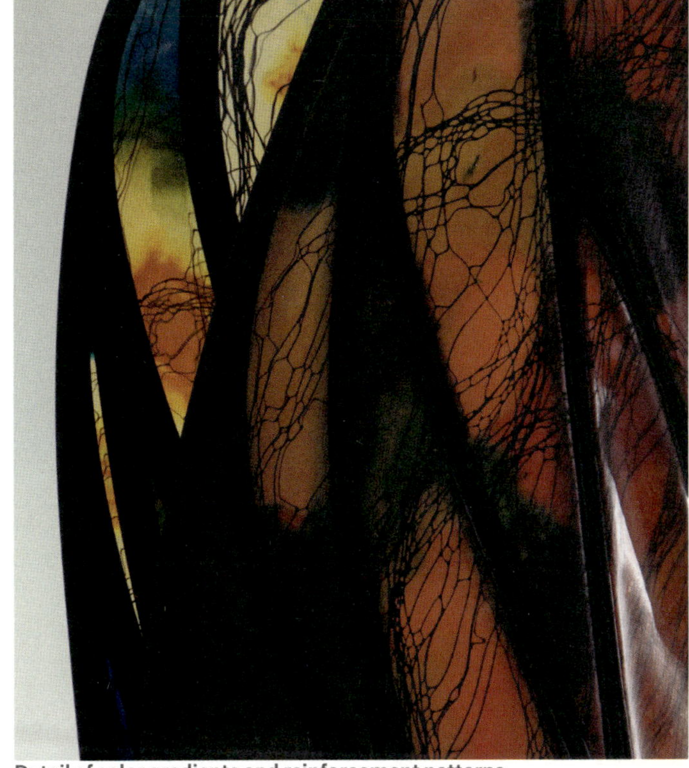

Detail of color gradients and reinforcement patterns.

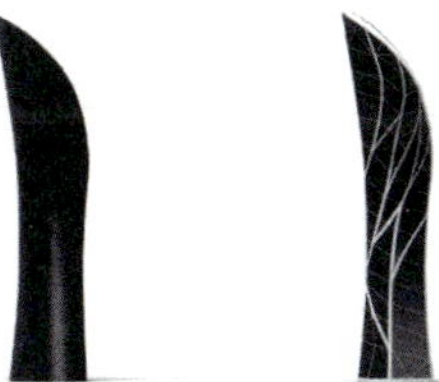

he base geometry was the tarting point from which the final orm, scaffold, and distribution of aterials were generated using mulations and environmental ata.

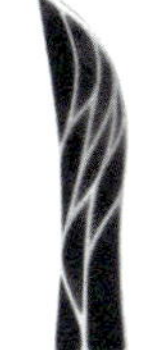

The base geometry was processed into a set of developable and printable surfaces. A developable surface can be flattened onto a plane without being distorted.

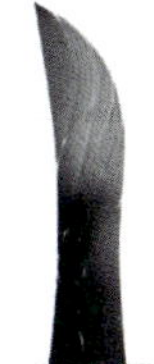

Low-resolution physics-simulation results were propagated to a high-resolution representation of the biopolymer skin, ultimately informing the thickness of rigid scaffolding elements.

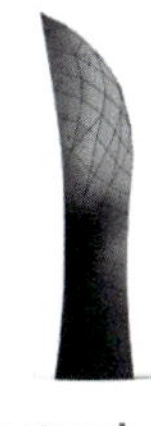

Structural stresses were simulated in a low-resolution representation of the skin.

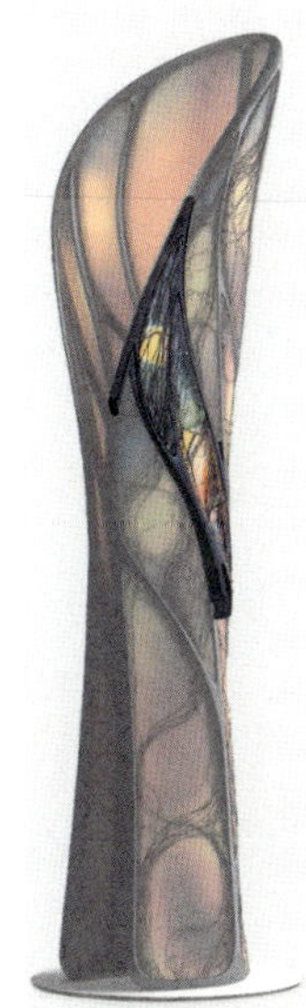

Prototype rendering showing the continuous gradation of biopolymers, creating fine gradients of material and optical properties.

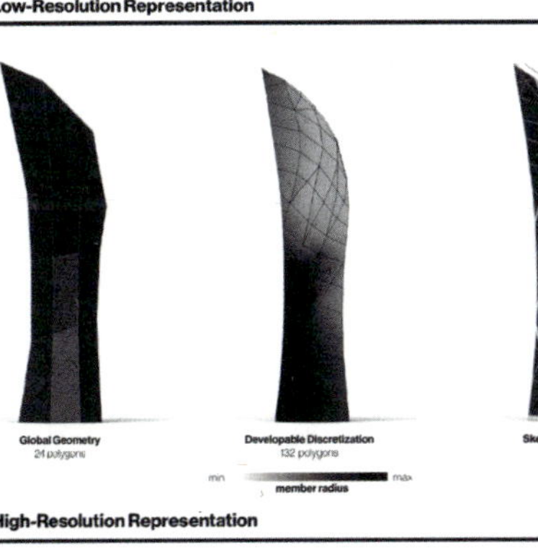
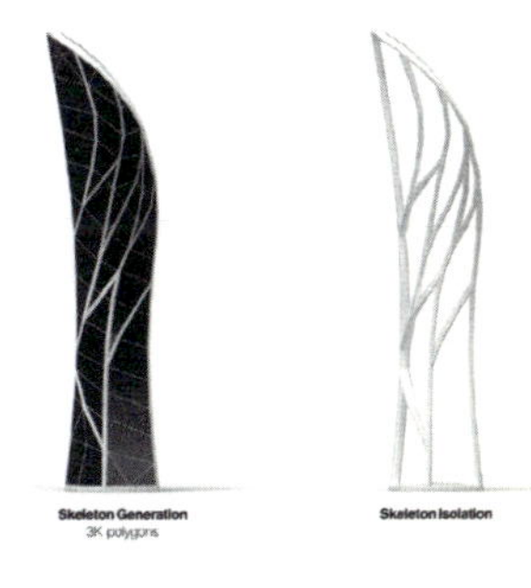

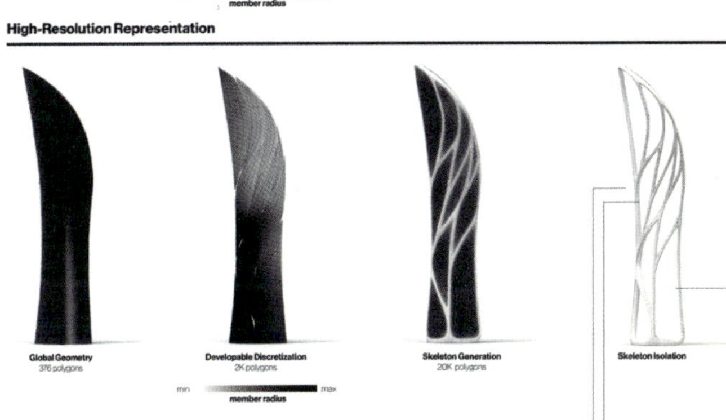

Low- and high-resolution representations of the structure's scaffold elements.

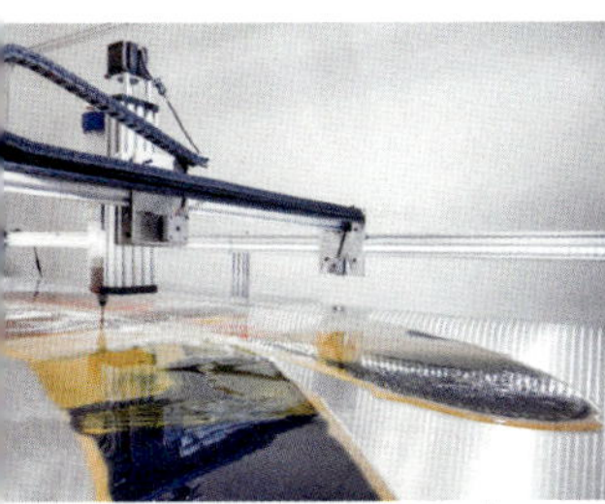

arge-scale numeric control ystem used for project abrication.

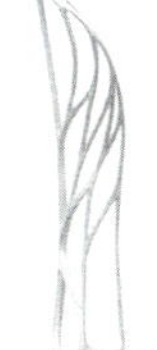

From simulated structural data, a rigid scaffold was formed to link the developable sections of biopolymer skin. These rigid elements were assigned thickness based on the amount of simulated stress at each point.

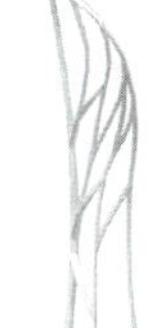

The low-resolution scaffold was processed into a high-resolution representation with channels for the friction fitting of printed panels.

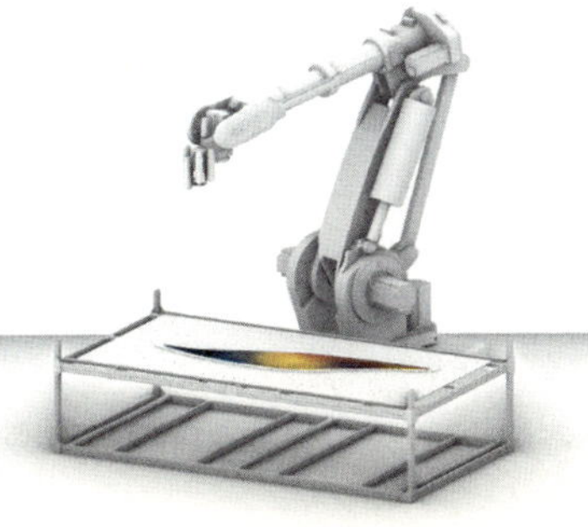

The extrusion system could be affixed to a CNC gantry or an industrial robot in order to access a larger fabrication space and make a larger object.

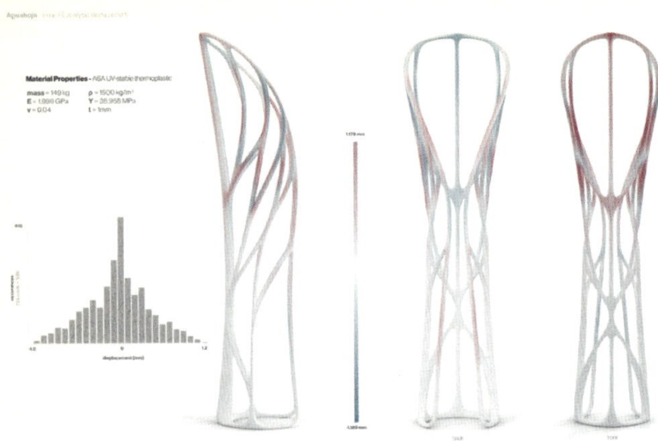

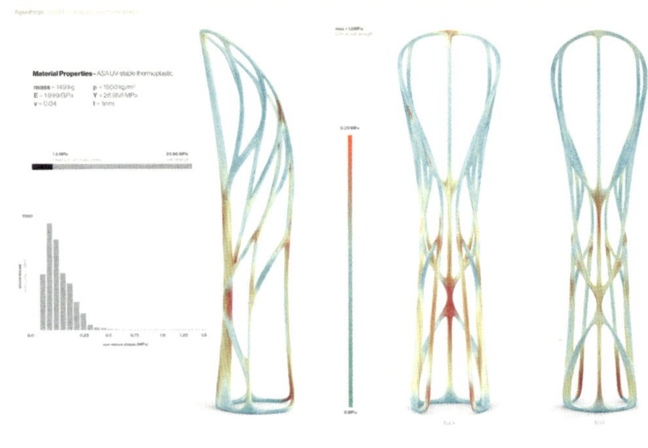

Intrinsic structural data, such as displacement and stress on the skeletal frame, governed the distribution of venation patterns.

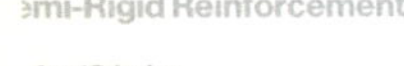

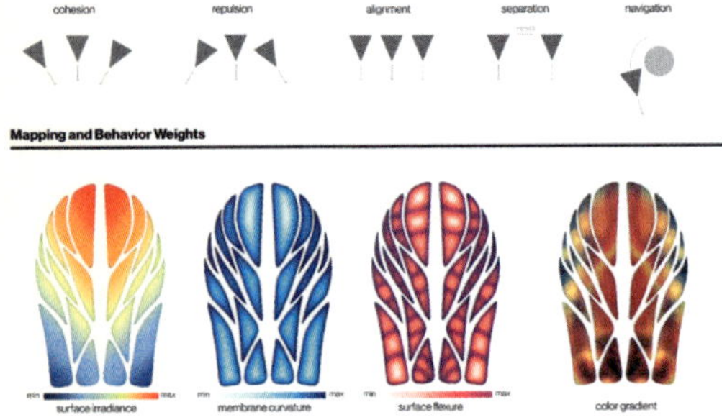

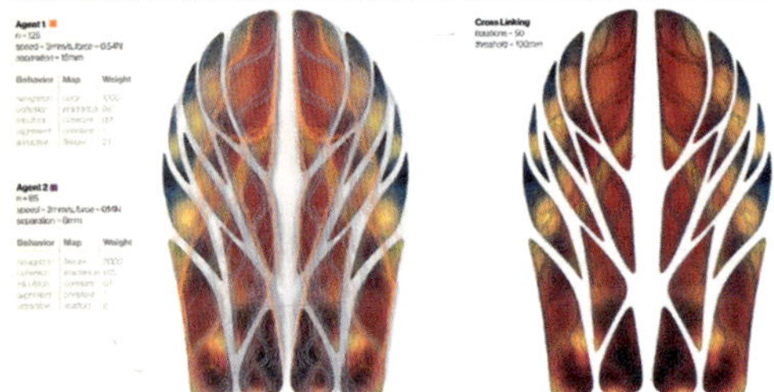

An agent-based swarm algorithm interpreted environmental and structural data maps to create a reinforcement pattern of rigid material that prioritized high-stress areas while avoiding the occlusion of high-irradiance sections.

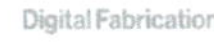

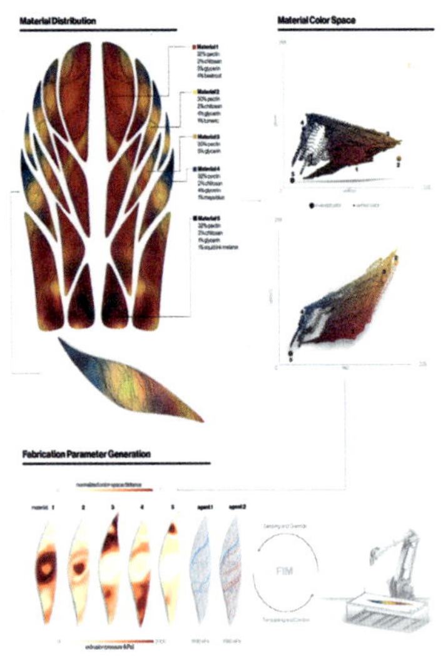

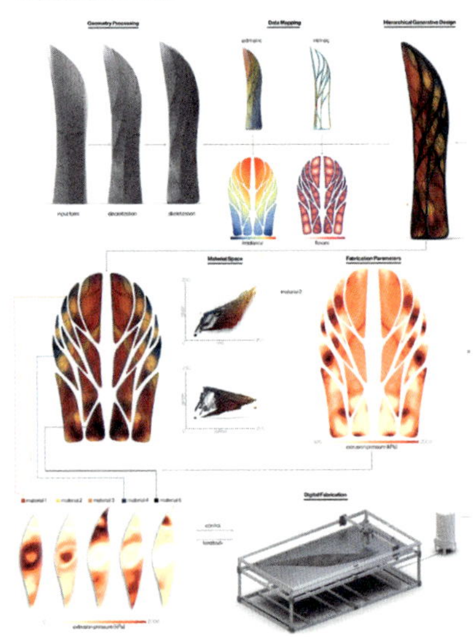

Generative workflow of base geometry processed into a constructable form and assigned material distributions based on a simulation. Material distributions were translated directly into fabrication information, creating a tight integration of design and fabrication. A feedback and override system enabled real-time iteration.

Stress simulation in the scaffold.

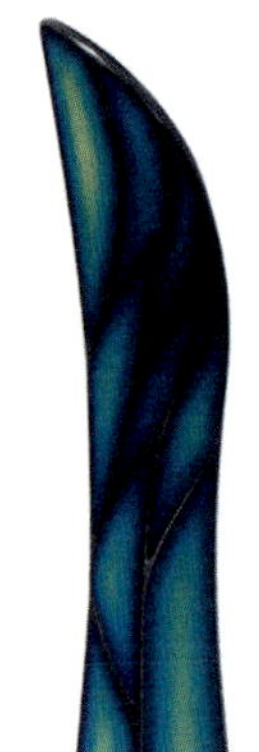

Rendering of the product of a generative workflow, from low-resolution base geometry to biopolymer structure.

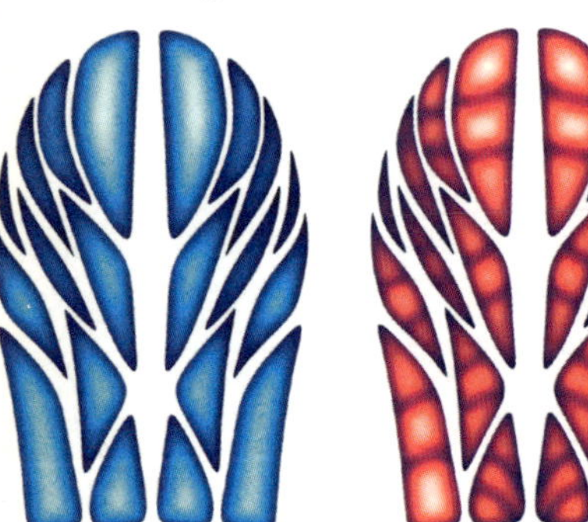

Unwrapped displacement map.

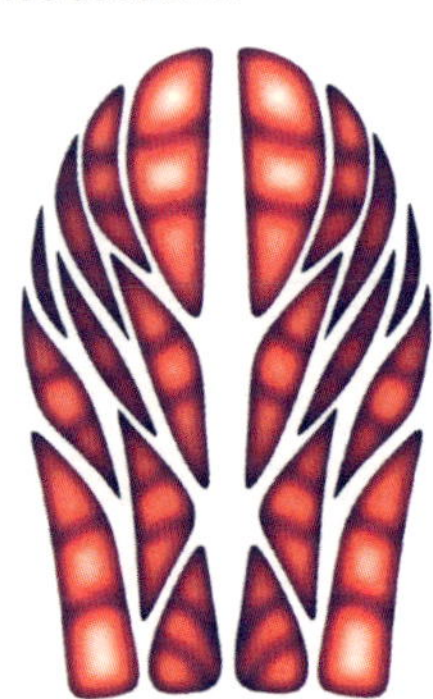

Flexure map generated by stress simulation and displacement mapping. The results suggest priority areas for reinforcement with rigid materials.

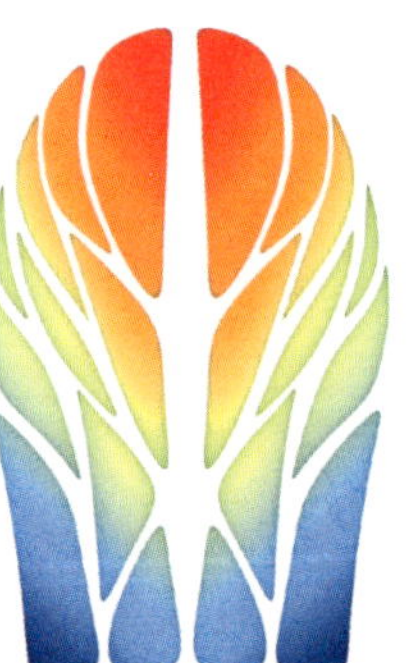

Irradiance map determining the distribution of transparent regions and color transitions based on their exposure to the sun. Irradiance measures the amount of sunlight and heat received by a structure.

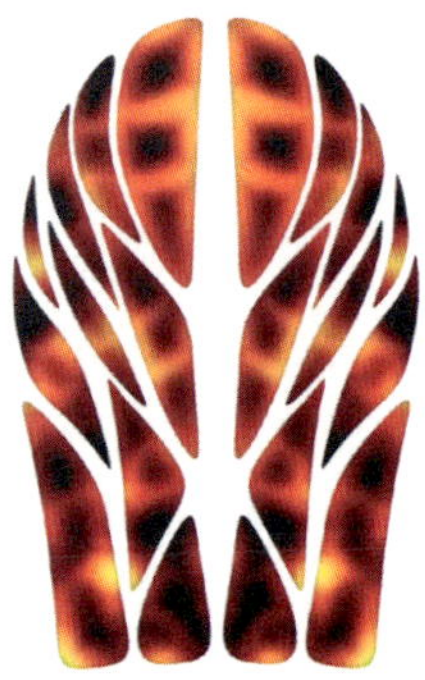

Aposematic pattern (a pattern used to repel predators) generated from irradiance, flexure, and displacement maps.

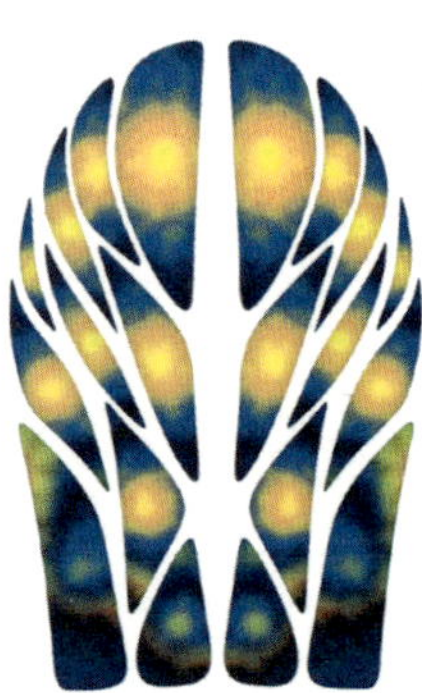

Pollinator-attraction pattern generated from irradiance, flexure, and displacement maps. The colors and patterns attract pollinators such as honeybees and butterflies but cannot be seen by many other insects.

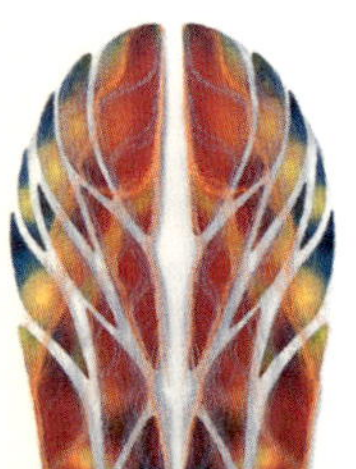

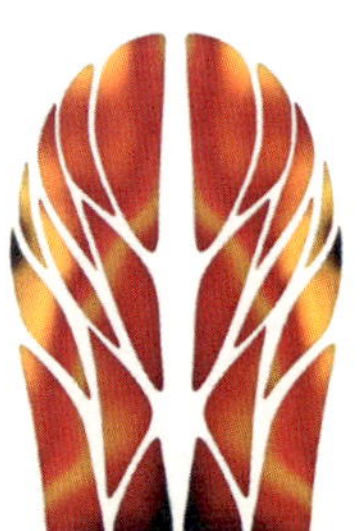

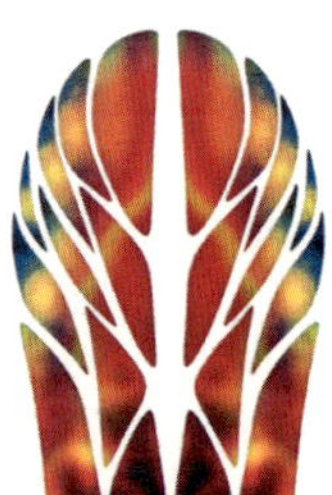

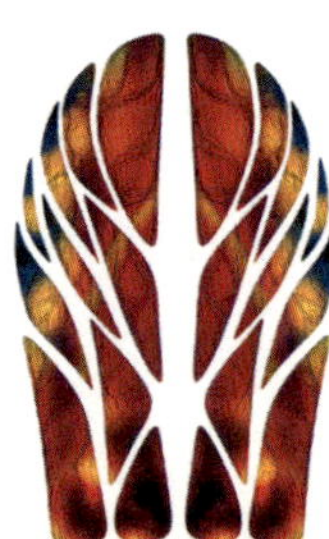

Pattern generated from irradiance, flexure, and displacement maps. The colors mimic environmental patterns that attract aphids and ladybugs but are not visible to predator organisms. Agent-based reinforcement patterns distributed rigid materials where structural stress and simulated flexure were highest.

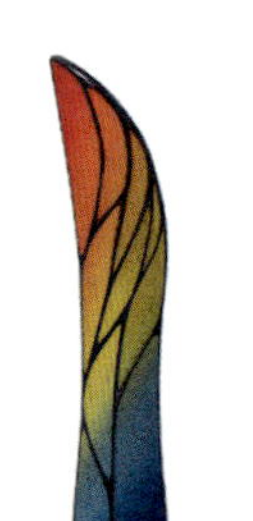

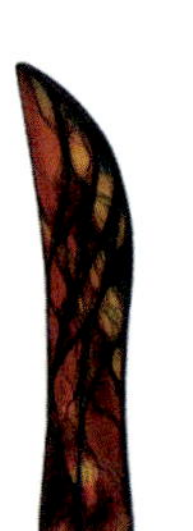

Simulations determined material makeup and structure: a displacement simulation informed the rigidity of the biopolymer skin; a site-specific irradiance simulation determined the distribution of materials.

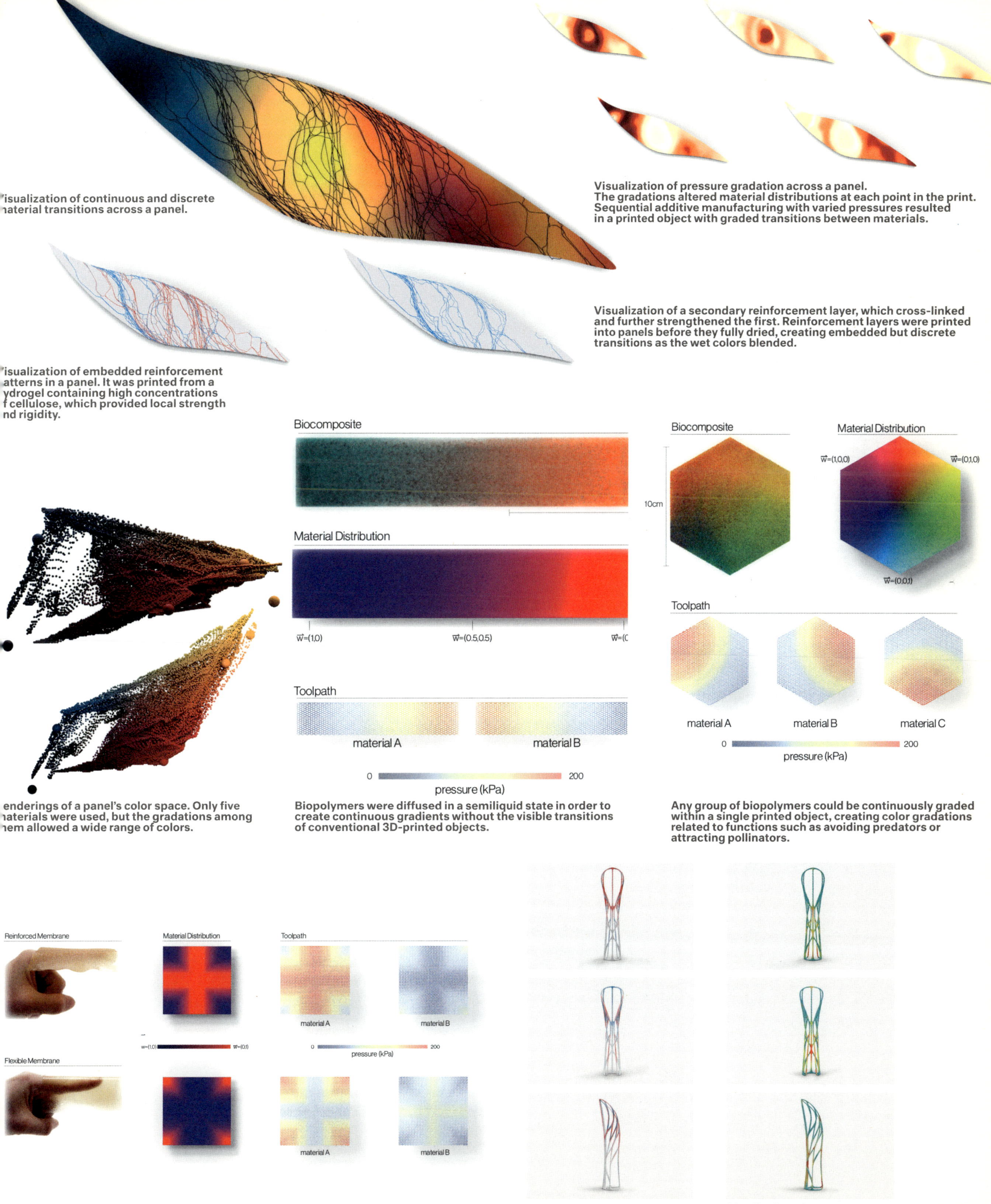

'isualization of continuous and discrete 'aterial transitions across a panel.

Visualization of pressure gradation across a panel. The gradations altered material distributions at each point in the print. Sequential additive manufacturing with varied pressures resulted in a printed object with graded transitions between materials.

Visualization of a secondary reinforcement layer, which cross-linked and further strengthened the first. Reinforcement layers were printed into panels before they fully dried, creating embedded but discrete transitions as the wet colors blended.

'isualization of embedded reinforcement atterns in a panel. It was printed from a ydrogel containing high concentrations f cellulose, which provided local strength nd rigidity.

enderings of a panel's color space. Only five 'aterials were used, but the gradations among 'em allowed a wide range of colors.

Biopolymers were diffused in a semiliquid state in order to create continuous gradients without the visible transitions of conventional 3D-printed objects.

Any group of biopolymers could be continuously graded within a single printed object, creating color gradations related to functions such as avoiding predators or attracting pollinators.

The gradation of materials in objects of identical geometry resulted in iterations united by a central pattern but with different degrees of flexibility or rigidity.

Distribution of displacement across the scaffold.

Distribution of stresses across the scaffold.

A

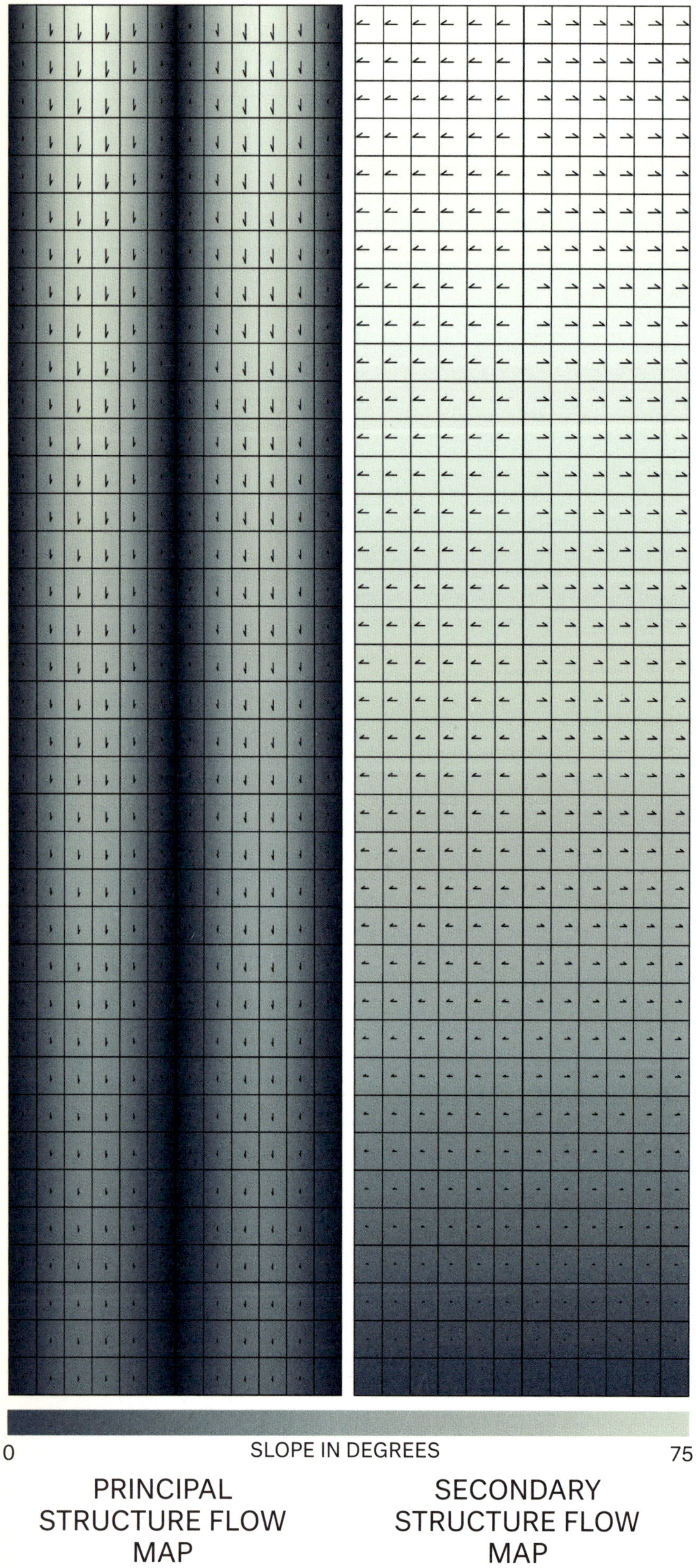

0 SLOPE IN DEGREES 75

PRINCIPAL STRUCTURE FLOW MAP

SECONDARY STRUCTURE FLOW MAP

B

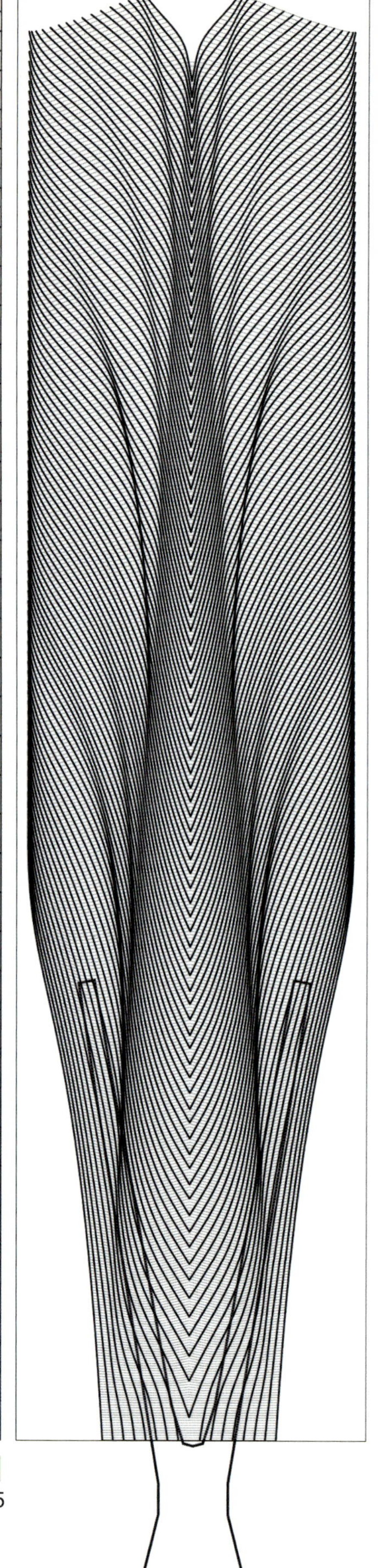

2D STREAMLINES

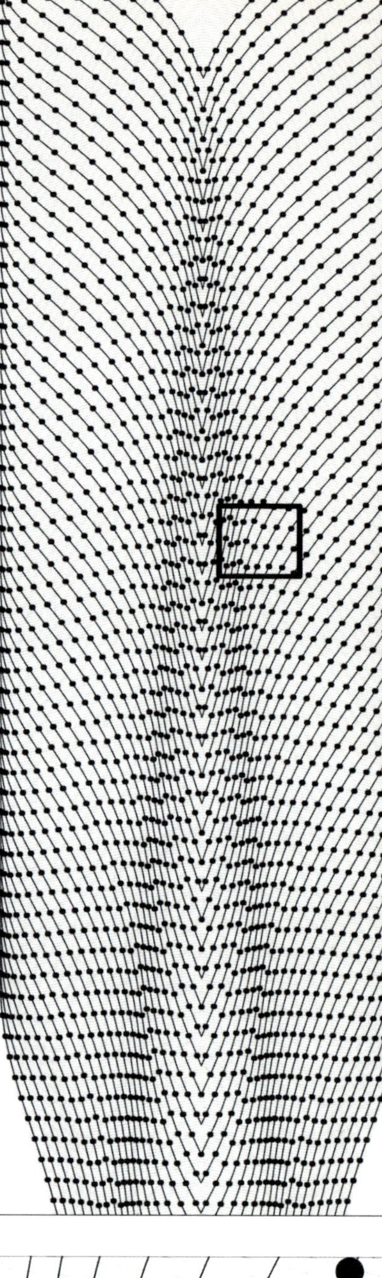

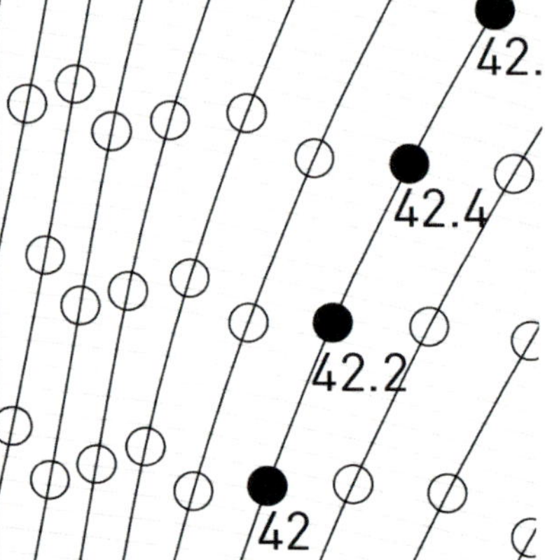

PRESSURE IN PSI

CONTINUOUS PRESSURE VARIATION MAP

Aguahoja I
The process of the design and robotic fabrication of large-scale chitosan composites. Intrinsic mechanical and extrinsic environmental properties drove the generated form. Vector fields encoded variation in principle and secondary structure and were interpolated into streamlines of material deposition trajectories (a); pressure, speed, and material concentration variations along structural streamlines were encoded into real-time instructions sent to the manufacturing platforms (b); hydration control maps on top of deposited structures provided evaporation-driven self-folding of the final three-dimensional structure (c).

C

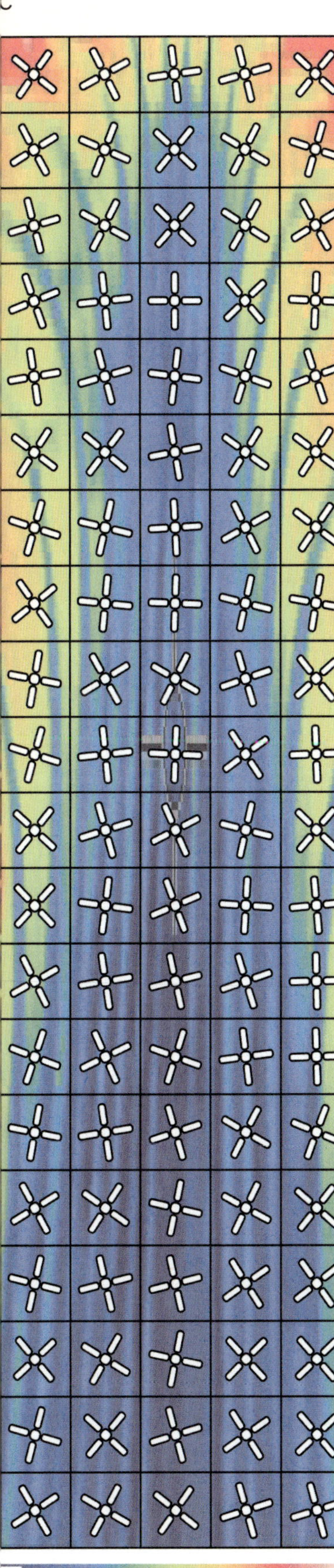

10C TEMPERATURE 24C

DIGITAL
HYDRATION CONTROL
MAP

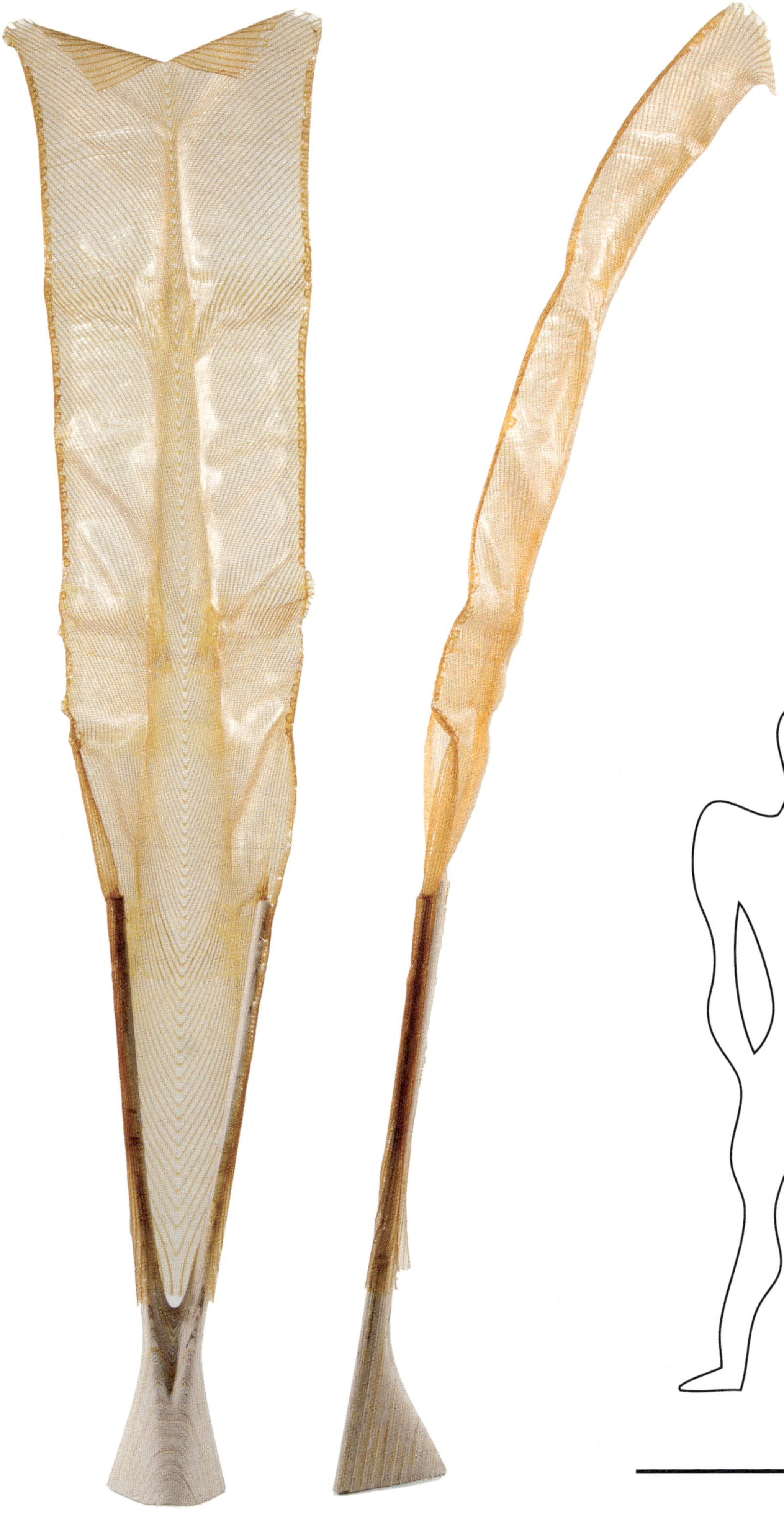

3D STRUCTURE

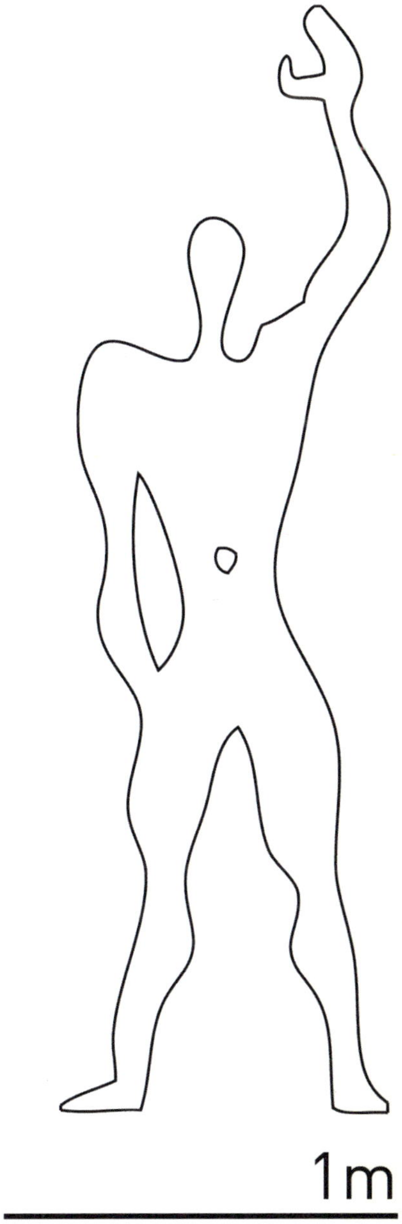

1m

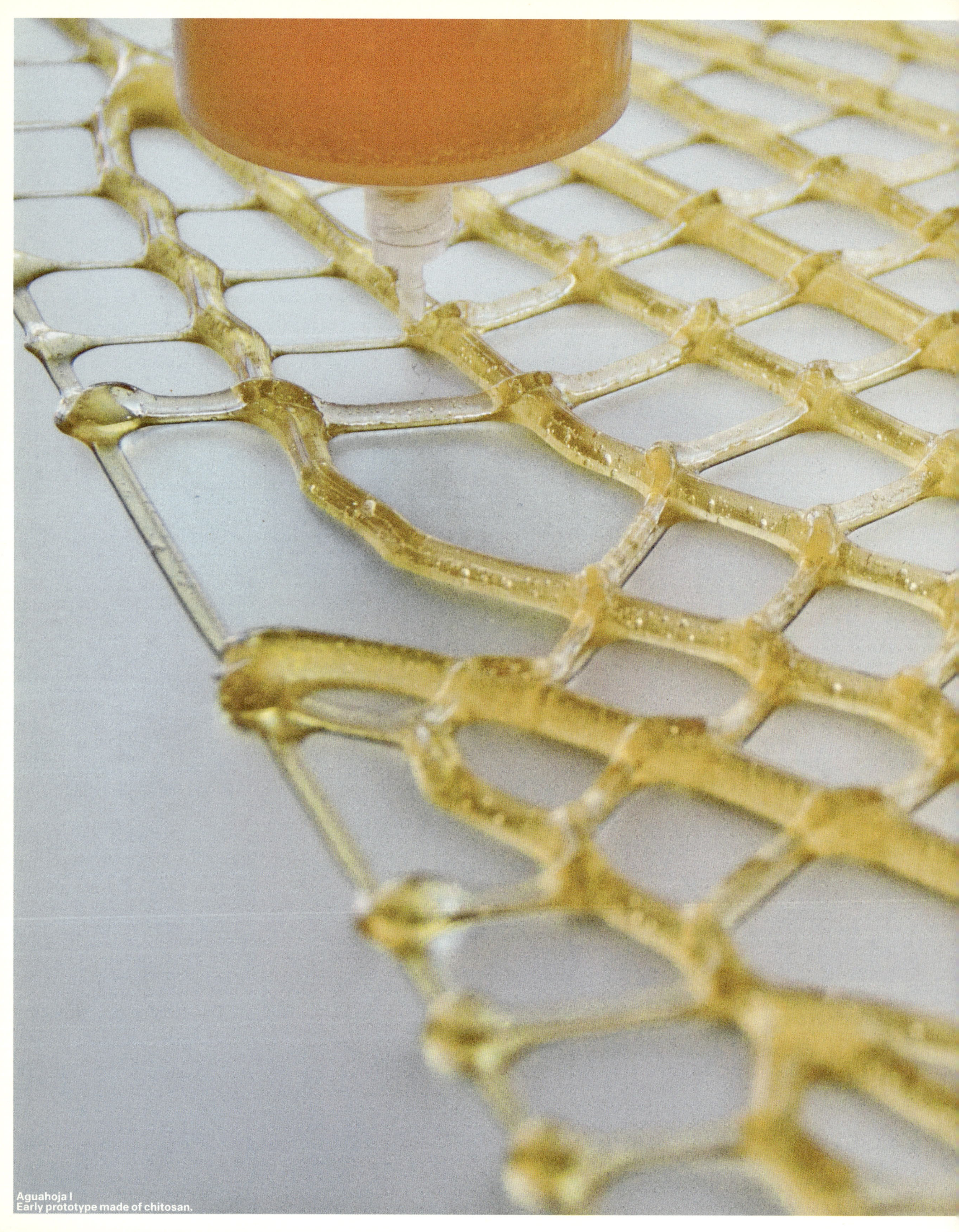
Aguahoja I
Early prototype made of chitosan.

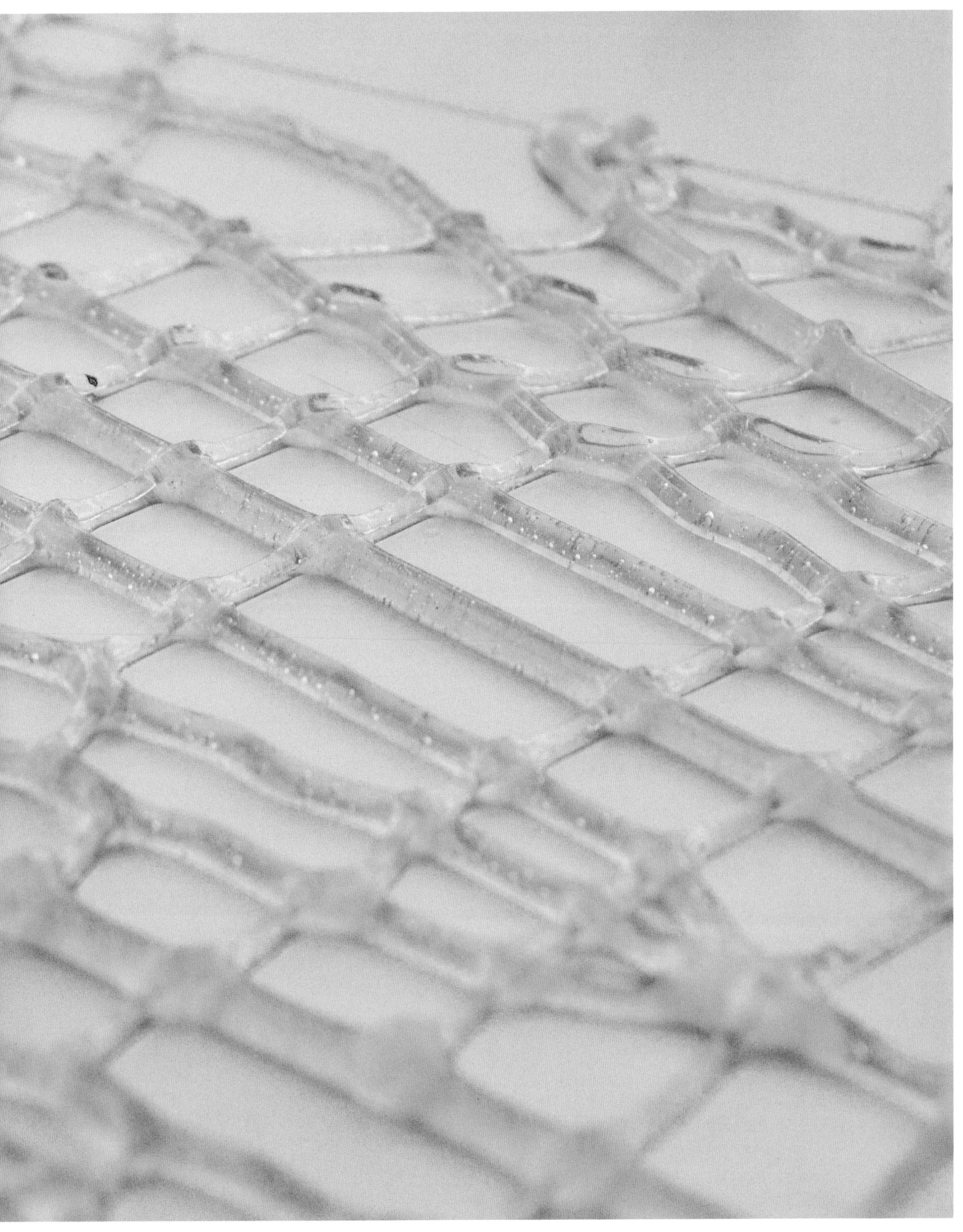

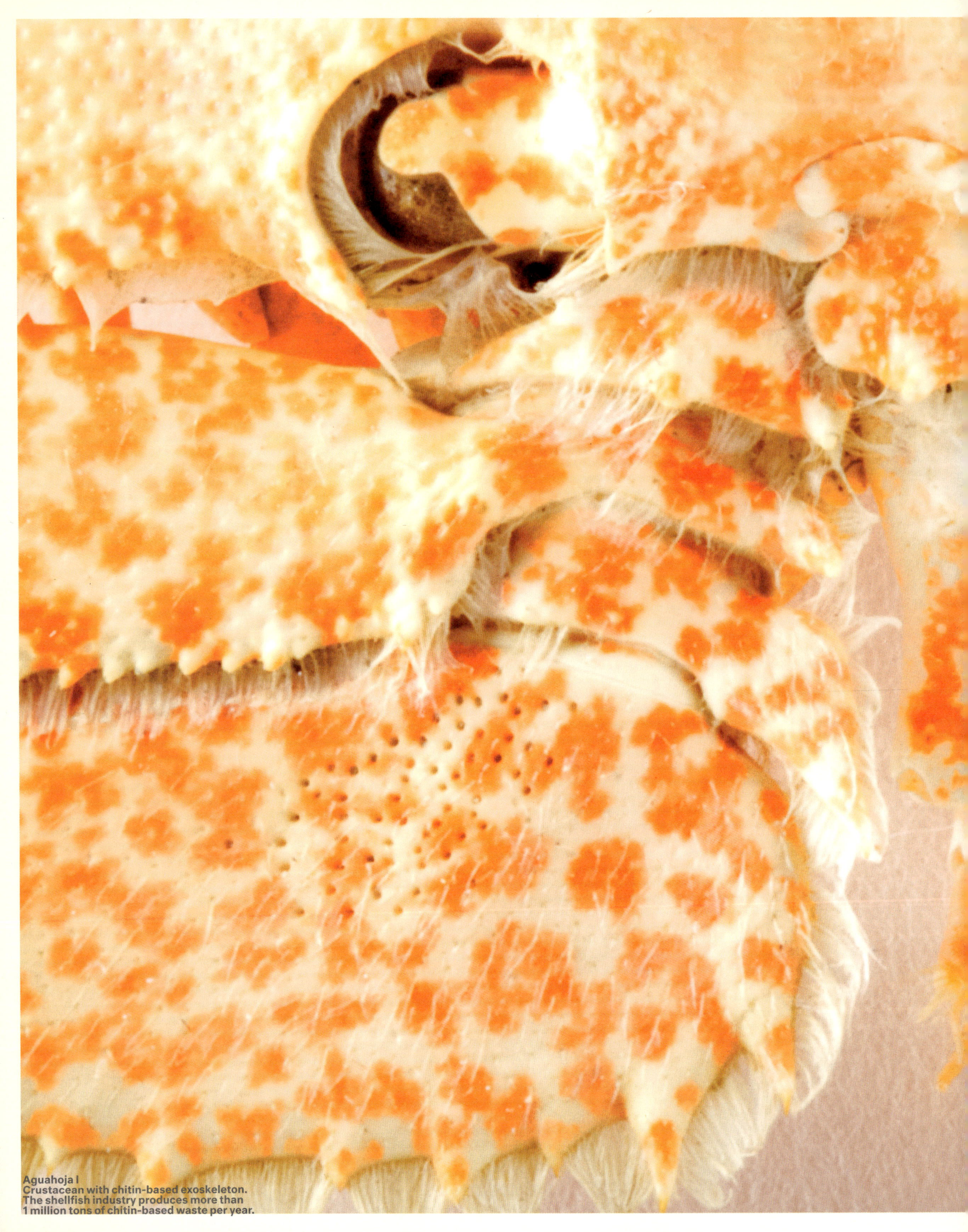

Aguahoja I
Crustacean with chitin-based exoskeleton.
The shellfish industry produces more than 1 million tons of chitin-based waste per year.

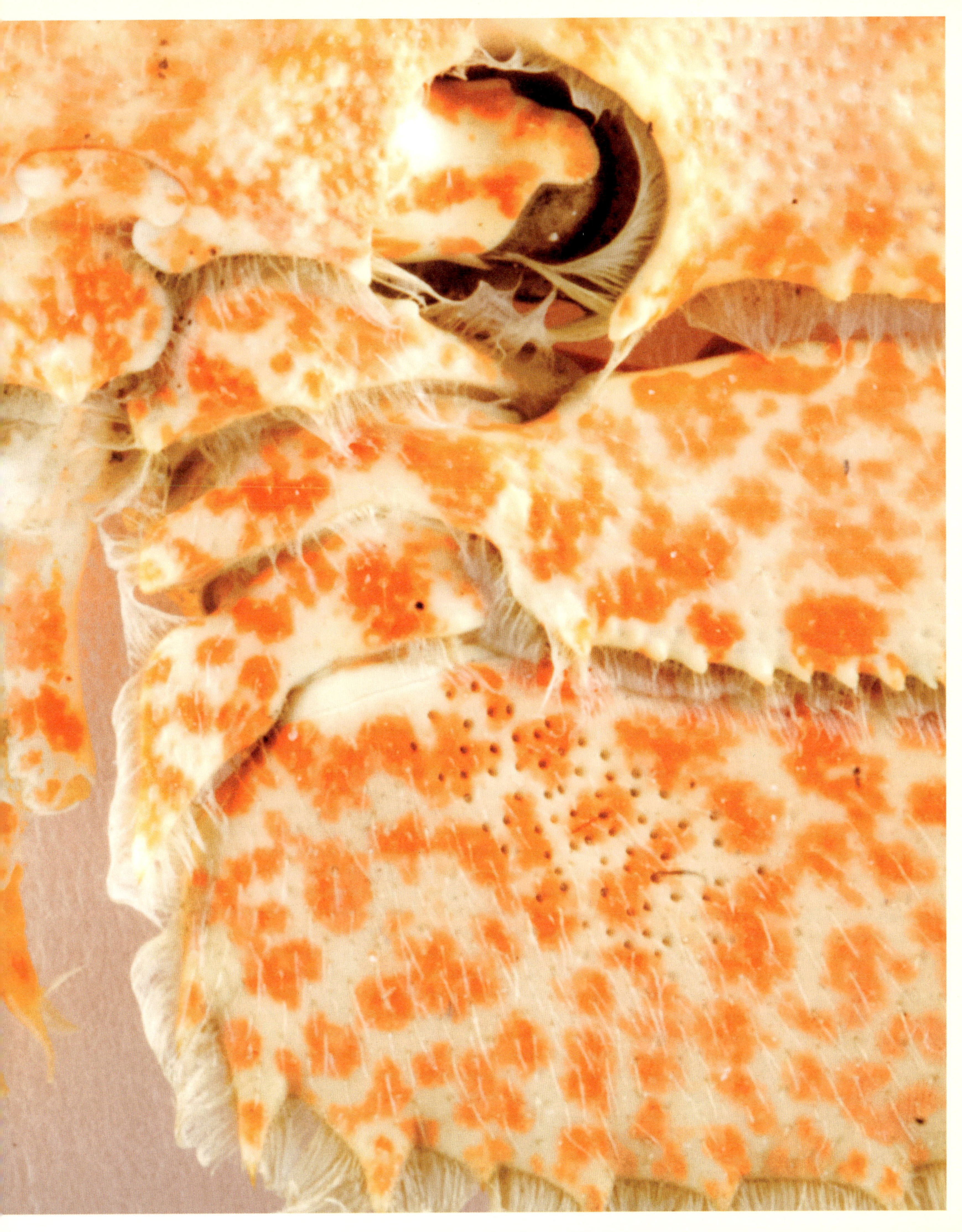

Aguahoja I
Rigid veins contained a higher concentration of chitosan, in contrast to the transparent cells.

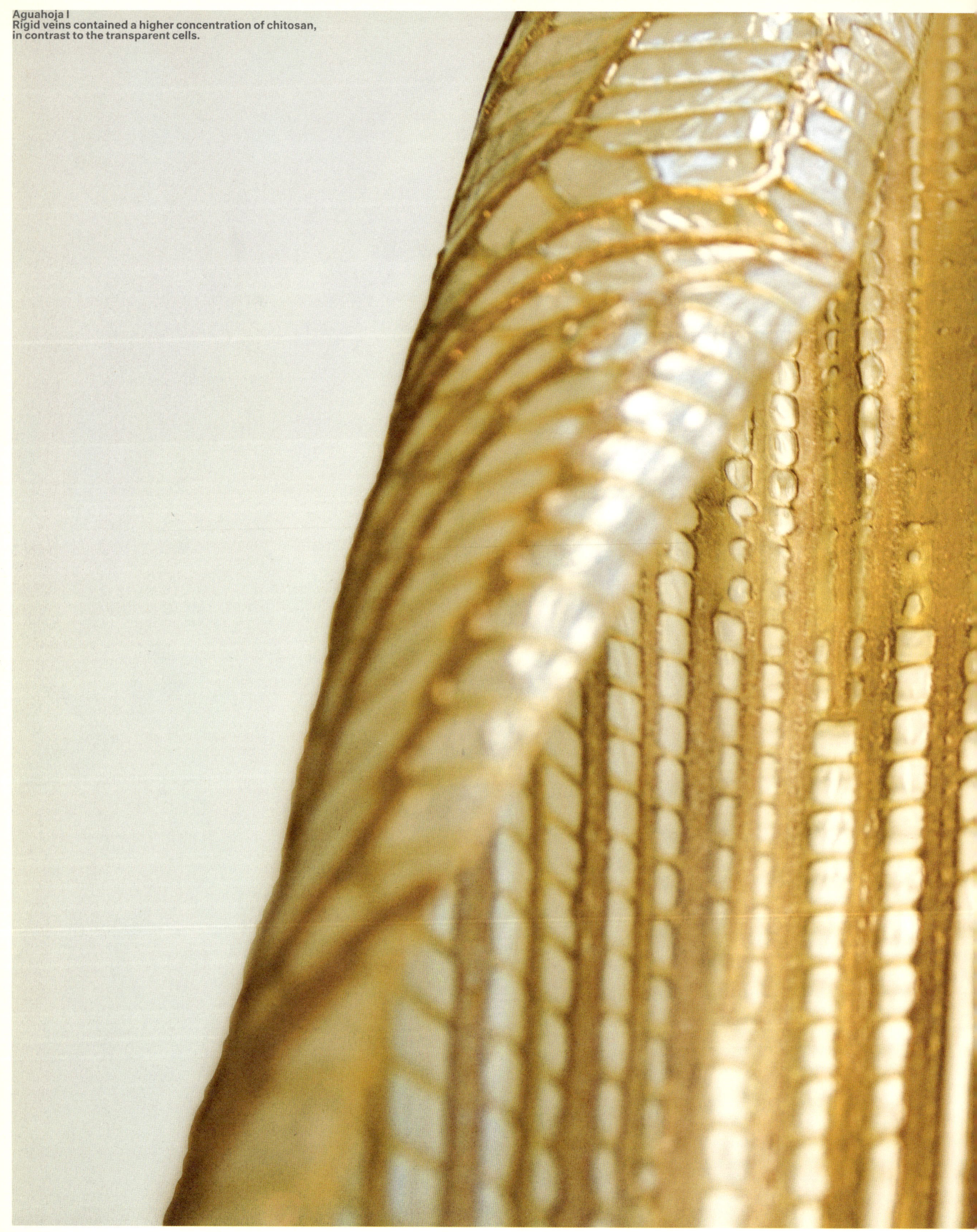

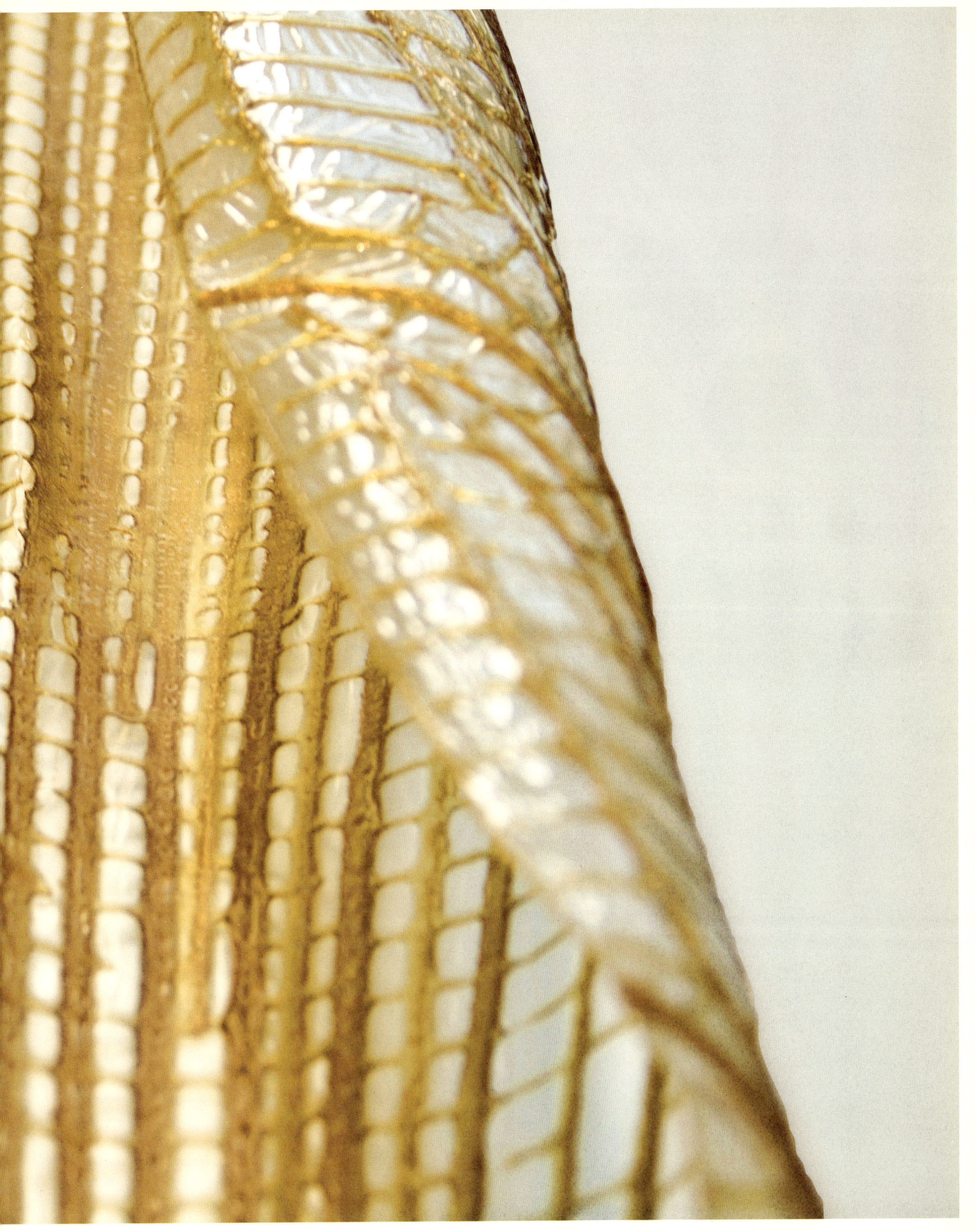

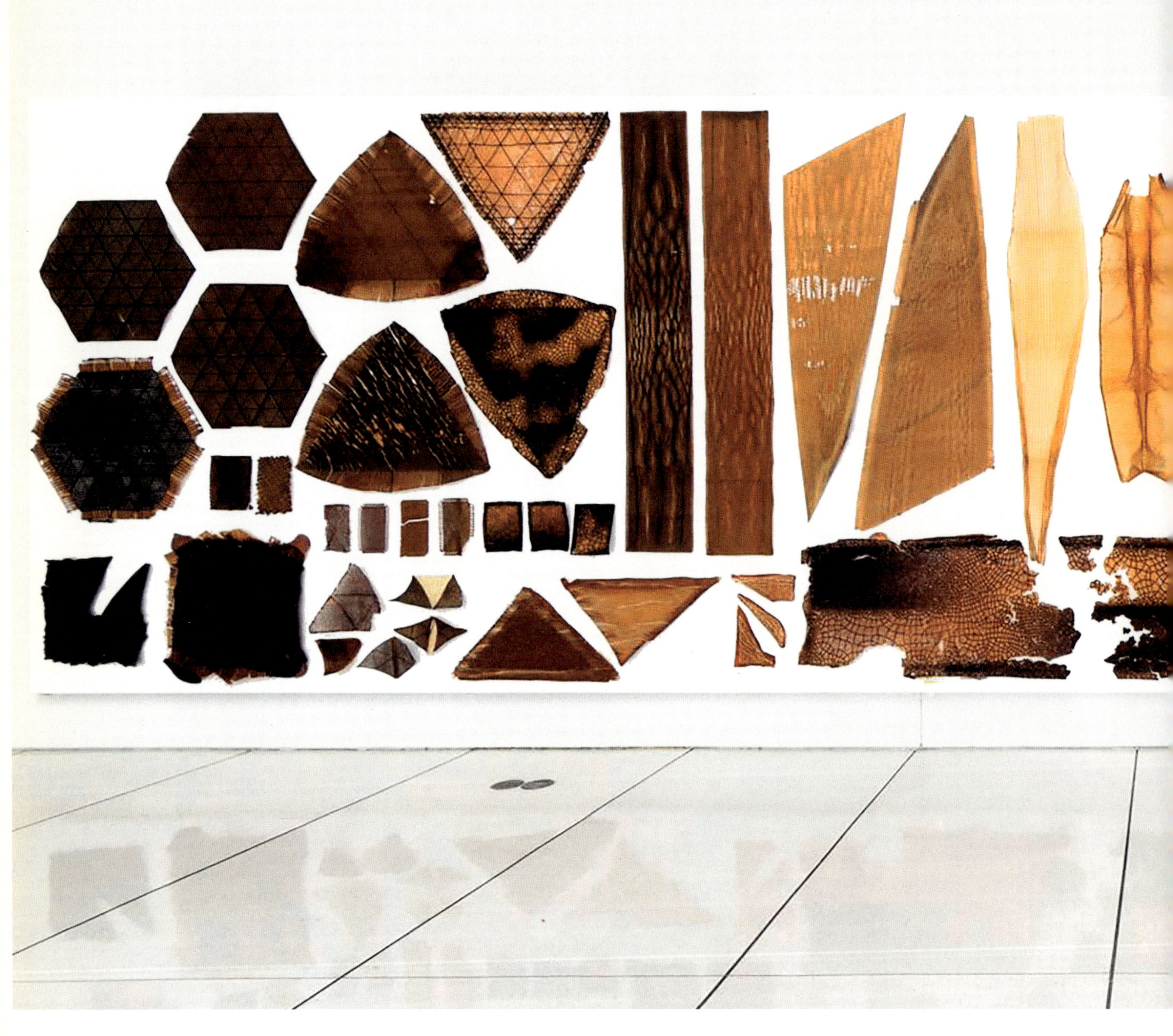

Aguahoja I
Material experiments made over four years demonstrated a range of aesthetics and behaviors elicited in medium- to large-scale printed objects.

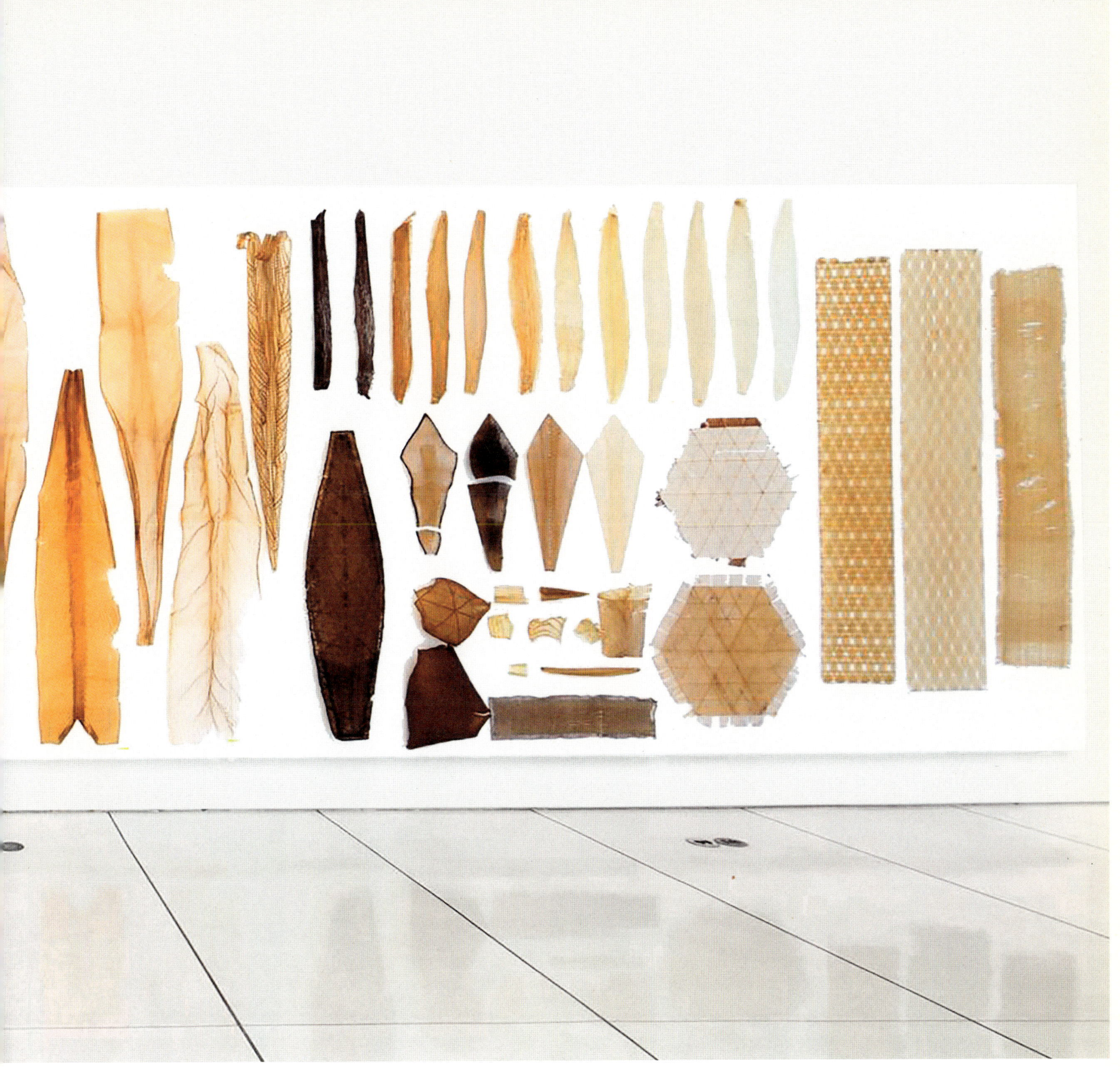

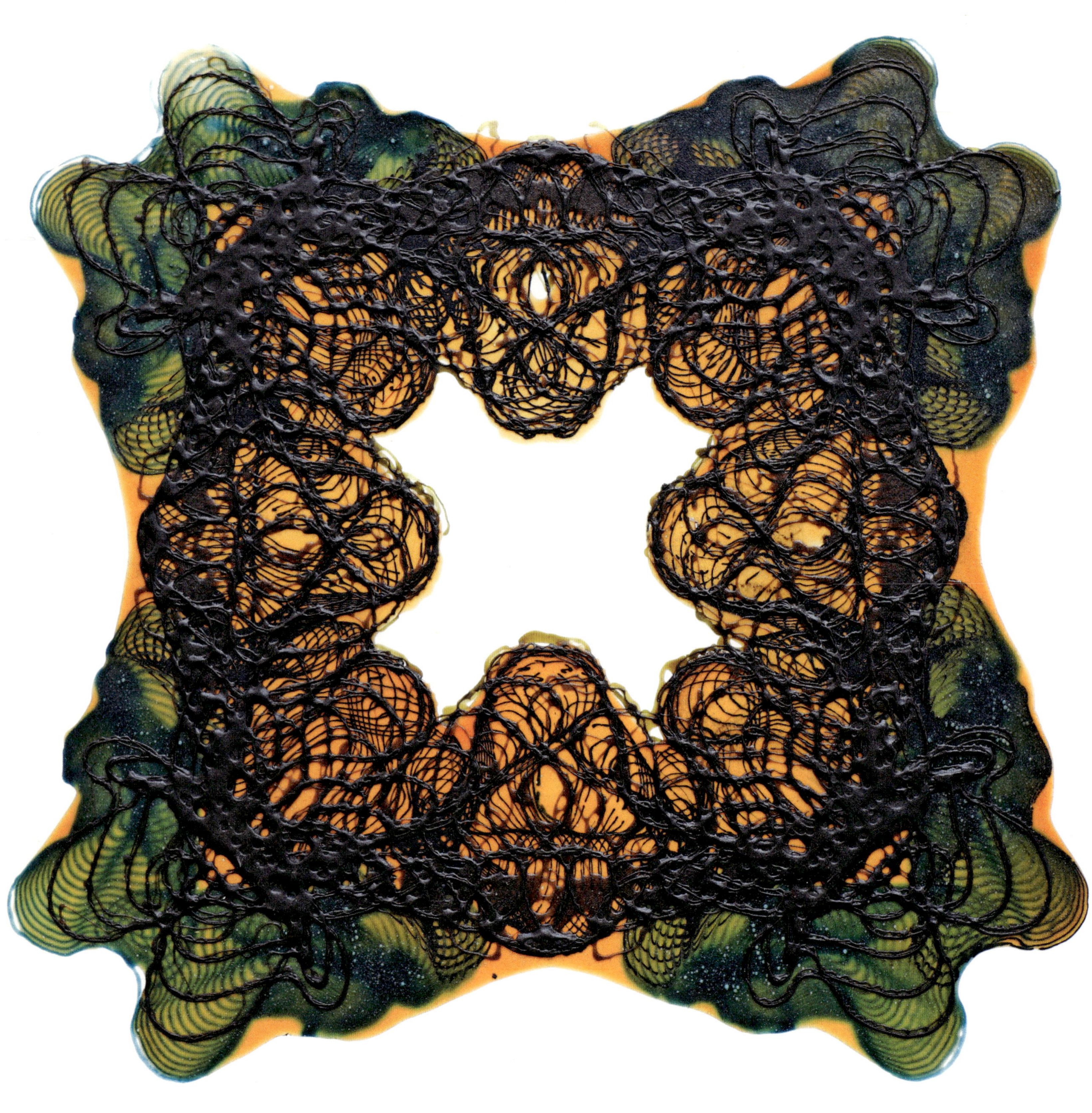

Aguahoja II
Prototype biocomposites formed of layers of aqueous materials (left) and containing turmeric and beet in a pectin membrane (right).

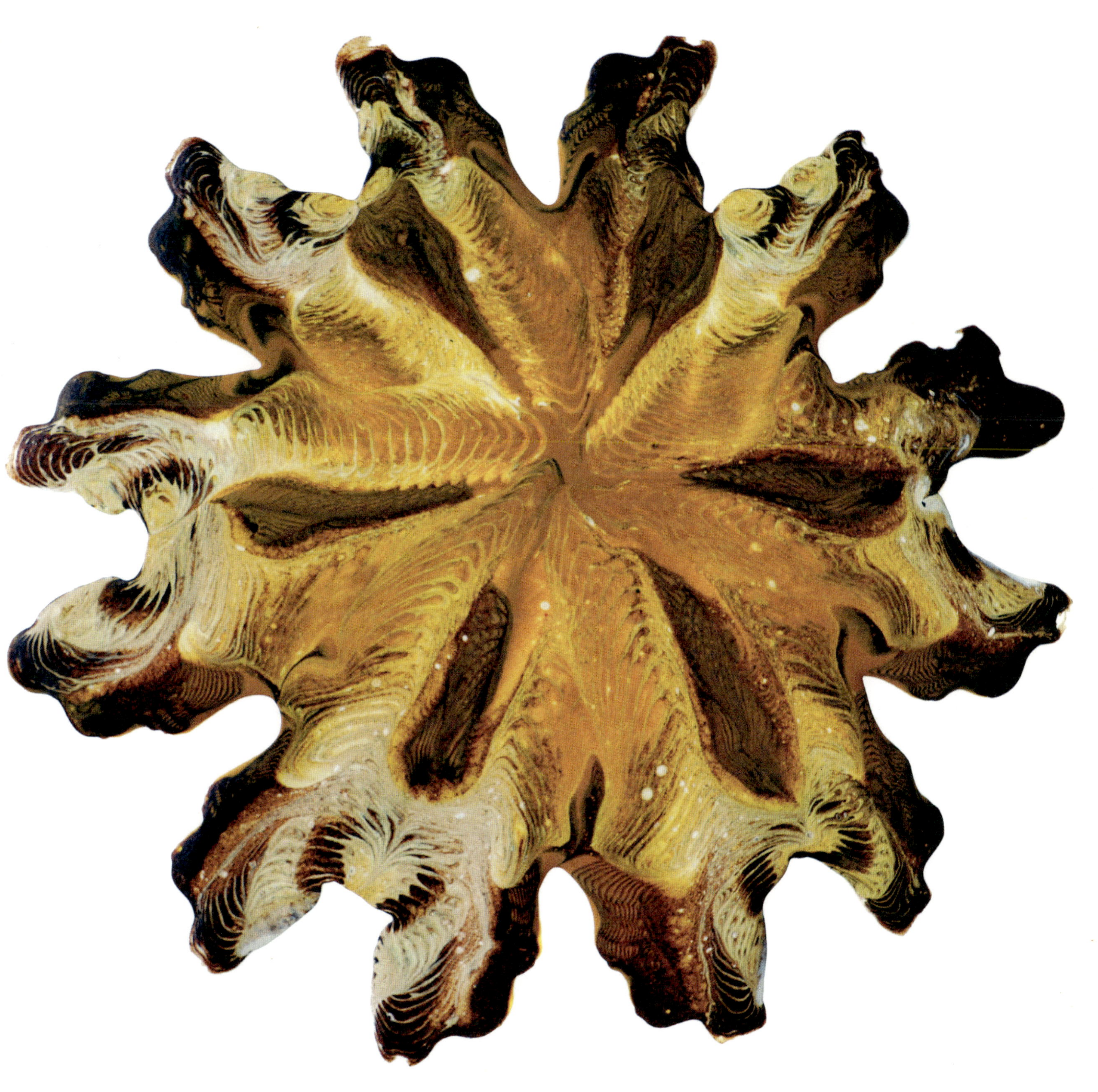

SILK PAVILION I

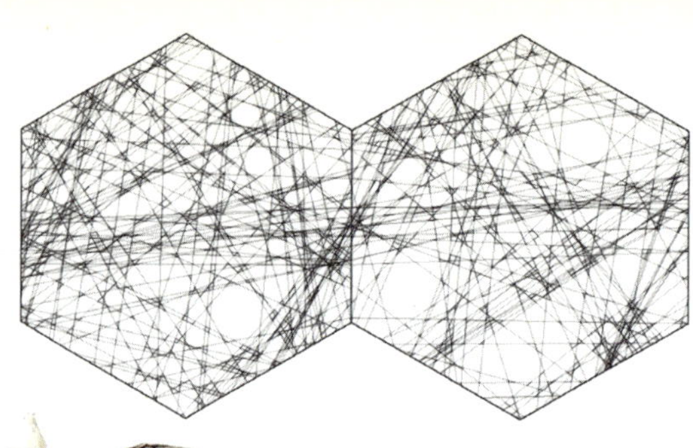

Schematic of the toolpath (the path followed by the tip of a spinning tool in order to produce a desired geometry) used to fabricate the structure's panels.

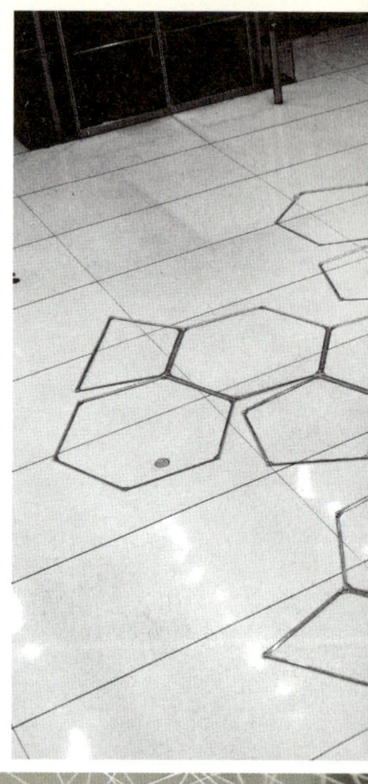

Neri Oxman and
The Mediated Matter Group
Silk Pavilion I. 2013
Stainless steel, CNC, and silkworm-spun silk thread
9 ft. 5⁄16 in. × 9 ft. 5⁄16 in. × 9 ft. 10 1⁄8 in.
(275 × 275 × 300 cm)
An MIT Media Lab project
Research team: Markus Kayser, Jared Laucks, Jorge Duro-Royo, Carlos David Gonzalez Uribe, Neri Oxman
Collaborators and contributors: Fiorenzo Omenetto, Tufts University; James C. Weaver, Wyss Institute, Harvard University

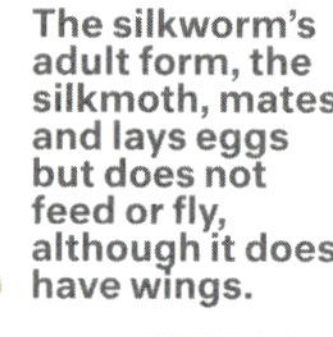

The silkworm's adult form, the silkmoth, mates and lays eggs but does not feed or fly, although it does have wings.

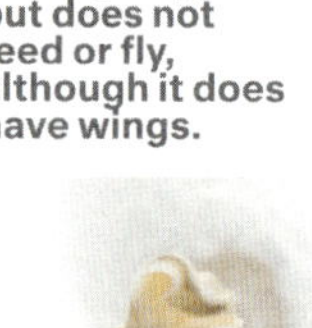

An adult silkmoth emerges from its cocoon.

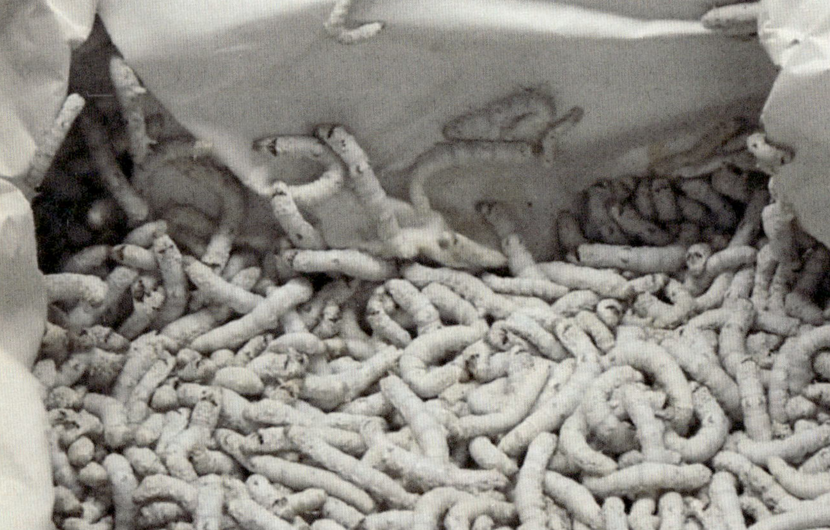

The larvae of the silkworm *Bombyx mori* were raised through their five instars (developmental stages) at the lab.

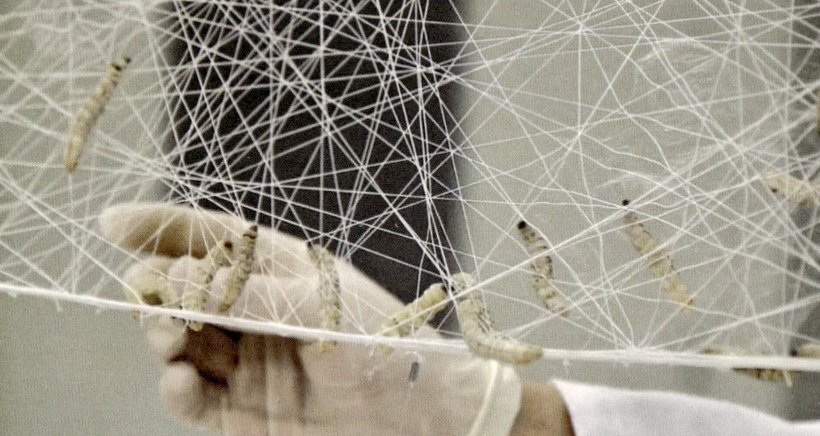

At the beginning of their spinning stage, the silkworms were placed on a scaffold of silk that had been robotically wound over stainless steel frames. Biologically spun silk thus covered robotically spun silk.

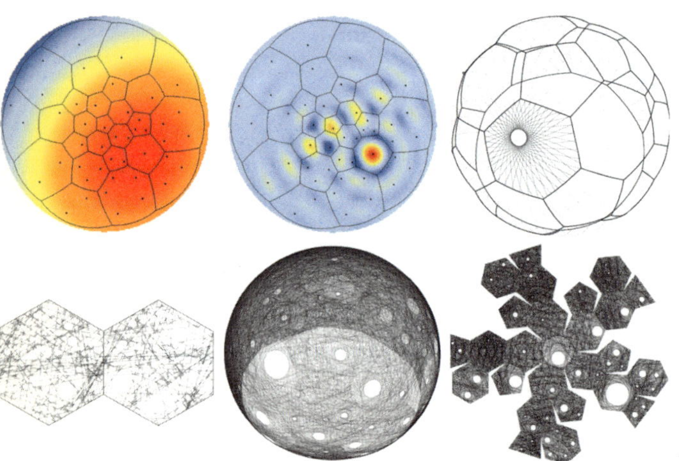

A diagram of the sun's path was used to determine the placement of apertures on the panels in order to modulate the distribution of light and heat on the surface and thus influence the deposition of silk.

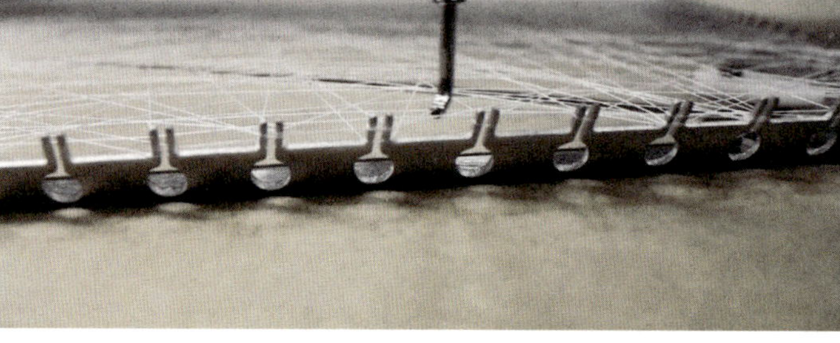

Custom-designed end effector for the robotic arm of the CNC machine that wound the thread on each stainless steel frame. The pattern was informed by the size of an average silkworm, in order to keep them from falling through the gaps.

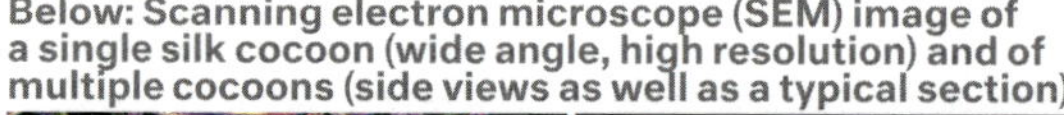

Below: Scanning electron microscope (SEM) image of a single silk cocoon (wide angle, high resolution) and of multiple cocoons (side views as well as a typical section).

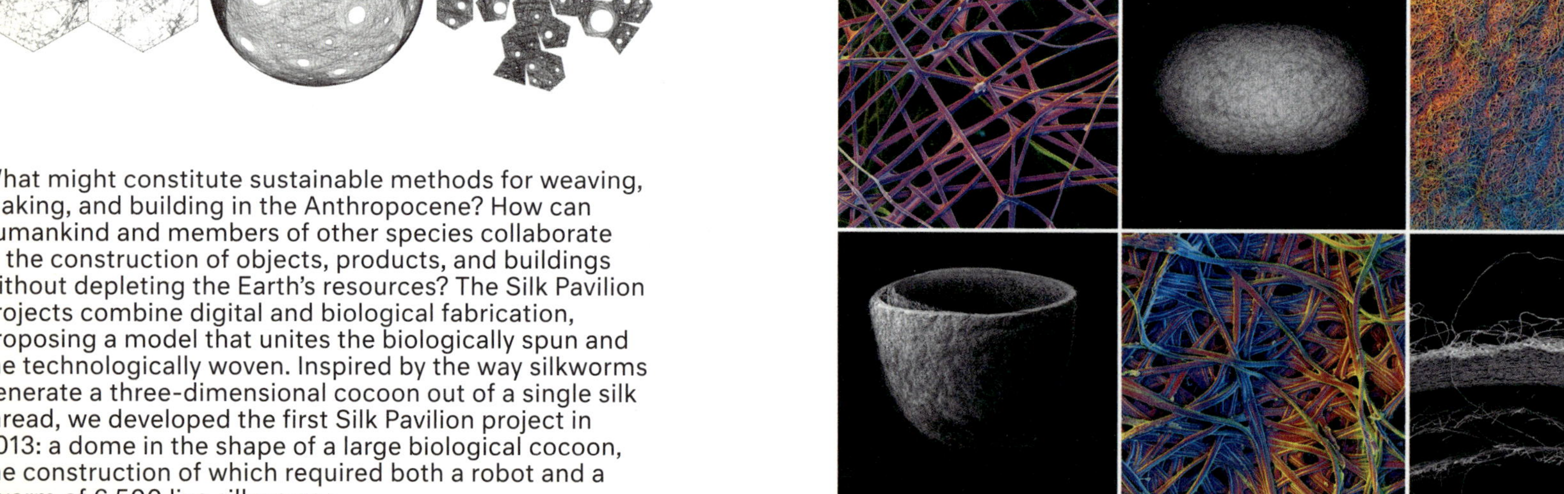

What might constitute sustainable methods for weaving, making, and building in the Anthropocene? How can humankind and members of other species collaborate in the construction of objects, products, and buildings without depleting the Earth's resources? The Silk Pavilion projects combine digital and biological fabrication, proposing a model that unites the biologically spun and the technologically woven. Inspired by the way silkworms generate a three-dimensional cocoon out of a single silk thread, we developed the first Silk Pavilion project in 2013: a dome in the shape of a large biological cocoon, the construction of which required both a robot and a swarm of 6,500 live silkworms.

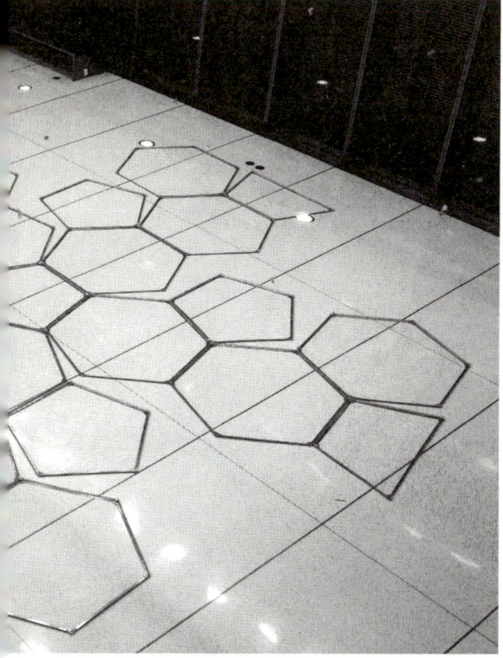

he pavilion's wenty-six flat olygonal anels laid out efore being ssembled into dome.

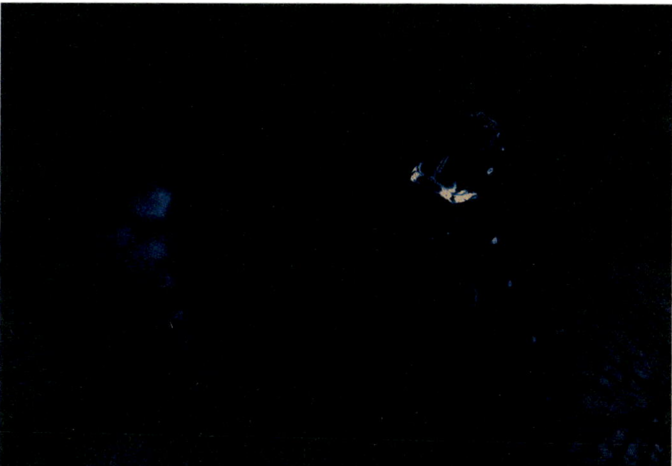

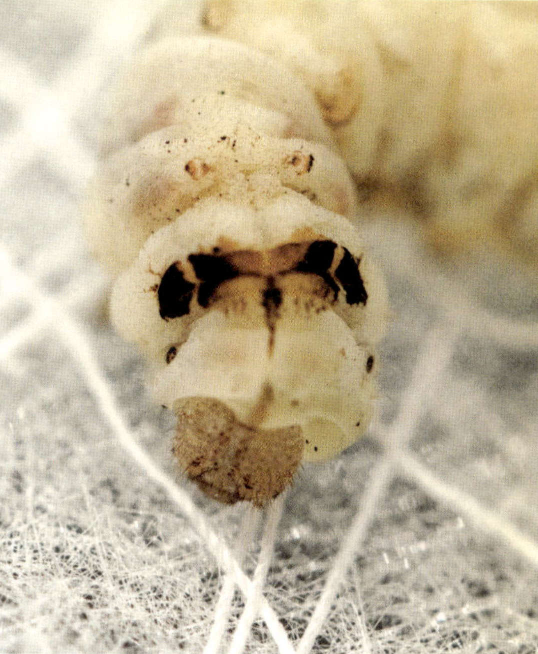

Over three weeks, 6,500 silkworms completed the pavilion, spinning, undergoing healthy metamorphosis, and laying eggs. Each worm spun a single silk thread about 0.6 mile (1 kilometer) long—all together, the length of the Silk Road.

In the first phase of construction, we used twenty-six polygonal stainless steel frames to create an underlying support structure. The frames were covered with a scaffolding of silk threads laid down by a computer numerically controlled (CNC) machine. An algorithm based on silkworm movement assigned the path of these continuous hreads across the frames, creating various degrees of density nd an overall surface geometry; the pattern also kept the ilkworms from falling through gaps.

nd effector of e CNC system inding silk read on a olygonal frame produce a affold for the orms.

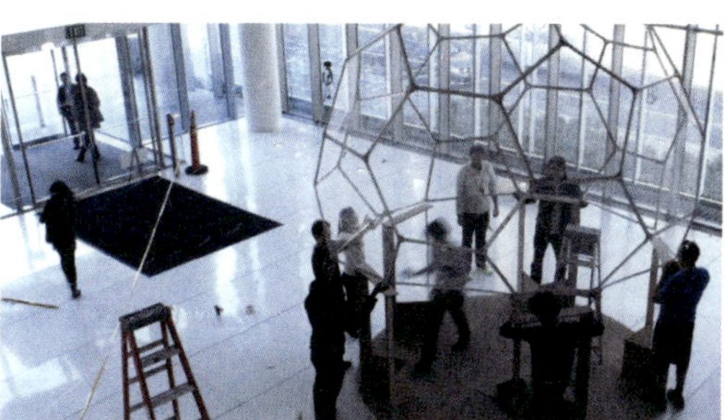

Silk Pavilion I in the lobby of the MIT Media Lab in 2013.

Holes in the panels were precisely designed to control and direct air and heat patterns within the structure, affecting the silkworms' movements.

or the pavilion's inner, secondary structure, we deployed e silkworms as biological printers, filling the gaps mong the machine-woven threads to create a dense, abitable shelter. The worms, each of which spun a read 0.6 mile (1 kilometer) long, were introduced into e bottom rim of the frame and then traveled over the tructure, spinning flat silk patches and reinforcing the nderlying structure. They underwent healthy etamorphosis during the spinning phase and were emoved from the structure following the pupation stage. s moths, they laid their eggs—about 1.5 million of em—enough for about 250 additional pavilions.

We found that the silkworms were affected by their spatial and environmental conditions, including changes in natural light and heat, and that they migrated to darker, colder areas. By varying these conditions, we were able to direct the silkworms to spin specific patterns. We used a seasonal sun map to guide the silkworms in the sun's path, with apertures in the underlying structure designed to lock in rays of light from the pavilion's south and east sides. The pavilion's main oculus was located against the east elevation and functioned as a sun clock.

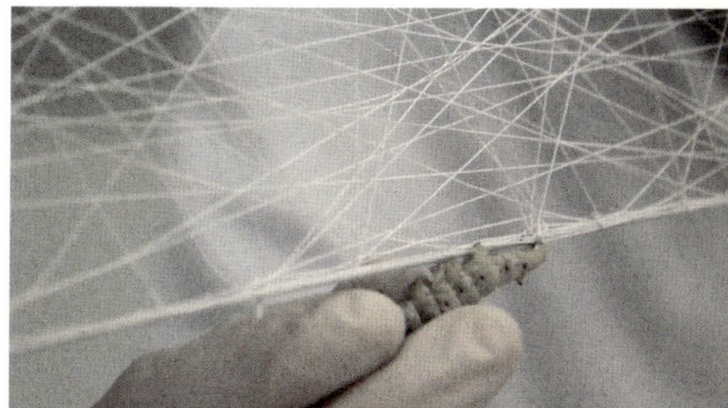

The silkworms were manually placed on the scaffold structure. Over time, they moved upward, spinning patches of silk and changing the overall form of the pavilion.

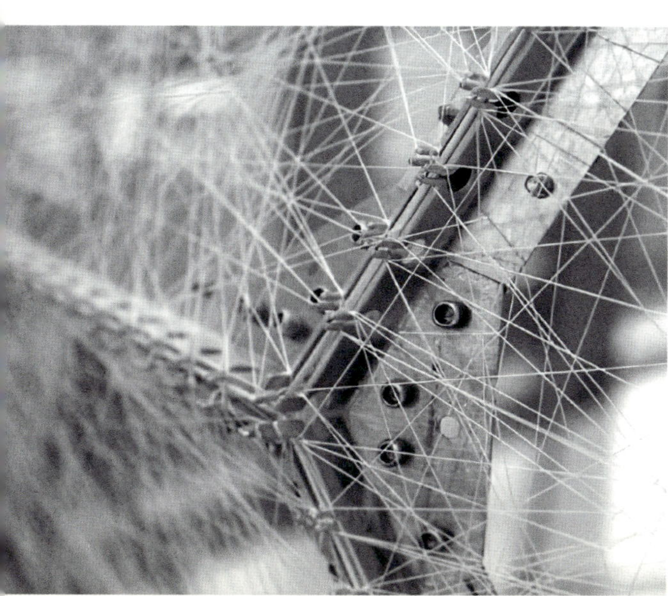

Scaffold detail. The silkworms naturally closed the gaps between the threads, creating a continuous surface.

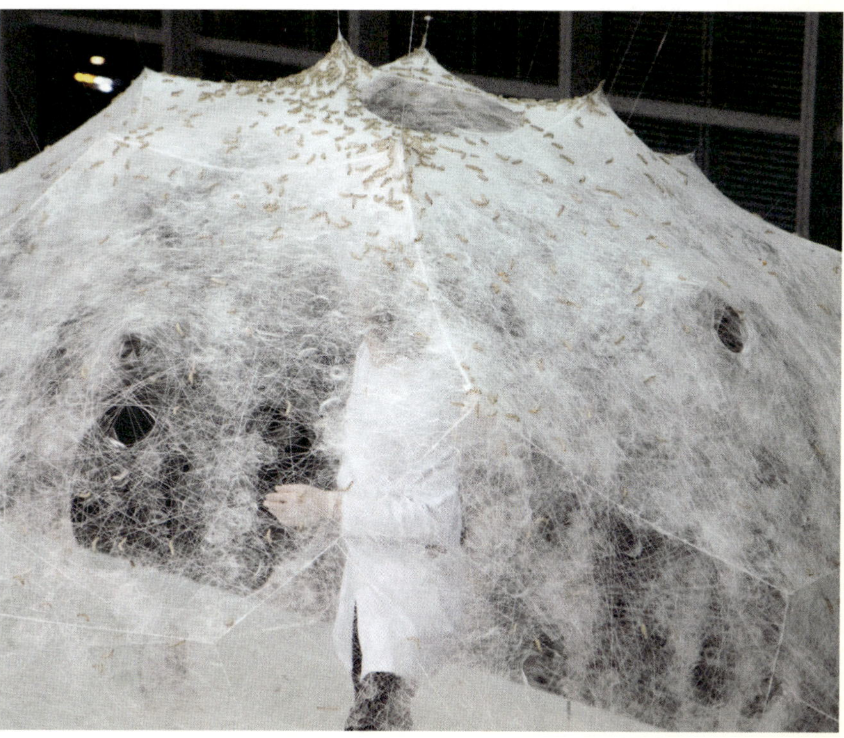

The structure became denser as the silkworms worked, and its form changed, demonstrating some of the ways in which biological processes affected and modified the original architectural design.

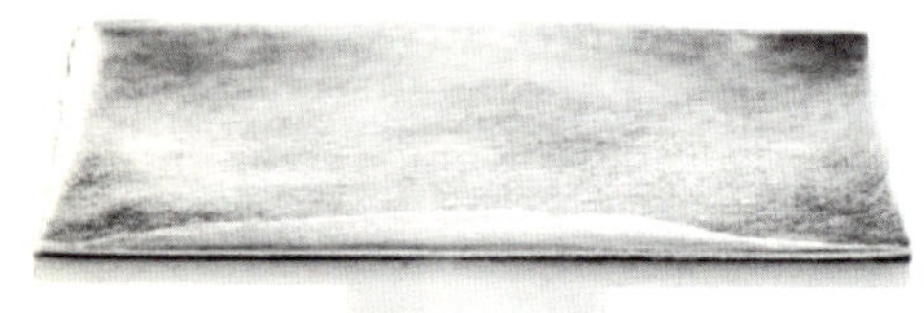

Test: Height
Dimensions: 80 × 80 mm
Cross section height: 0 mm

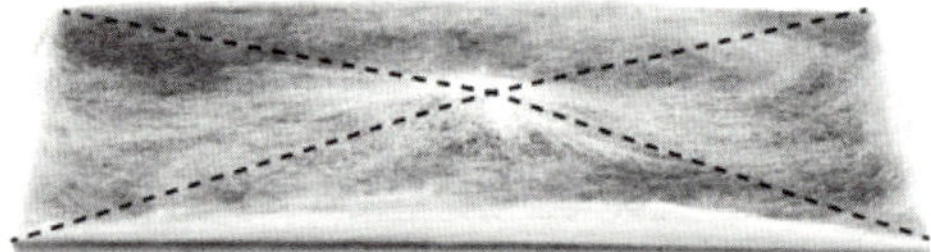

Test: Height
Dimensions: 80 × 80 mm
Cross section height: 3 mm

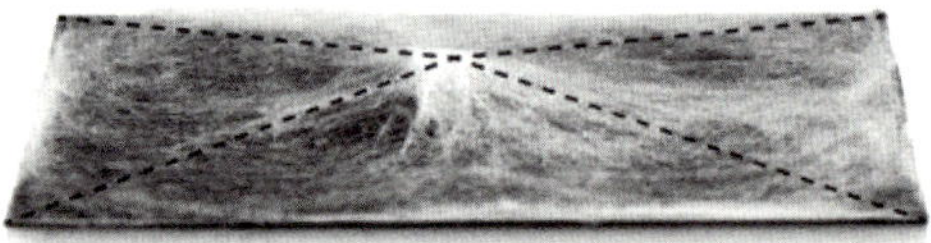

Test: Height
Dimensions: 80 × 80 mm
Cross section height: 6 mm

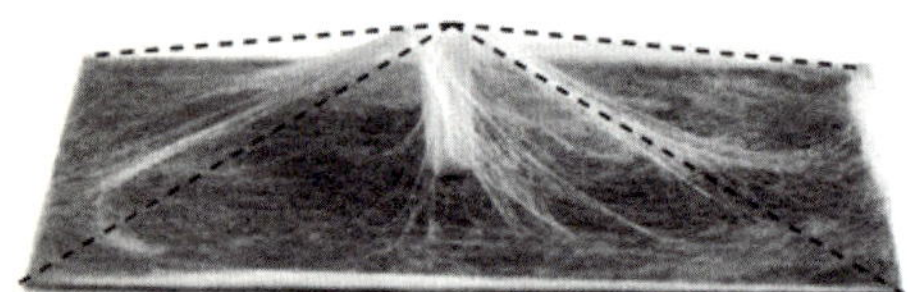

Test: Height
Dimensions: 80 × 80 mm
Cross section height: 15 mm

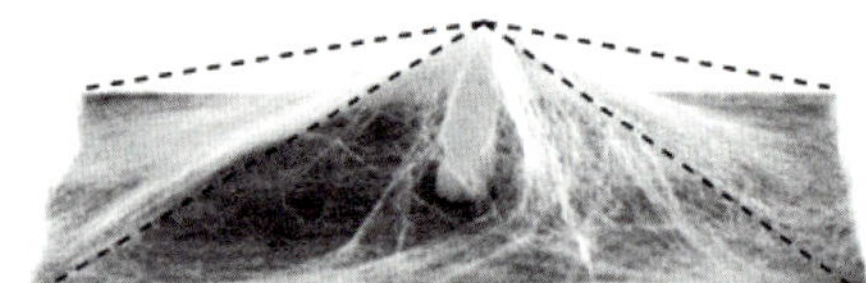

Test: Height
Dimensions: 80 × 80 mm
Cross section height: 18 mm

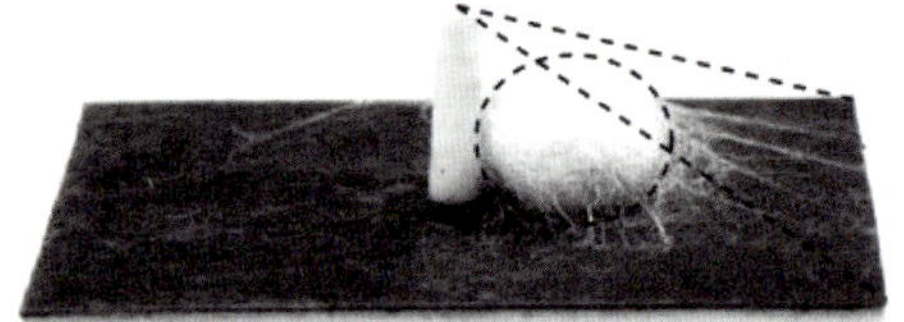

Test: Height
Dimensions: 80 × 80 mm
Cross section height: 21 mm

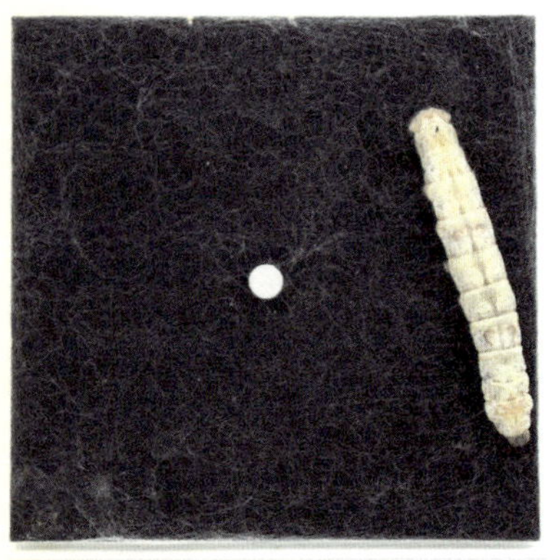
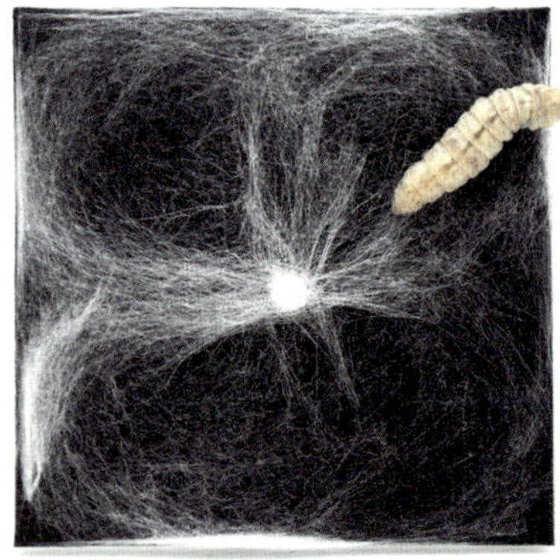
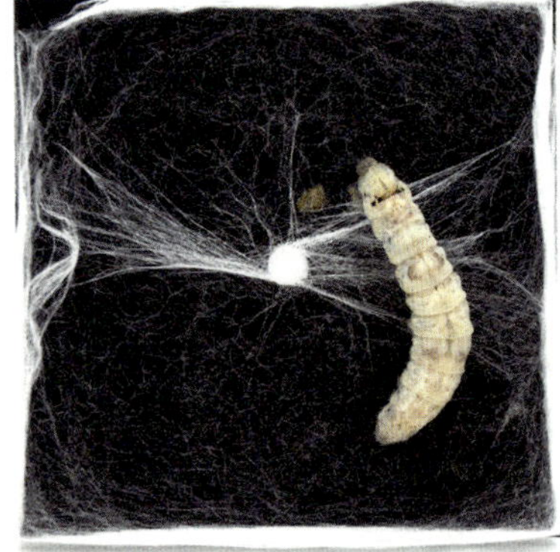
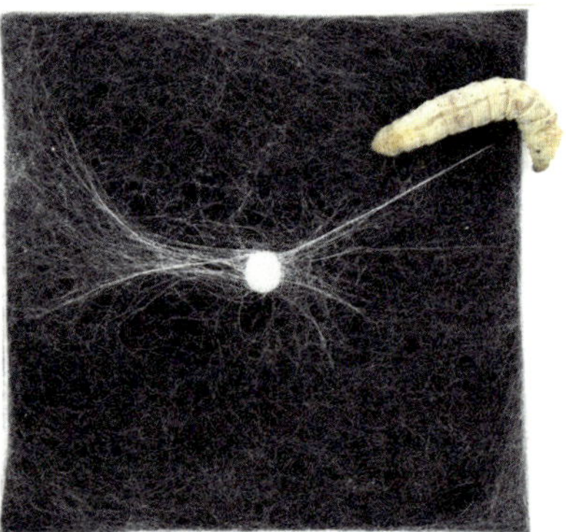

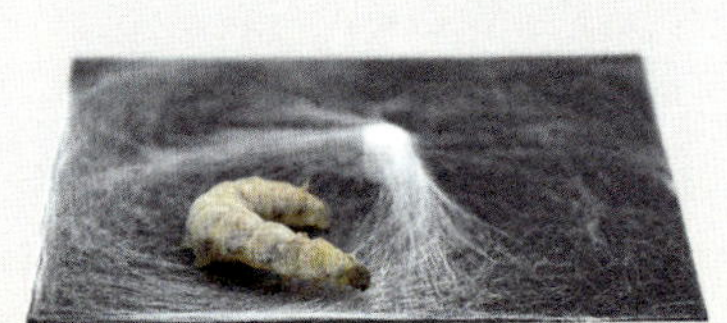
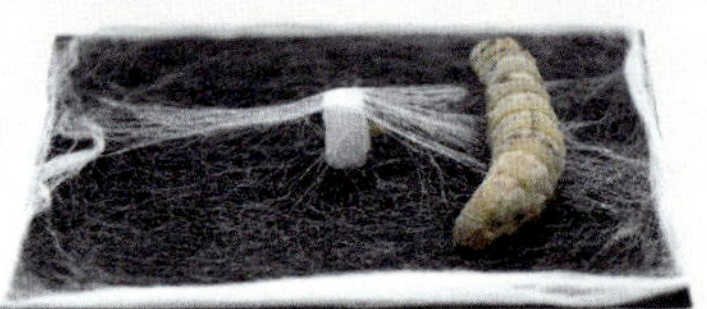
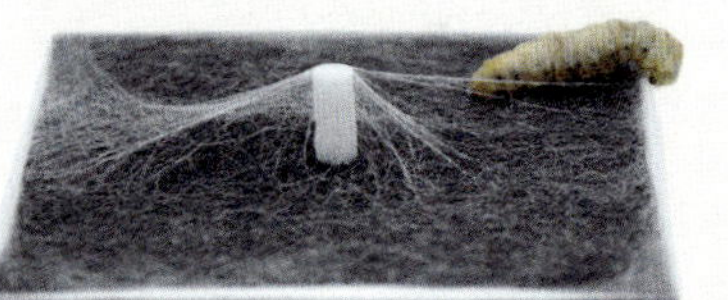

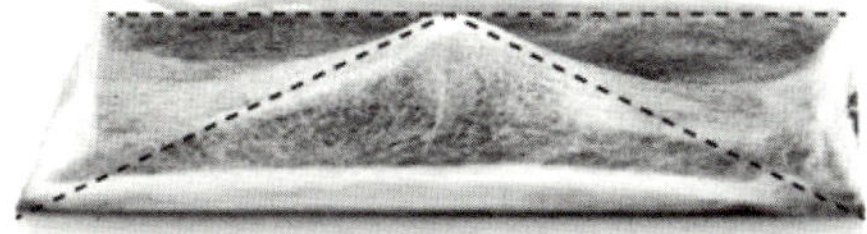

Test: Height
Dimensions: 80 × 80 mm
Cross section height: 9 mm

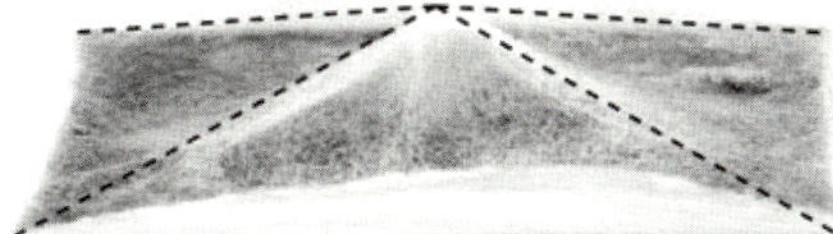

Test: Height
Dimensions: 80 × 80 mm
Cross section height: 12 mm

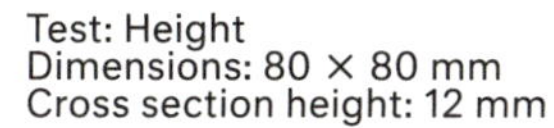

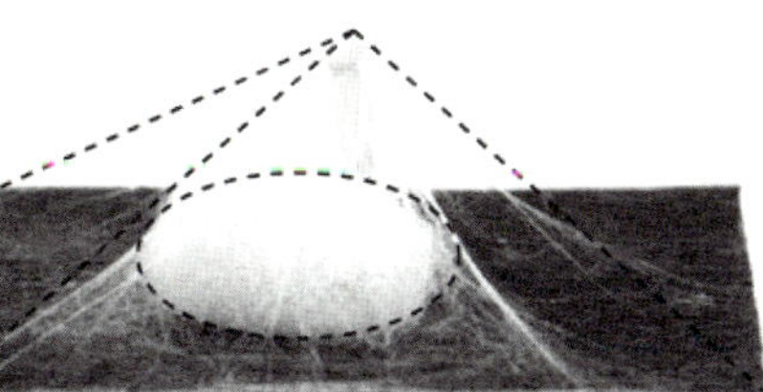

Test: Height
Dimensions: 80 × 80 mm
Cross section height: 24 mm

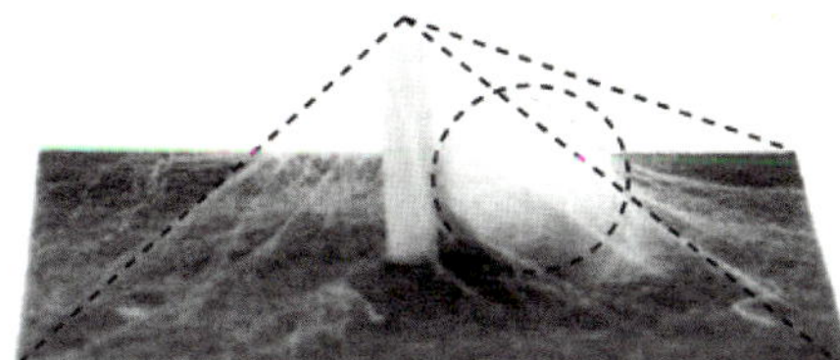

Test: Height
Dimensions: 80 × 80 mm
Cross section height: 27 mm

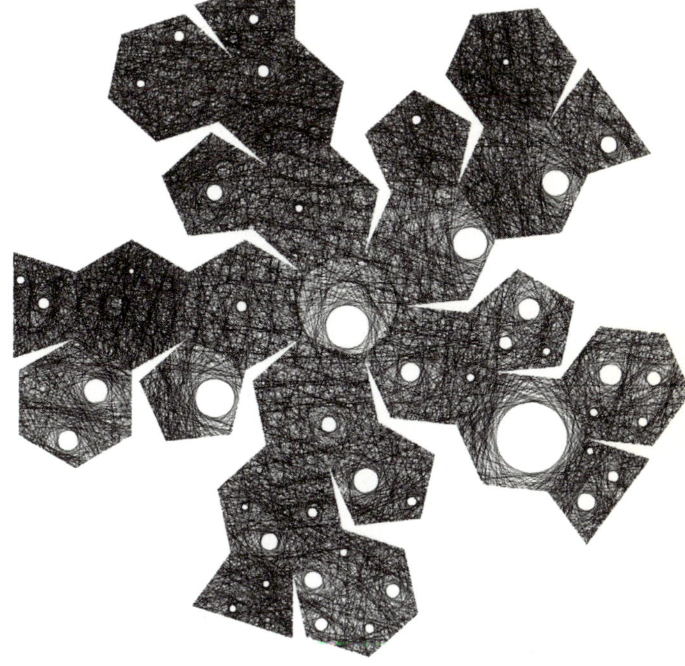

Diagram of the pavilion as a flat, two-dimensional structure with panels unfolded, with a total span of about 1,830 square feet (170 square meters). The panels functioned as an underlying support structure for the silkworms.

An experiment placing the silkworms on surfaces with central rods demonstrated that when the rod was taller than about 13⁄16 inch (21 millimeters), the worms spun three-dimensional cocoons, and at lower heights they produced flat patches. Over the course of a few days, each worm spun a flat patch, a cocoon, or a transitional form, depending on the geometry of the underlying structure.

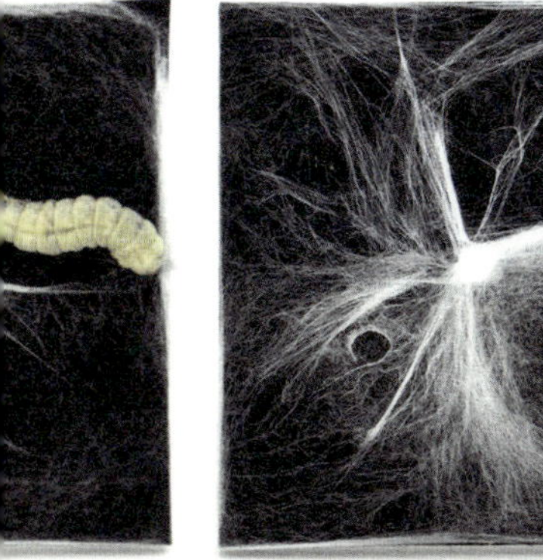

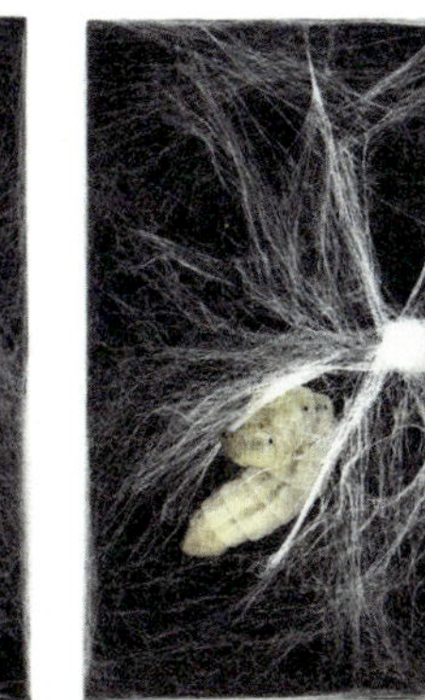

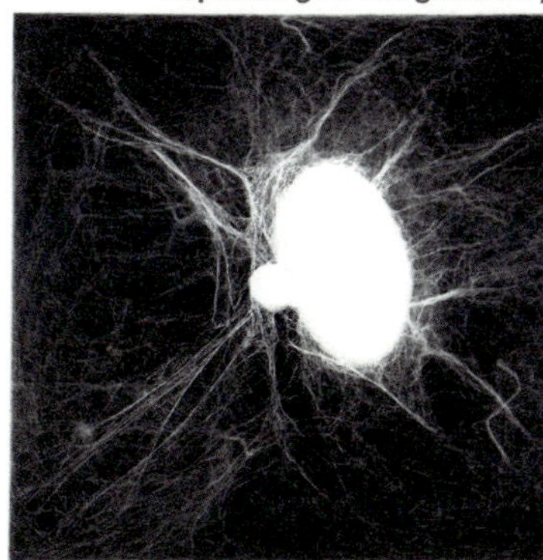

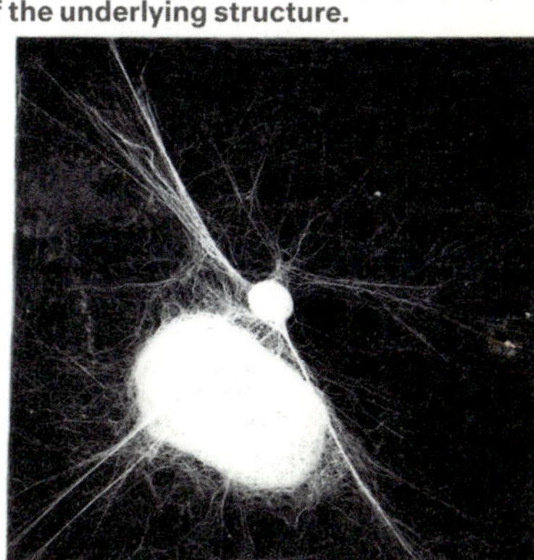

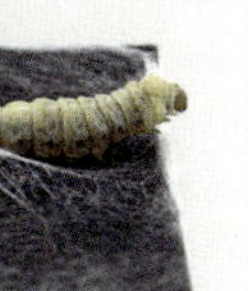

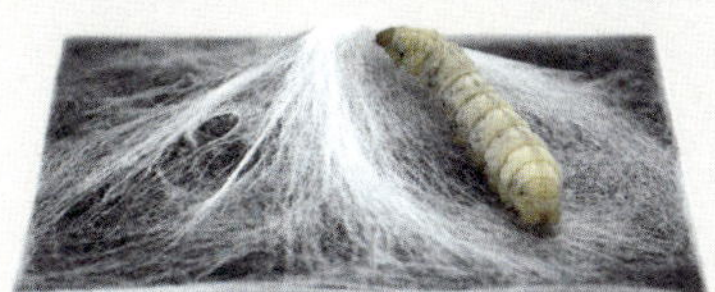

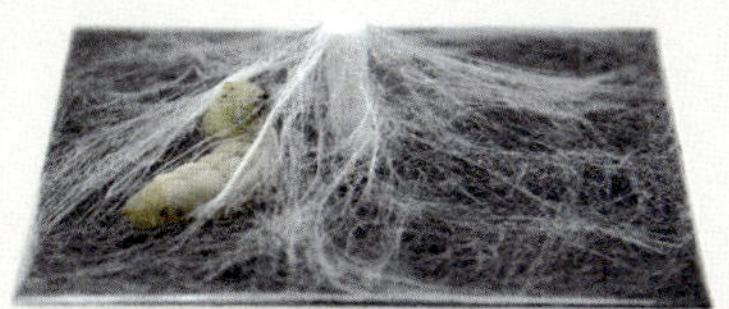

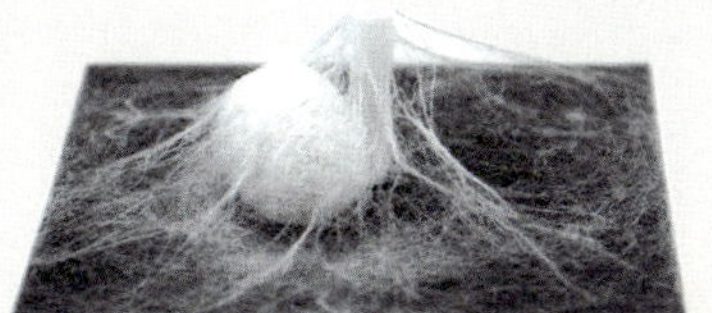

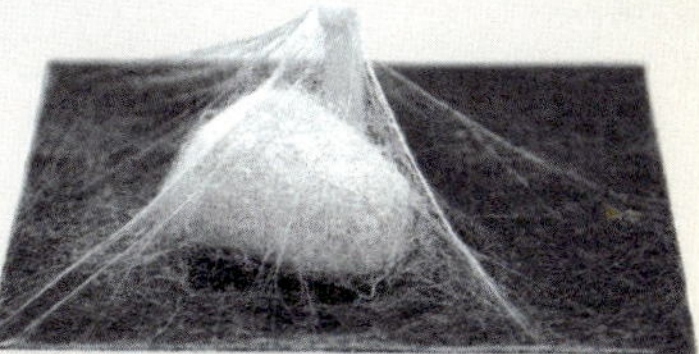

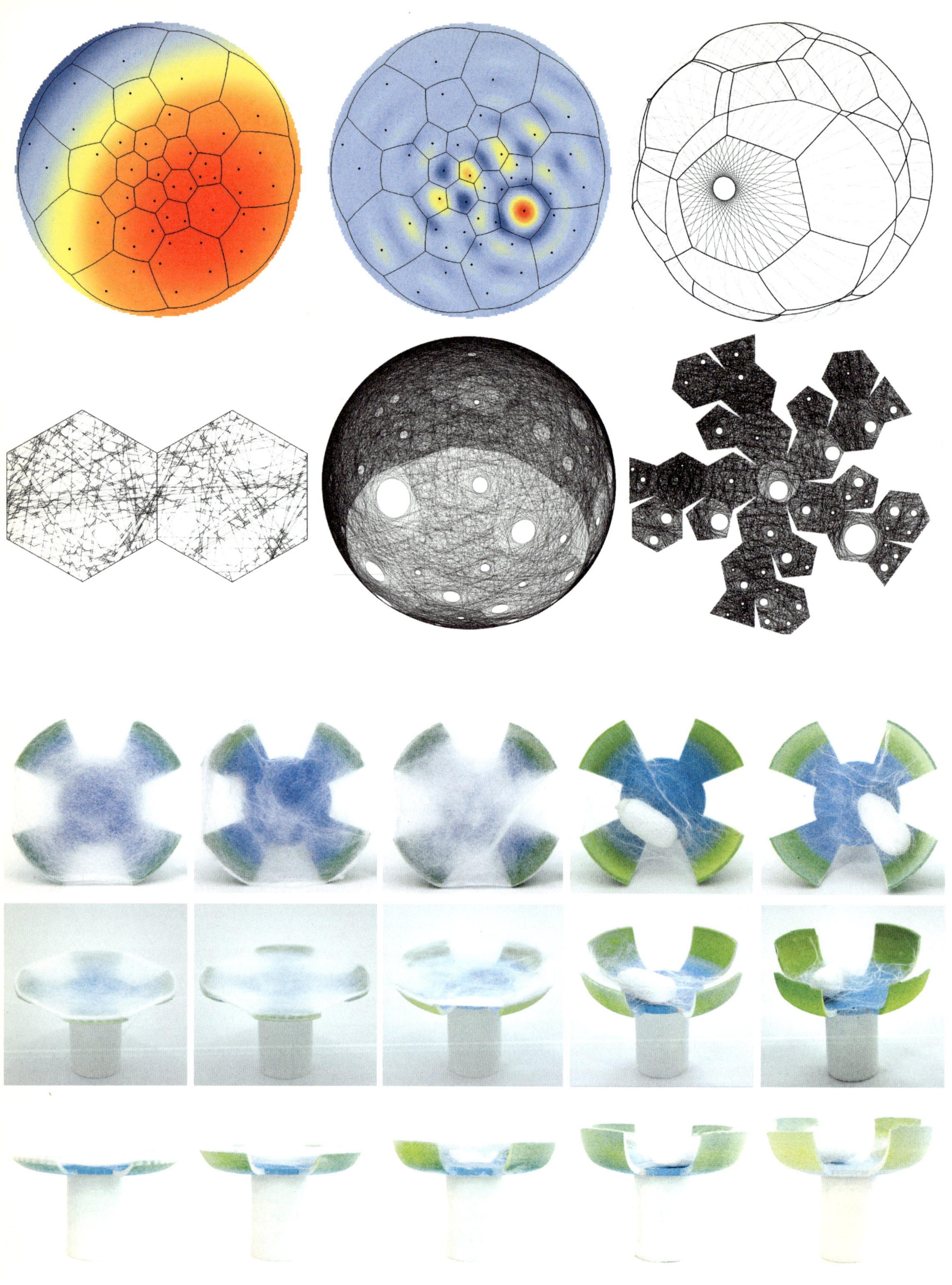

Silk Pavilion I
Variations in light and heat exposure and the shape of the underlying structure influenced the silkworms' behavior.

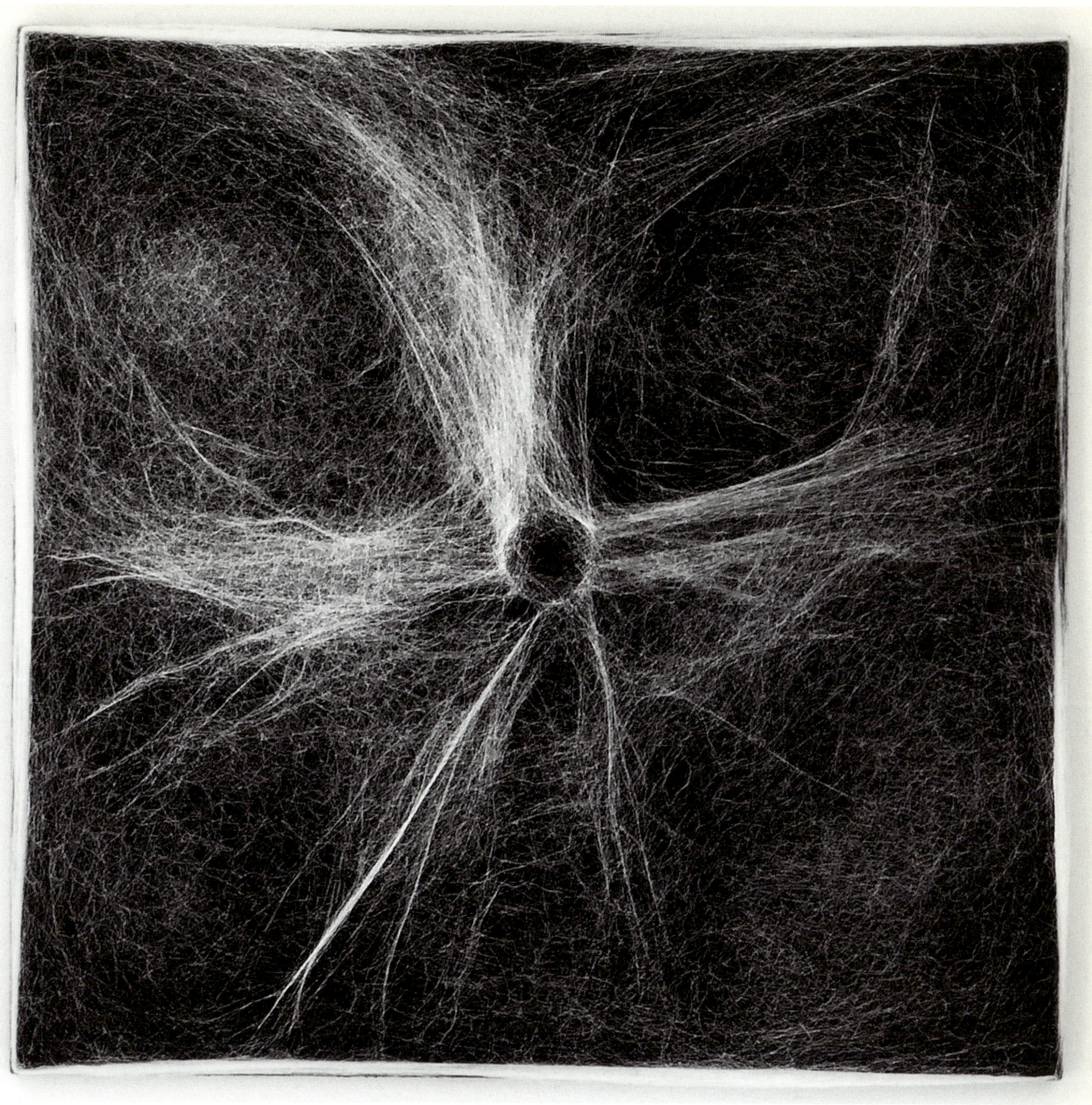

Silk Pavilion I
Over three weeks, 6,500 silkworms completed the pavilion.

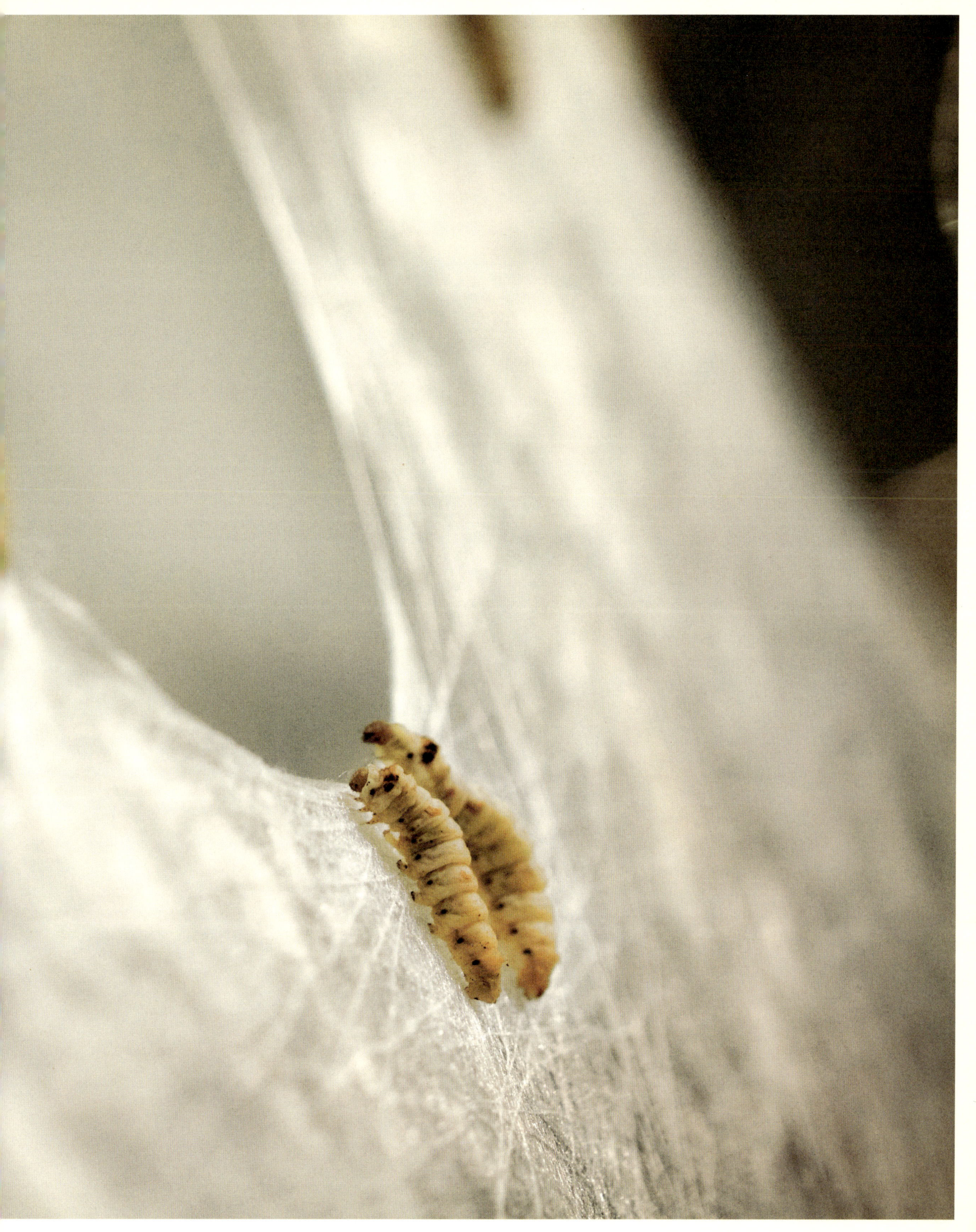

Silk Pavilion I
Holes in the panels were precisely designed to control and direct air and heat patterns within the structure.

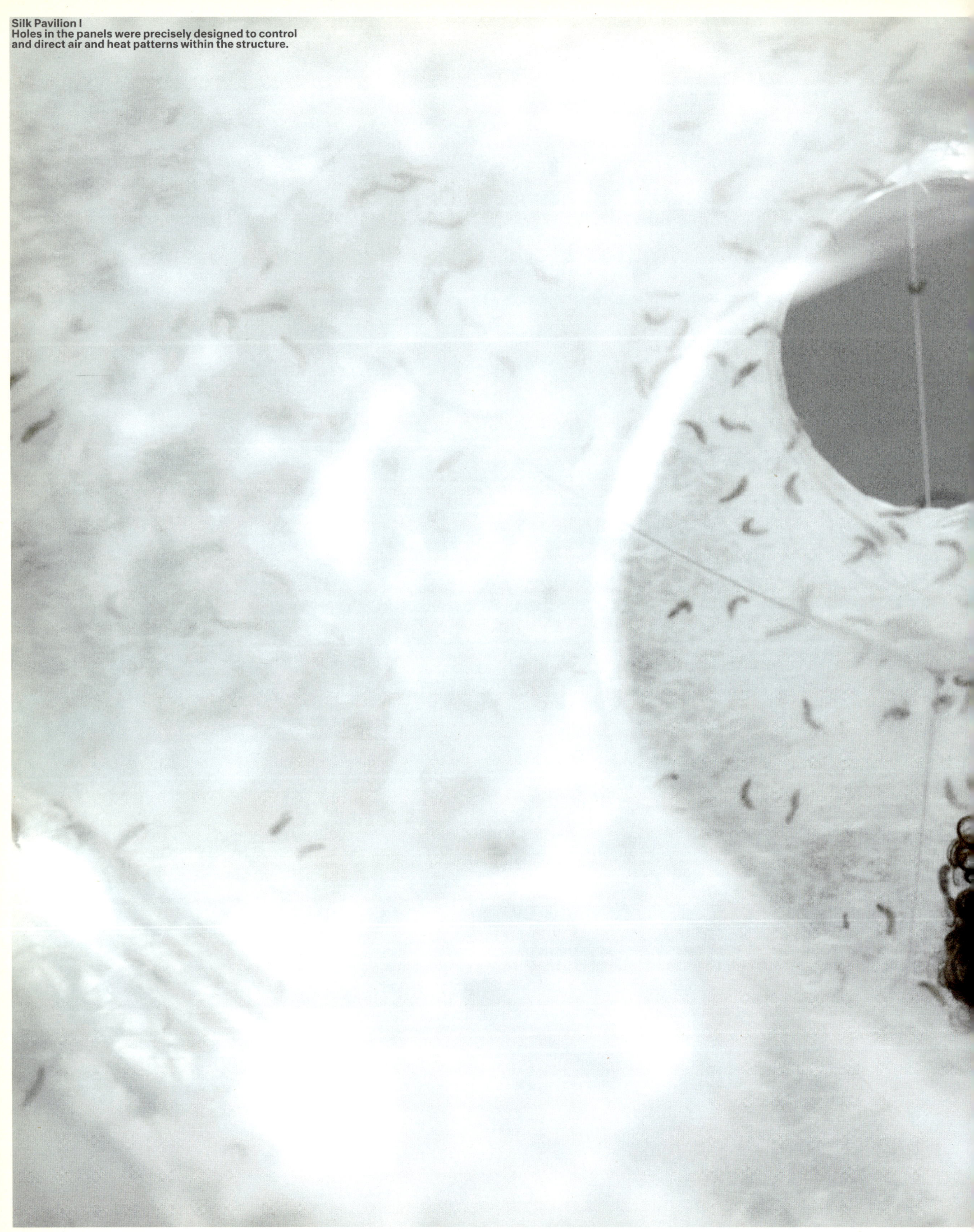

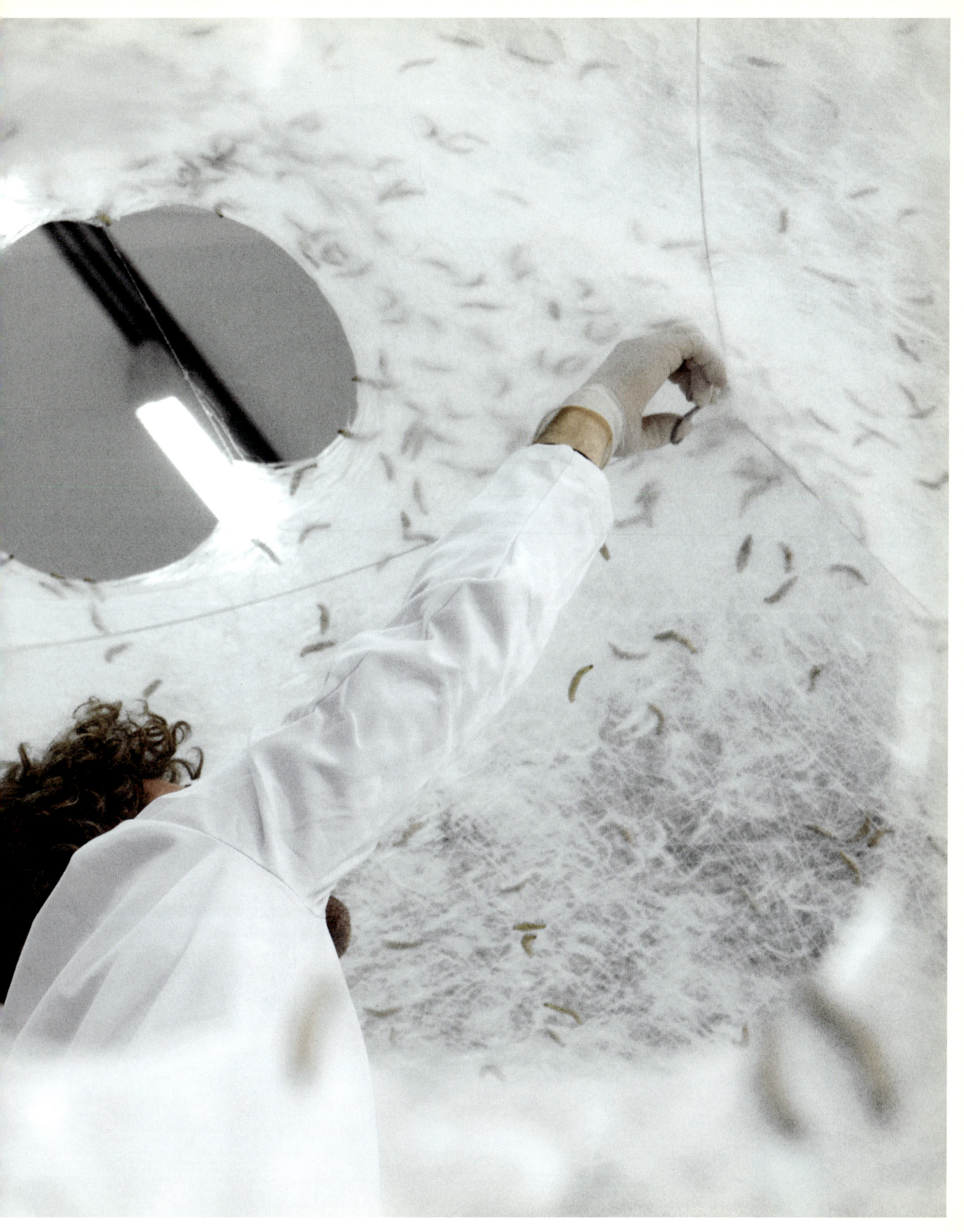

SILK PAVILION II

Neri Oxman and
The Mediated Matter Group
Silk Pavilion II. 2019
Stainless steel cables, aluminum, polyvinyl alcohol knit textile, and silkworm-spun silk
16 ft. 4 13⁄16 in. × 16 ft. 4 13⁄16 in. × 19 ft. 8 3⁄16 in.
(500 × 500 × 600 cm)
An MIT Media Lab project
Research team: João Costa, Christoph Bader, Sunanda Sharma, Felix Kraemer, Susan Williams, Jean Disset, Neri Oxman
Undergraduate researcher: Sara L. Wilson
Collaborators and contributors: Davide Biasetto, Il Brolo Società Agricola S.r.l.; Levi Cai; Silvia Cappellozza and Alessio Saviane, Council for Agricultural Research and Agricultural Economics Analysis (CREA-AA); Natalia Casas; Kelly Egorova; Fiorenzo Omenetto, Tufts University; Sol Schade, Advanced Functional Fabrics of America (AFFOA); James C. Weaver, Wyss Institute, Harvard University; Nitzan Zilberman; Bodino; Front Inc.; Robert Wood Johnson Foundation; The Museum of Modern Art, New York

Commissioned for the exhibition *Neri Oxman: Material Ecology*, The Museum of Modern Art, New York, 2020

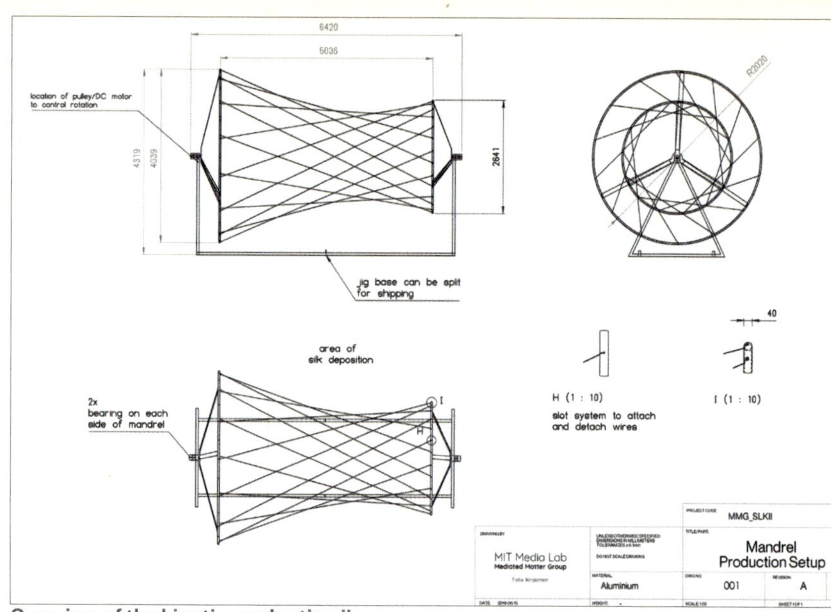

Overview of the kinetic production jig.

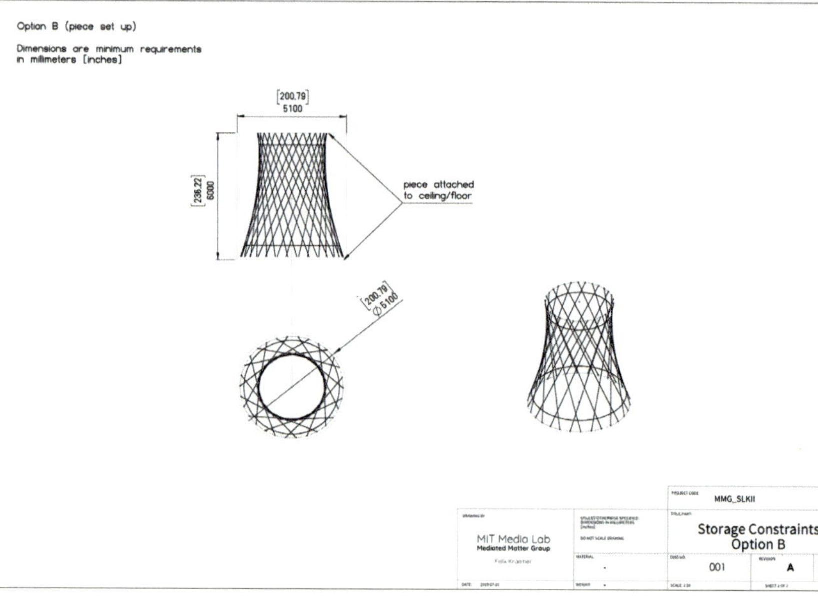

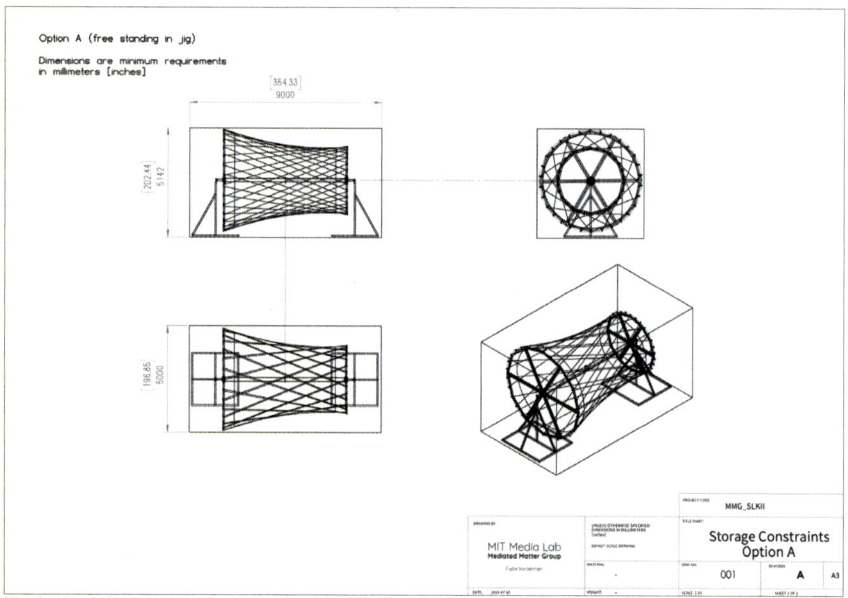

Technical drawings of storage and shipping constraints for pavilion elements, including the production jig. The constraints informed design decisions such as size, compactness, and mechanical operation.

To produce the installation for MoMA, a mesh of stainless steel wire rope was mounted on a kinetic jig at the silkworm-rearing facility and reinforced with water-soluble knit fabric, forming the pavilion's base geometry. After the silkworms spun over this structure, the pavilion was removed from the jig and shipped to New York.

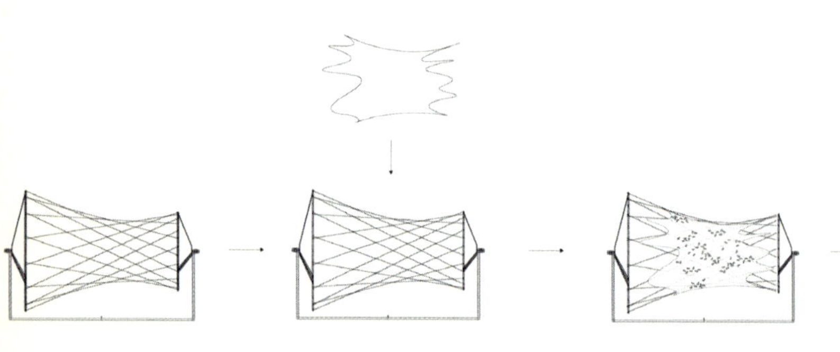

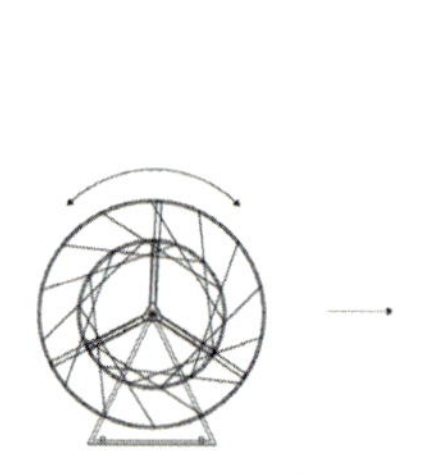

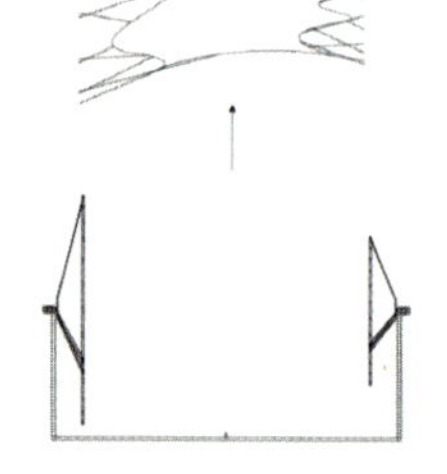

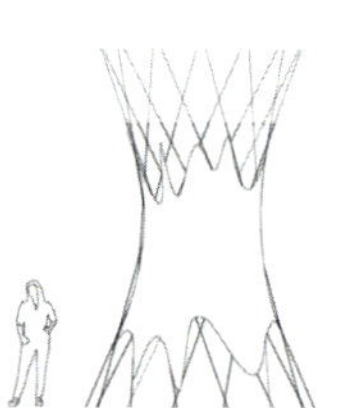

Initial setup.

A soluble knit is added to the wire structure to form base for spinning phase.

Deposition of organic silk layer by silkworms.

Density of silk deposition is controlled by rotation of mandrel.

Silk structure is removed from mandrel setup.

Resulting structure is installed in gallery space.

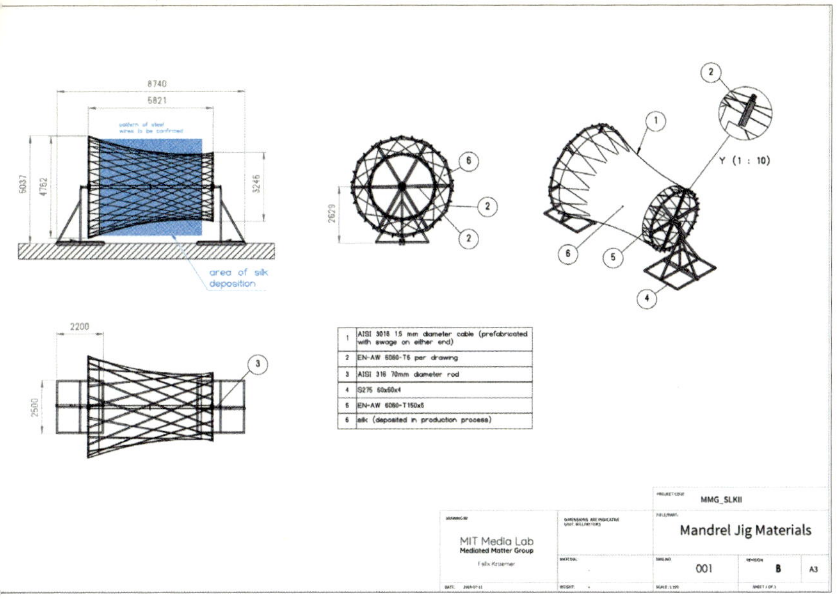

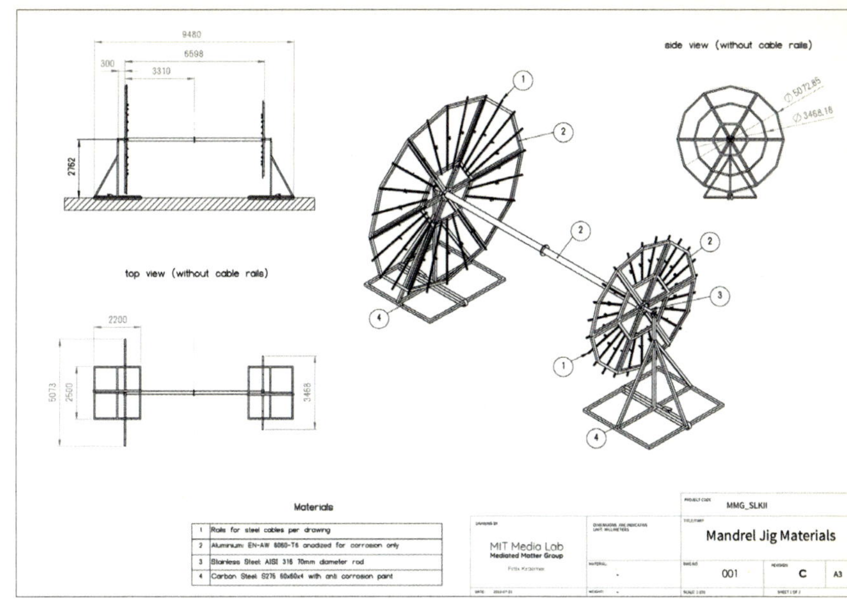

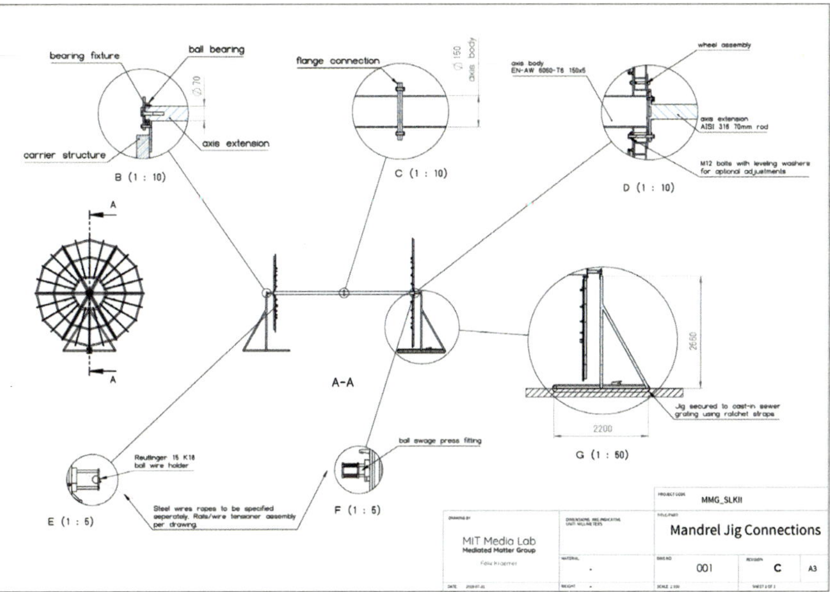

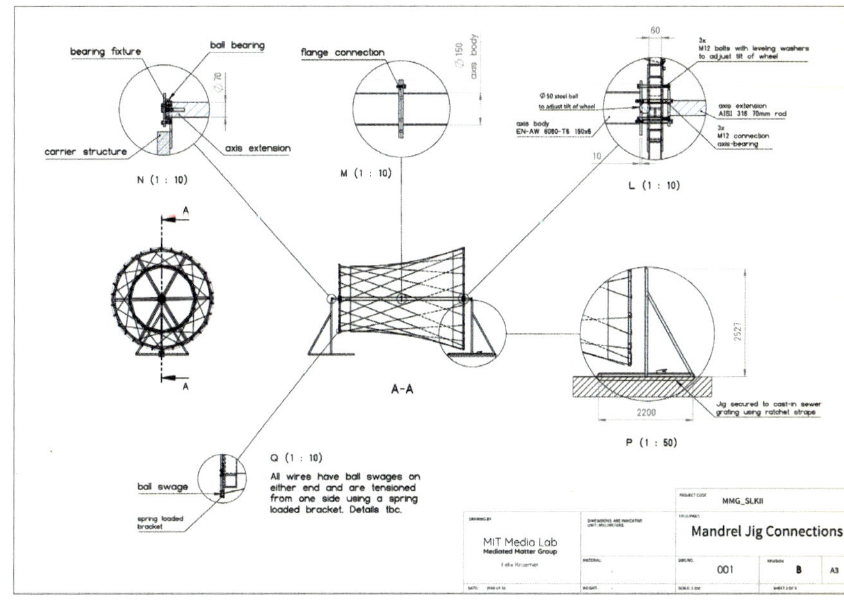

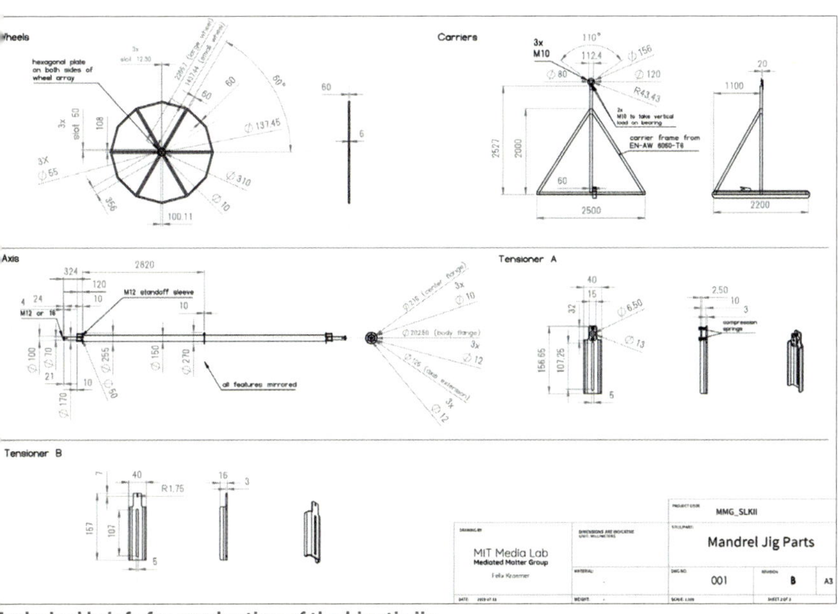

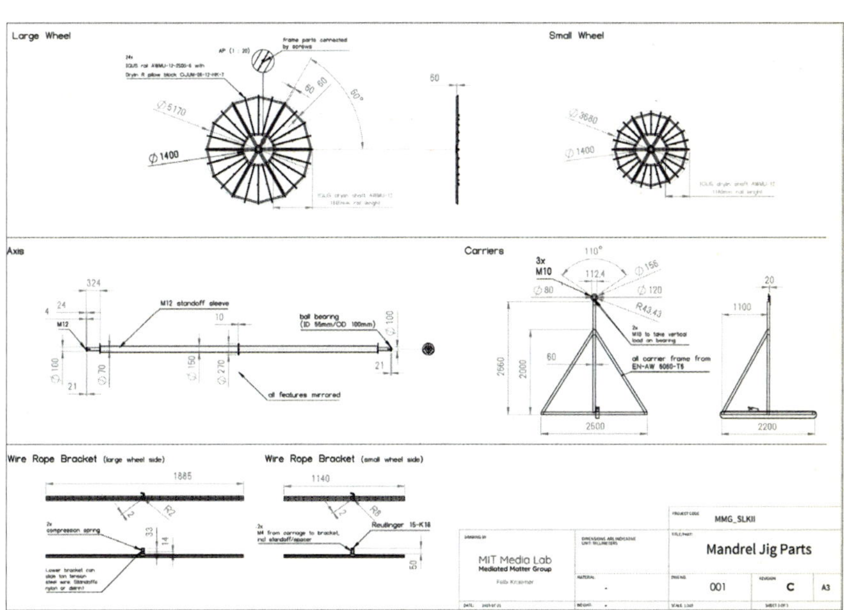

Technical briefs for production of the kinetic jig structure: materials (top), connections (center), and details (bottom).

Silk Pavilion II, developed for the *Material Ecology* exhibition at The Museum of Modern Art in 2020, added kinetic manufacturing to the mix of technological and biological construction methods explored in Silk Pavilion I. Both pavilions bridge the deep disjuncture created by the ways in which the natural is destroyed in service of the made. Silk is traditionally harvested by boiling the cocoon to dissolve the adhesive that attaches the silk to the layers below, thus allowing a single strand to be unrolled. This process, of course, kills the larva, disrupting the development of the organism. The pavilions bypass this problem by using structures designed to influence silkworms to spin in sheets rather than cocoons. Thus this crew of small, remarkable insects became a crew of construction workers and designers, collaborating with human- and robot-made forms.

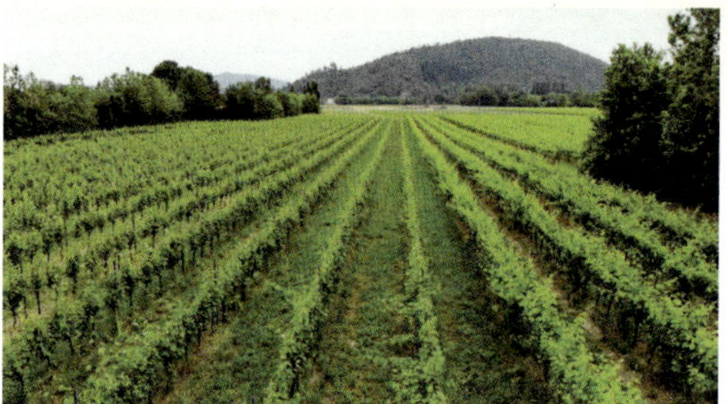

The farm primarily produces organic wine. The silkworm-rearing facilities are part of an effort in Northern Italy to expand the silk industry.

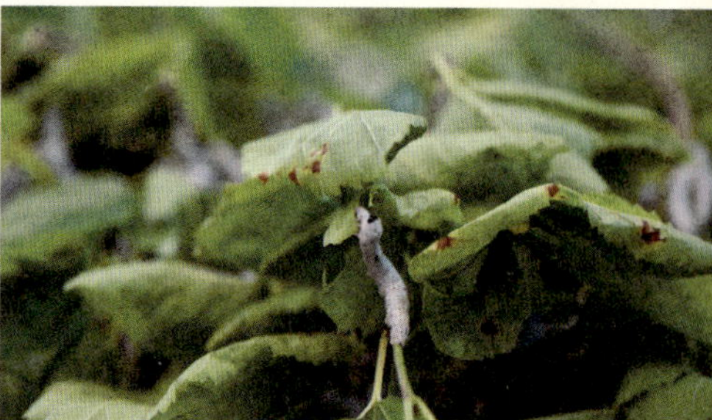

The space at the facility was maintained at 71.6–77 degrees Fahrenheit (22–25 degrees Celsius) and approximately 70 percent relative humidity. The worms were kept in a bed of mulberry leaves and were ready to spin a few days after arrival.

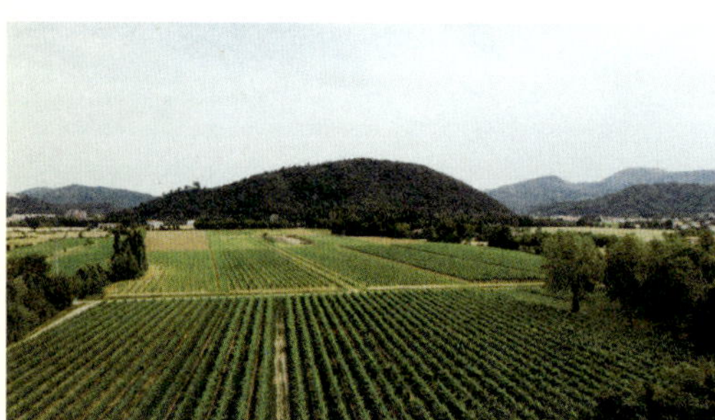

The outskirts of Abano Terme, Italy, the location of the farm where the silkworms for Silk Pavilion II were raised, with the Euganean Hills in the background.

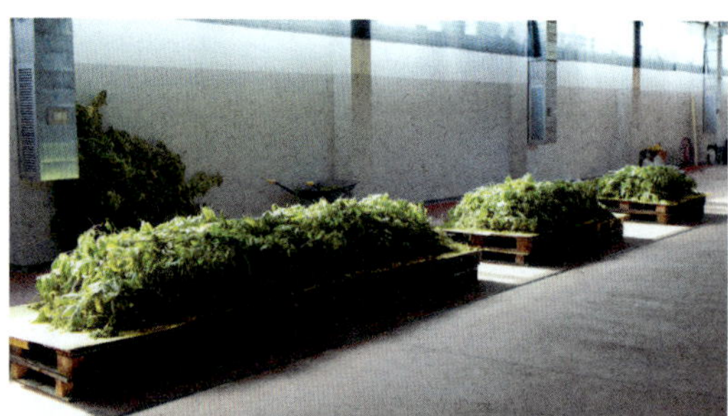

The silkworms were in a late larval stage when they arrived at the farm, where they were fed fresh mulberry leaves cut from trees in a Teolo farm facility.

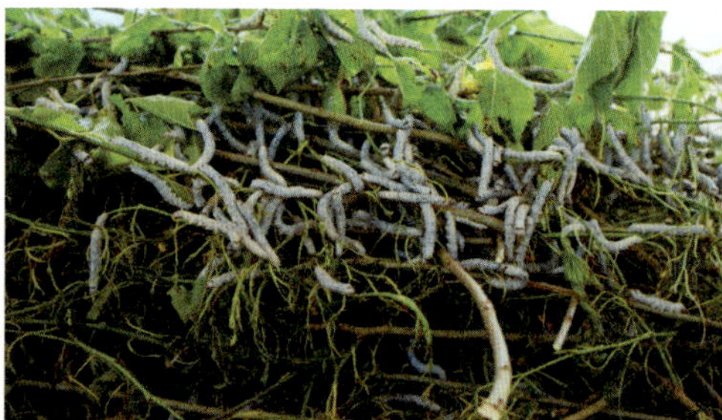

When a silkworm is ready, it climbs upward, looking for the ideal height and the right spot to start spinning.

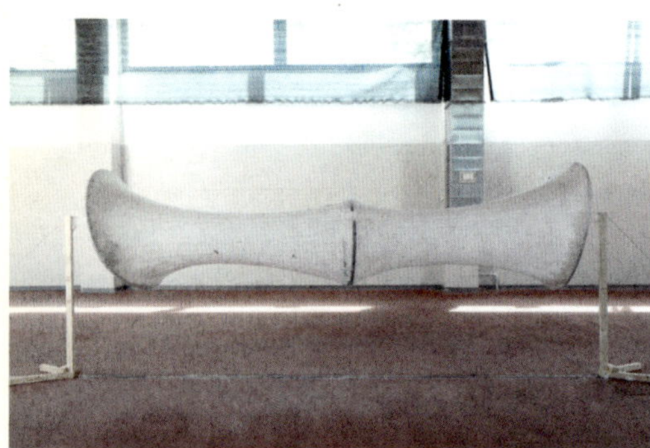

Rotating mandrel prototype used to investigate the upward path of the silkworms on a three-dimensional rotating form.

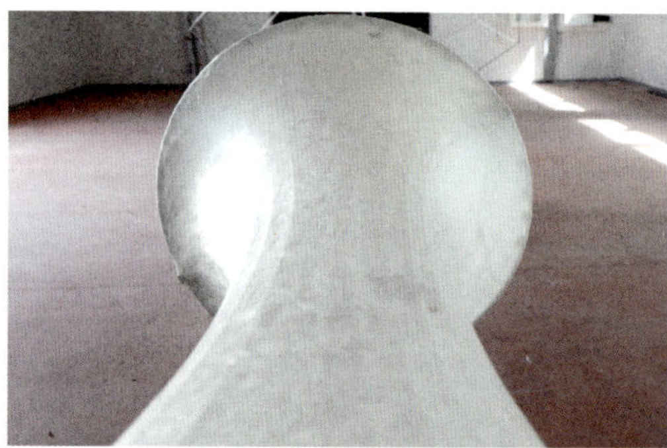

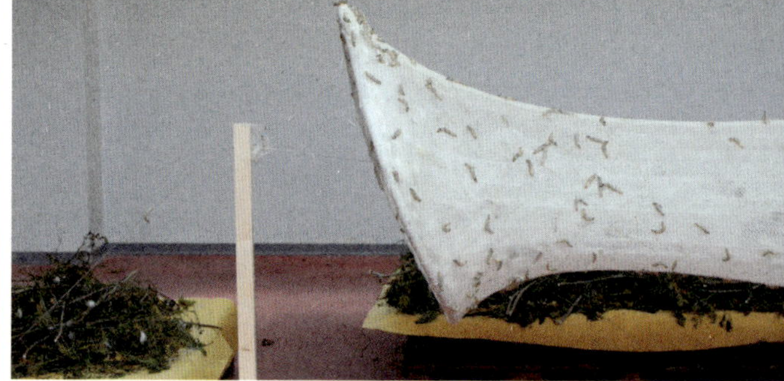

In a departure from the materials and geometry of Silk Pavilion I, the second version had three interrelated layers. The primary, innermost structure was made of braided stainless steel wire ropes. The secondary structure was two-dimensional: a surface of water-soluble knit fabric that served as a scaffolding on which the silkworms were positioned. The tertiary three-dimensional structure was spun by more than 17,000 silkworms in Abano Terme, in Teolo, near Padua, Italy, at one of the largest silkworm-rearing facilities in Europe. We turned to this well-equipped European facility as the source of our animal collaborators for several reasons, one of which was that the silk industry has never quite developed in the United States the way it has in Italy or China. The result is that most of the few silkworm-rearing facilities in the U.S. raise the worms as food for reptiles or as classroom specimens, and therefore in fairly small batches; a considerable amount of labor is required to care for a large number of worms. In addition, nuclear polyhedrosis virus (BmNPV) continues to plague *Bombyx mori* in the U.S., and mulberry leaves—the worms' sole diet—are difficult to find.

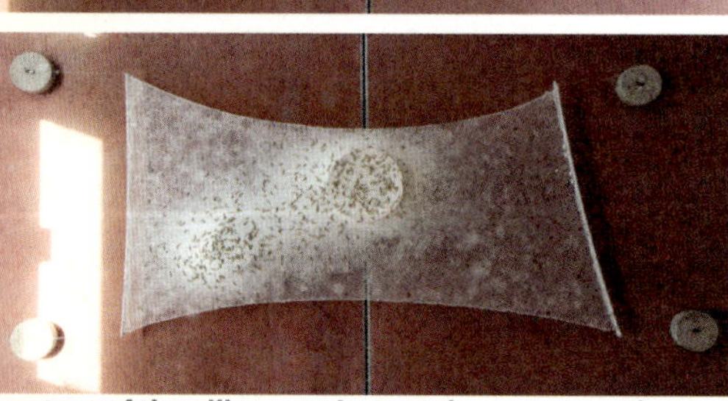

In a test of the silkworms' upward movement, they were placed on knitted mesh that had been stretched over elevated hoops to create uneven surfaces. Over the course of five days, the silkworms gravitated toward the structure's higher points, where they deposited denser layers of silk.

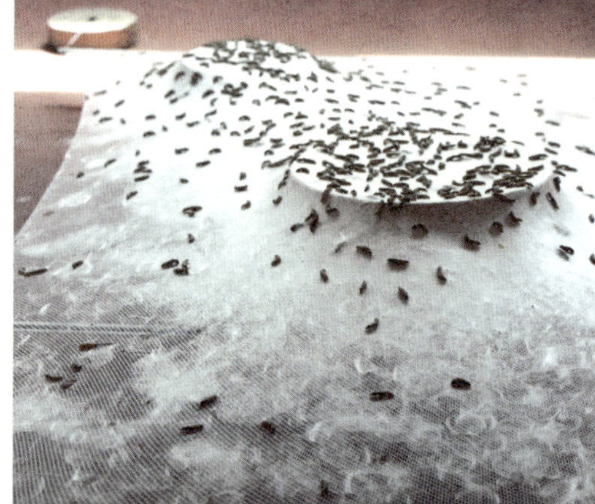

Worms spinning on an elevated hoop prototype.

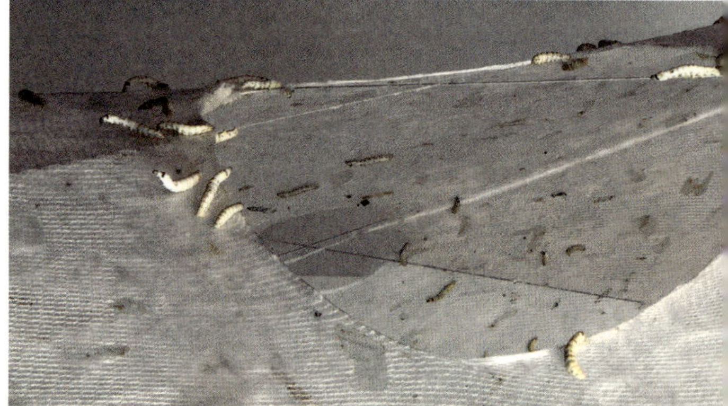

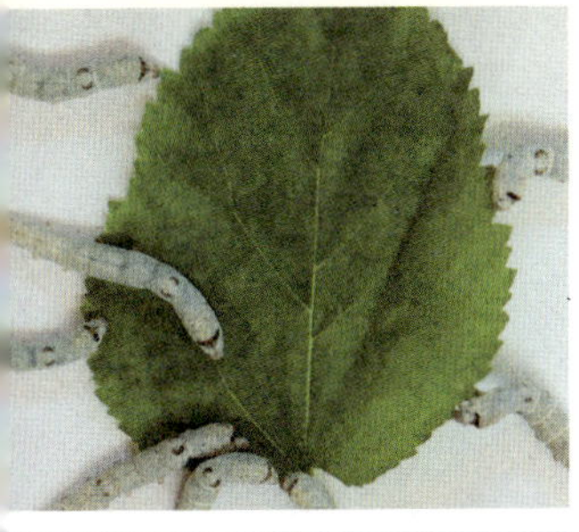

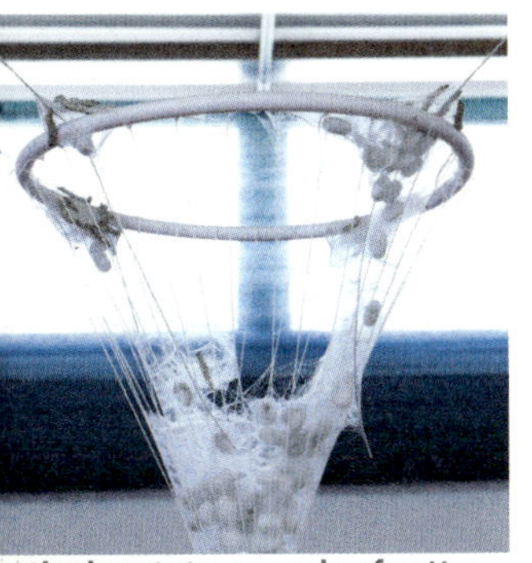

ertical prototype made of cotton wine. The tests on this structure emonstrated the need for spatial onstraints—and therefore rchitectural design—without hich the silkworms would spin ocoons instead of patches.

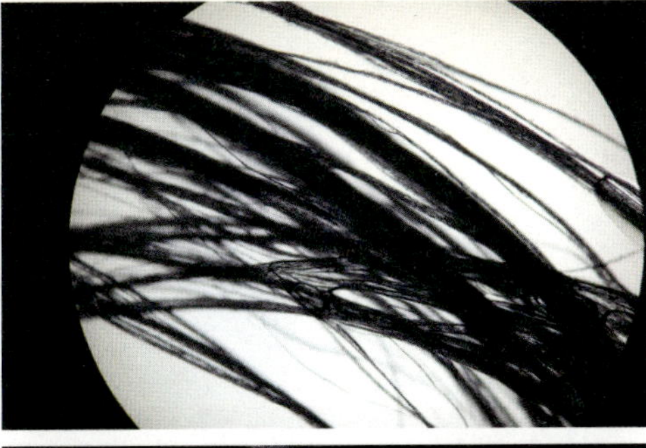

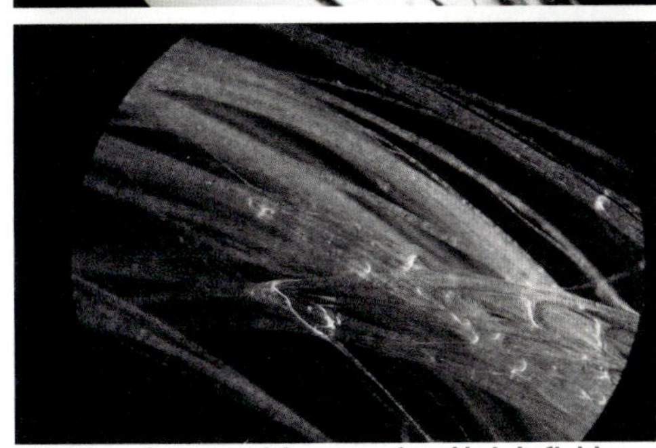

Silk strands seen with transmitted brightfield microscopy.

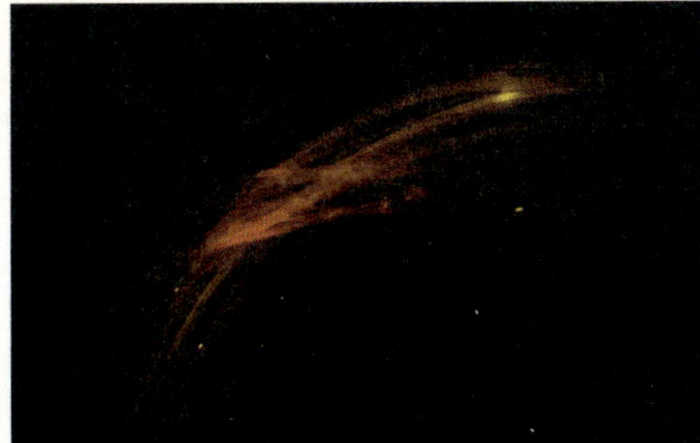

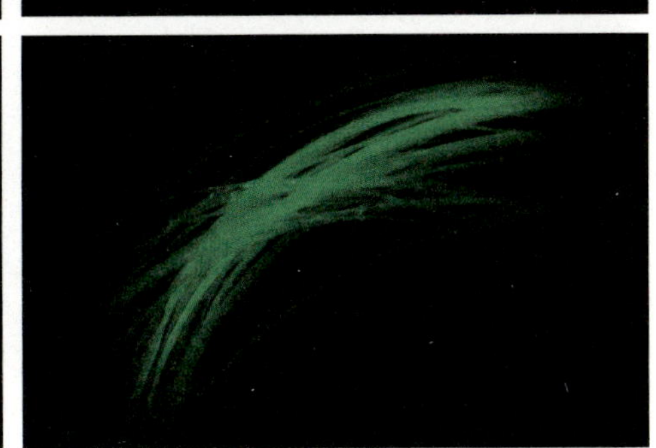

Silk strands seen with fluorescence microscopy.

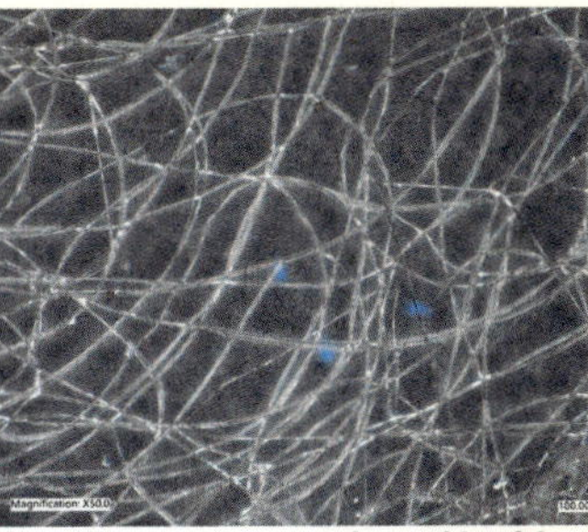

The silkworms spun thread in thin layers, creating a fine mesh.

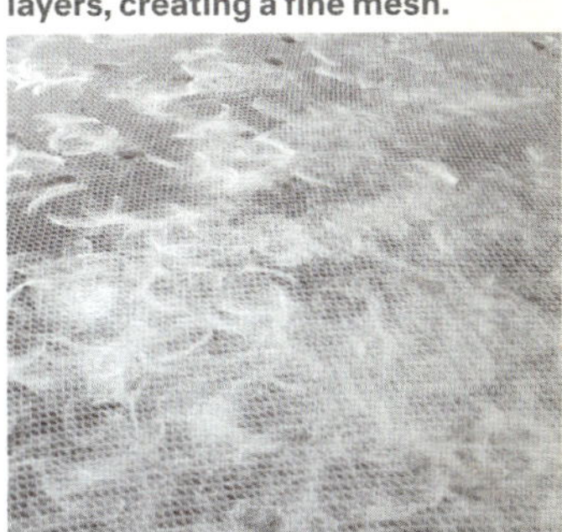

For ten days, the worms spun enough silk to reach from the surface of the Earth to the center and back again. They moved horizontally over the underlying structures, their upward spinning motion helped by a rotating mandrel driven by a top-down kinetic motor. The density of the resulting silk layer varied depending on the environmental factors affecting different parts of the structure; heat and light influenced the worms' movement, as did the topology of the kinetic hyperboloid. A chemical reaction between the silkworms' excretions and the fabric created holes in the knit layer, which released some of the structure's tensile stress and created a metabolic canvas of organic waste. Thus were structural forces influenced biochemically, suggesting broader possibilities in the synthesis of the natural and the designed.

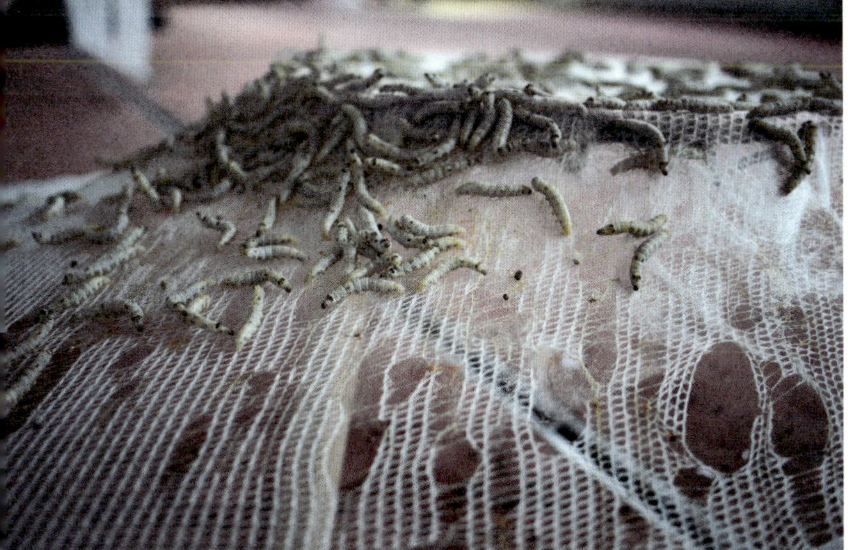

he mesh on the scaffold structure was spun with a custom-designed nit that dissolved on contact with the silkworms' liquid excretions. he worms' last excretion before spinning created holes and spaces ney then filled in with silk.

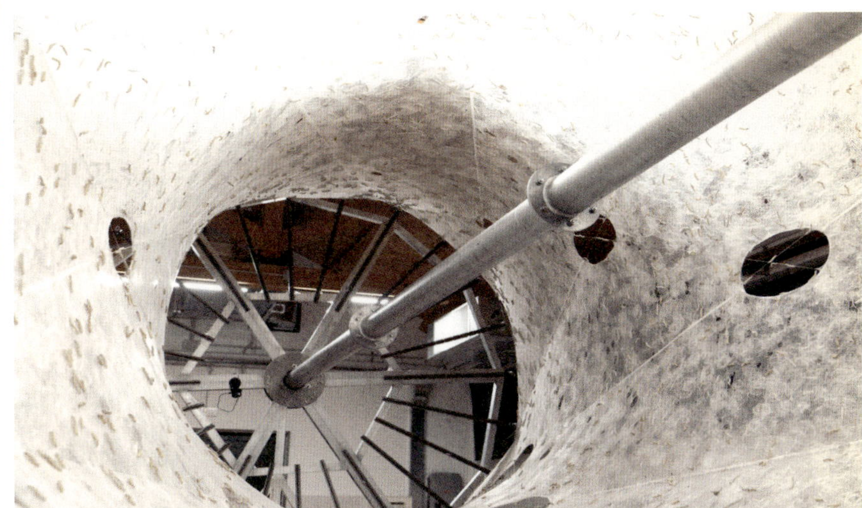

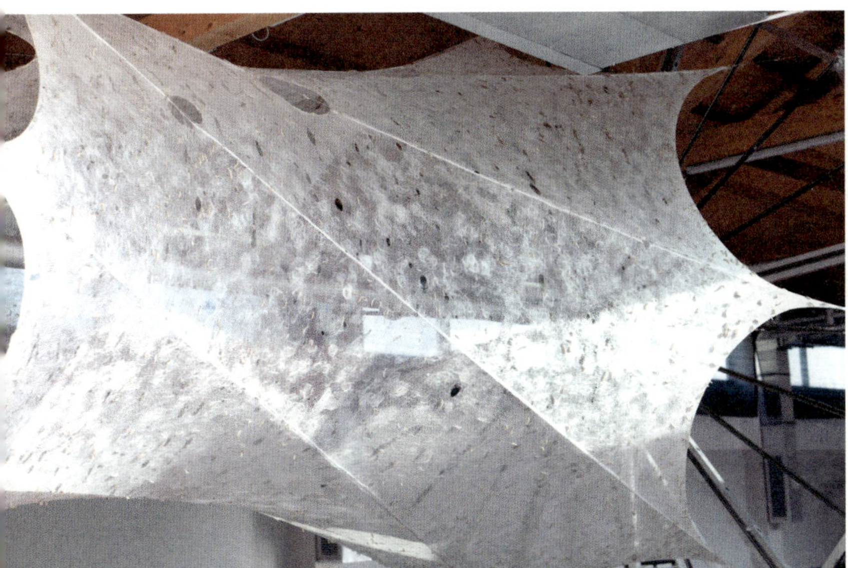

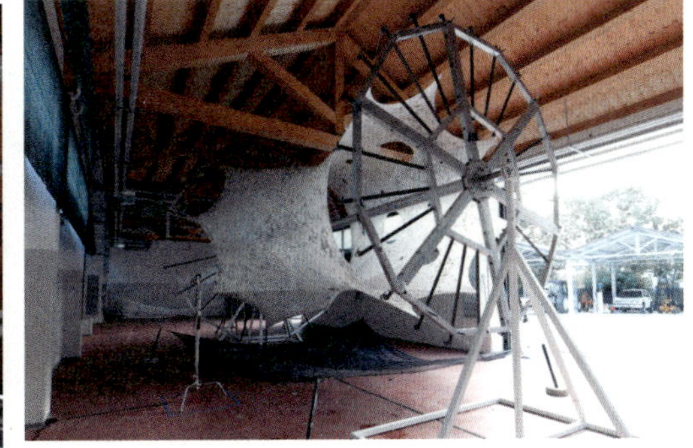

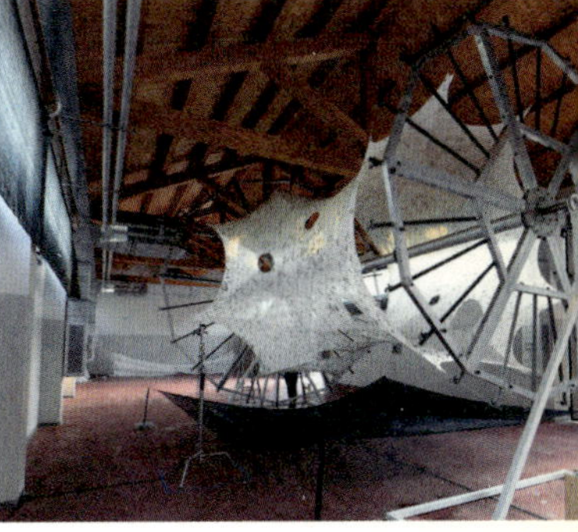

Kinetic jig structure at the silkworm-rearing facility in Abano Terme. For Silk Pavilion II, 17,532 silkworms were placed on the structure over a ten-day period.

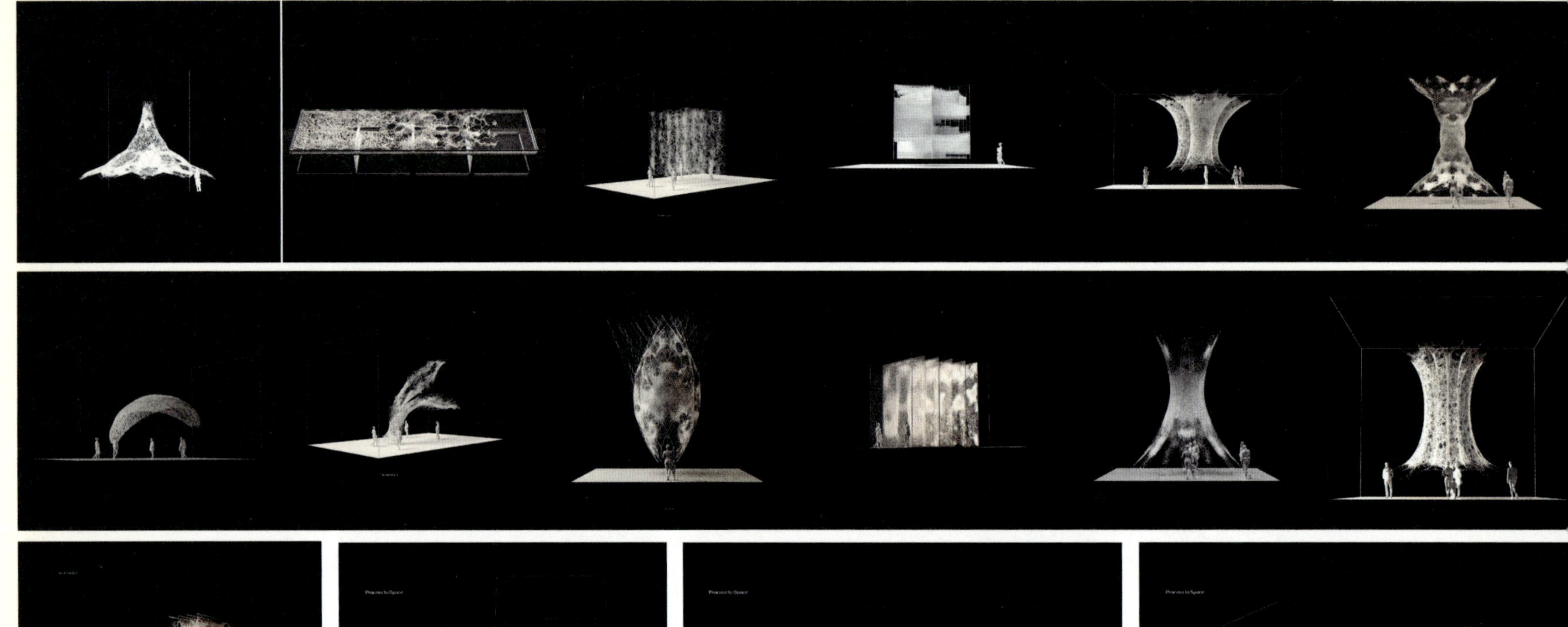

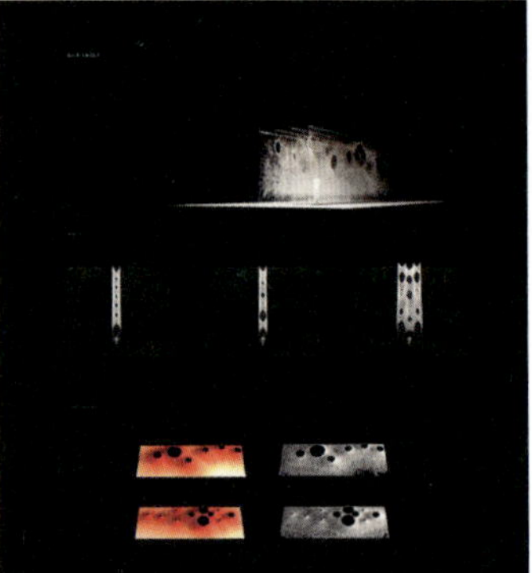

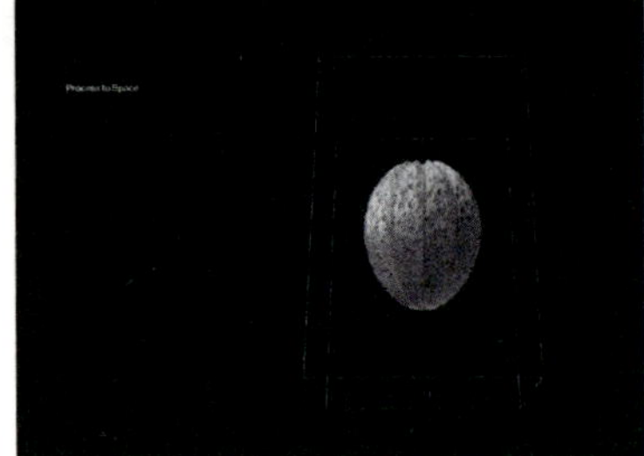

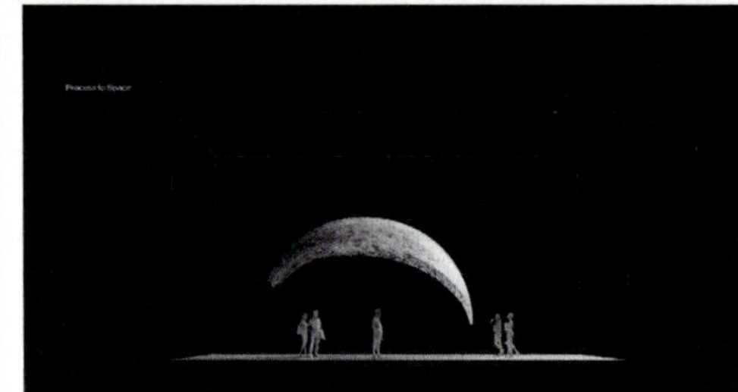

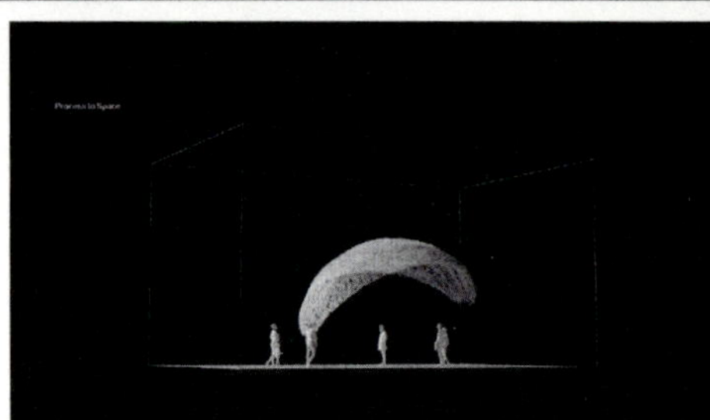

Conceptual design iterations of the pavilion for MoMA's 1 North gallery, testing different tensile, wire, and tessellated structures.

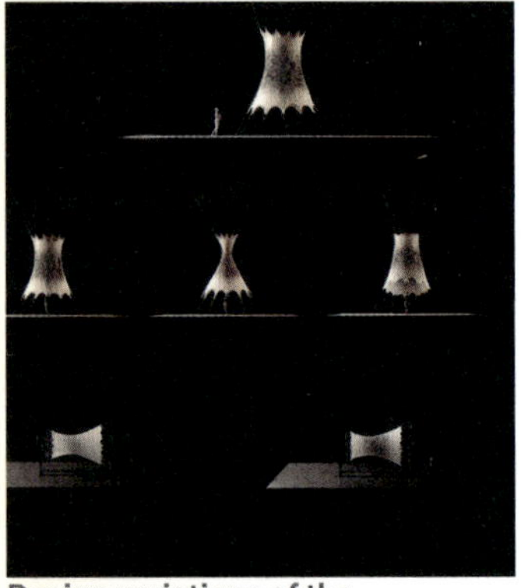

Design variations of the hyperboloid structure, with silk-production rig.

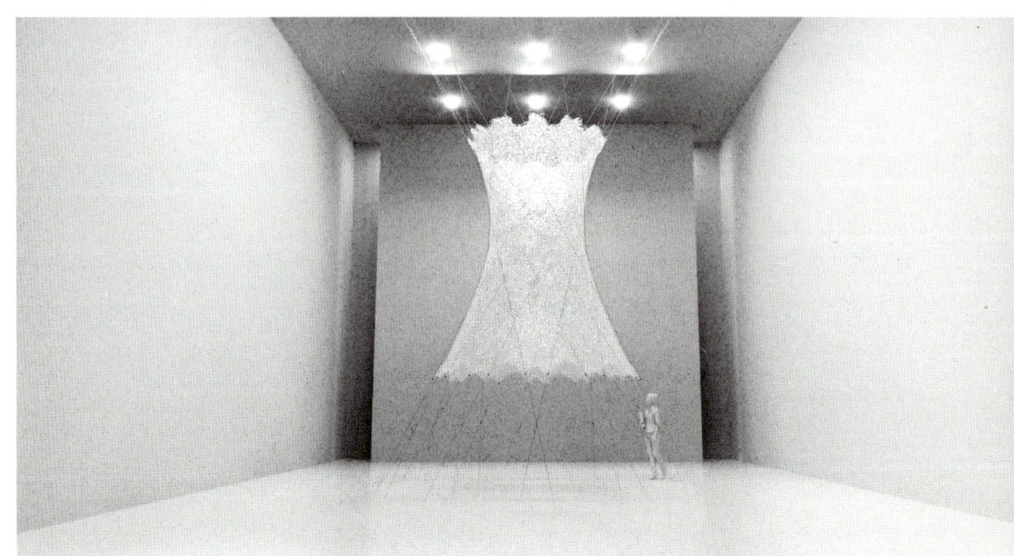

Early rendering for MoMA's 1 North gallery.

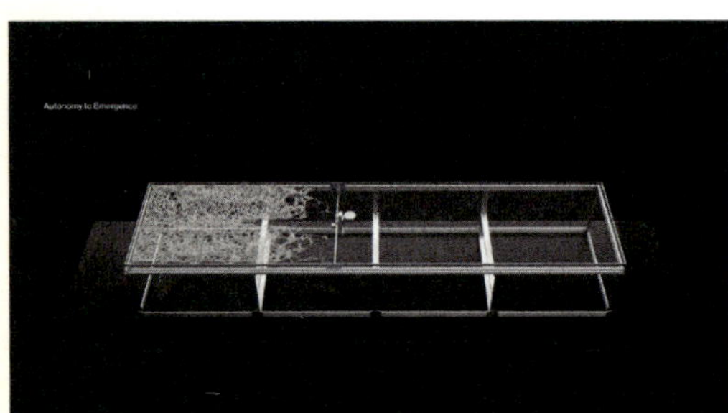

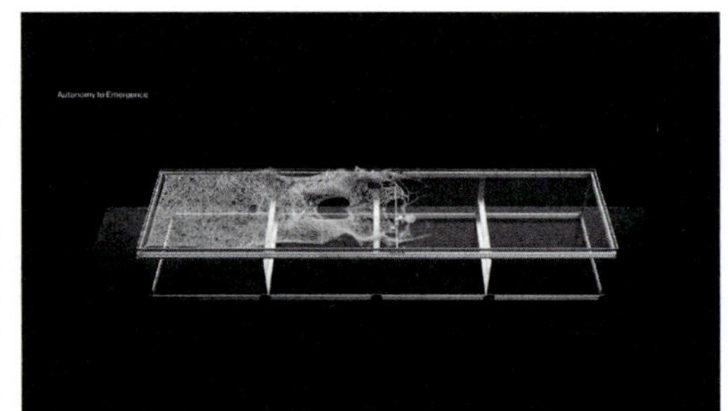

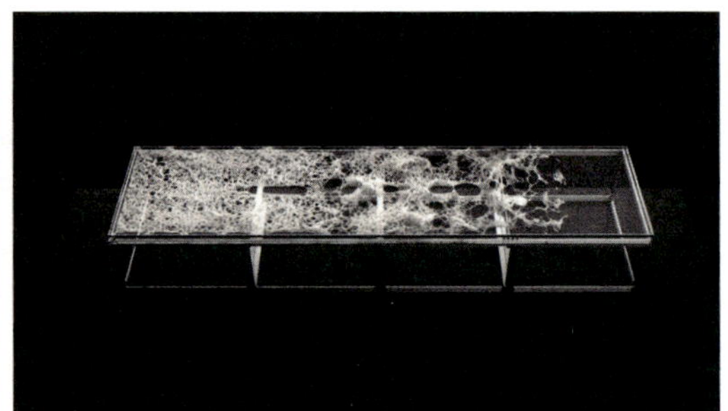

Conceptual rendering of a hybrid artificial-natural collaborative system, showing how an intelligent system can develop a self-learned symbiosis by mediating autonomously between fabrication and biological agents. The actions of the robotic system influence those of the biological organisms, which the robotic system then models, maps, and memorizes. The result is a new techno-organic system designed to take information from an emergent system and self-construct a final architecture.

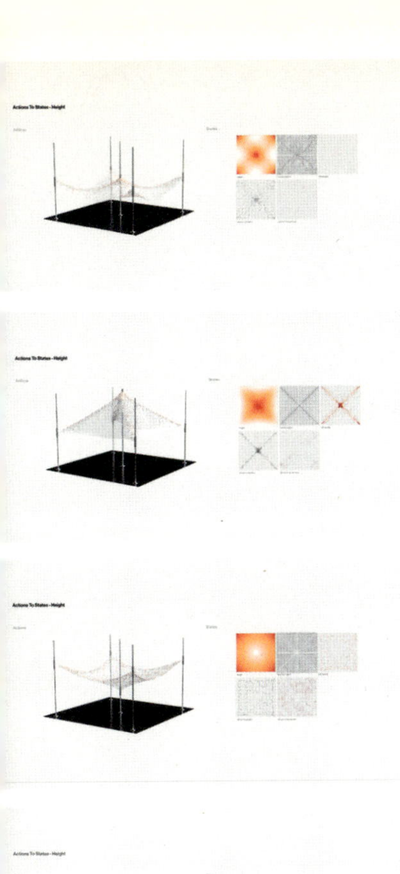

Conceptual renderings of a digital fabrication system designed to enable forms of collaboration between hardware (3D gantry), software (algorithms), and wetware (silkworms and other living organisms). These design studies explore synergies between hard and soft control and top-down and bottom-up construction; they also question the prevailing notion that design artifacts are the outcome of linear and streamlined systems. In process-based interactions, by contrast, fabrication tools provide a set of structural and environmental conditions for the organisms deployed; in turn, the organisms generate feedback that informs the environmental conditions generated by the fabrication tools. The fabrication system and the organism(s) combine to contribute to new forms of fabrication "intelligence."

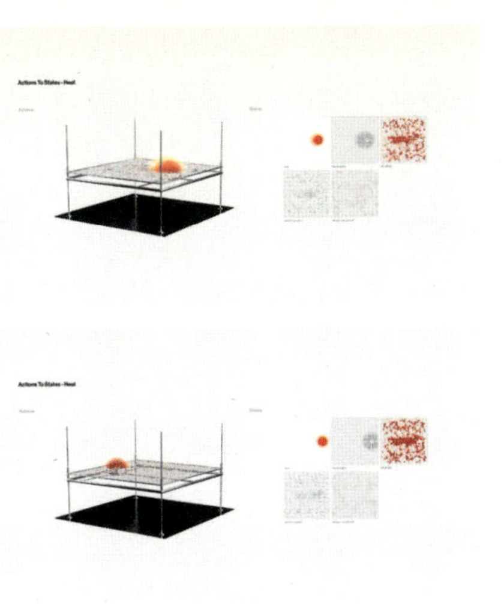

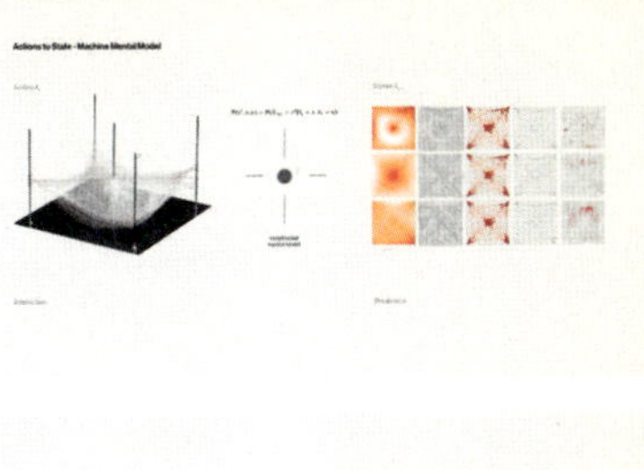

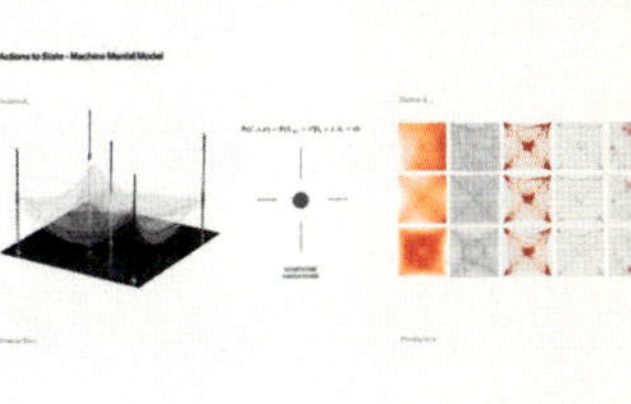

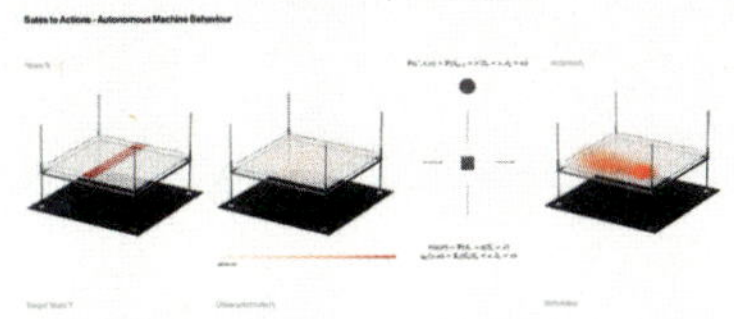

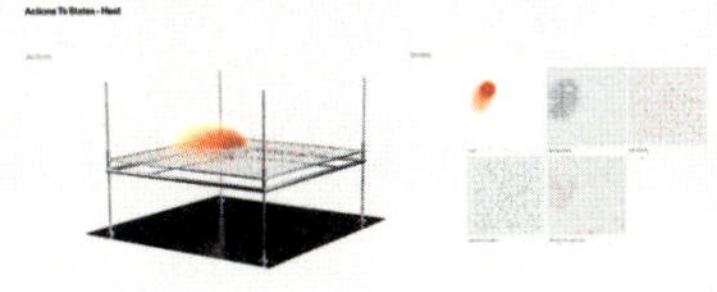

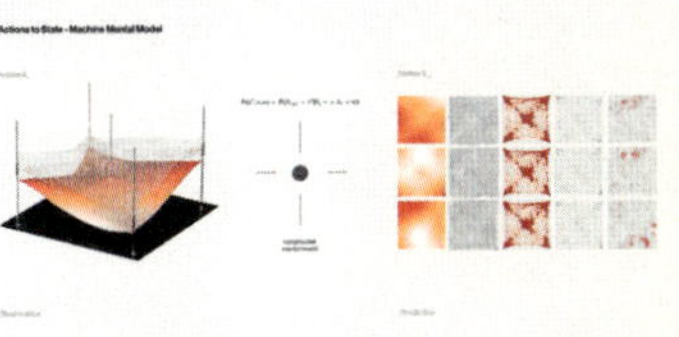

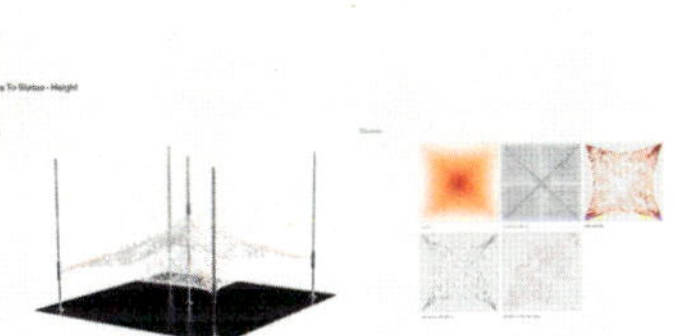

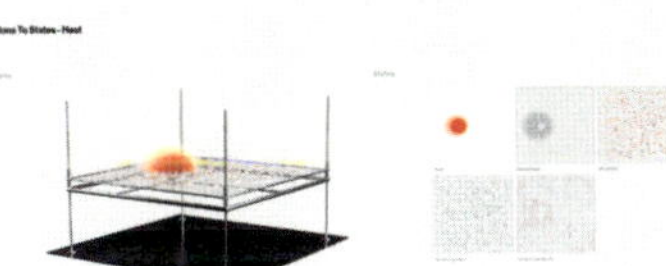

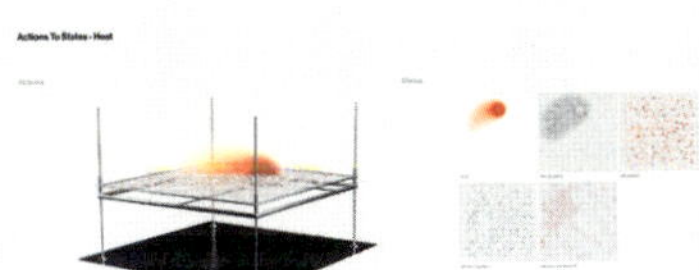

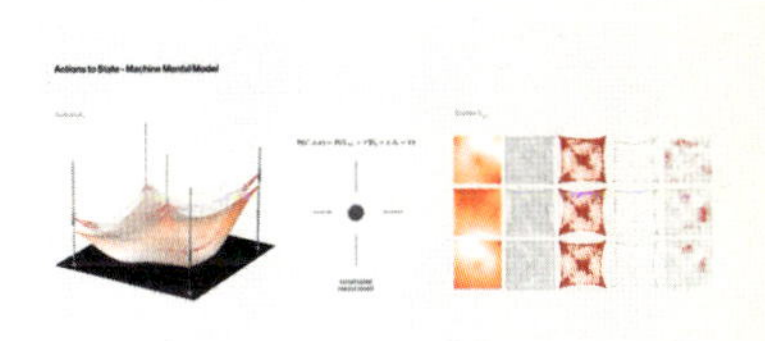

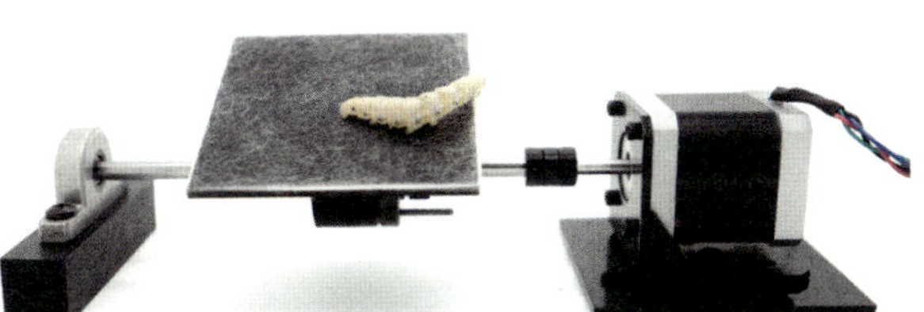

rotating mechanism changed the angles of the spinning urface to test the silkworms' spinning behavior in motion.

n example of heat templating used to uide spinning trajectories: map of heating nodules mounted under a spinning surface eft); overhead view of the fiber distribution n the heat-templated platform (right).

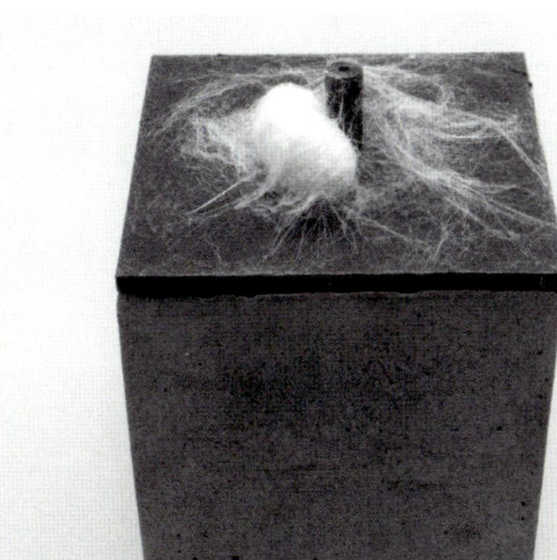

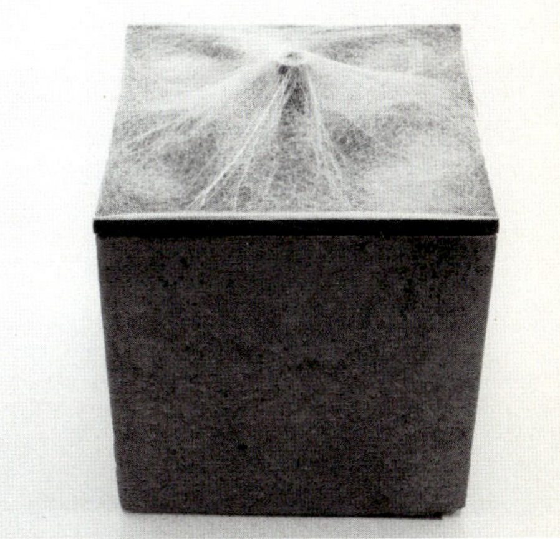

Investigation into the structural height at which a silkworm will stop producing three-dimensional cocoons. When the central element in the setup was less than 13/16 inch (21 millimeters) tall, the silkworm generated flat sheets; when the central element was taller, the worm spun a cocoon.

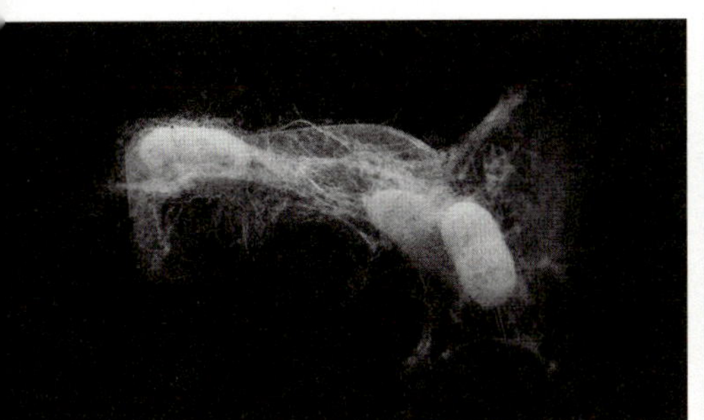

tudy of the aggregation and nterconnection of cocoons.

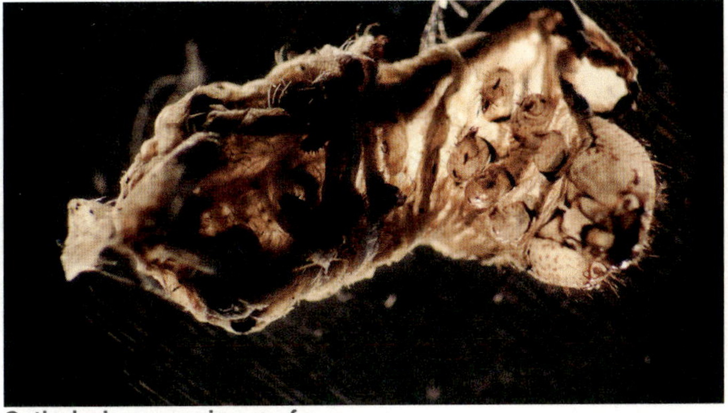

Optical microscope image of a silkworm's shed skin. As worms grow over their larval period, they shed and renew their skin four times.

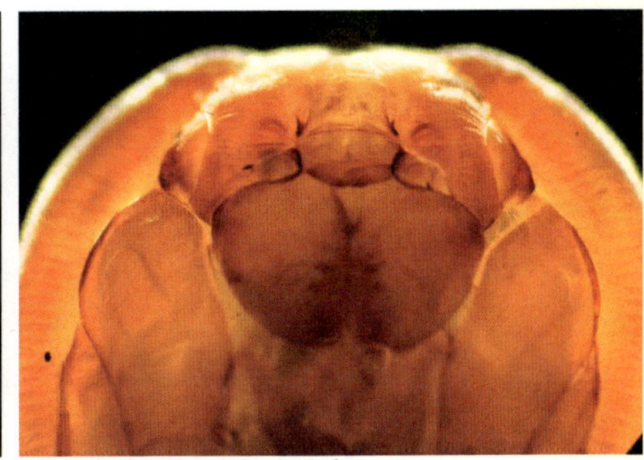

Optical microscope image of a pupating silkworm, the stage in which it metamorphoses.

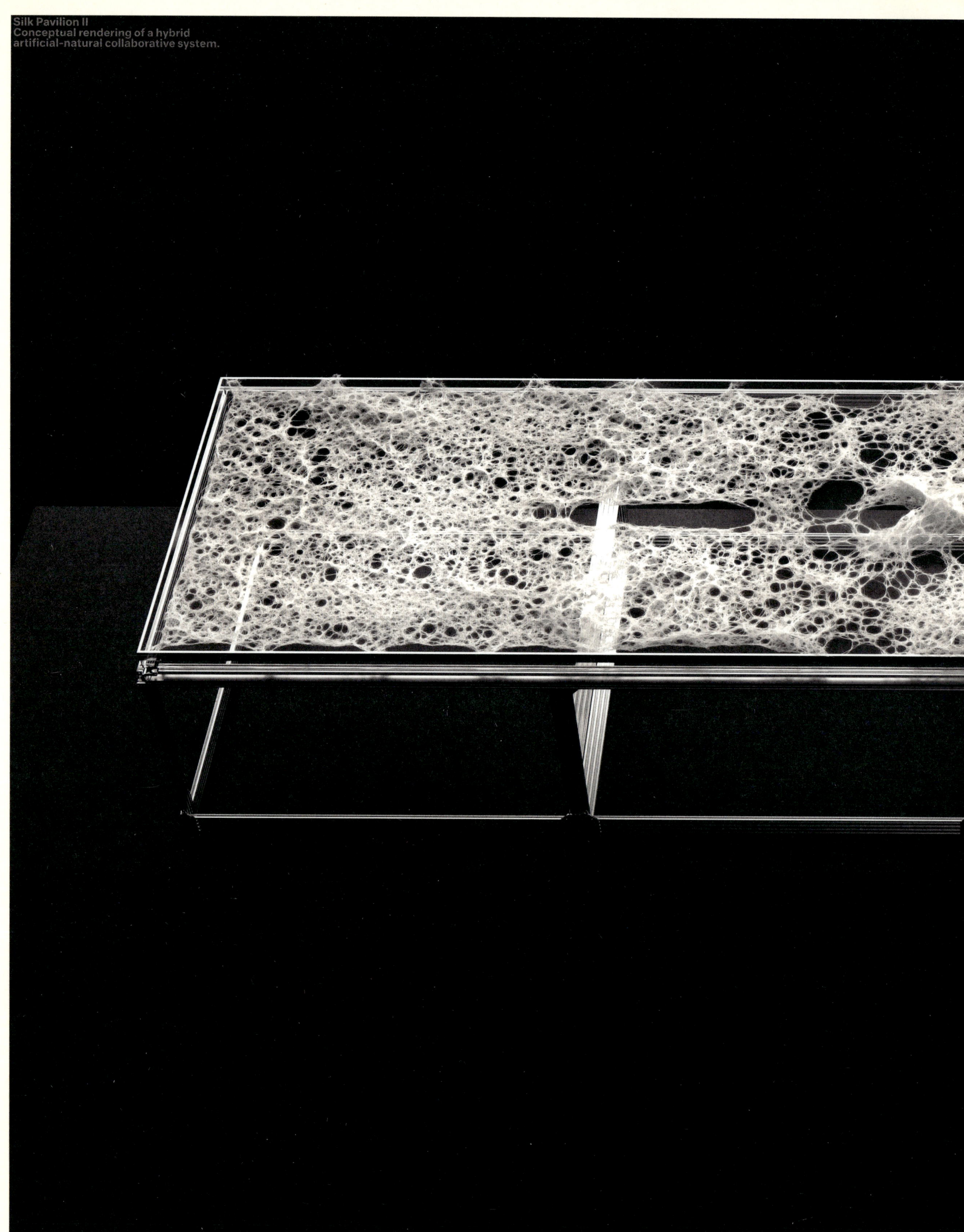

Silk Pavilion II
Conceptual rendering of a hybrid artificial-natural collaborative system.

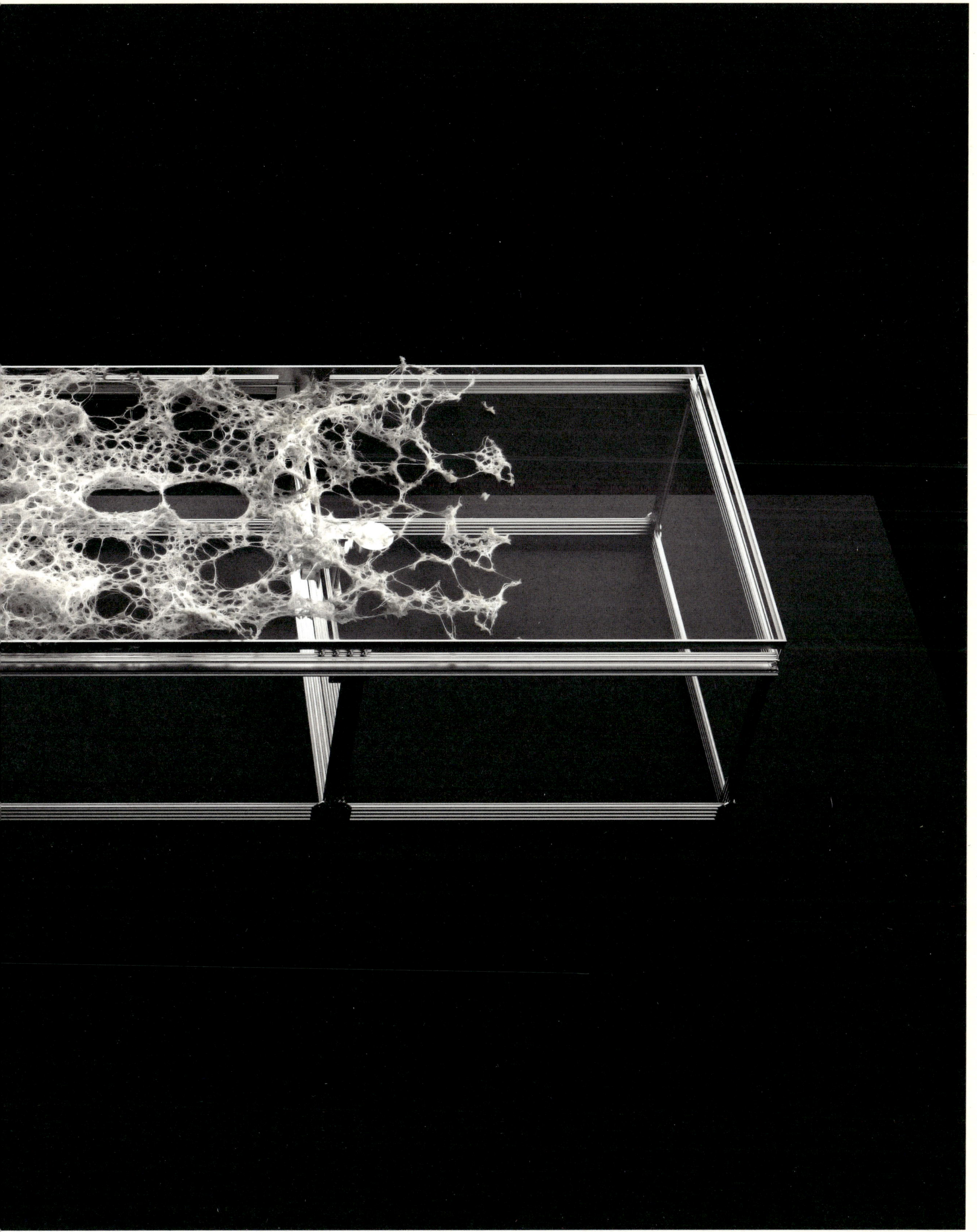

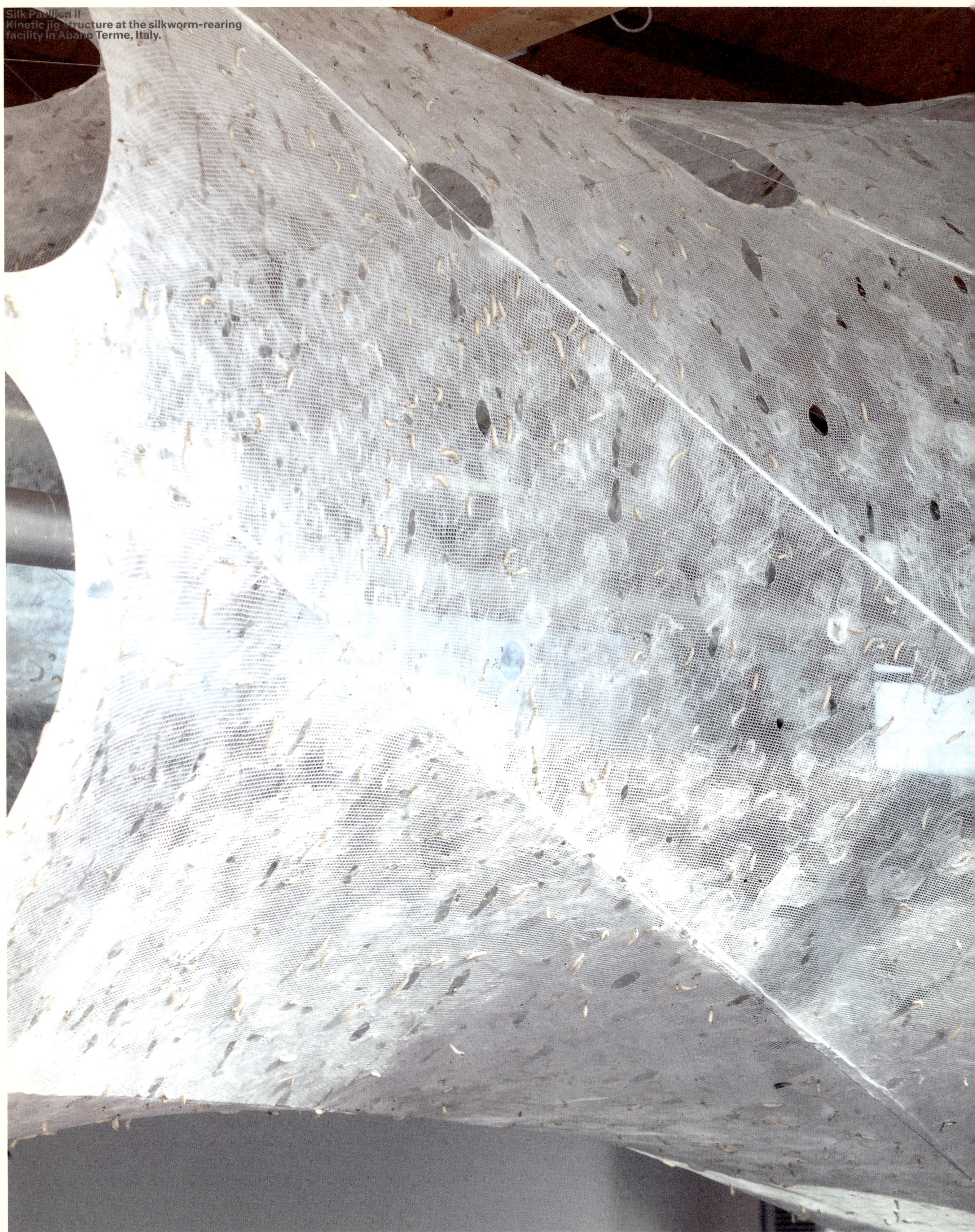

Silk Pavilion II
Kinetic jig structure at the silkworm-rearing facility in Abano Terme, Italy.

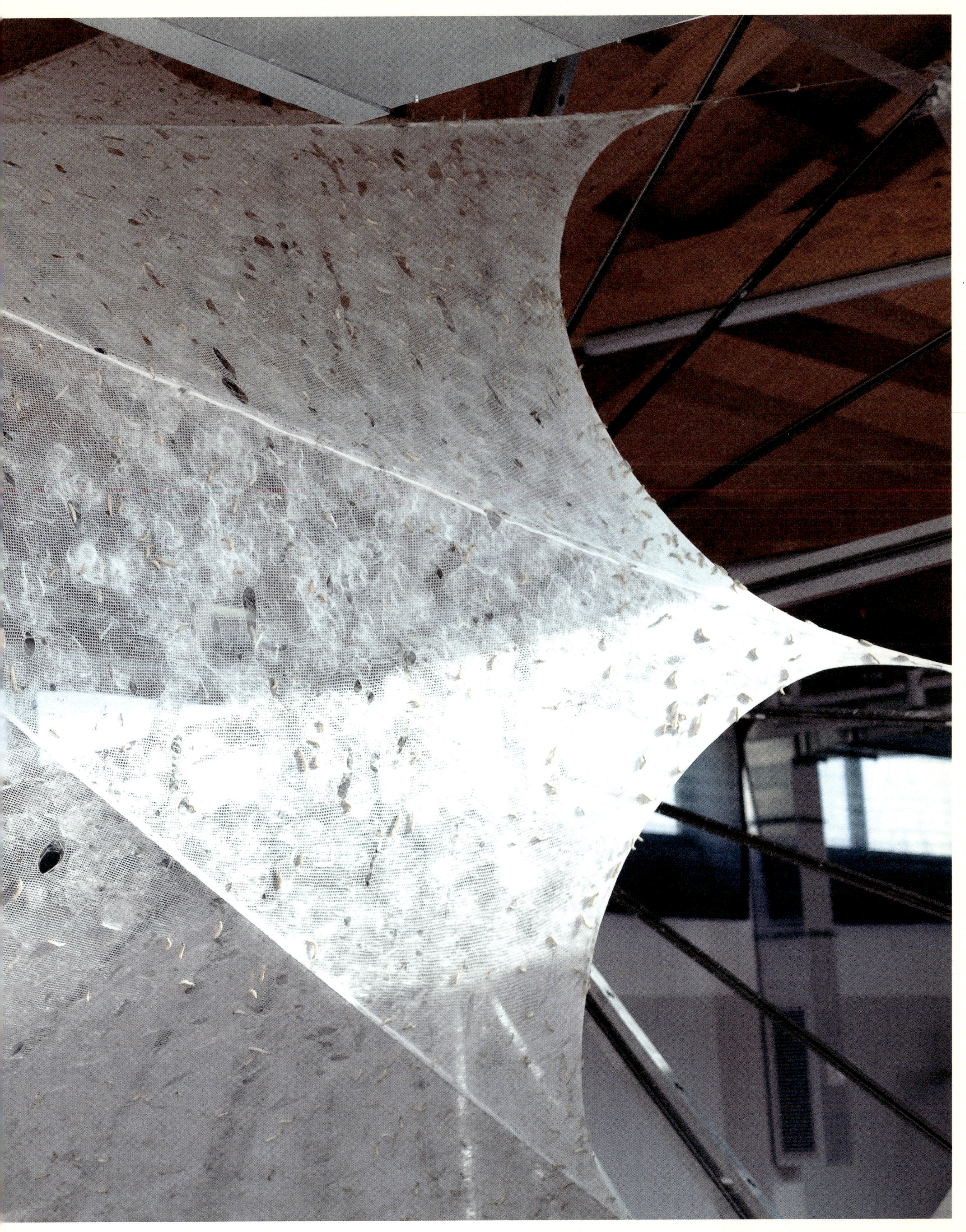

GLASS I

Glass has long played a significant role in the evolution of product and architectural design—there are records of Egyptians using it for decorative objects before 3000 BCE—with its performance and functionality increasing as more effective processing methods, such as blowing, pressing, and forming, have been developed. Through industrialization it has become a ubiquitous and defining feature of the built environment, yet the control and customization of geometrically complex shapes has remained elusive.

Neri Oxman and
The Mediated Matter Group
Glass I. 2015
3D-printed glass
Dimensions variable, up to
7 ⅞ × 7 ⅞ × 5 ⅞ in. (20 × 20 × 15 cm)
An MIT Media Lab project
Research team: John Klein, Michael Stern, Markus Kayser, Chikara Inamura, Giorgia Franchin, Shreya Dave, Daniel Lizardo, Peter Houk, Neri Oxman
Collaborators and contributors: Mary Ann Babula; P. T. Brun; Jeremy Flower; Wyss Institute, Harvard University; Rubix Composites; Skutt Kilns; Glass Art Society; MIT Center for Bits and Atoms; MIT Edgerton Center; MIT Central Machine Shop; MIT Mechanical Engineering Department; MIT Glass Lab

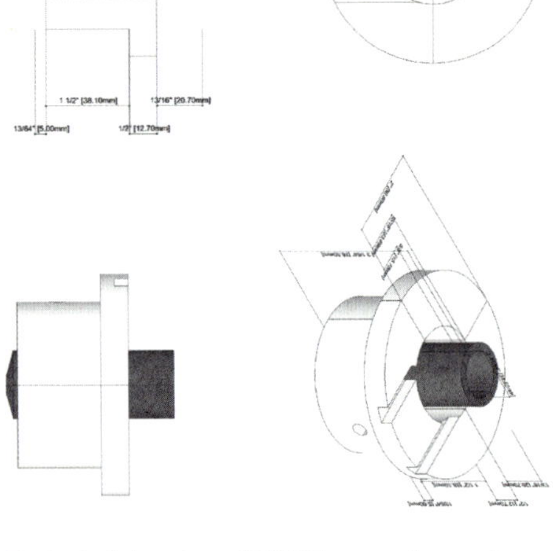
Technical drawing of G3DP's ceramic nozzle.

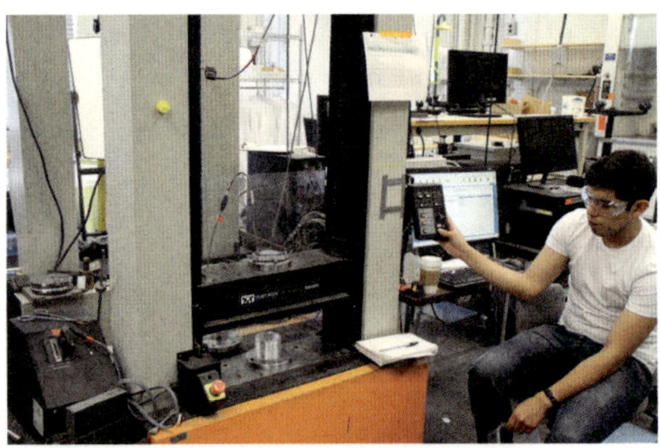

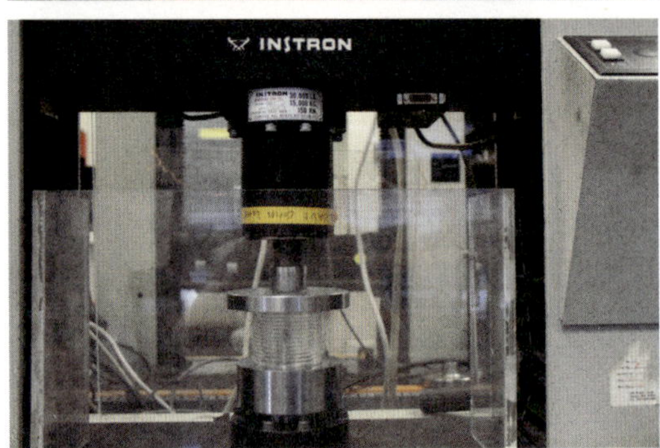

Compression test of a glass sample.

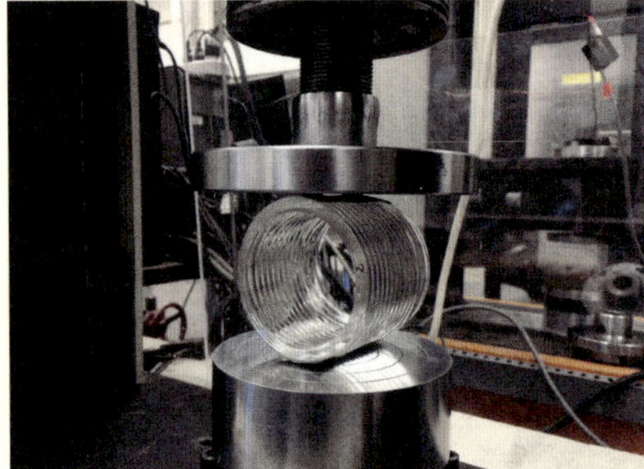
Structural one-point bending test of a cylindrical print.

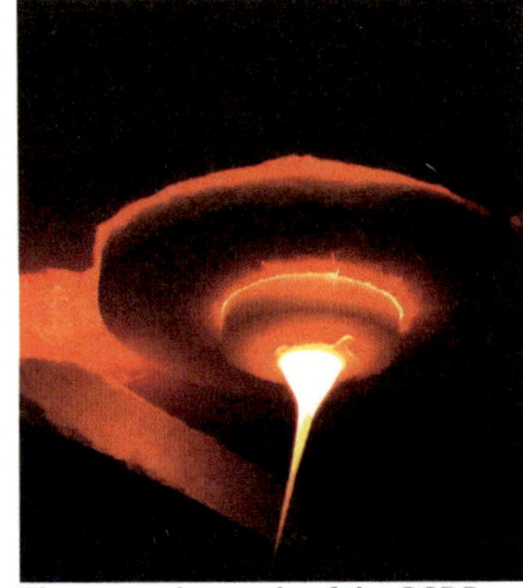
The ceramic nozzle of the G3DP system, coated in molten glass.

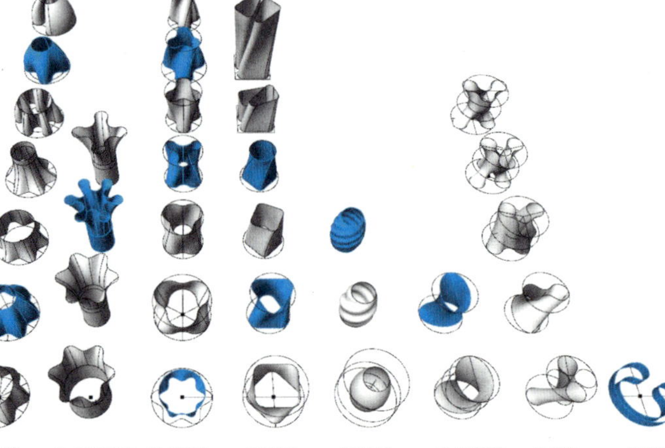

Above: A structural family of experimental objects varying in geometric complexity. Objects shaded in blue went on to be digitally fabricated with the G3DP.

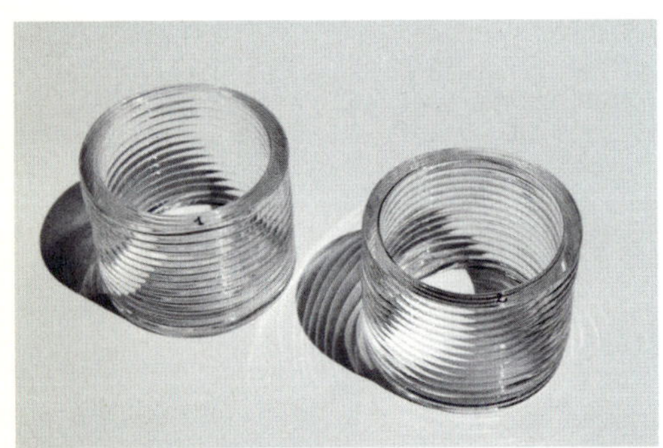
3D-printed glass cylinders.

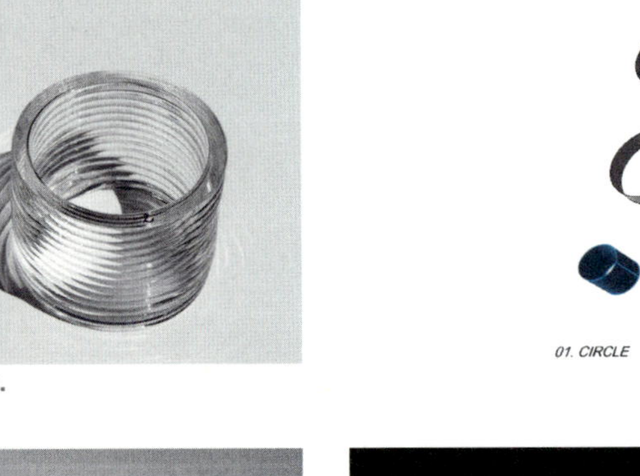

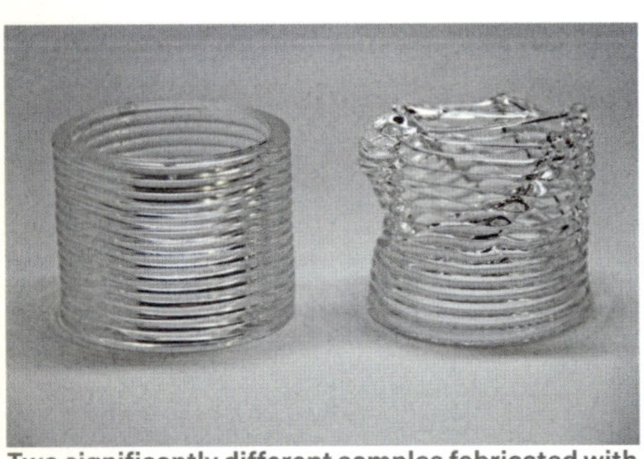
Two significantly different samples fabricated with standard printing parameters: one showing typical results, the other the outcome of atypical system behavior.

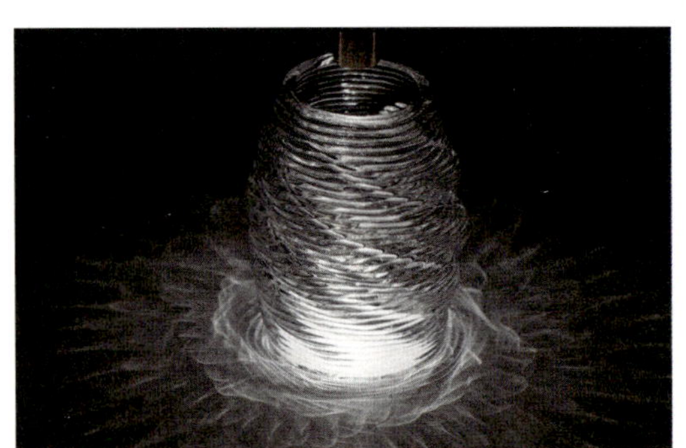
A glass object illuminated from above, generating a complex pattern of caustics (light rays reflected and refracted to create areas of light and shadow).

In 2015, we developed G3DP (or, very simply, Glass 3D Printer), a technology that allowed us to shape the material's optical properties, creating the possibility of glasswork that is functional, technically controlled, and true to glass's physical properties. The printed structures made of extruded layers of molten glass that adhere to each other, have complex variations in shape. G3DP's dual-heating chamber functioned in the upper chamber as a kiln and in the lower chamber as an annealer; the kiln ran at approximately 1,900 degrees Fahrenheit (1,137.8 degrees Celsius).

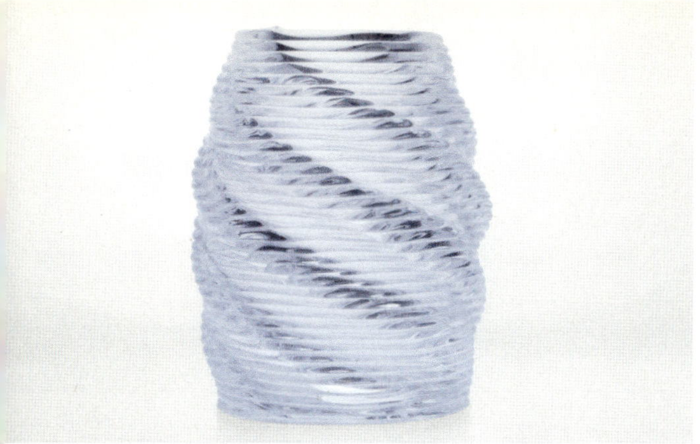

)bject printed to explore changes in concavity/ onvexity out of plane.

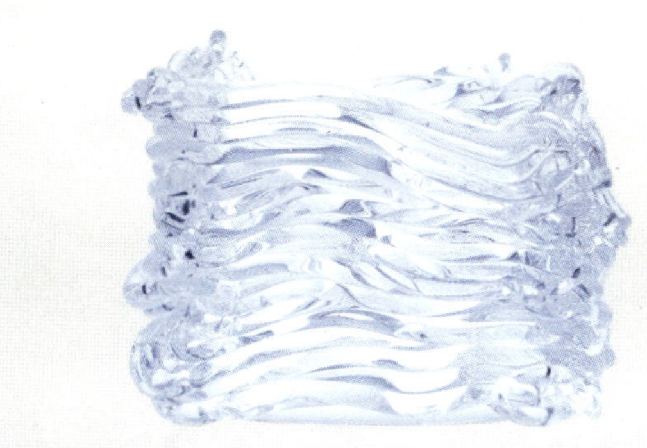

The first object printed during G3DP development, pre-CNC, with the object path still traced out by hand.

G3DP-generated object exploring the effects of curvature.

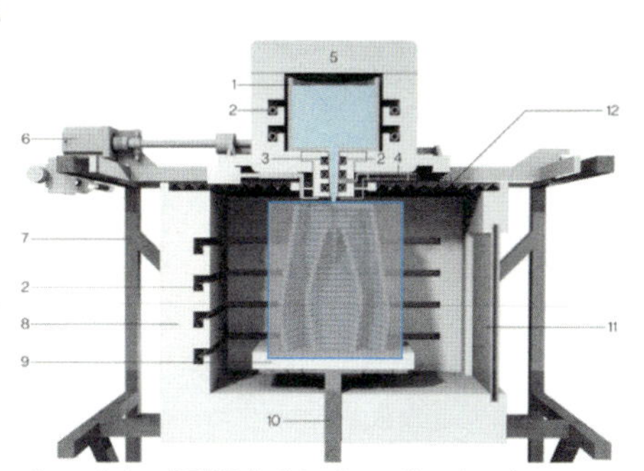

ection of the G3DP (left); glass flowing from he crucible to the printed object (right).

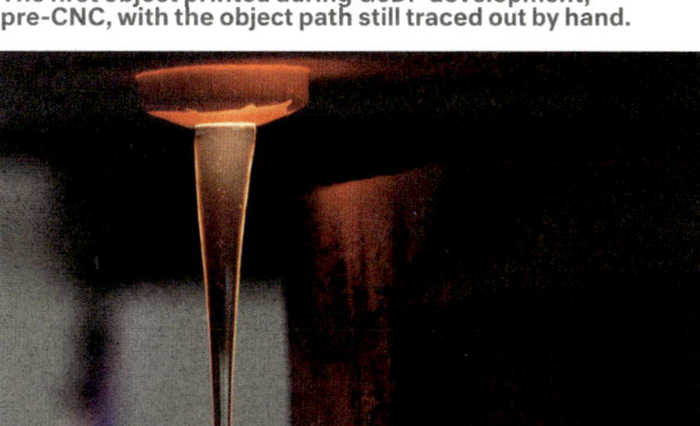

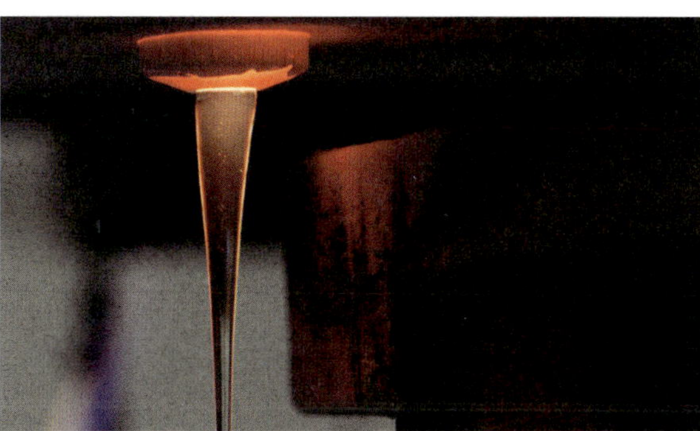

Molten glass beginning to flow from the nozzle of G3DP as a print is initiated.

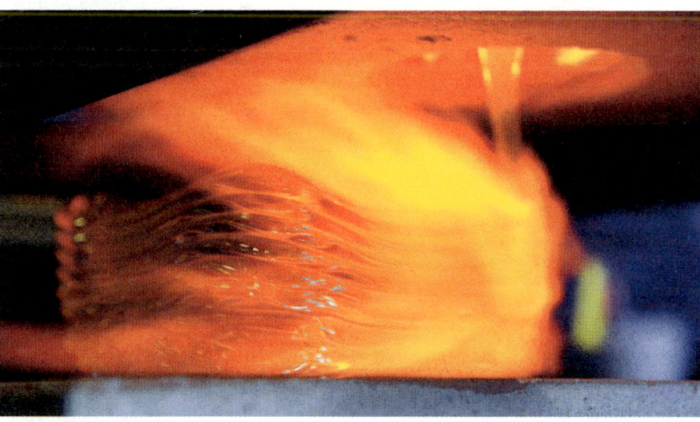

Experimental object generated during G3DP development. The printed object was kept from cracking by manual torching.

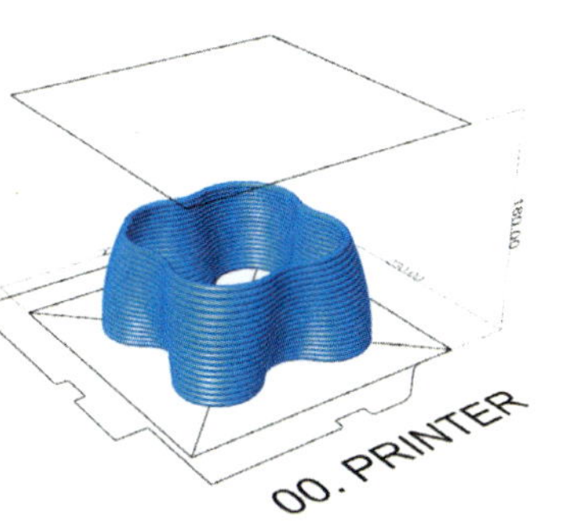

Above: Structure simulated by G-code (a programming language used to control an automated machine) with a shape based on the code generated in the software Rhinoceros with the plug-in Grasshopper.

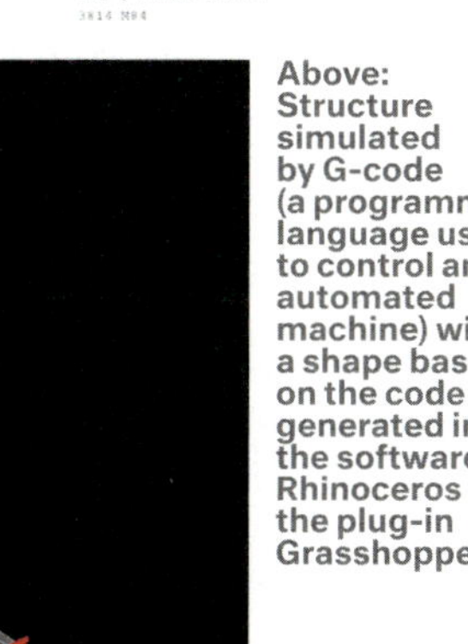

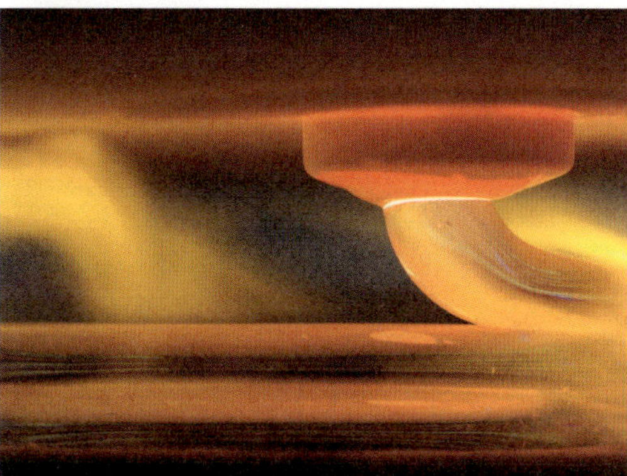

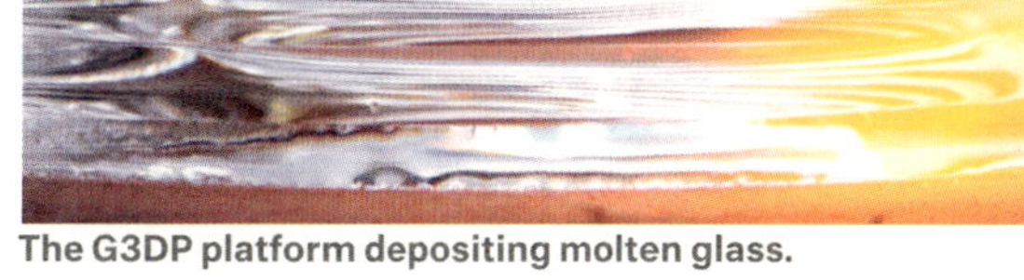

The G3DP platform depositing molten glass.

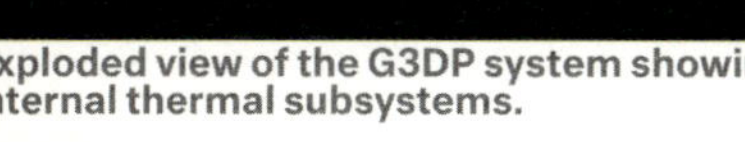

xploded view of the G3DP system showing ıternal thermal subsystems.

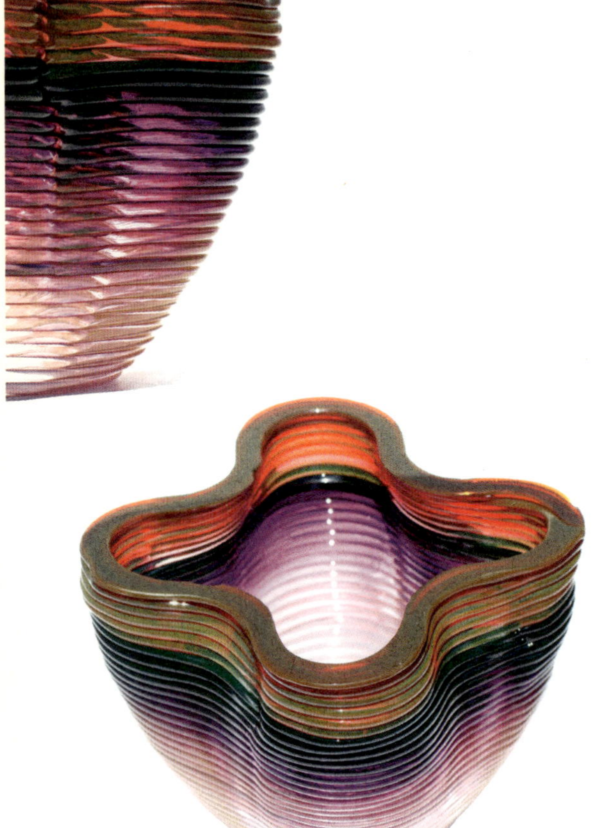

G3DP-generated object exploring the use of color.

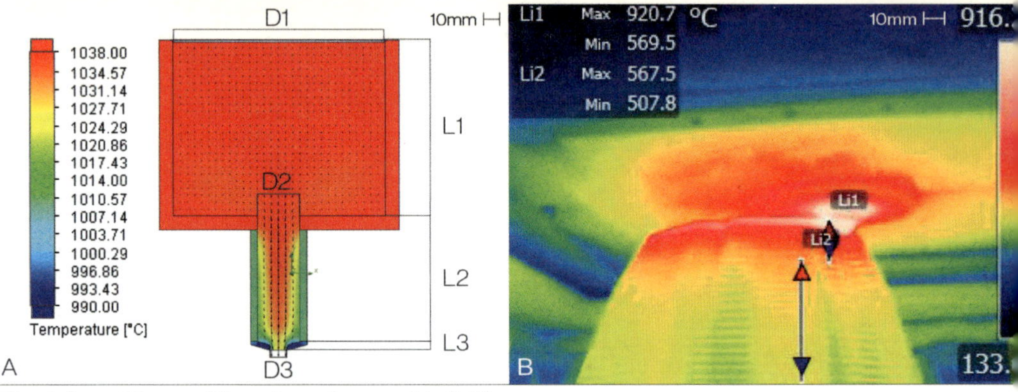

Simulation of the crucible and nozzle assembly (left) showing temperature distribution of the printing platform and thermal image of an object being printed (right).

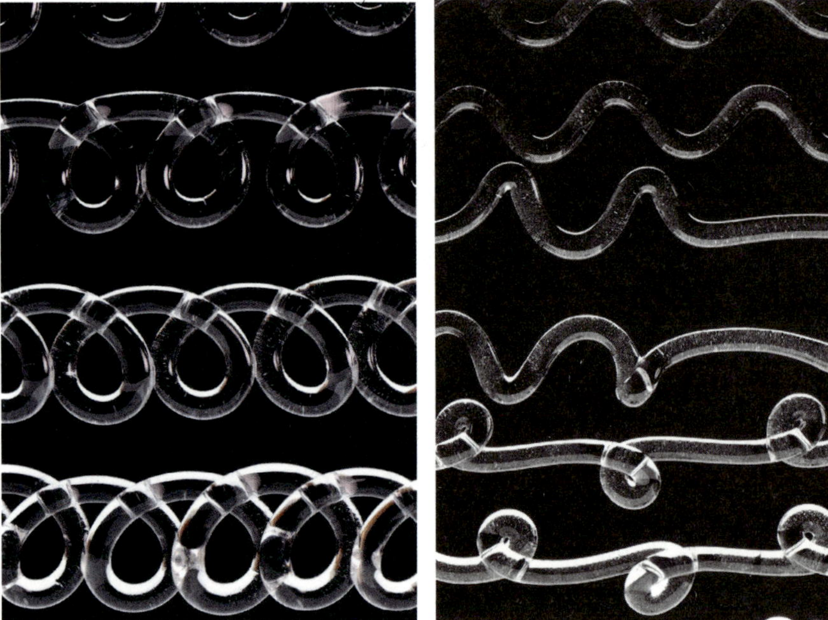

Above and below: Meandering paths and alternating loops were created when the nozzle deposited glass 3 15⁄16 inches (100 millimeters) above the build plate.

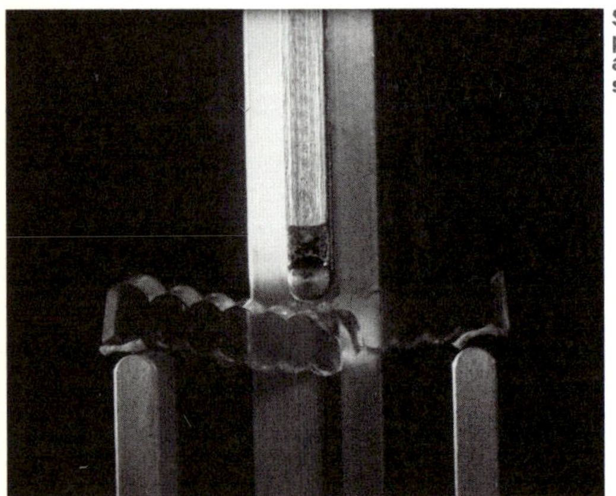

Structural three-point bending test of a printed glass sample.

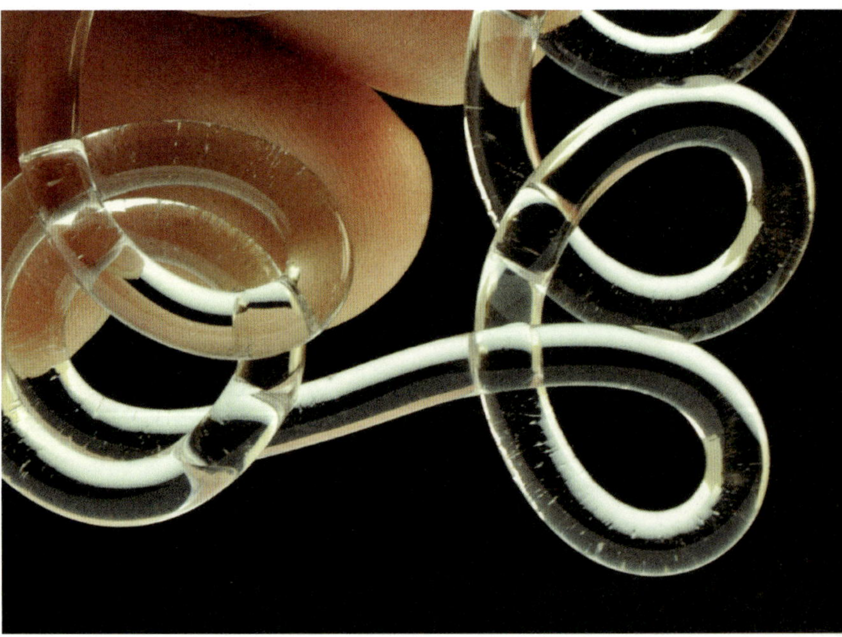

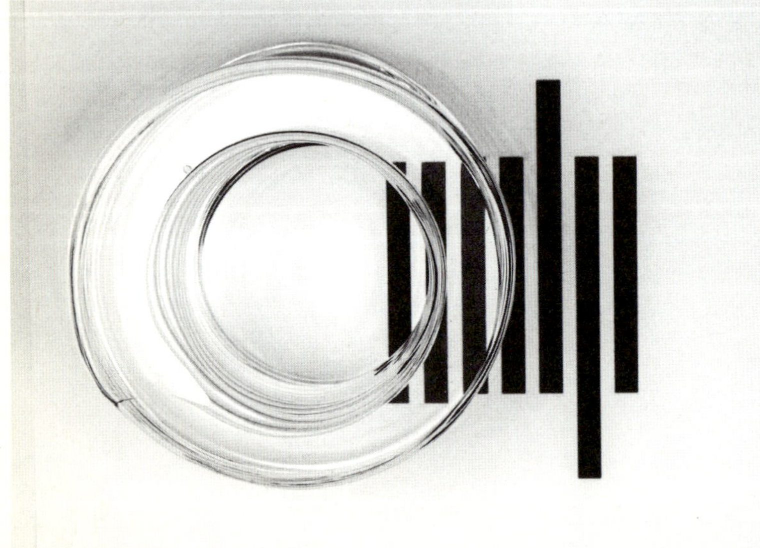

The high degree of optical transparency in the printed glass, demonstrated in a top-down view of a 2 ¾-inch (70-millimeter) cylinder.

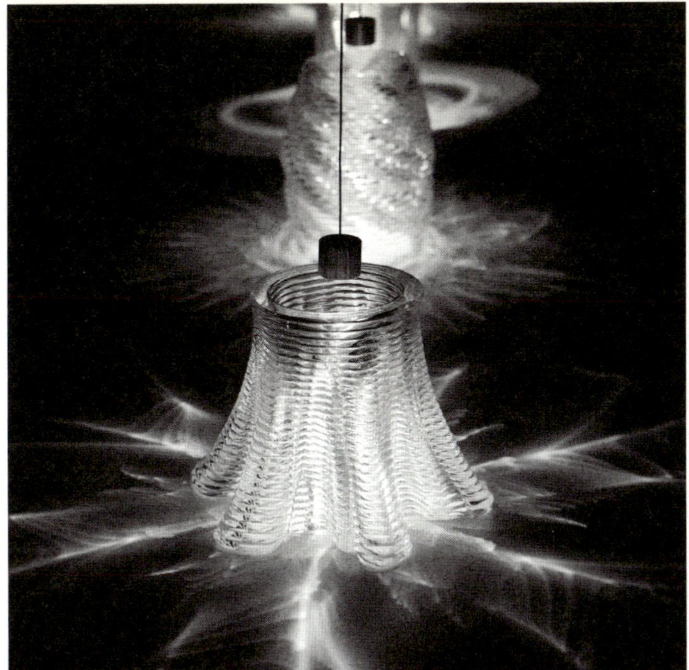

Glass objects illuminated from above generated a complex pattern of caustics (light rays reflected and refracted to create areas of light and shadow).

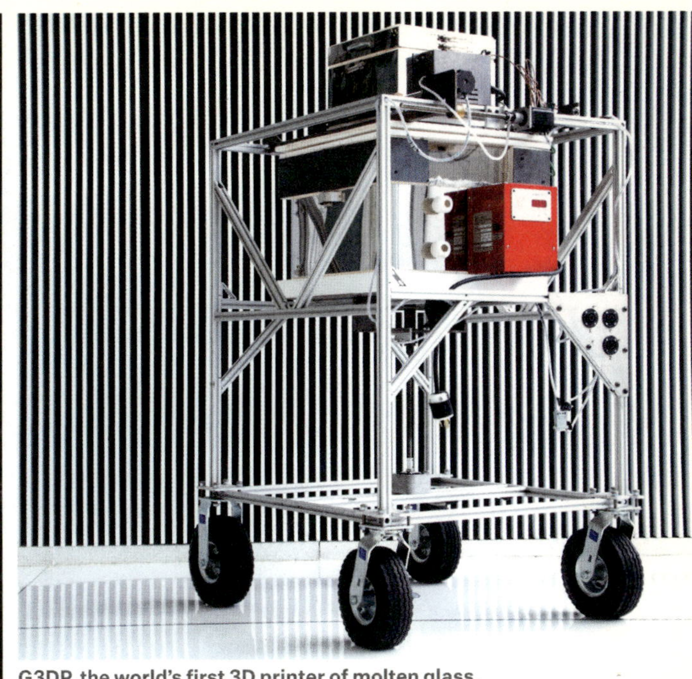

G3DP, the world's first 3D printer of molten glass.

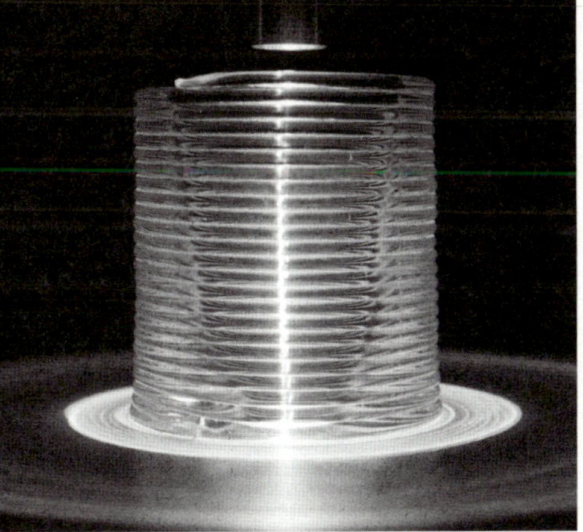

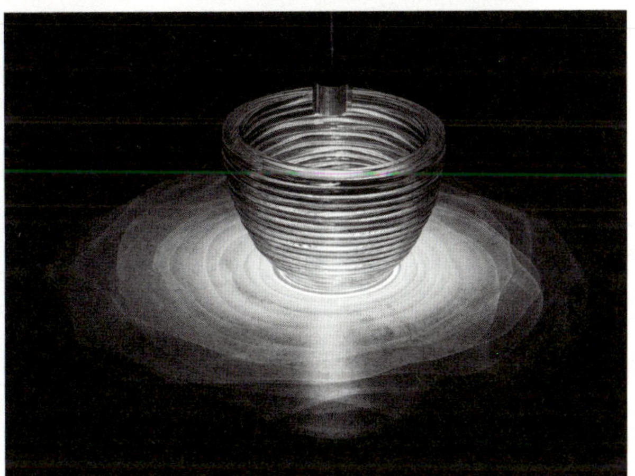

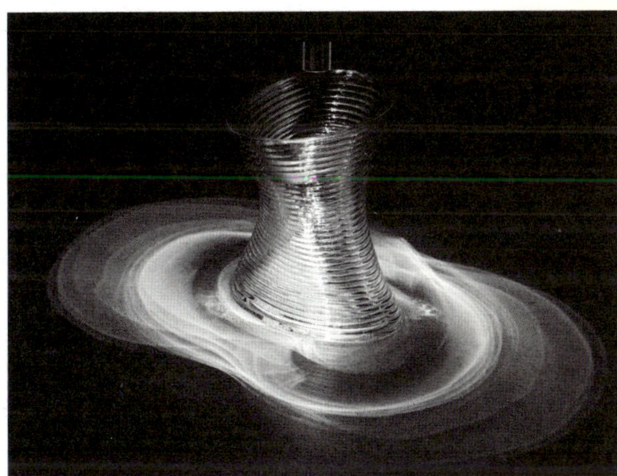

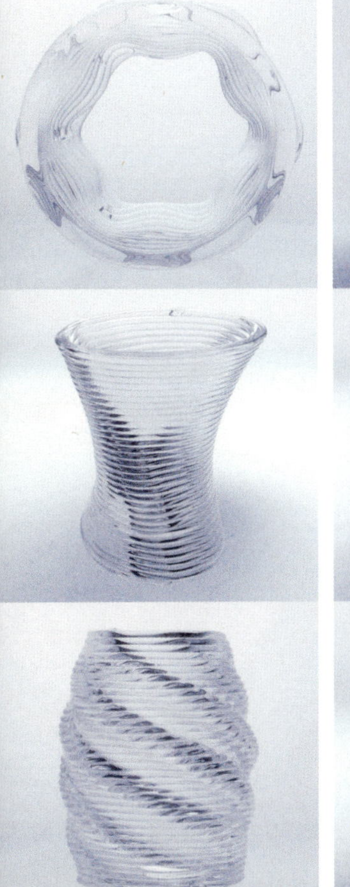

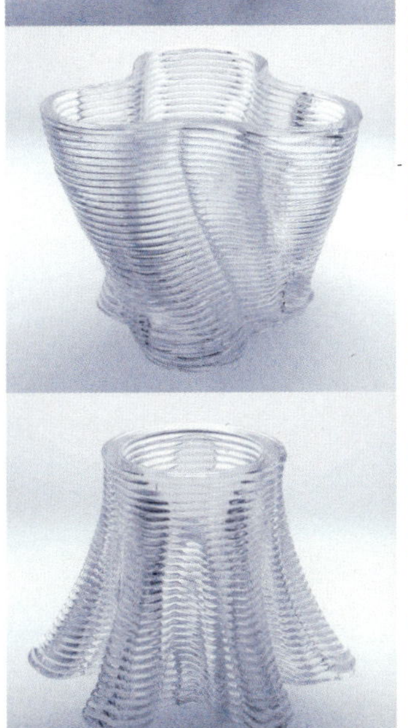

The first G3DP-printed objects.

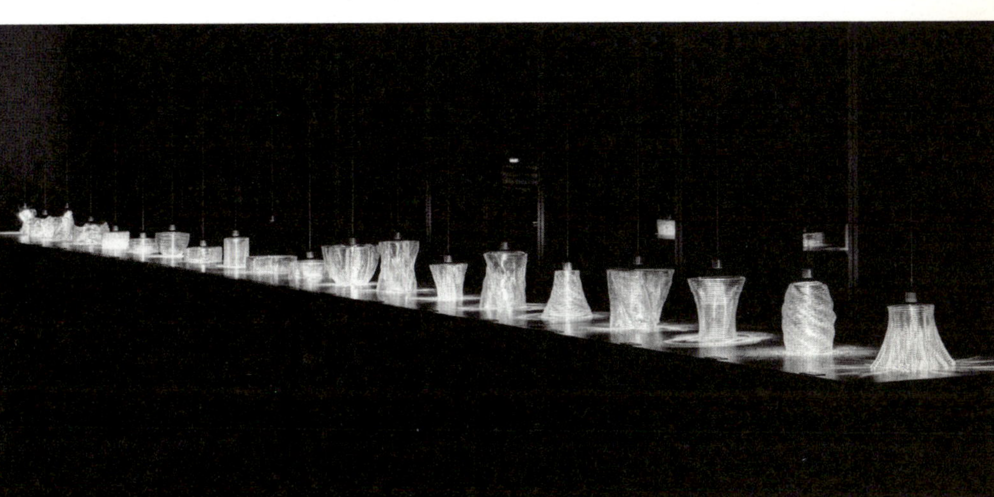

An installation of G3DP-printed objects at the MIT Media Lab, 2015.

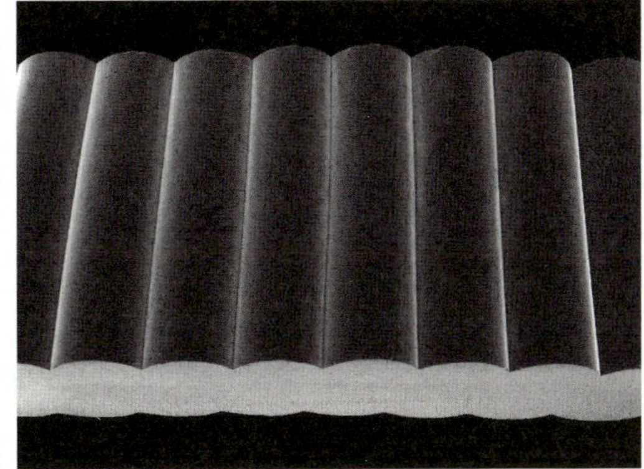

Scanning electron microscope (SEM) image showing a cross section of a printed glass wall.

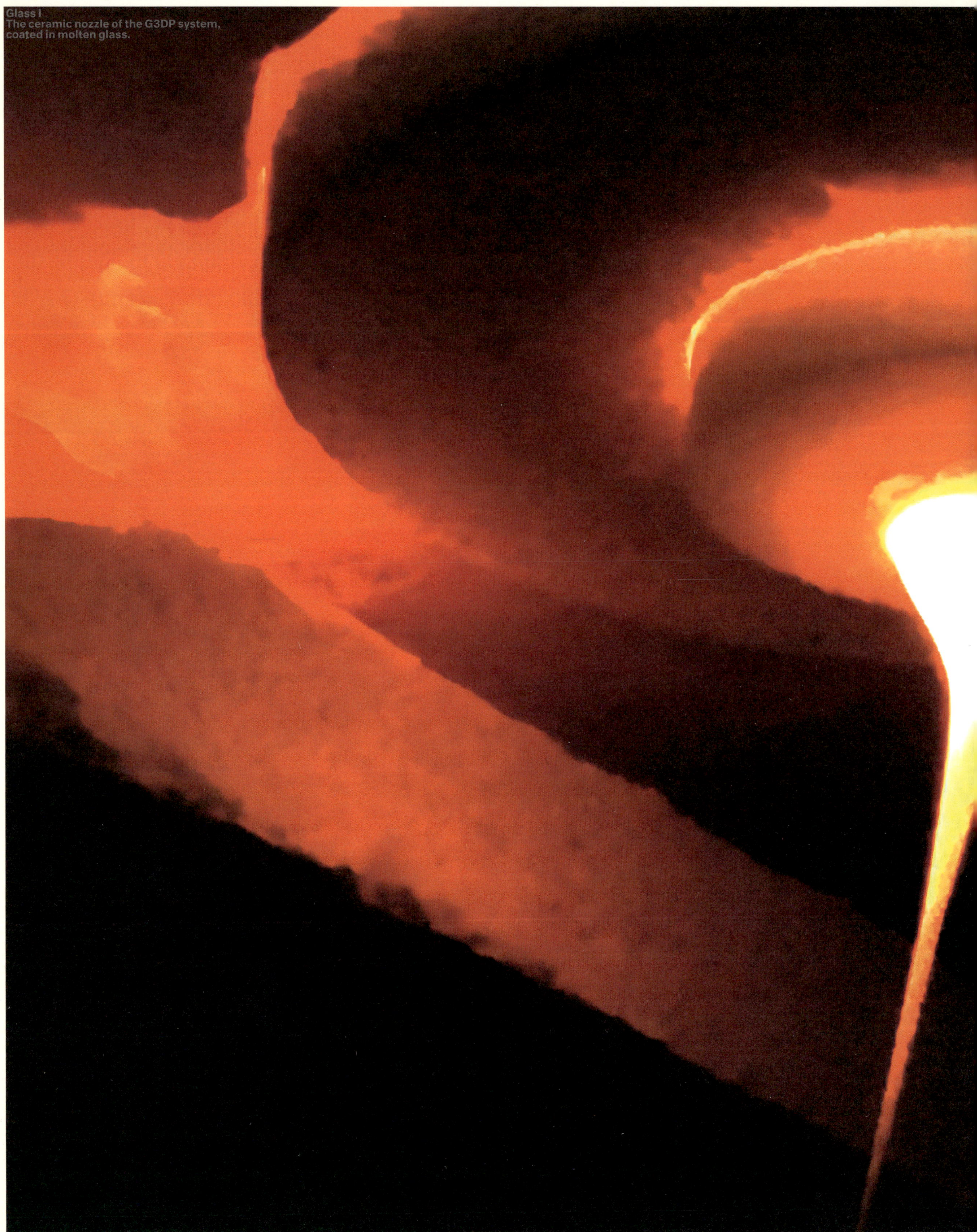
Glass I
The ceramic nozzle of the G3DP system, coated in molten glass.

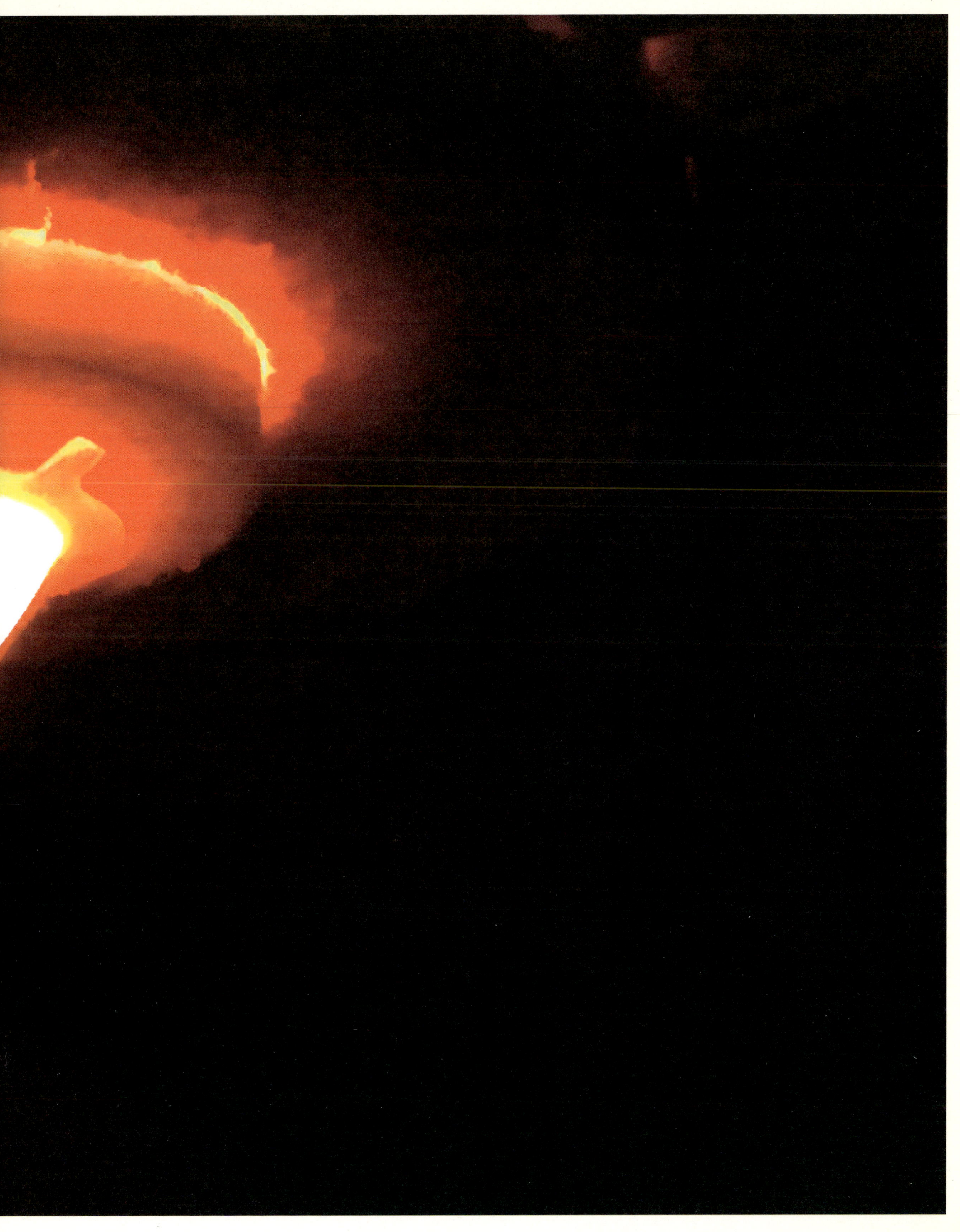

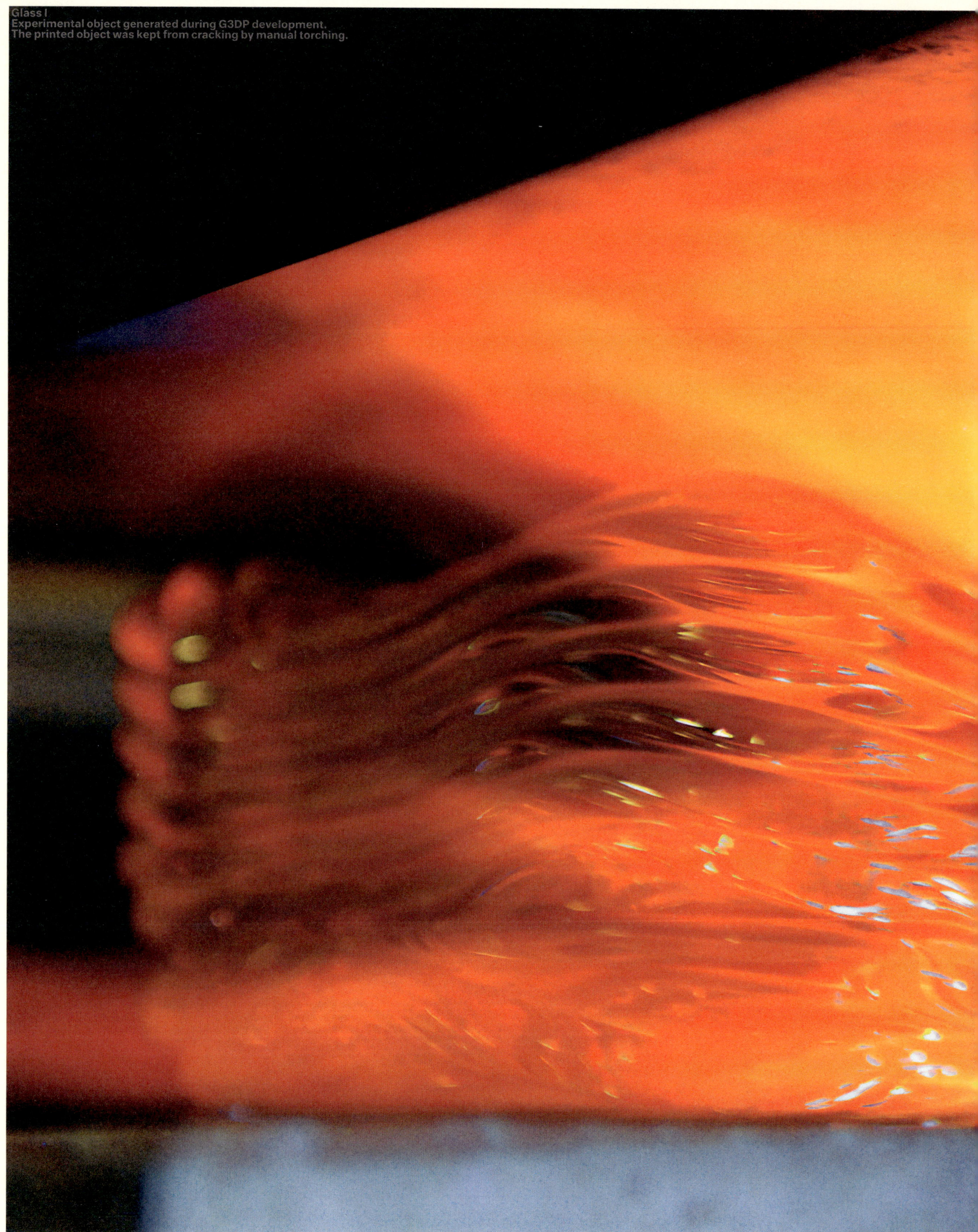

Glass I
Experimental object generated during G3DP development.
The printed object was kept from cracking by manual torching.

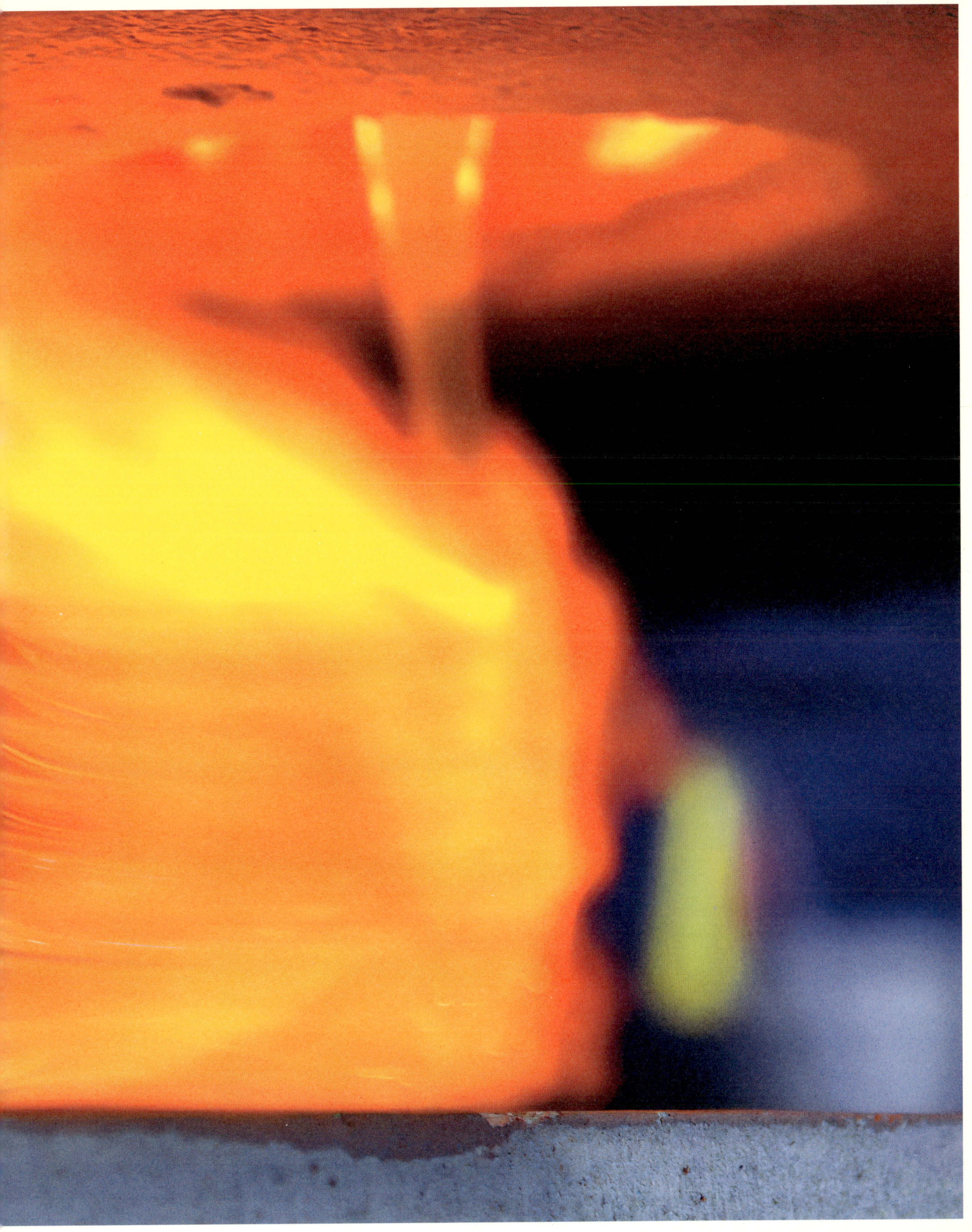

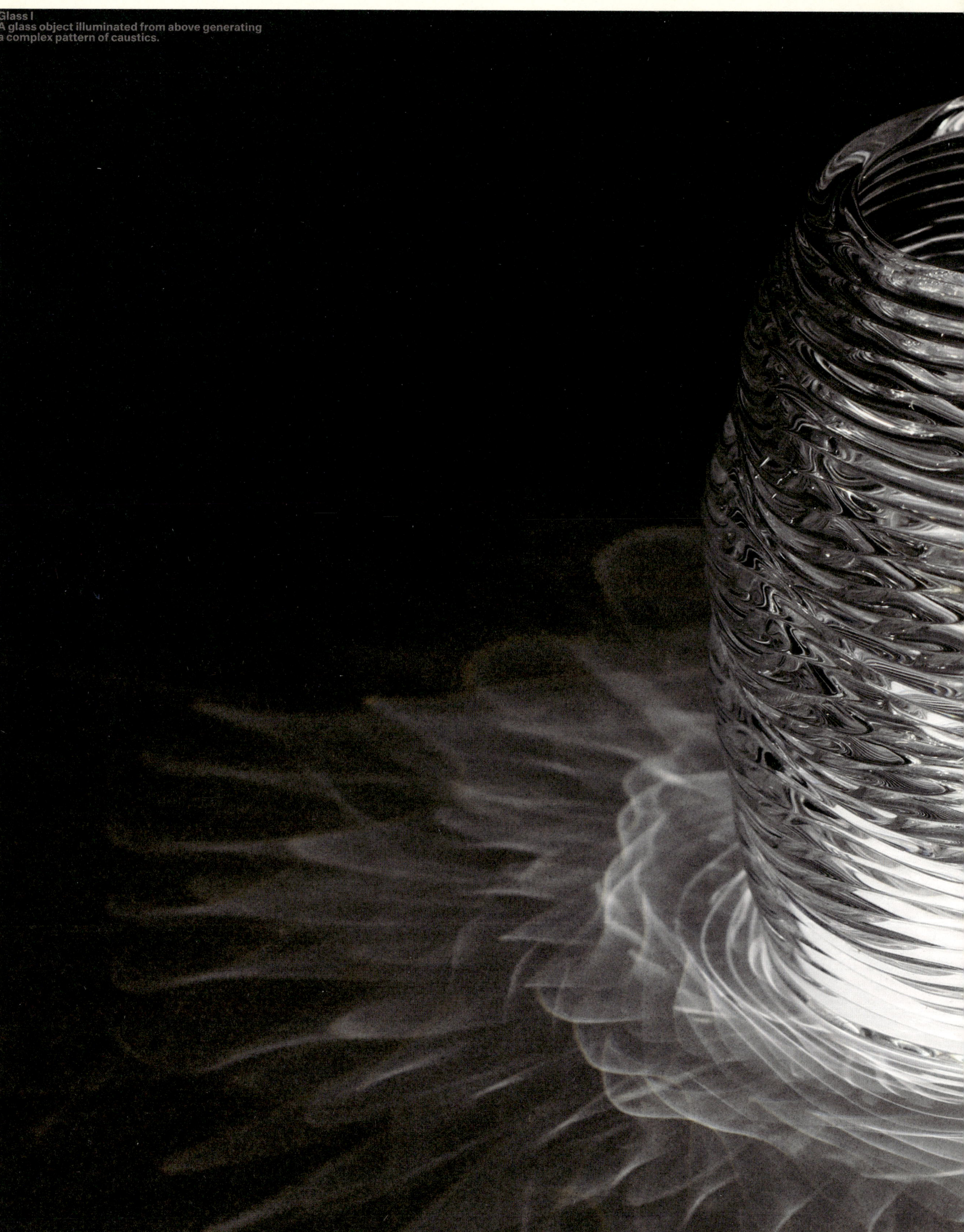

Glass I
A glass object illuminated from above generating a complex pattern of caustics.

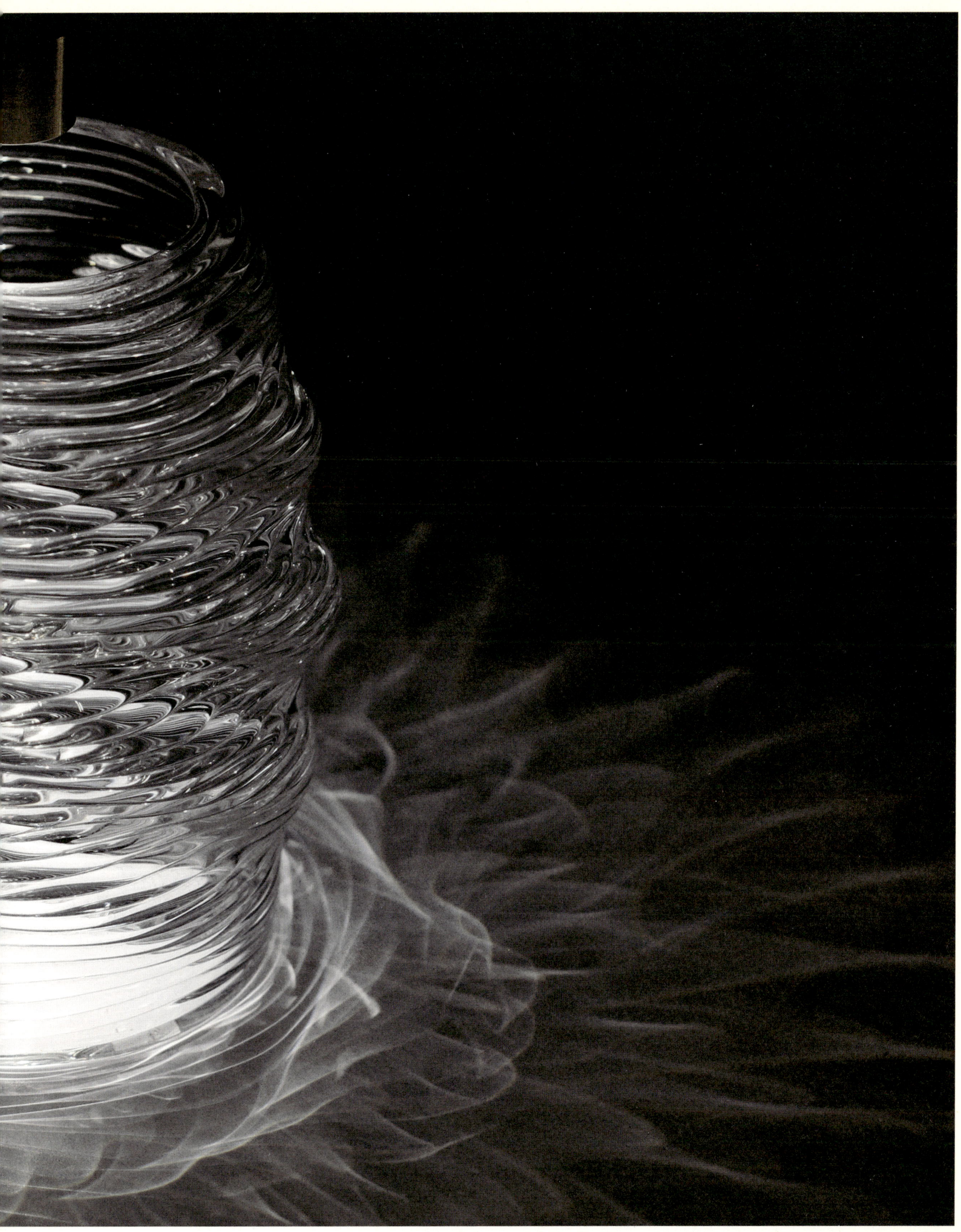

Glass I
The first object printed during G3DP development, pre-CNC, with the object path still traced out by hand.

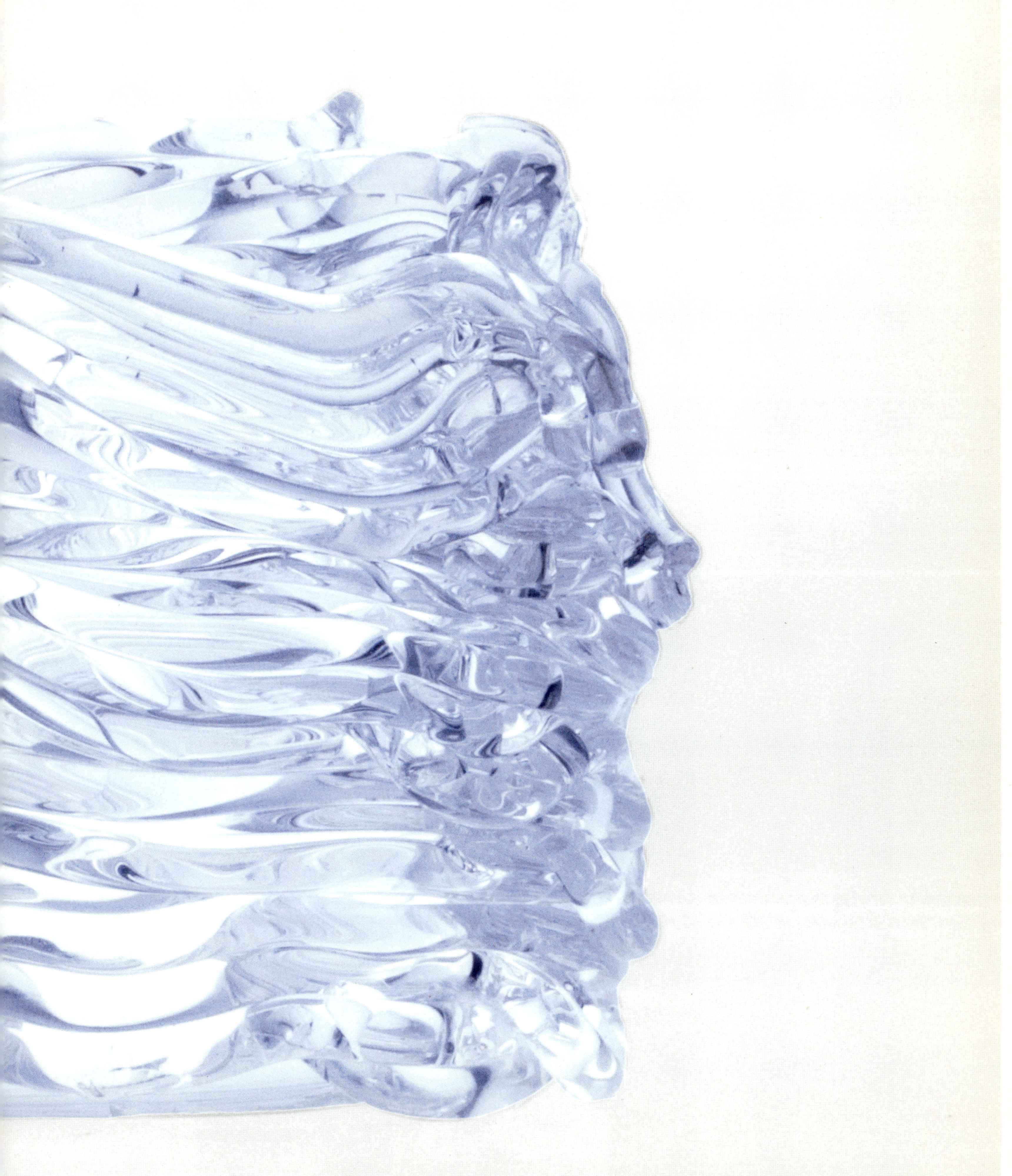

GLASS II

Neri Oxman and The Mediated Matter Group
Glass II. 2017
3D-printed glass, steel, brass, and silicone
11 13⁄16 in. × 11 13⁄16 in. × 9 ft. 10 1⁄8 in.
(30 × 30 × 300 cm)
An MIT Media Lab project
Research Team: Chikara Inamura, Michael Stern, Daniel Lizardo, Tal Achituv, Tomer Weller, Owen Trueblood, Nassia Inglessis, Giorgia Franchin, Marianna Gonzalez, Yinong Liu, Kelly Egorova, Peter Houk, Neri Oxman
Collaborators and contributors: Paula Aguilera, Mary Ann Babula, David J. Benyosef, Jeremy Flower, Sadie Forbes, Skutt Kilns, Andrew Magdanz, Robert Philips, Andy Ryan, Susan Shapiro, Neils La White, Jonathan Williams, Pentagram, Simson Gumpertz & Heger, Front Inc., Almost Perfect Glass, Spiral Arts Inc., Rubix Composites, Deltech Furnaces, Mori Building Co. Ltd. and Mori Building Group, Lios, NOE LLC, MIT Center for Bits and Atoms, MIT Central Machine Shop, Lexus

Commissioned by Lexus for *Yet*, 2017 Lexus Design Awards, Milan Design Week

Structural simulation of a column, used to validate the installation's response to a lateral load.

G3DP2 is our most recent and advanced version: a high-fidelity, large-scale additive printer for creating glass structures at architectural scales. The printer's architecture and operations were both restructured, with a digitally integrated three-zone thermal-control system and a four-axis motion-control system that together allow us to have a higher degree of control over shape and optical capabilities, and to produce larger structures with finer details. G3DP2 can process up to 33 pounds (15 kilograms) of 2,000-degree Fahrenheit (1,093-degree Celsius) molten glass in a single build.

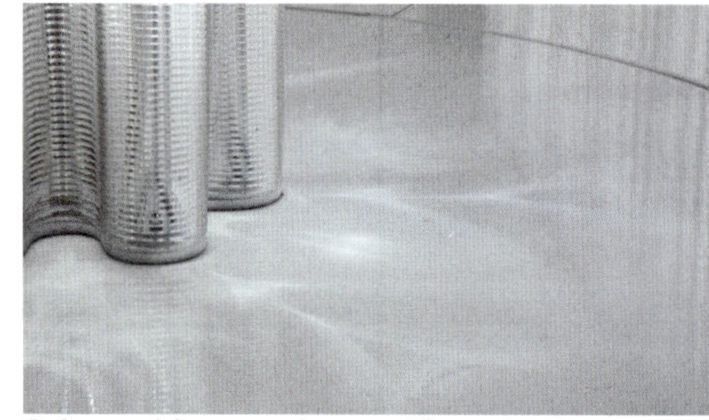

Column with supporting steel floor-plate system, Ark Hills Sengokuyama Mori Building, Tokyo, 2018.

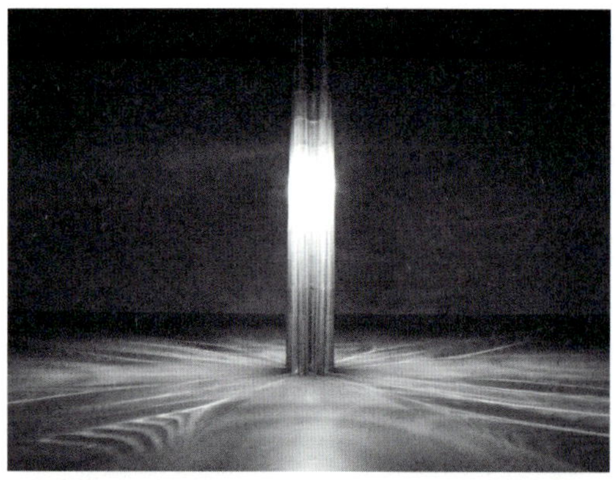

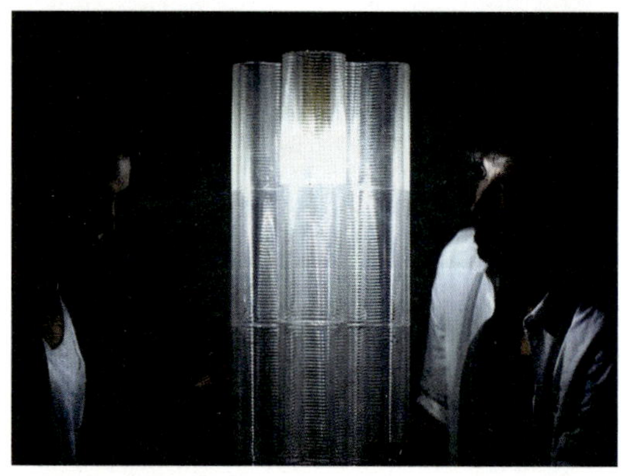

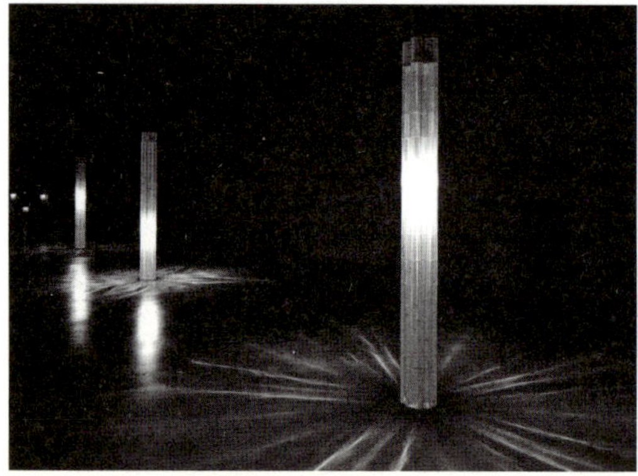

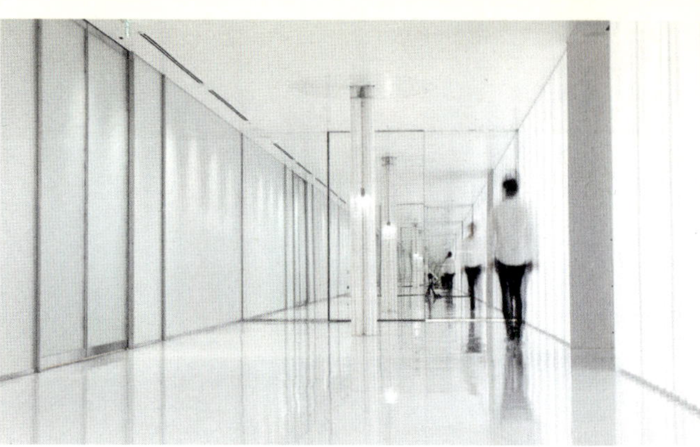

With G3DP2 we produced Glass II, which consists of three columns, each about 10 feet (3 meters) tall, each made of fifteen elements with cross sections that are radially symmetric and flowerlike, containing lobes that look like petals branching along its height. Inspired by the columns of Antoni Gaudí's Basílica de la Sagrada Família in Barcelona, the columns vary in shape from top to bottom, becoming narrower as they grow taller in order to reduce the weight at the top; the column is thus stronger at its base, supporting the structural load. As a result, the optical properties also vary, with curved surfaces creating reflections and refractions, or caustics. These optical properties let the columns function as lenses at an architectural scale, concentrating and dispersing light outside and in. By further developing these properties, we intend to create a model for passive thermal regulation and the harnessing of solar energy.

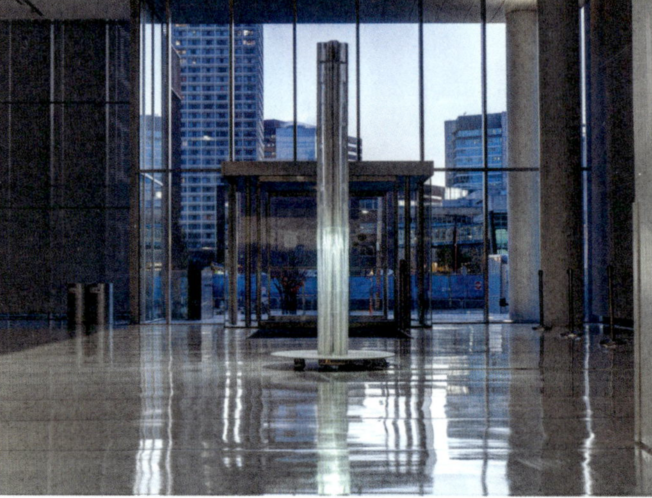

Morning view of a full-scale installation test in the east lobby of the MIT Media Lab, 2017.

The computational process used to generate each column's shape was influenced by the constraints of the manufacturing platform and the requirements of the object being printed. For example, the higher the load created by the column's volume at any given height, the greater its total surface area would be, with more lobes in its section—requiring the printer to operate at a tighter turning radius toward the column's base. In other words, the more load the column was required to bear, the more intricate its cross section would be. Such interactions between form and formation, between an object's shape and the technology necessary to create it, are further examples of the productive tension between nature-made materials and novel technologies that can shape them in environmentally responsive ways.

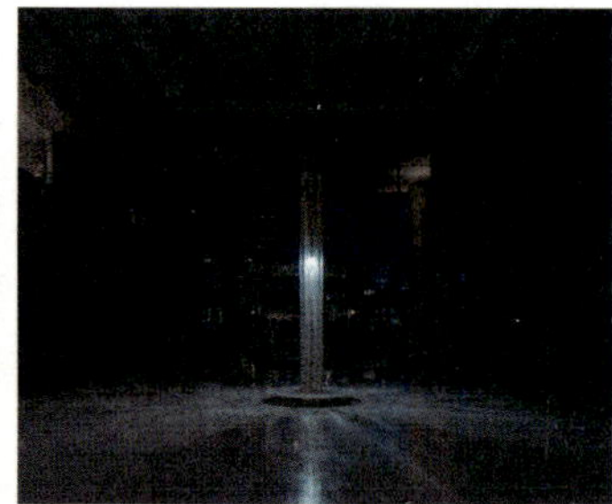

Installation views of the first 10-foot (3-meter) glass columns, Milan Design Week, 2017.

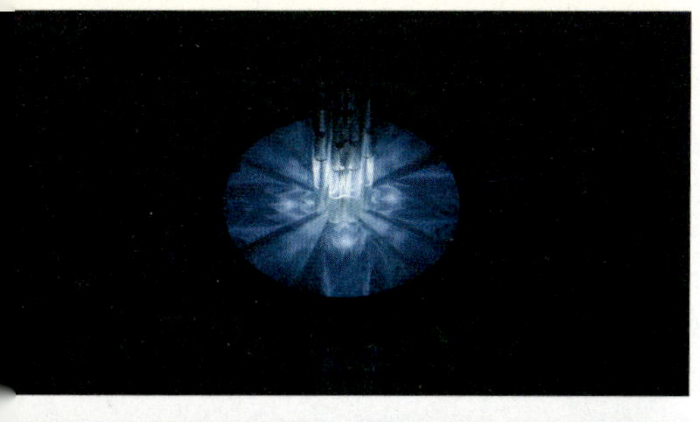

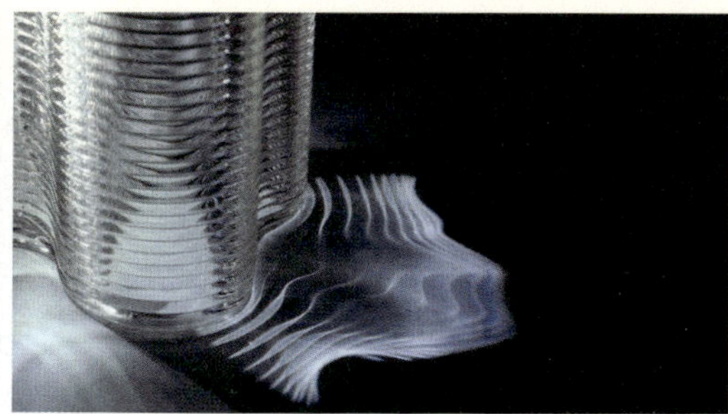

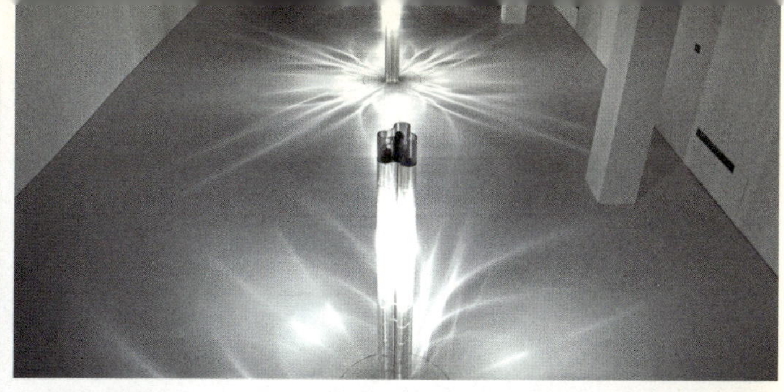

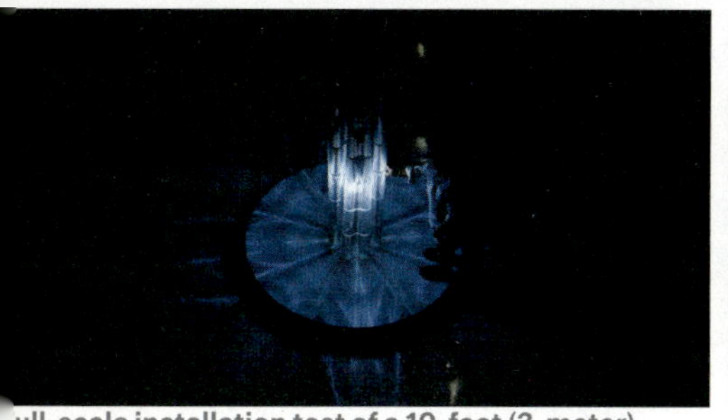

ull-scale installation test of a 10-foot (3-meter) ;lass column in the lobby of the MIT Media Lab, 2017.

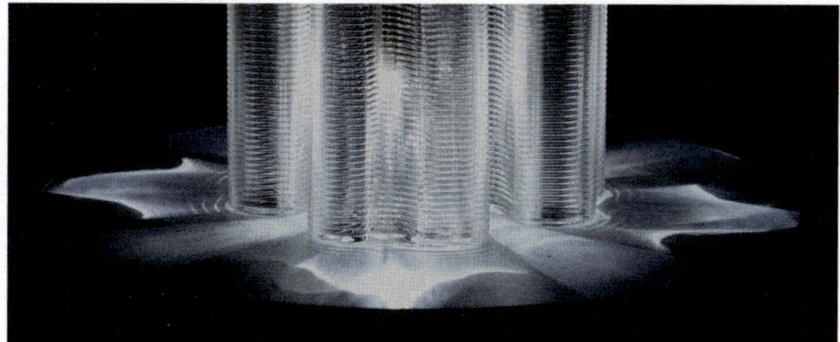

Caustic patterns.

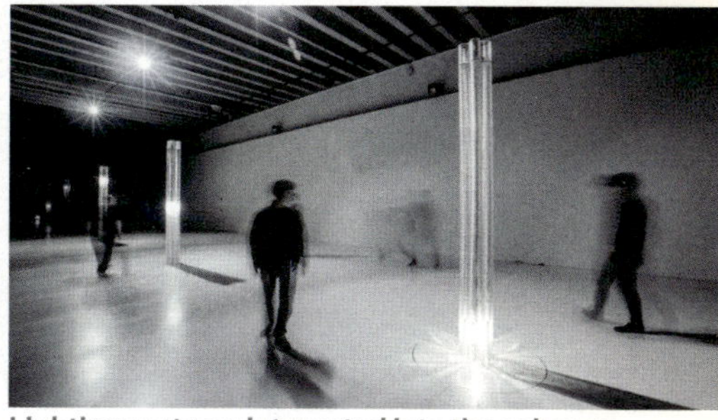

Lighting systems integrated into the columns creating an immersive field of caustics, Milan Design Week, 2017.

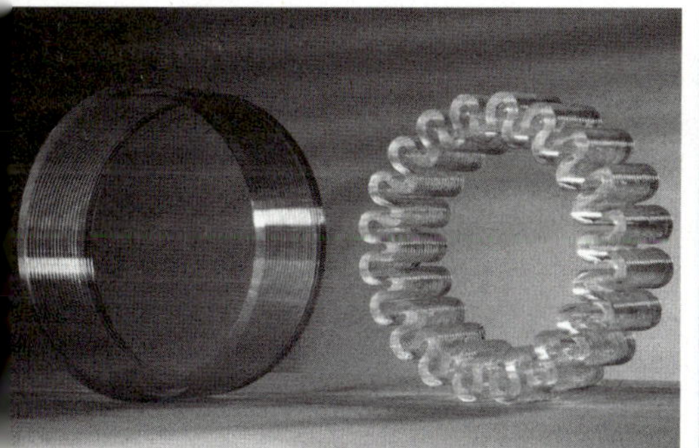

:urvature test components. These tests perfected he fidelity and repeatability of glass production.

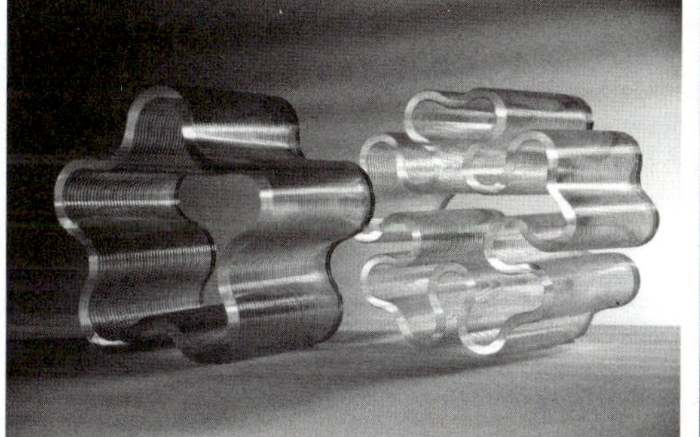

Top and bottom components of the five-lobe column, demonstrating a significant change in the cross section shape.

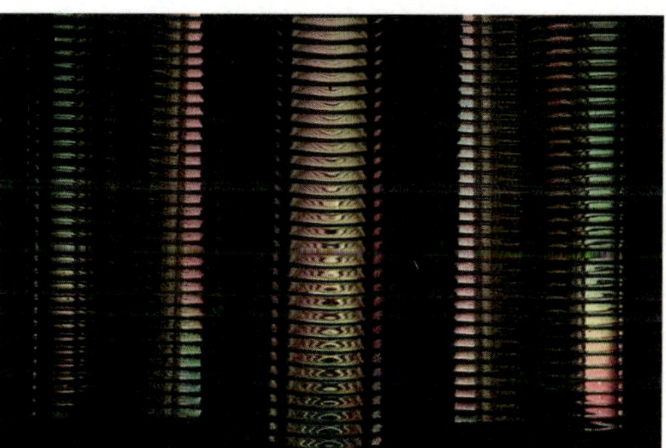

Structural testing of a column component in compression, viewed through a polariscope. Stresses appear as colors in the object.

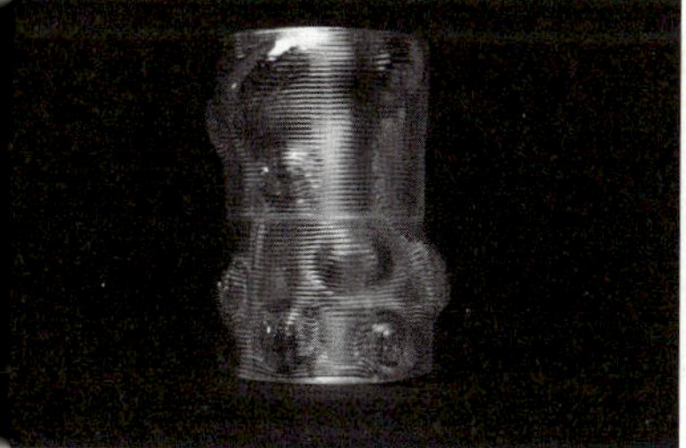

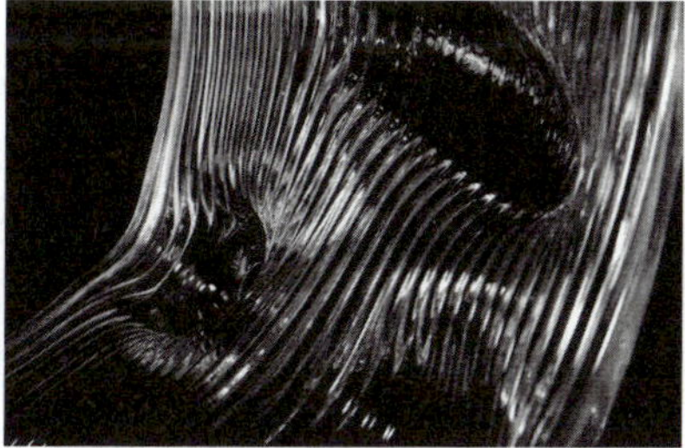

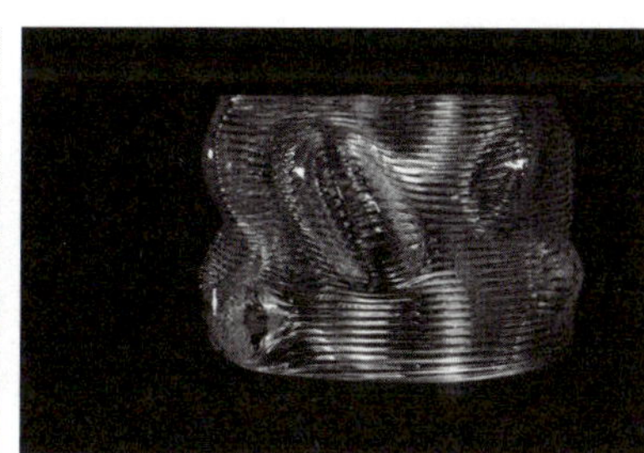

Experiments for Glass III, made of printed, reheated, blown glass.

3elow: Assembly drawings of printed glass column nd structural pre-tension system (upper left), drive nodule subsystem for lighting housed within the column (lower left), column baseplate (upper right), and motion system for lighting (lower right).

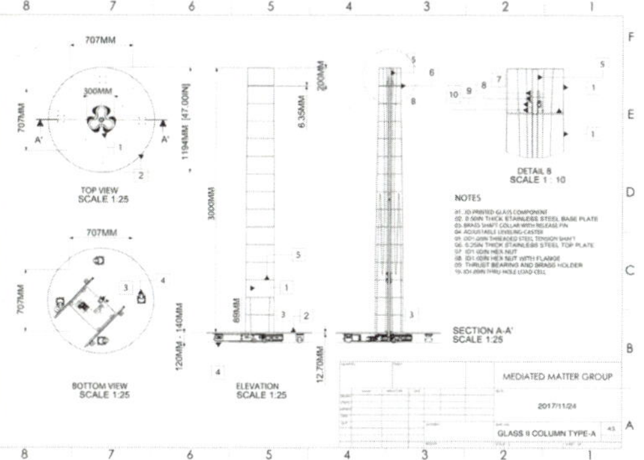

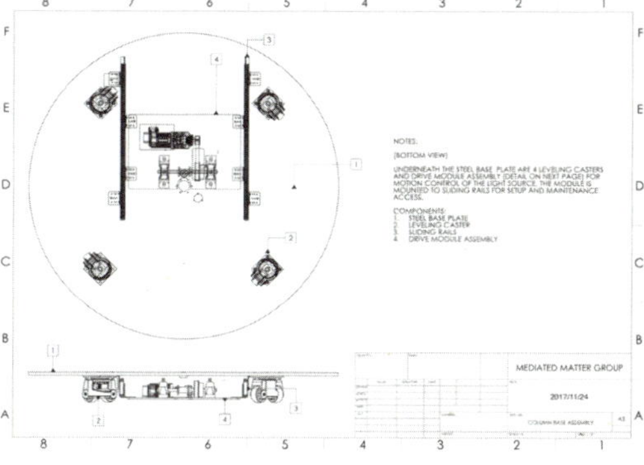

Rendering of G3DP2 control station containing motion- and thermal- control electrical systems.

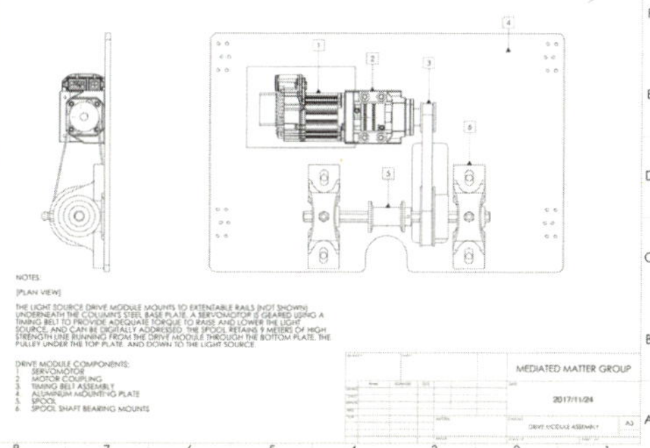

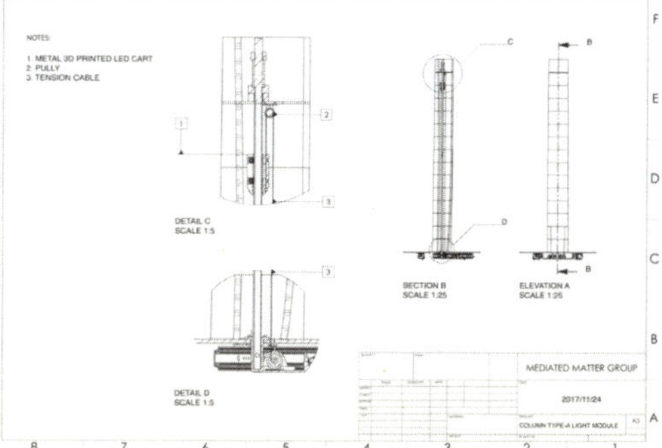

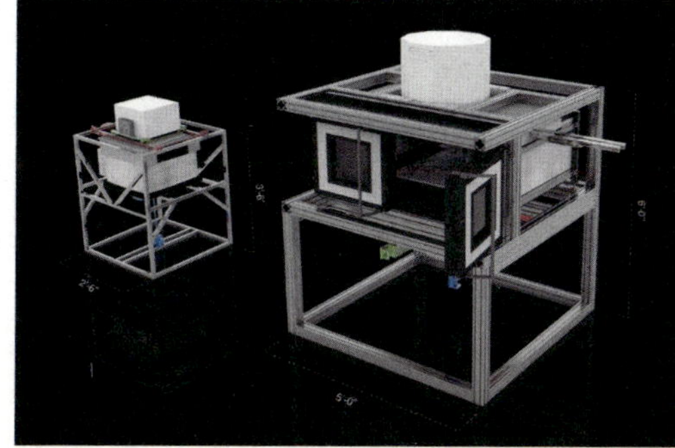

Comparative rendering of G3DP and G3DP2 platforms.

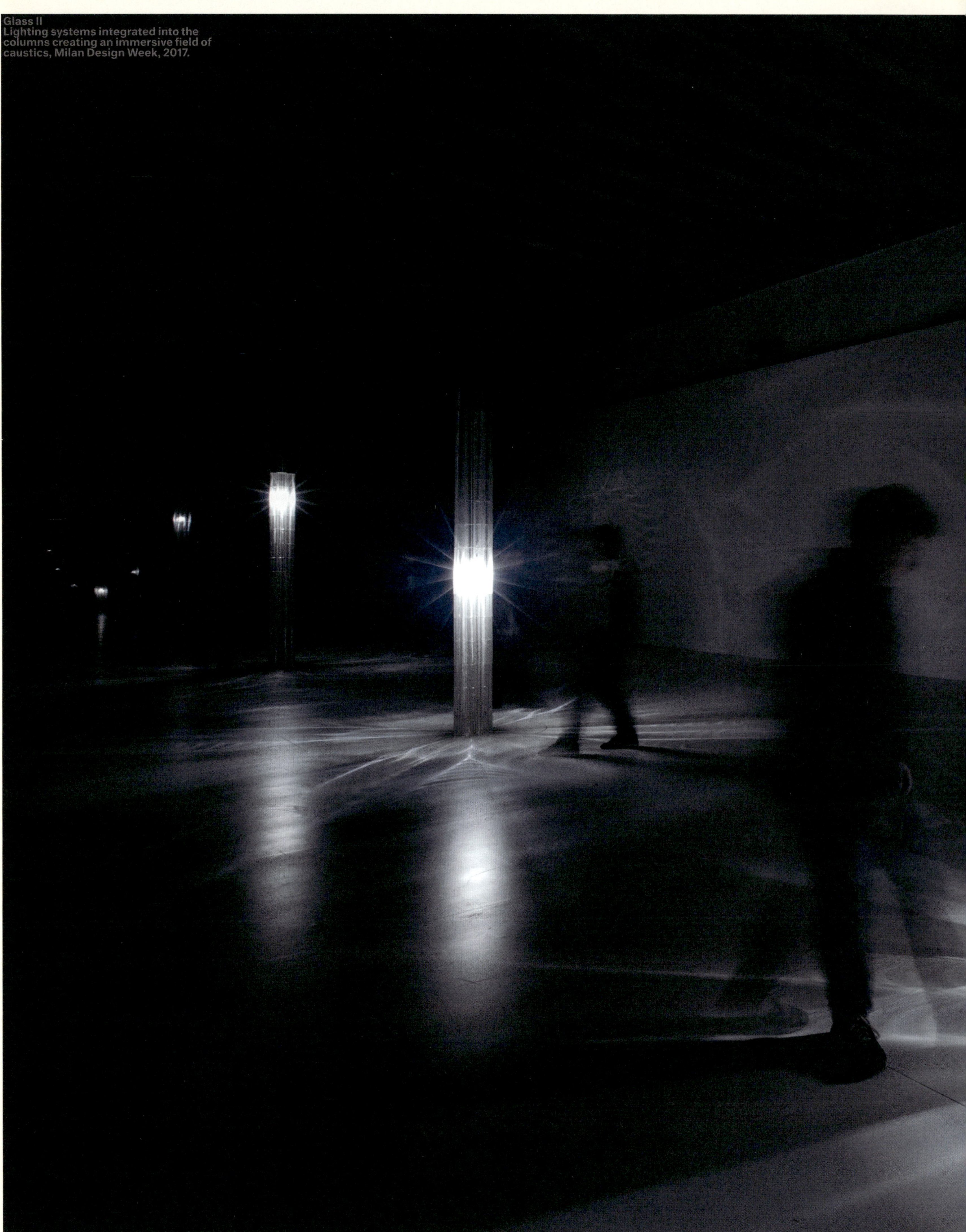

Glass II
Lighting systems integrated into the columns creating an immersive field of caustics, Milan Design Week, 2017.

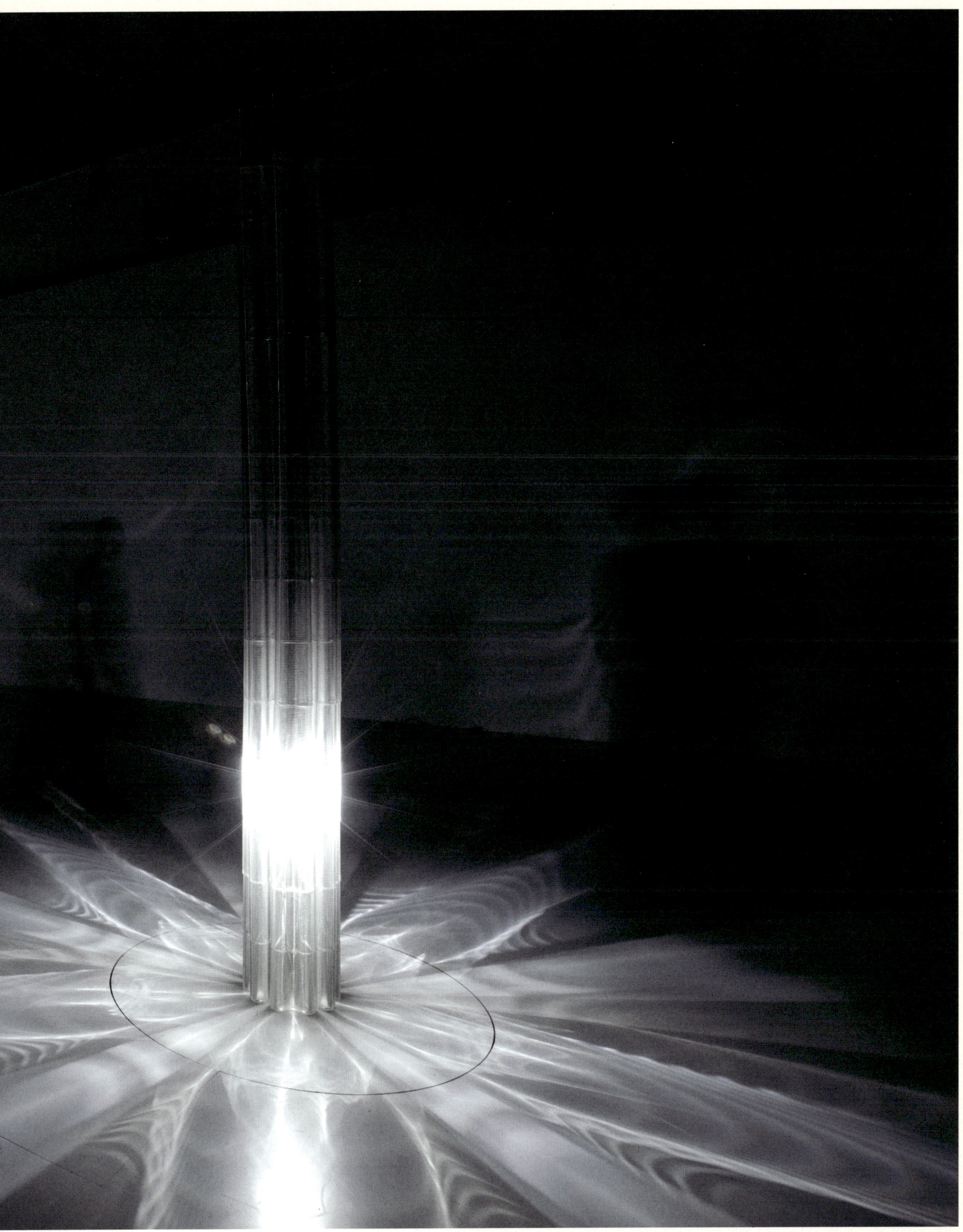

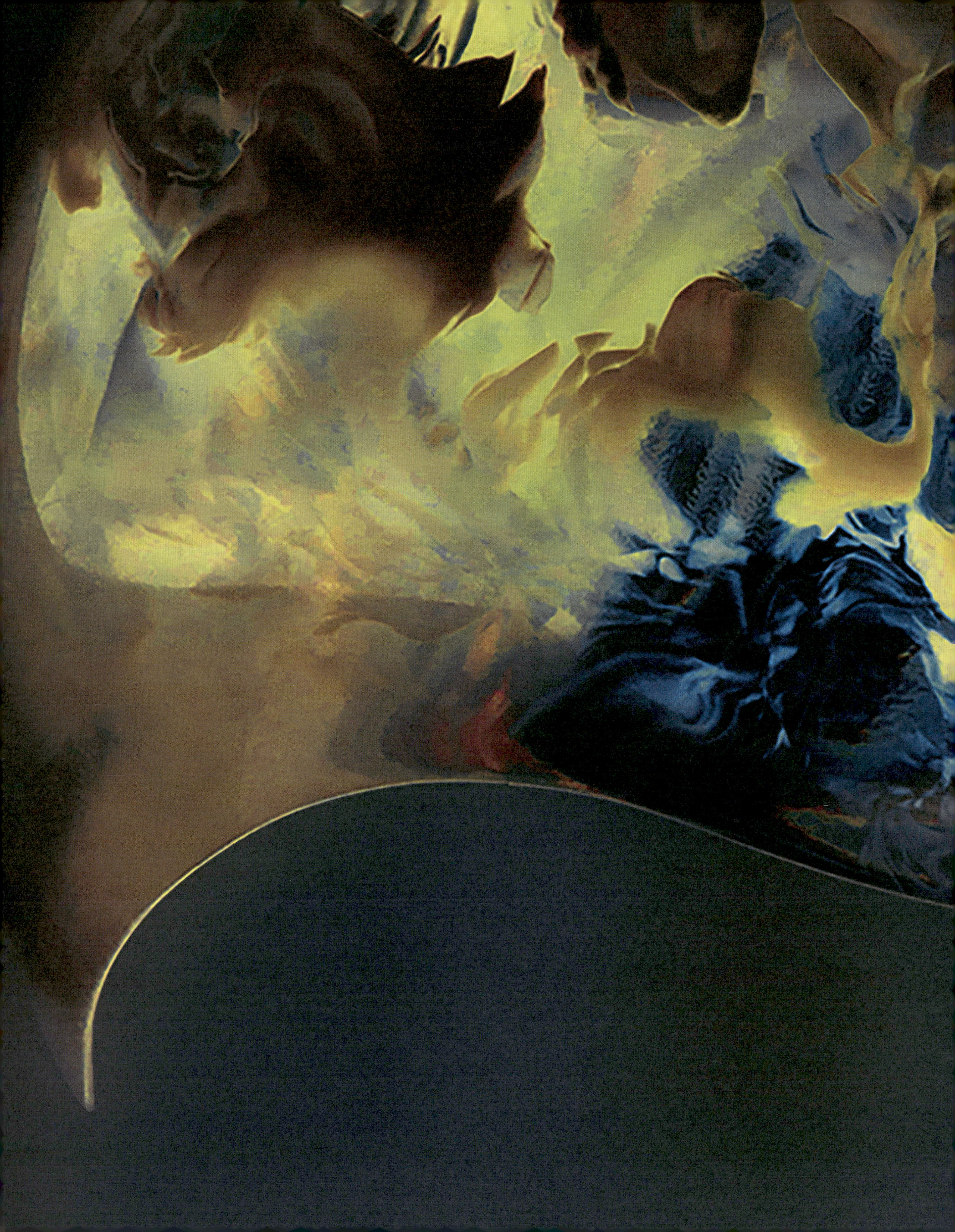

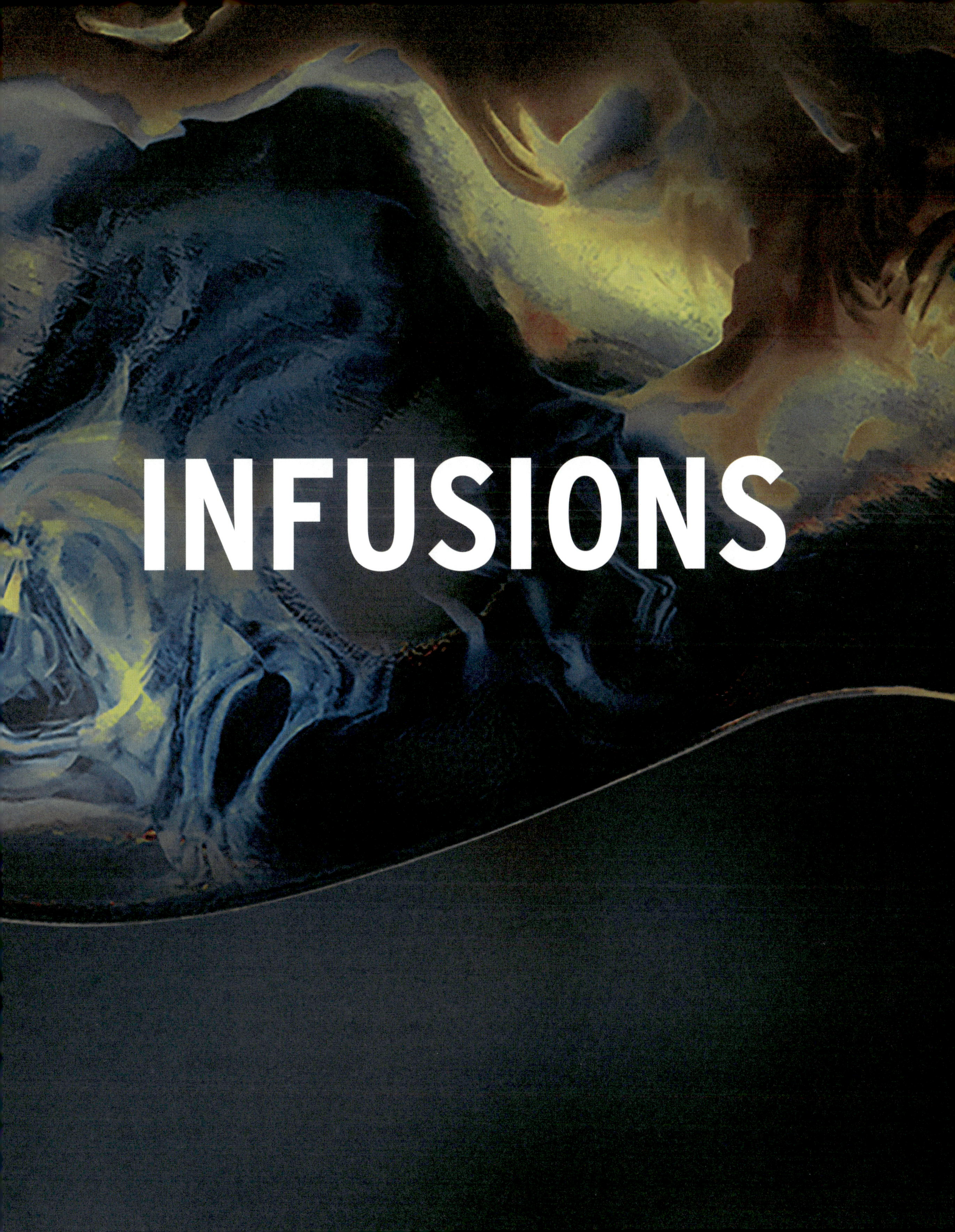

INFUSIONS

IMAGINARY BEINGS

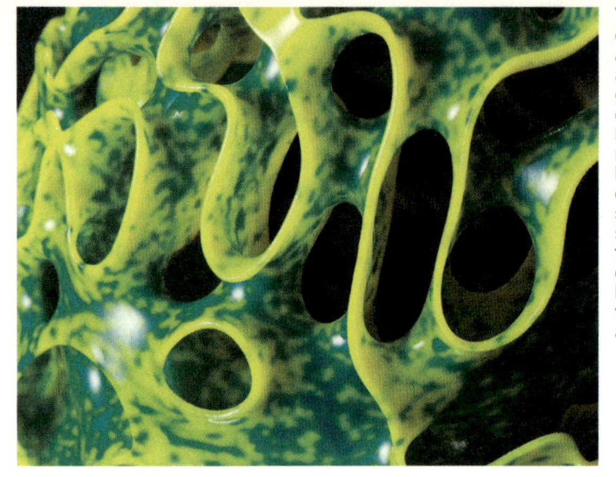

Two different textures were combined to create Pneuma 2, a lightweight, shock-absorbing armor. A cellular, volumetric, geometry-based texture enables expansion and contraction during breathing, while a dotted, material-based texture enables local variations in stiffness. The outer boundaries of the cellular structure are stiff, to maintain its structural integrity, while soft tissue is located within. This was one of our first projects to use bitmap printing for the design and fabrication of material properties at human-hair resolution.

Neri Oxman in collaboration with W. Craig Carter and Joe Hicklin
Imaginary Beings. 2012
Photopolymers
Dimensions variable
Produced by Stratasys Ltd.
Contributors: Turlif Vilbrandt; James C. Weaver, Wyss Institute, Harvard University; Stratasys Ltd.

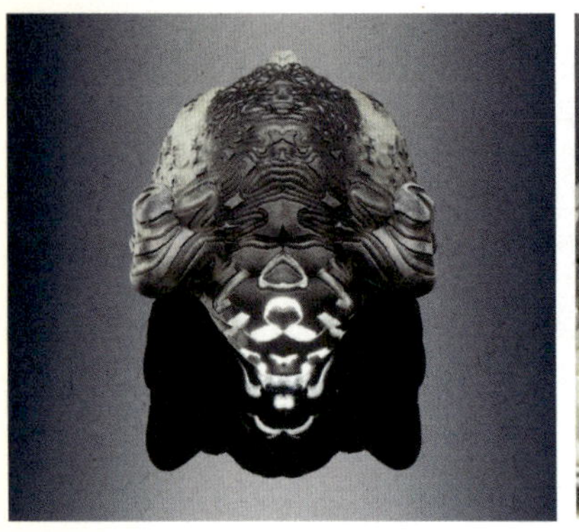
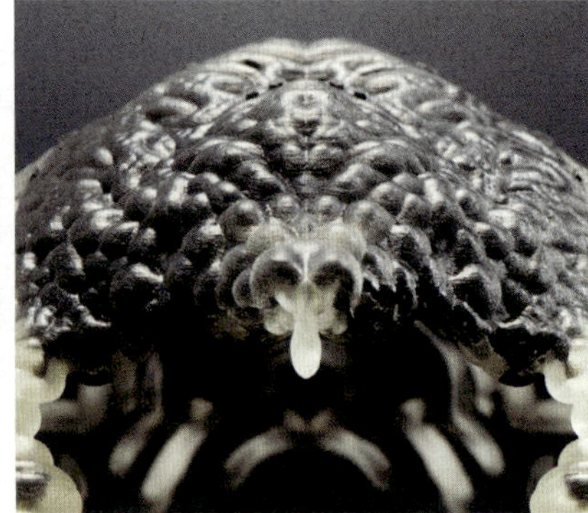
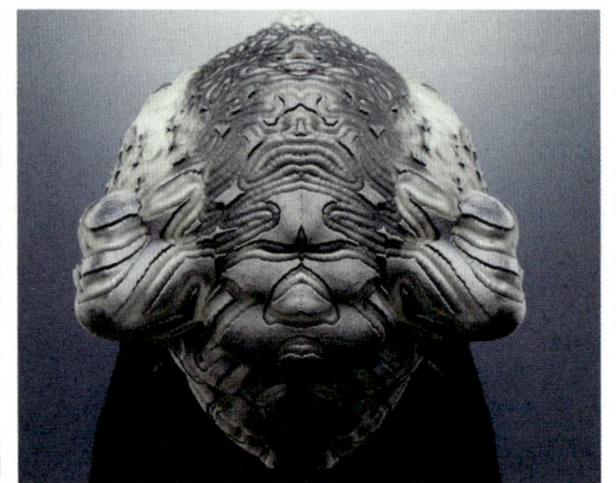
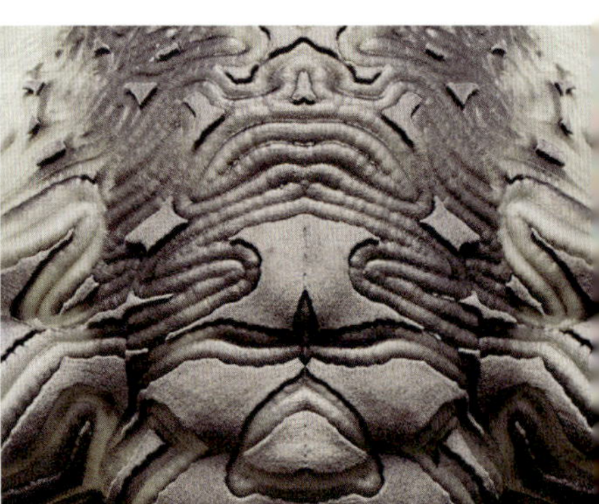

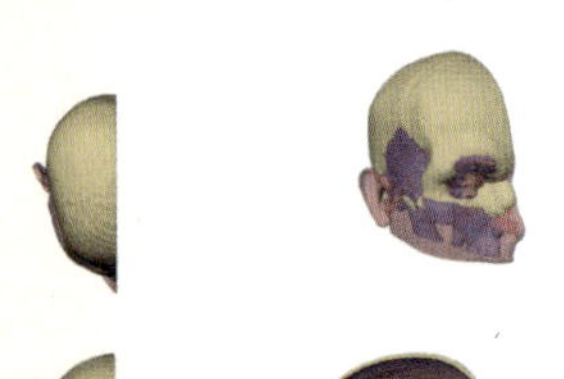
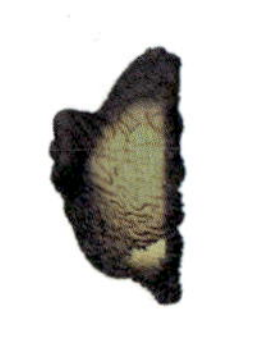
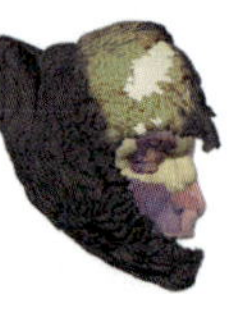
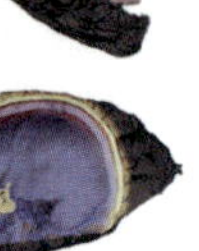
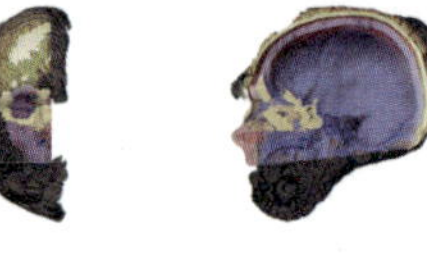
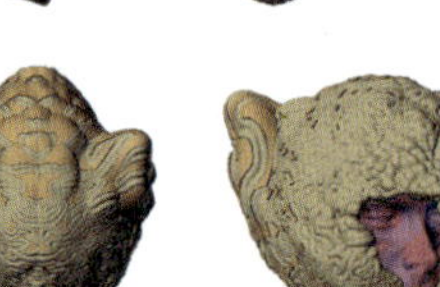
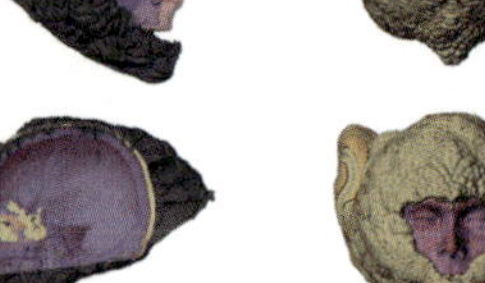
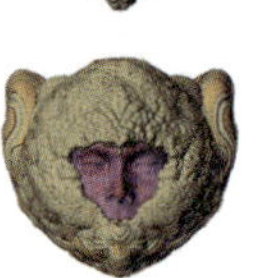
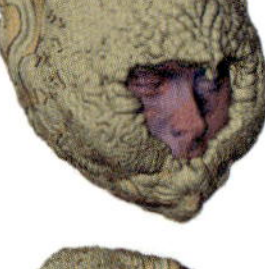
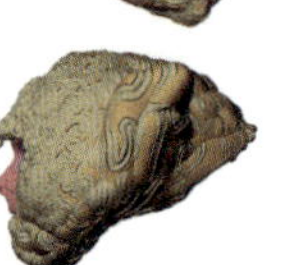
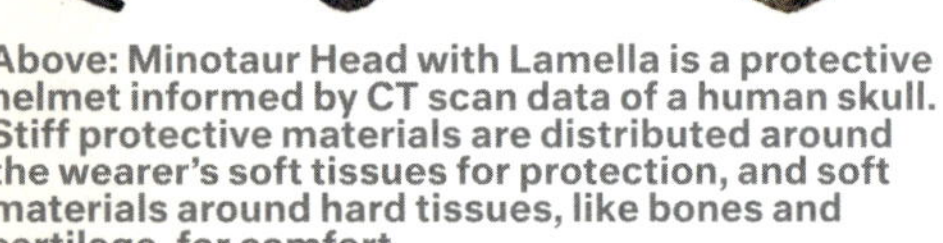

Above: Minotaur Head with Lamella is a protective helmet informed by CT scan data of a human skull. Stiff protective materials are distributed around the wearer's soft tissues for protection, and soft materials around hard tissues, like bones and cartilage, for comfort.

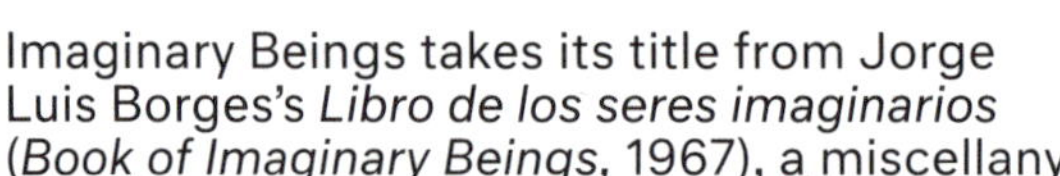

Imaginary Beings takes its title from Jorge Luis Borges's *Libro de los seres imaginarios* (*Book of Imaginary Beings,* 1967), a miscellany of more than a hundred fantastical beasts from folklore and literature. These creatures exist on the cusp of the natural and the mythical; they augment, fuse with, and adapt nature, exploring the limits of human biology and experience.

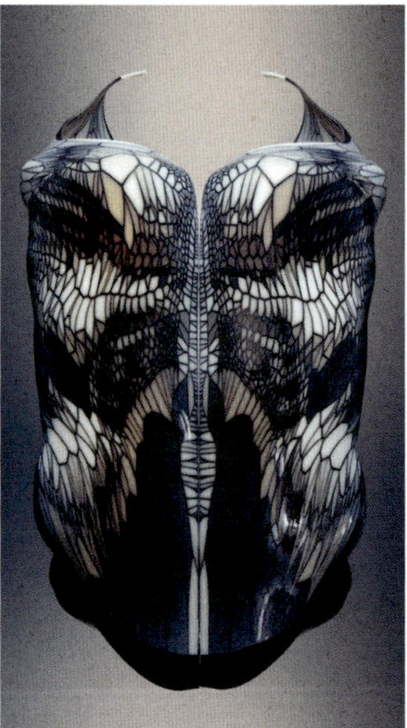
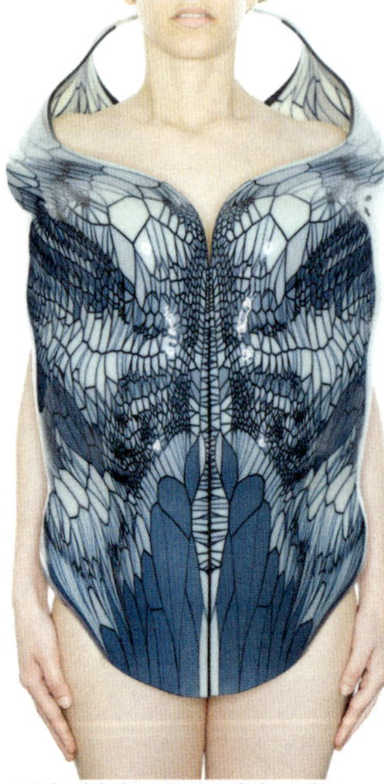

Above: Arachne is armor for the torso, with thickness and density corresponding to the anatomy of the wearer's rib cage. It is filled with stiff materials where it needs to shield bony tissue; softer, more flexible cells cover intercostal muscles.

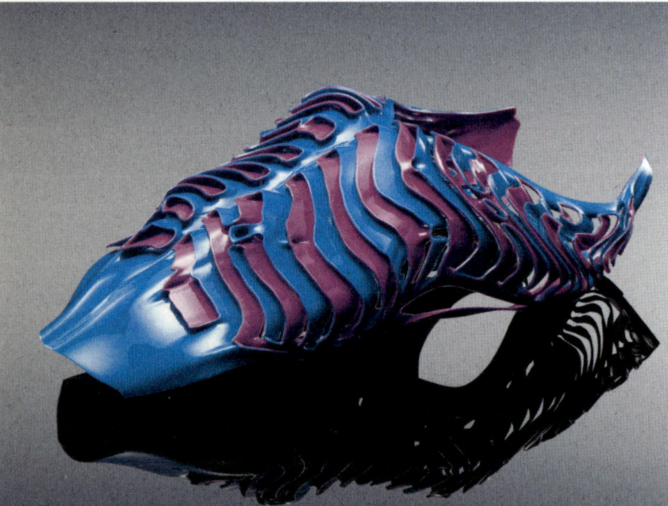

Leviathan 1 is armor made of a thin, stiff shell with double-sided cuts through it. The cuts make the object flexible and allow the wearer a greater range of movement. The inner side of each one is lined with soft materials, for comfort, while the shell provides protection.

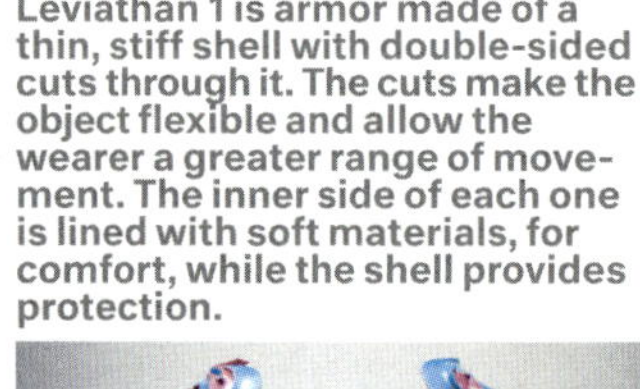
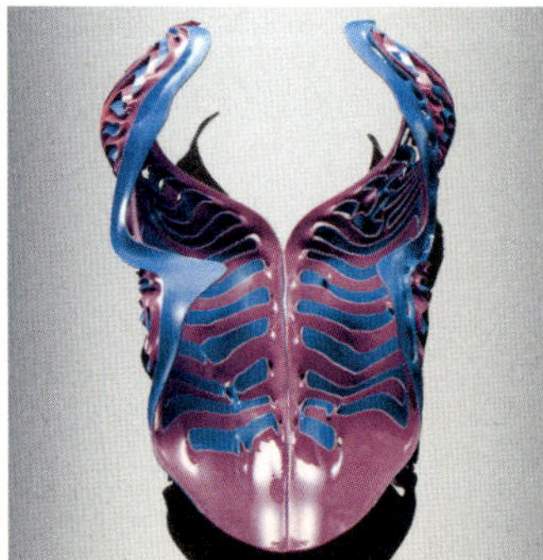
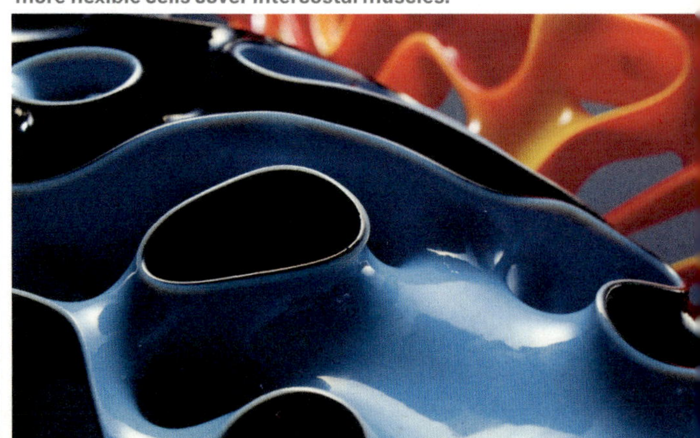

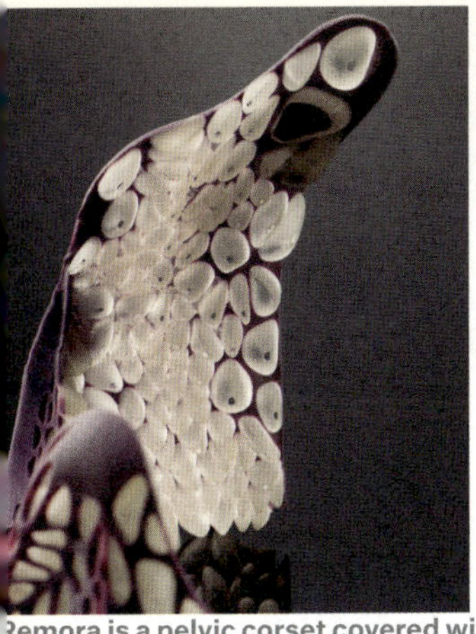

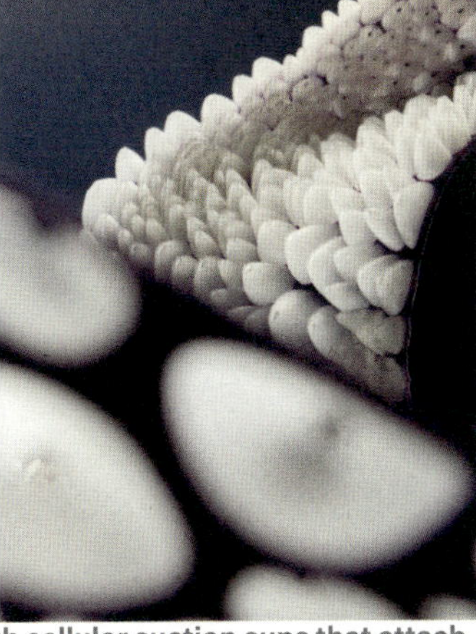

Remora is a pelvic corset covered with cellular suction cups that attach to the skin and enhance the wearer's circulation. Their varied size and density were informed by the body's natural curvature.

For those of us working at the intersection of technology and biology, of the made and the natural, Borges's stories are also tales of a possible future, one inhabited by creatures born with superpowers such as the ability to disappear or to walk through fire. Behind these visions of the possible lies a history of discovery and storytelling, from Leonardo da Vinci's human-powered aircraft, inspired by the wings of Icarus, to mythical occurrences of material self-repair and regeneration, as far back as Prometheus and his liver.

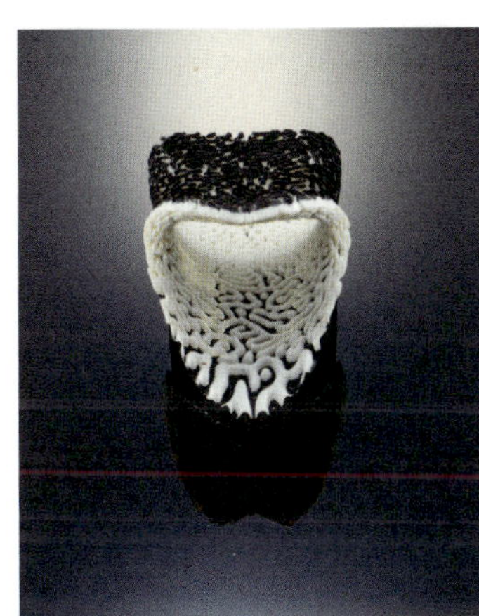

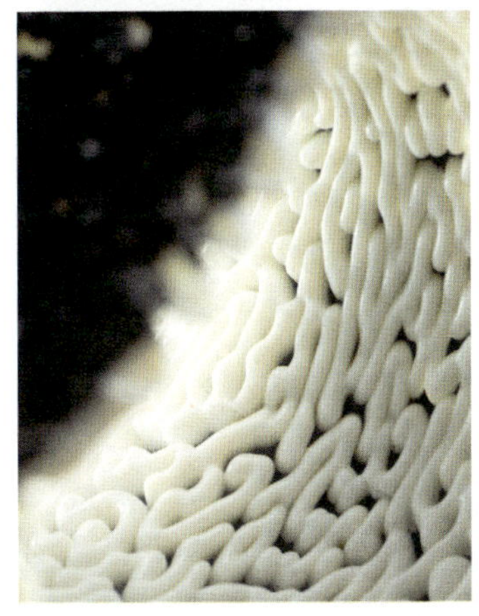

Gravida is a pregnancy corset with an external armor shell and an internal soft membrane. Its relatively high ratio of surface area to volume maximizes heat retention while providing comfort, support, and flexibility.

Design's history also features the amplification of human powers and compensation for human limitations, from basic technologies such as the wheel and everyday tools to high-tech prosthetics and wearables. The Imaginary Beings series works in that tradition, proposing improvements to the skeletal, pulmonary, and muscular functions of the human body. At its core is the premise that biological organs and organisms will someday be digitally designed and developed through technologies that will allow humans the abilities of Borges's imaginary beings. We will be able to fly, breathe underwater, and turn invisible.

In this speculative project, eighteen prototypes turn Borges's mythical qualities into prosthetics that could, as technology improves, enhance human abilities. For their design and digital fabrication, we drew on a library of algorithms inspired by natural forms and developed 3D-printing technologies, including a novel bitmap-printing technique, to support material performance and expression. Minotaur Head with Lamella, for example, is a protective helmet designed from the data provided by a CT scan of a human skull. Like other prototypes in the series, it explored the process of additive manufacturing using more than one material—both soft and stiff. The former make up the internal structure, providing comfort to the skull's hard tissues, while the latter make up the exterior protective shell. The helmet's elastic modulus was calculated for a high-resolution printer (16 microns, or 0.016-millimeter layers) at 700 dpi, allowing us to create stress/strain profiles for multiple points across the object's surface area.

What was once considered magic and captured in legend will be reality, as design and material technologies combine to offer more than meets the skin—spider suits, wing contraptions, and ultralight helmets—in what we call Mythologies of the Not Yet.

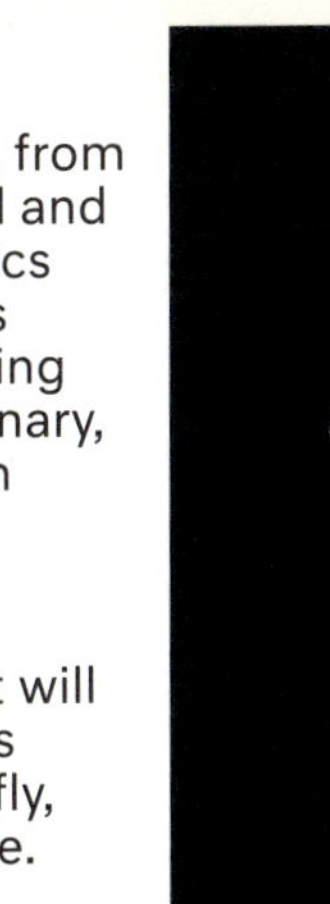

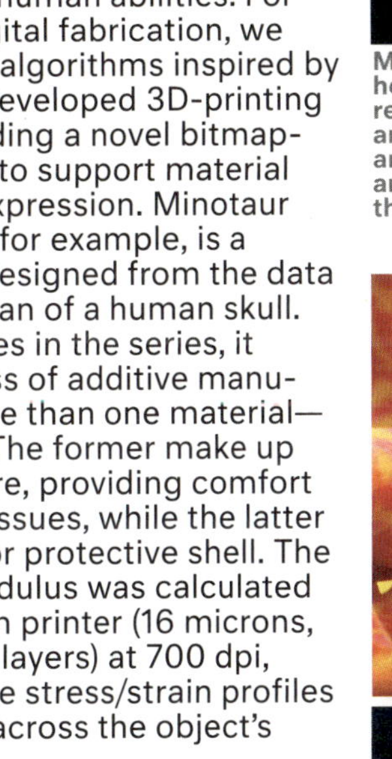

Minotaur Head with Sutures is a shock-absorbing helmet 3D printed in one swoop, with no assembly required. The sutures of varied thickness, density, and curvature were inspired by horn morphology and the suture calcification of the human skull, and they ensure flexibility without compromising the level of support for the wearer.

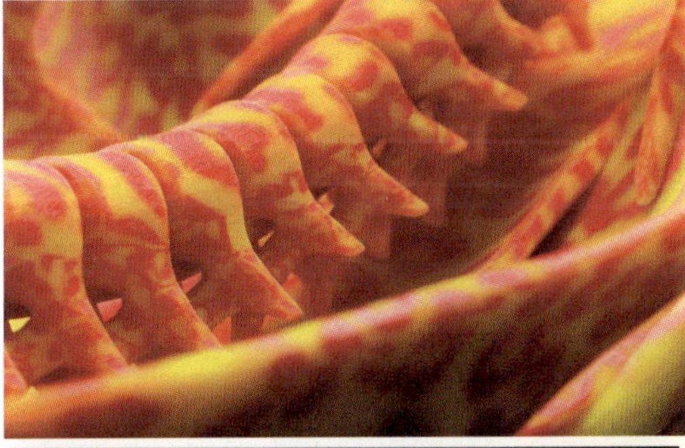

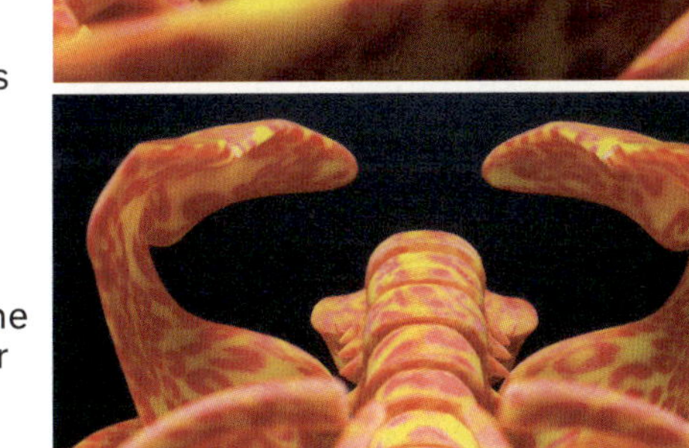

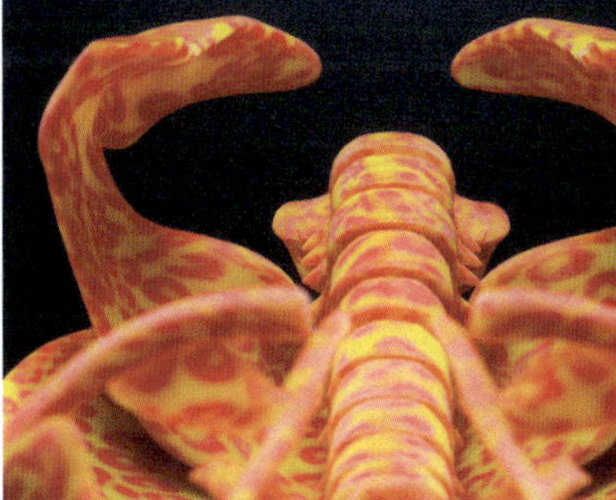

Kafka is armor with a soft spine-like element wrapped in a stiffer material to provide flexibility, protection, and support. It culminates in a semi-flexible neck element. The ornate soft-printed texture is more densely patterned in regions of high curvature, such as those located around spinal joints, providing for increased flexibility. The chimerical nature of this torso, and of Imaginary Beings in general, is embodied in this composition combining soft internal insect shell, stiff human armor, and semistiff leopard spots.

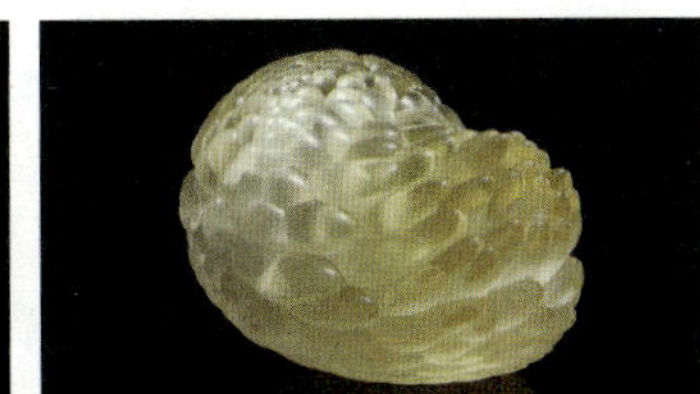

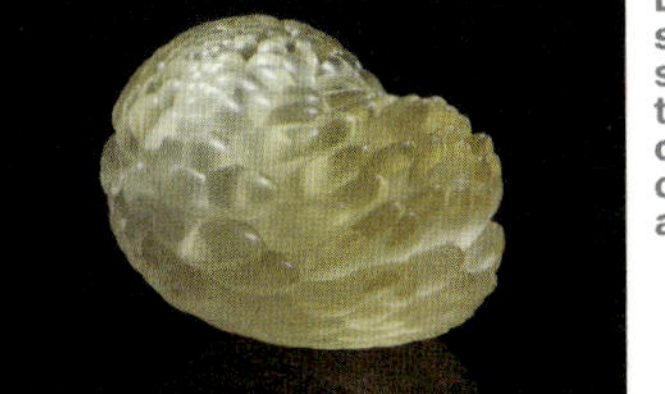

Daphne is a prosthetic knee socket that combines a robust shell structure with a soft inner tissue to provide protection and comfort. Its elastic moduli are calculated according to the anticipated load.

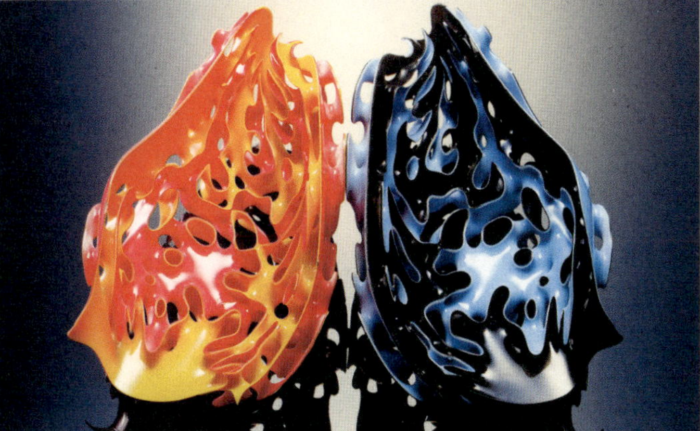

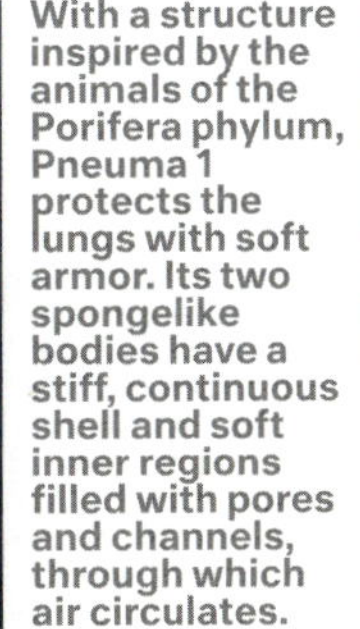

With a structure inspired by the animals of the Porifera phylum, Pneuma 1 protects the lungs with soft armor. Its two spongelike bodies have a stiff, continuous shell and soft inner regions filled with pores and channels, through which air circulates.

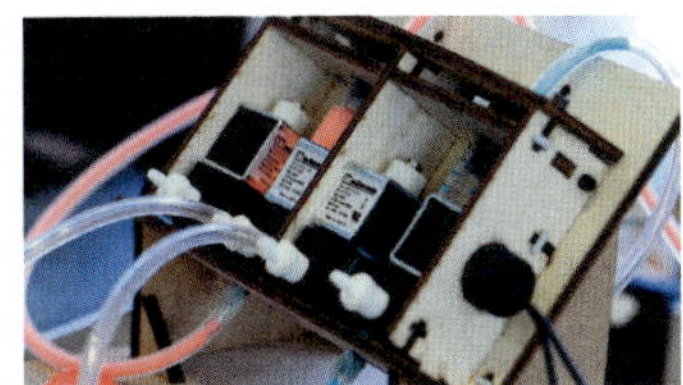

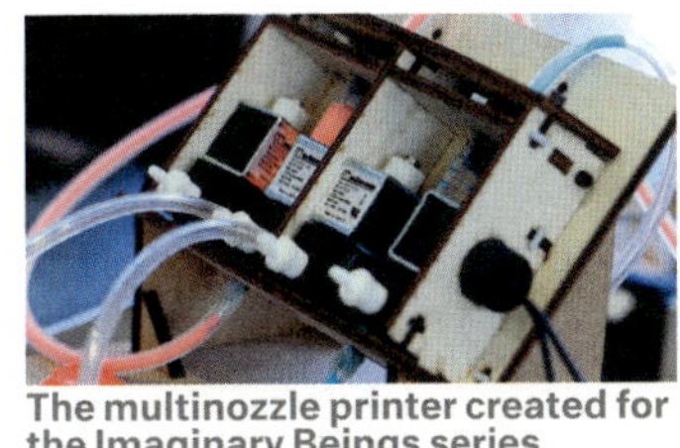

The multinozzle printer created for the Imaginary Beings series increases gradient resolution without sacrificing printing speed. The gradients were generated internally, in each nozzle reservoir, before extrusion.

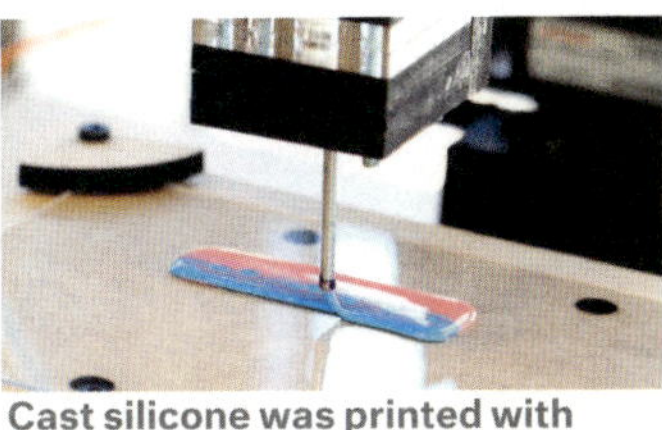

Cast silicone was printed with color gradients using a 3D-printing platform. Red and blue adhesives were mixed in different controlled ratios: the softer blue silicone (Shore 00-10) was combined with a harder red silicone (Shore 00-50) to produce gradients in color and hardness.

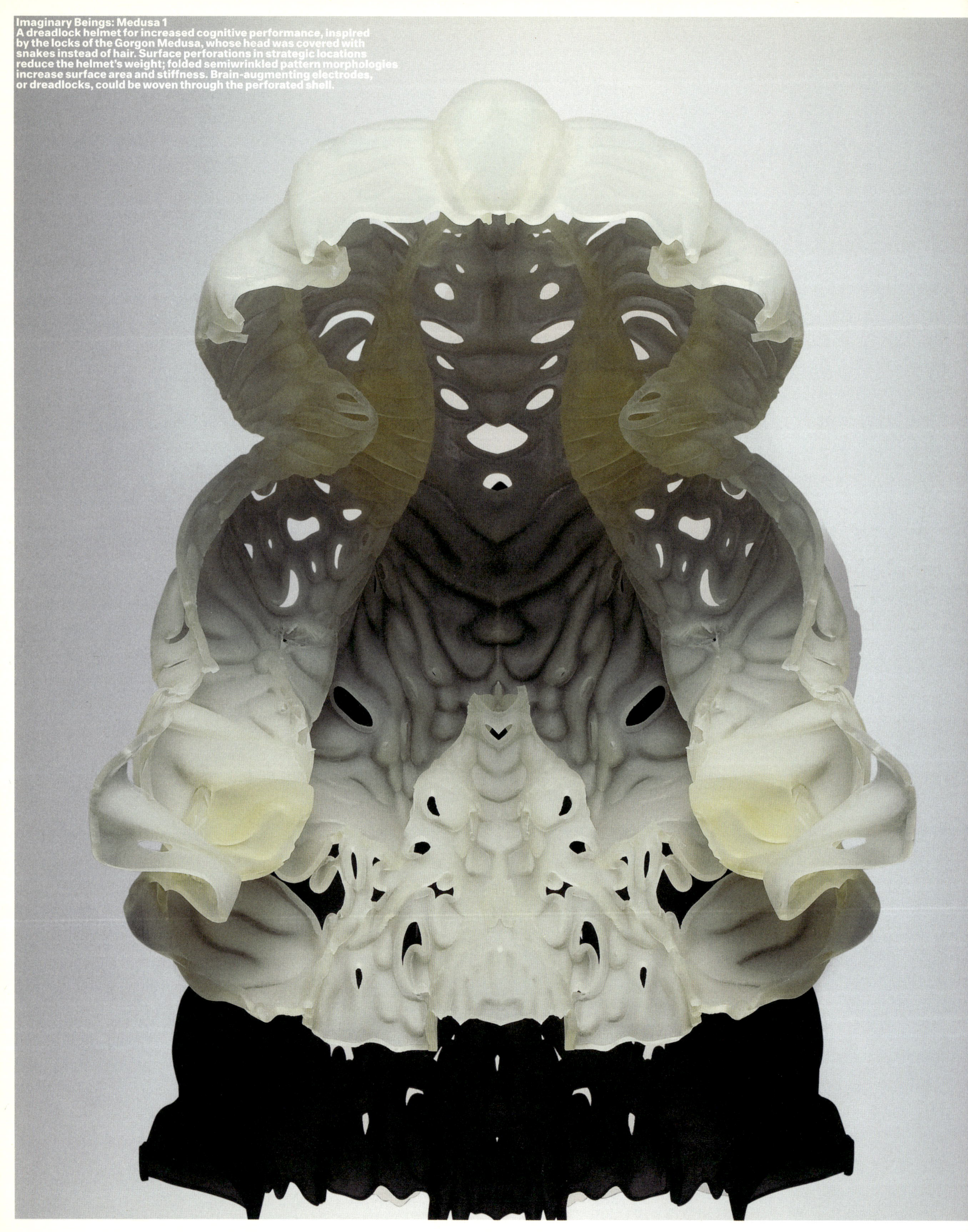

Imaginary Beings: Medusa 1
A dreadlock helmet for increased cognitive performance, inspired by the locks of the Gorgon Medusa, whose head was covered with snakes instead of hair. Surface perforations in strategic locations reduce the helmet's weight; folded semiwrinkled pattern morphologies increase surface area and stiffness. Brain-augmenting electrodes, or dreadlocks, could be woven through the perforated shell.

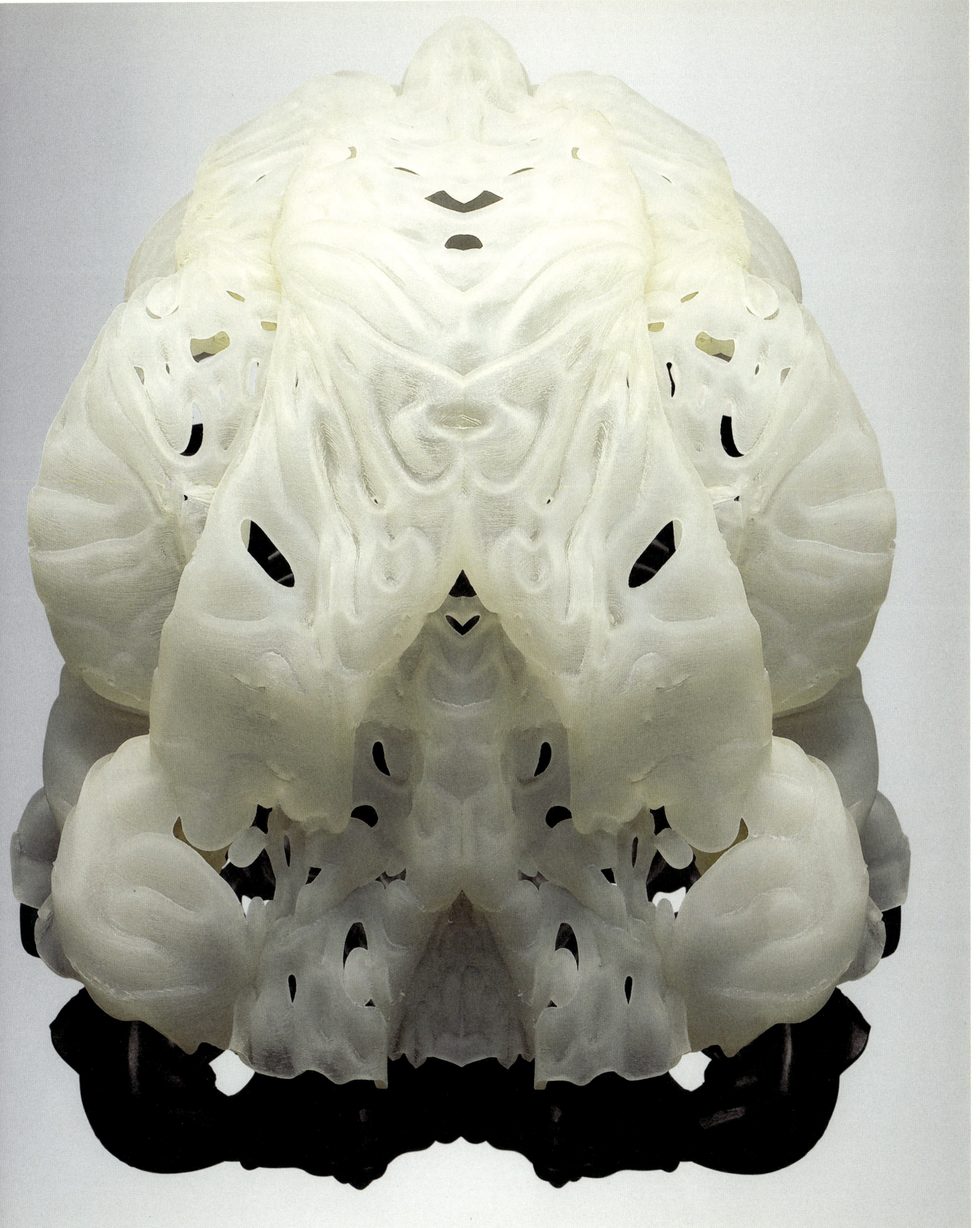

Imaginary Beings: Doppelgänger
A winged corset created through an algorithmic faceting technique. The white facets are visible from above (to attract mates), and the black facets are visible from below (to hide from predators).

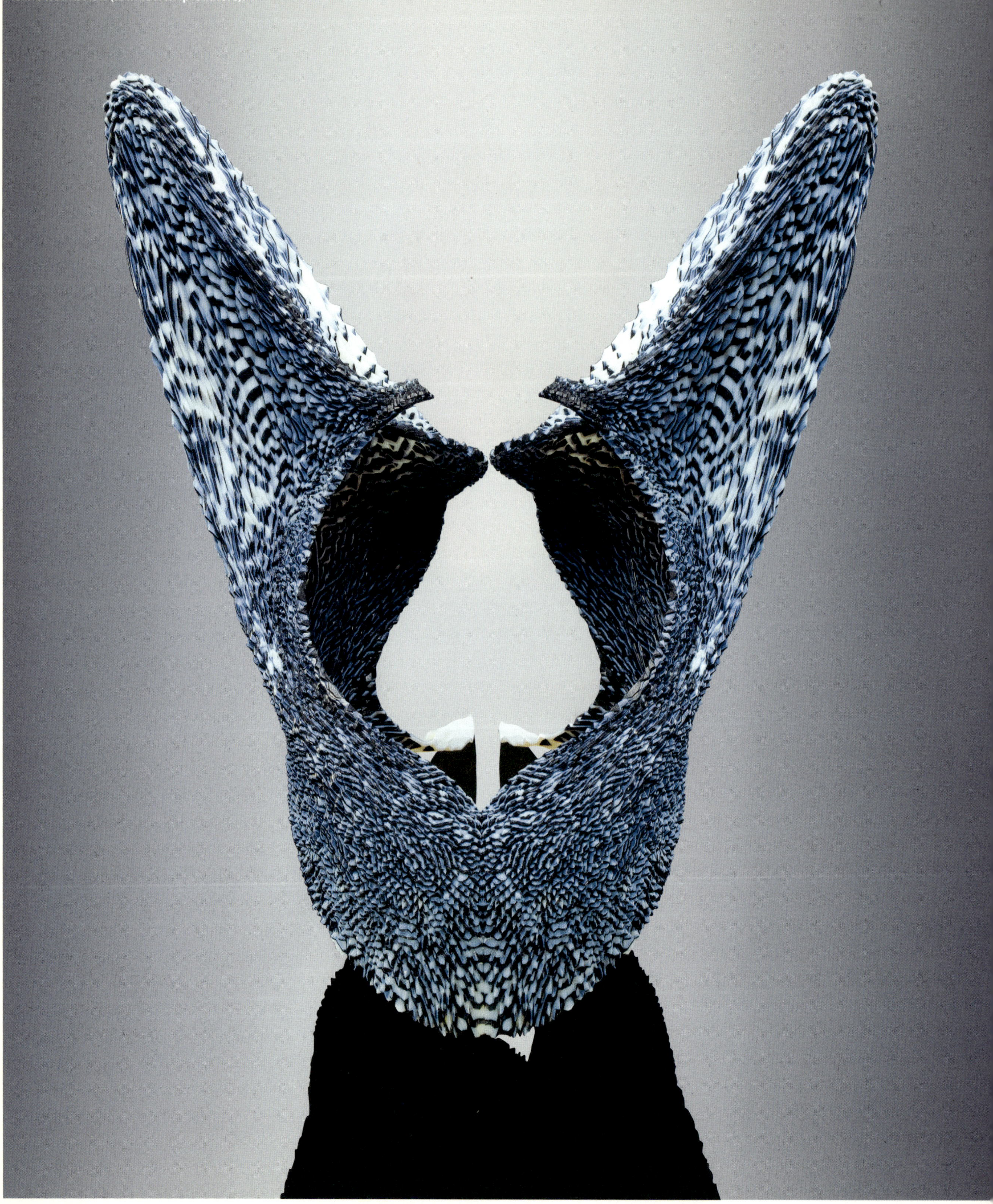

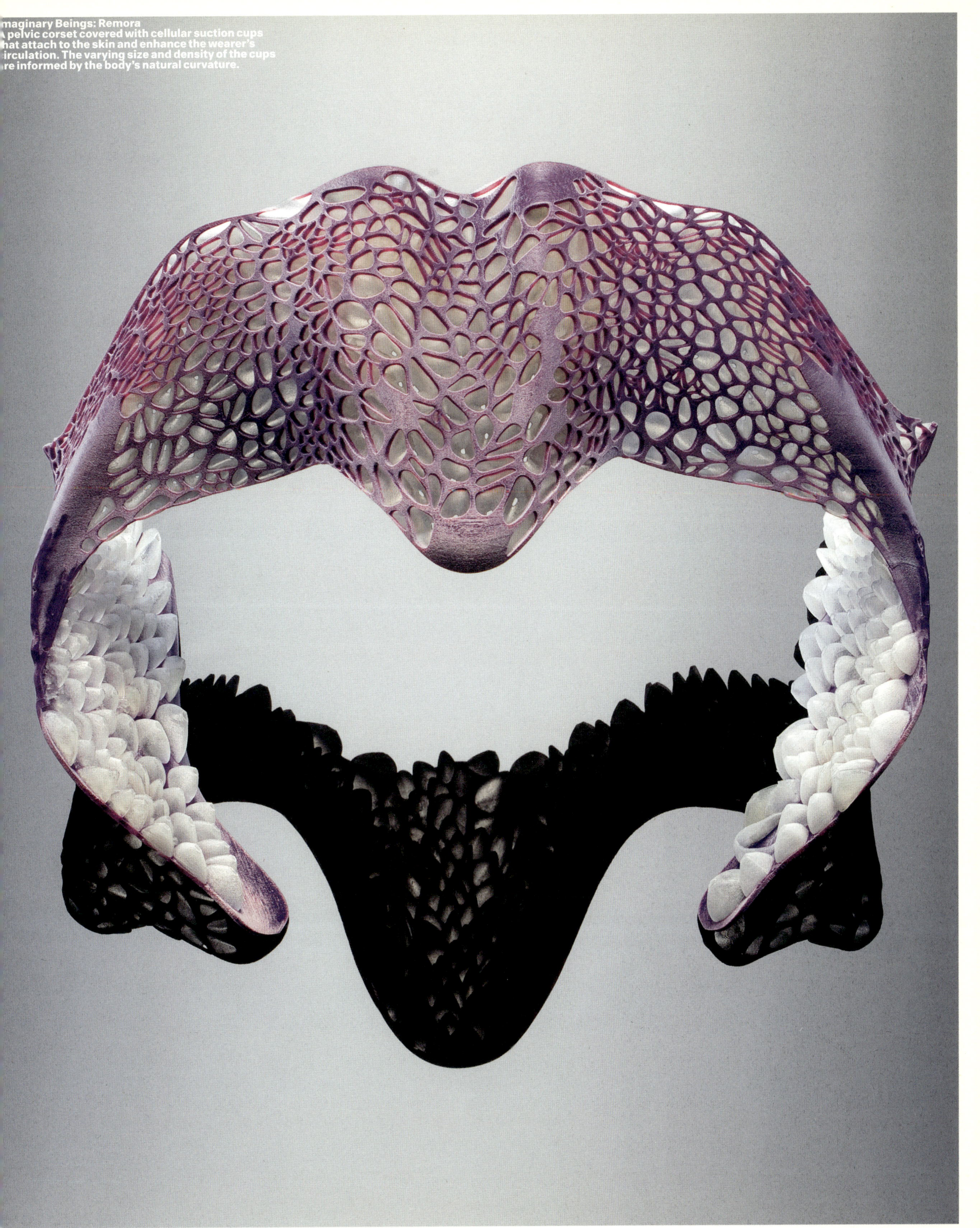

maginary Beings: Remora
A pelvic corset covered with cellular suction cups that attach to the skin and enhance the wearer's circulation. The varying size and density of the cups are informed by the body's natural curvature.

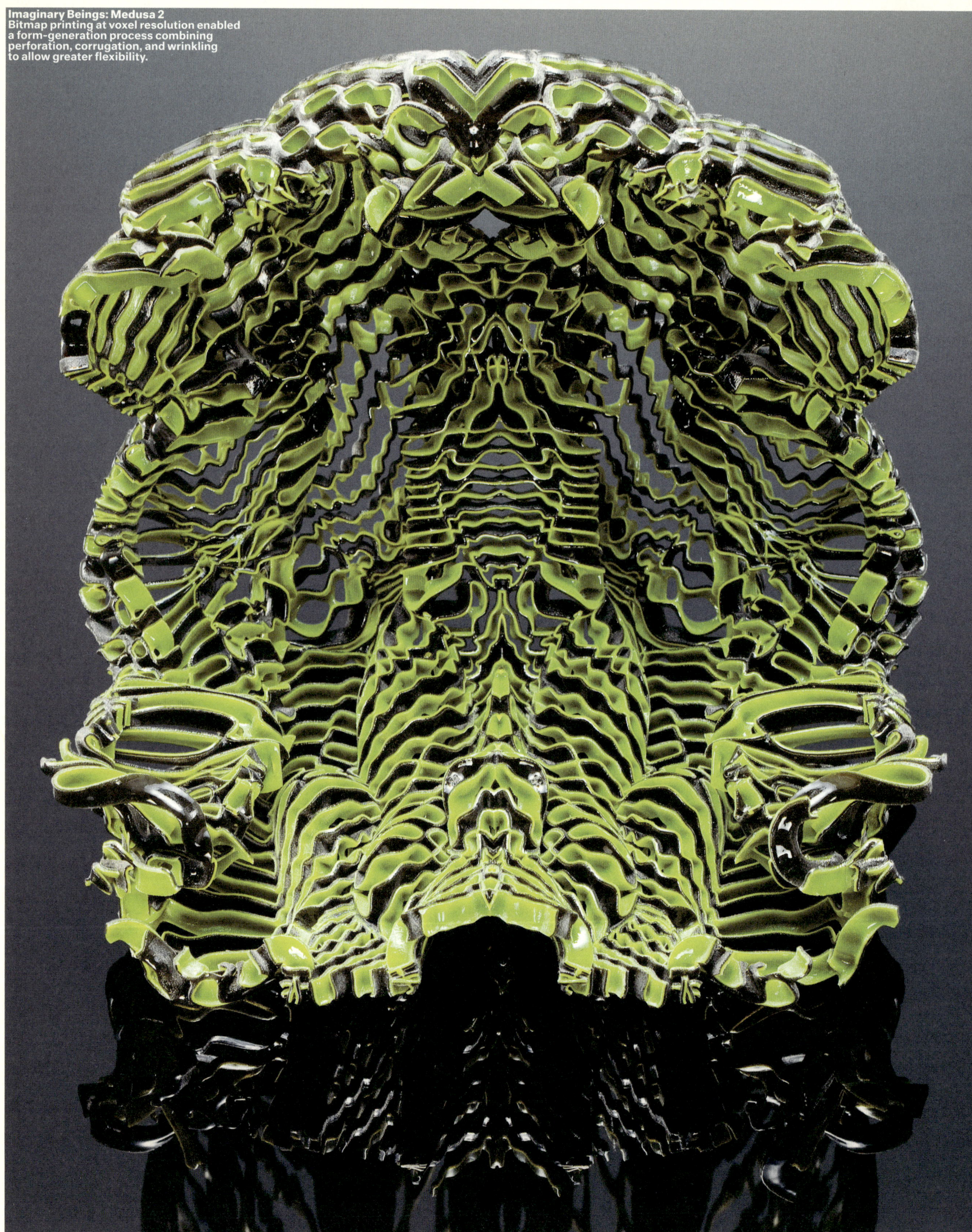

Imaginary Beings: Medusa 2
Bitmap printing at voxel resolution enabled a form-generation process combining perforation, corrugation, and wrinkling to allow greater flexibility.

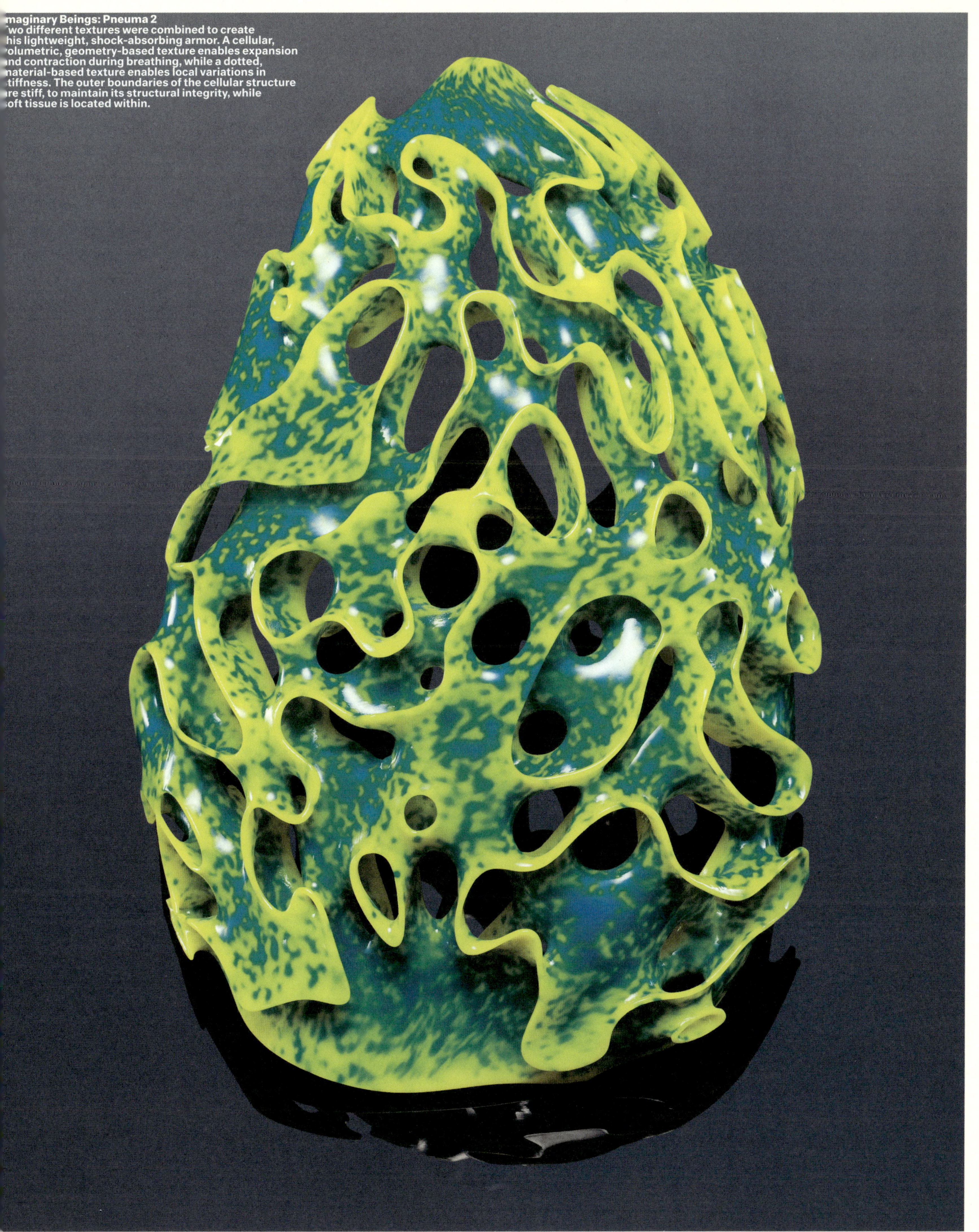

maginary Beings: Pneuma 2
wo different textures were combined to create his lightweight, shock-absorbing armor. A cellular, olumetric, geometry-based texture enables expansion nd contraction during breathing, while a dotted, naterial-based texture enables local variations in tiffness. The outer boundaries of the cellular structure re stiff, to maintain its structural integrity, while oft tissue is located within.

VESPERS / LAZARUS

Neri Oxman and
The Mediated Matter Group
Lazarus. 2016
Photopolymers
10 3⁄16 × 6 5⁄16 × 5 7⁄8 in.
(25.9 × 16 × 15 cm)
Produced by Stratasys Ltd.
Research team: Christoph Bader, Dominik Kolb, James C. Weaver, Neri Oxman
Collaborators and contributors: Gal Begun, Boris Belocon, Naomi Kaempfer, Danielle van Zadelhoff

Created in collaboration with Stratasys Ltd. for the New Ancient Collection

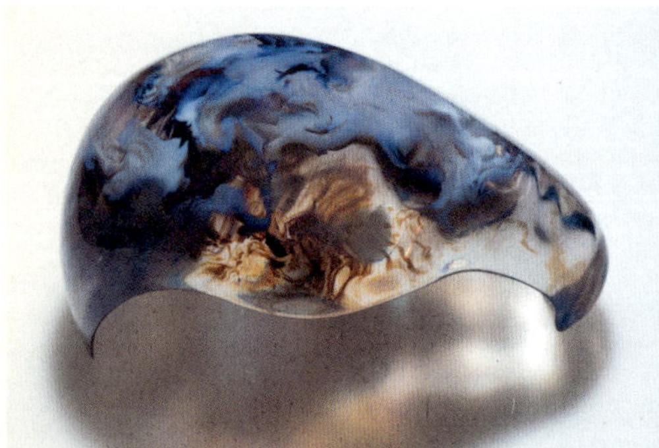

Lazarus.

Vespers is a collection of masks that explore what it means to design (with) life. From the relic of the death mask to contemporary living devices, the collection charts a trajectory from ritual to form to rebirth: from an ancient typology to a novel technology, from a conceptual piece to a tangible set of tools and techniques for combining programmable matter and programmable life. This project points toward an imminent future of customized wearable interfaces and building skins that fit not only a particular shape but also specific material, chemical, and genetic makeups—a future in which 3D-printed interfaces will guide the formation of drugs calibrated to fit a user's genetic makeup or will create architecture that responds and adapts to environmental cues.

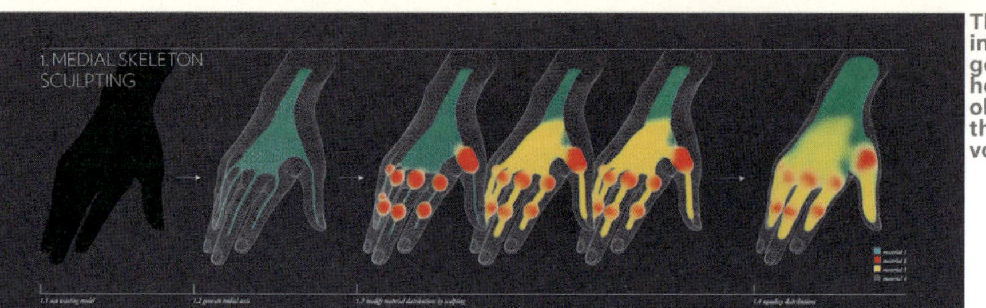

Thanks to advancements in additive manufacturing, geometrically complex, materially heterogeneous, high-resolution objects can now be fabricated. For this project, different methods of volumetric modeling were explore

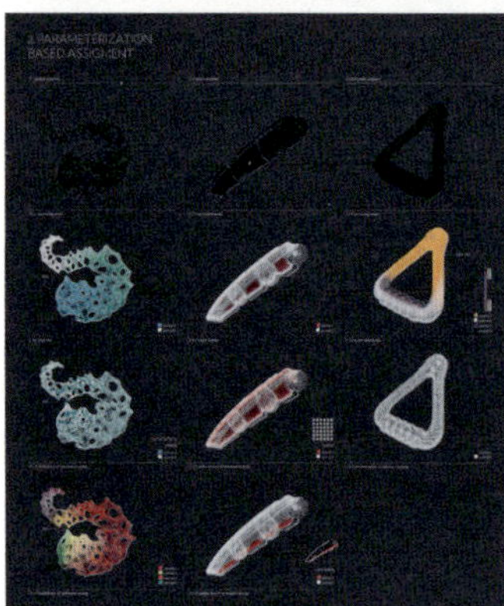

Objects and building components designed to promote functionally graded properties that enhance overall structural and environmental performance. The project's hybrid approach to heterogeneous material modeling (combining user-driven and generative model methods) enabled the augmentation of current 3D-modeling methods.

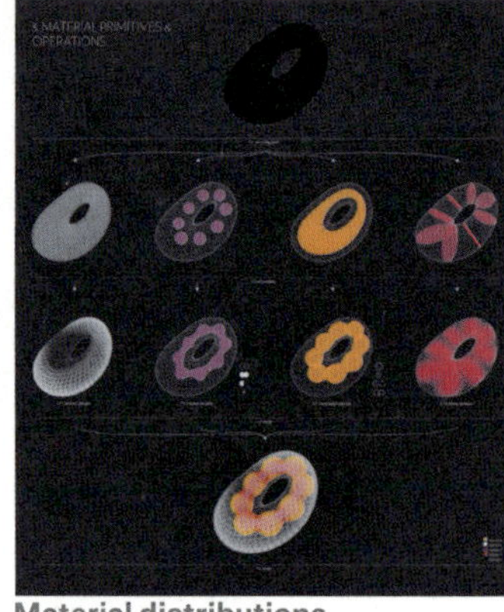

Material distributions "sculpted" from the medial axis of a three-dimensional object.

Before Vespers, there was Lazarus, our first reinterpretation of an ancient death mask. The death mask is traditionally a wax or plaster impression taken directly from a corpse; our version, which we call an "air urn," is designed to contain a person's last breath. Unlike the handmade mask, Lazarus was our first project to use a data-driven material modeling (DdMM) approach, creating high-resolution, geometrically complex, materially heterogeneous 3D-printed objects. The surface area of an individual mask is modeled on the face of a person near death; its material composition is then informed by the physical flow and distribution of air across that surface—data that could be gathered from the wearer or digitally generated by a computational process. The mask is then additively manufactured to create a unique artifact for a person's final exhale.

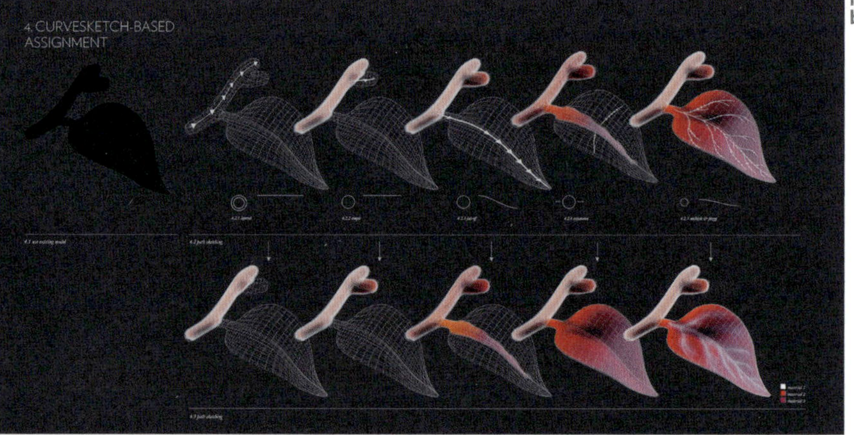

Material distributions determine by a network of curves.

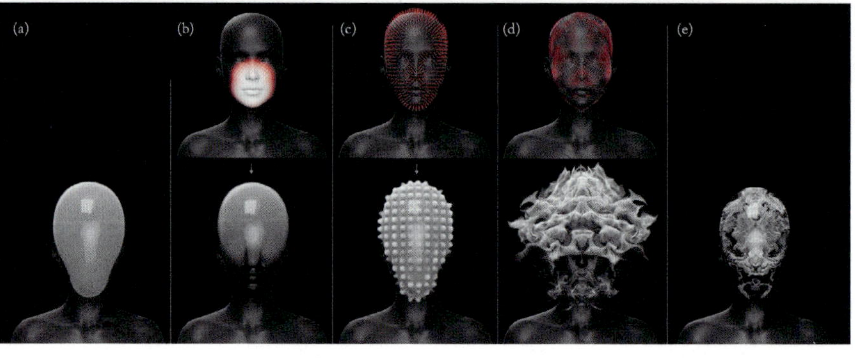

Data-source contributions to the generation of internal material distributions, with gray areas representing VeroWhite (RGD835) resin and transparent areas representing VeroClear (RGD810) resin.

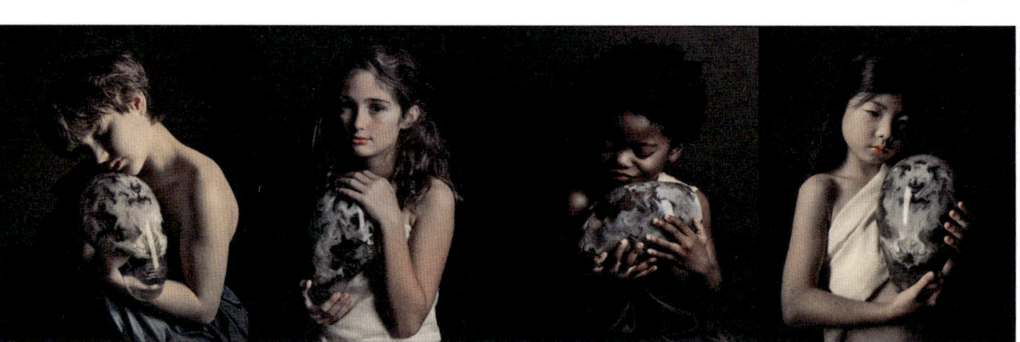

The air urn as a memento of the departed—a 3D-printed vessel for their last breath.

omputational framework for the roduction of bitmap-printable arts: object representation and ata generation (a), optional eneration of a polygonal slice (b), iority ordering of specified aterial domains (c), rasterization), on-slice-time data transfer and odification (e), composition (f), aterial dithering and direct tmap-printing system (g).

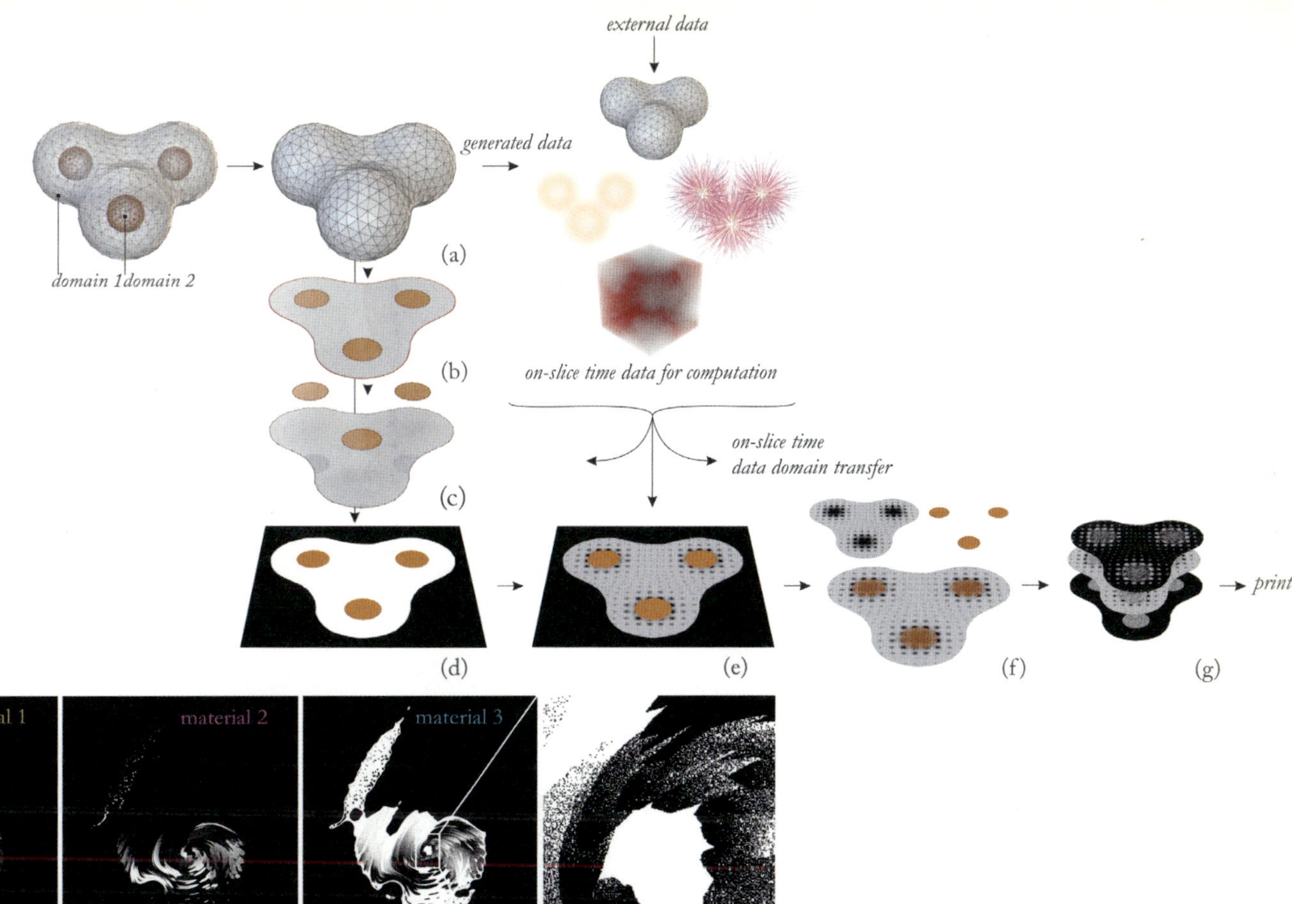

oss section showing variations in the material composition an object printed in three different materials.

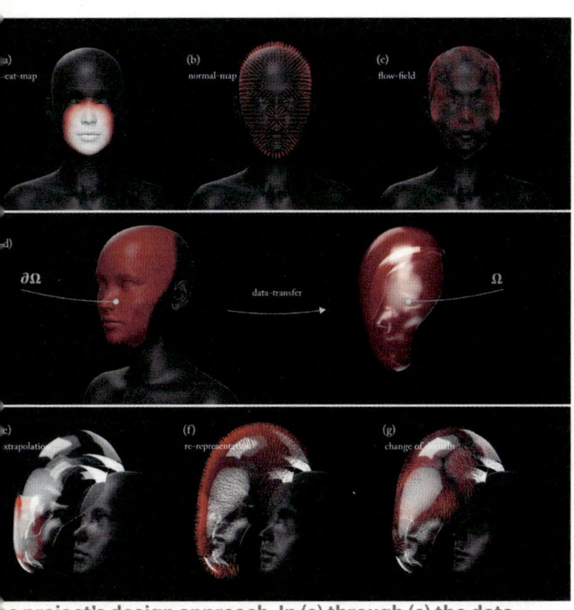

e project's design approach. In (a) through (c) the data urces are generated: heat map (a), normal map (b), velocity eld computed over the surface of the face (c). In (d) through), the data sources are transferred to the object of interest.

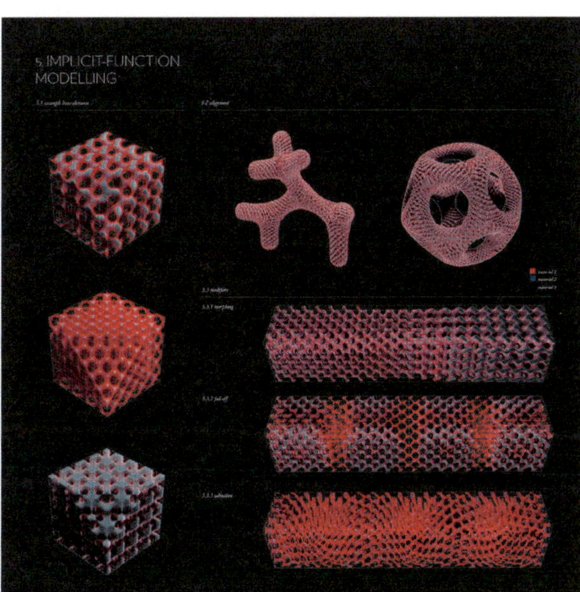

One of the data-driven material-modeling methods developed for the project. The parameters of a range of material compositions are determined by a given geometry; they then inform the material distributions.

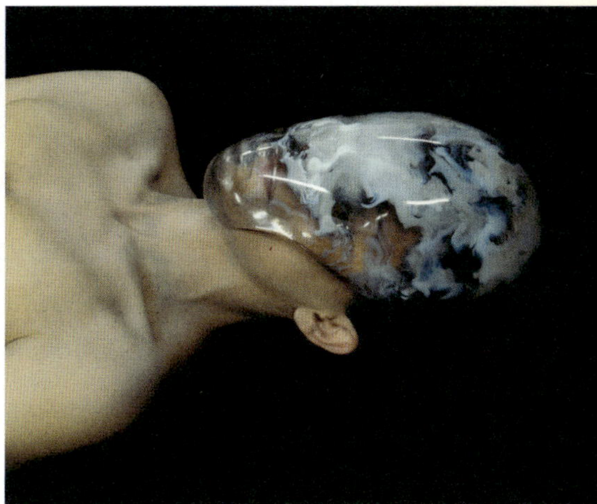

The mask's material composition is informed by the physical flow of a person's last breath over the surface of the face. Unlike the design of a traditional handmade mask, Lazarus's is data driven, digitally generated, and additively manufactured.

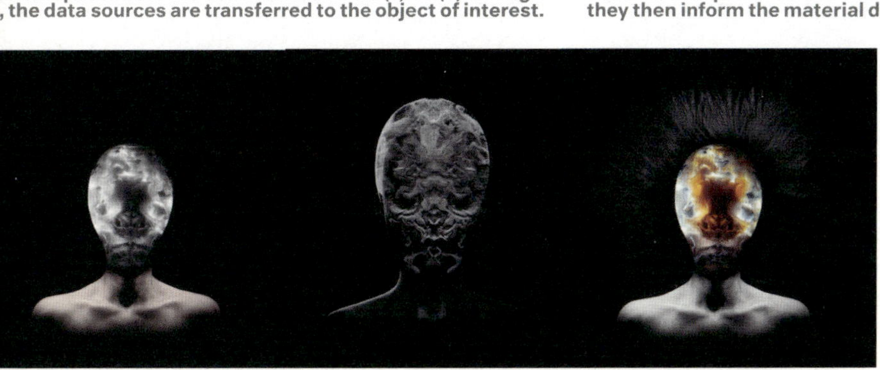

The mask is the product of a functional advection workflow, 3D printed from rigid white and transparent materials.

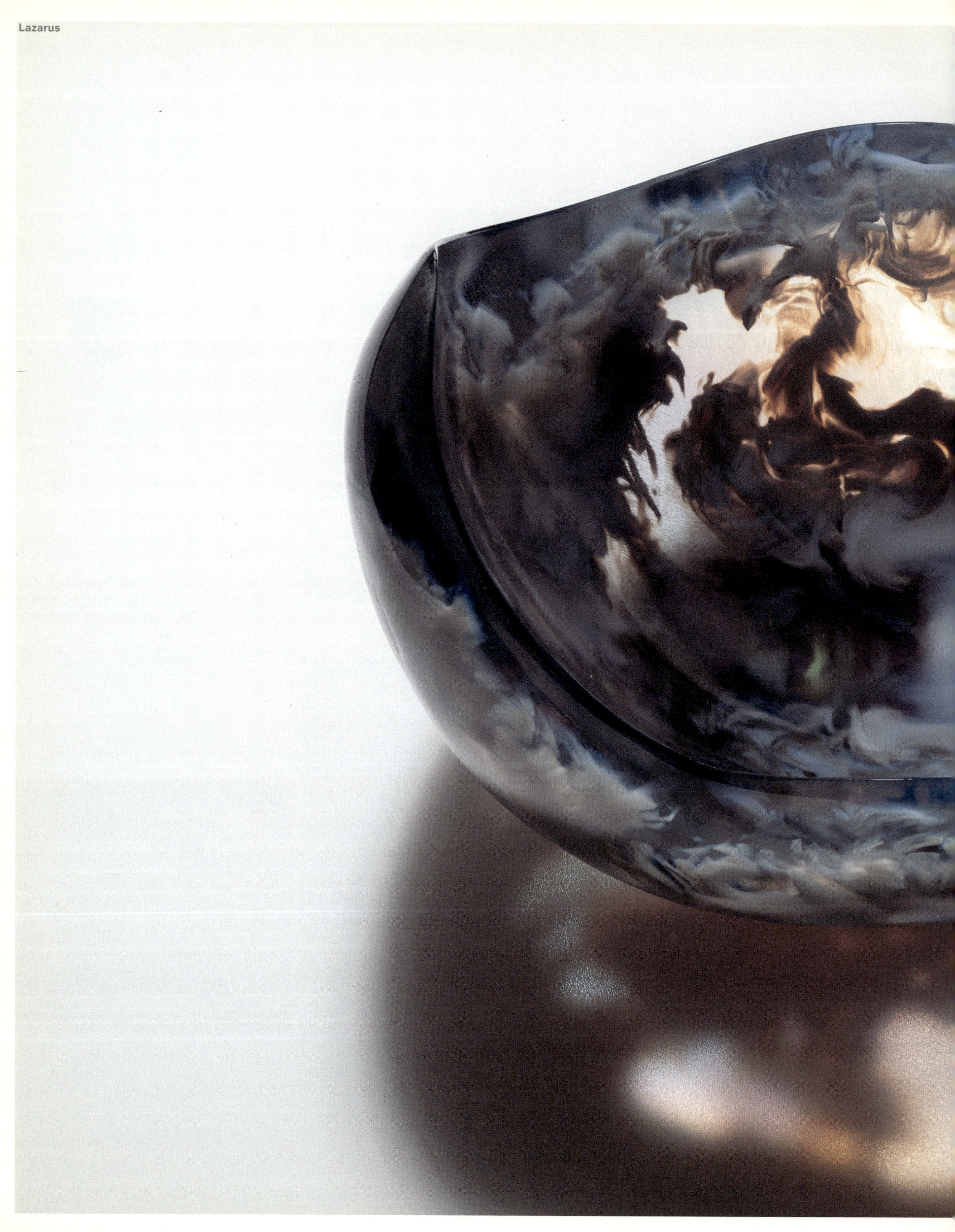

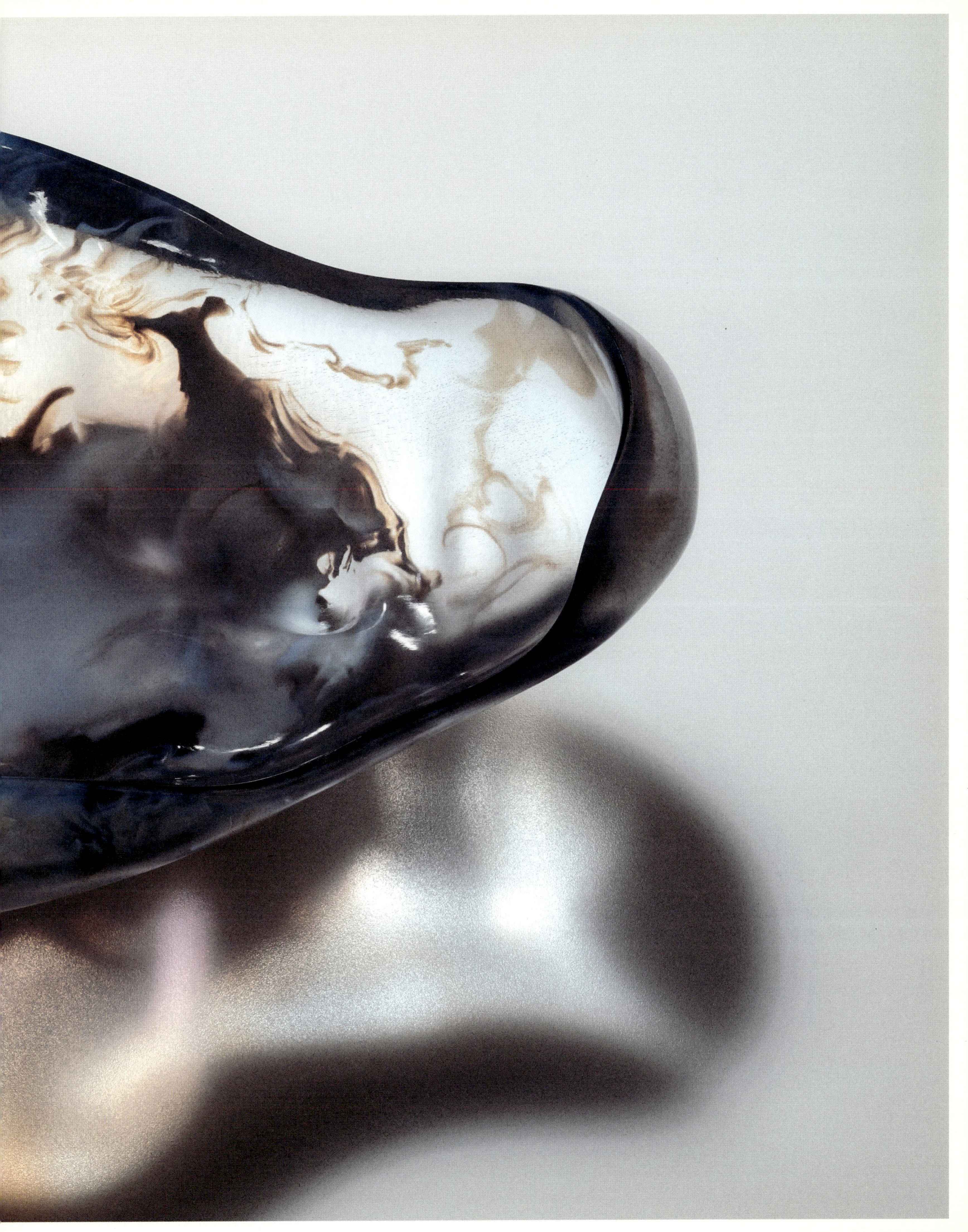

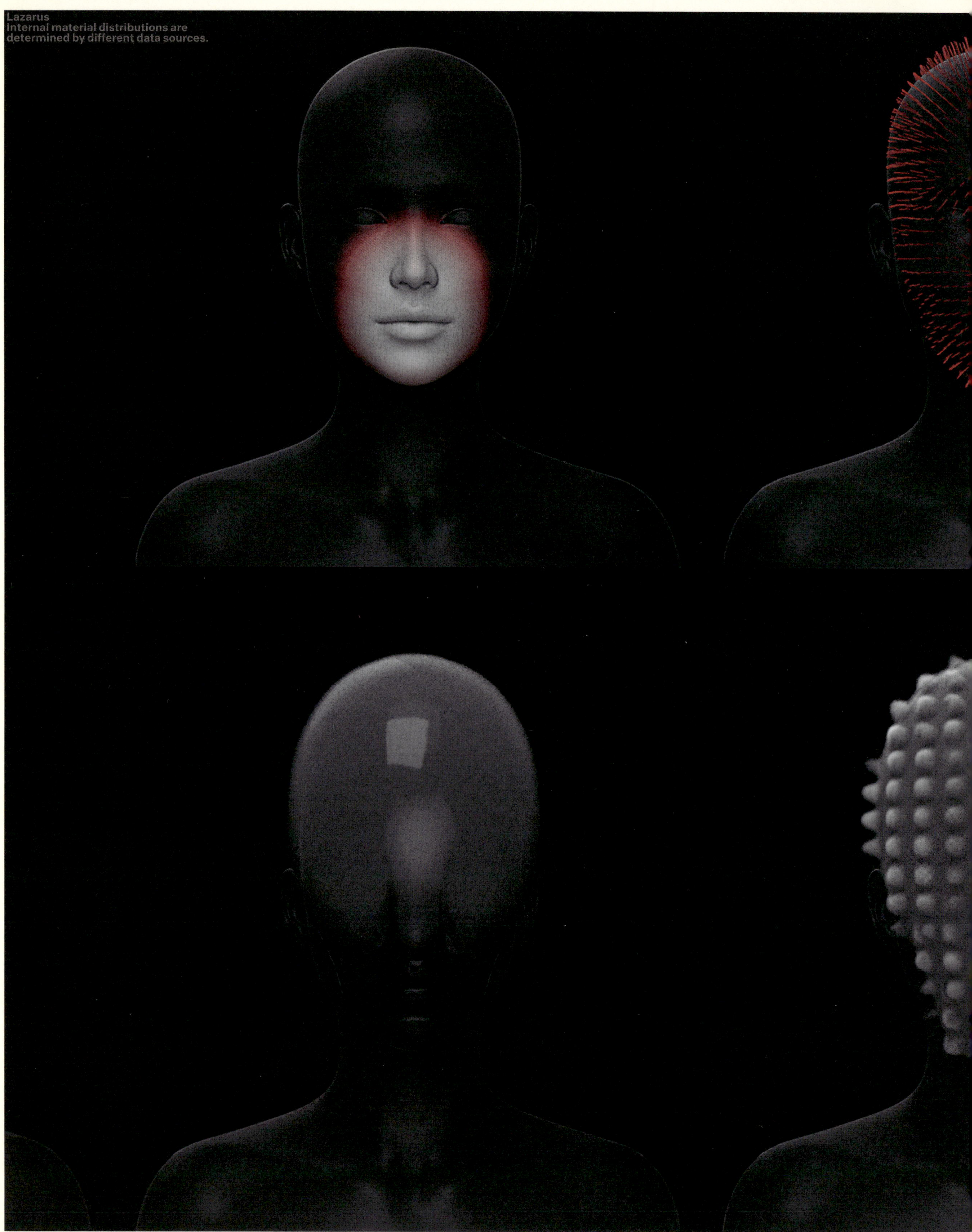

Lazarus
Internal material distributions are determined by different data sources.

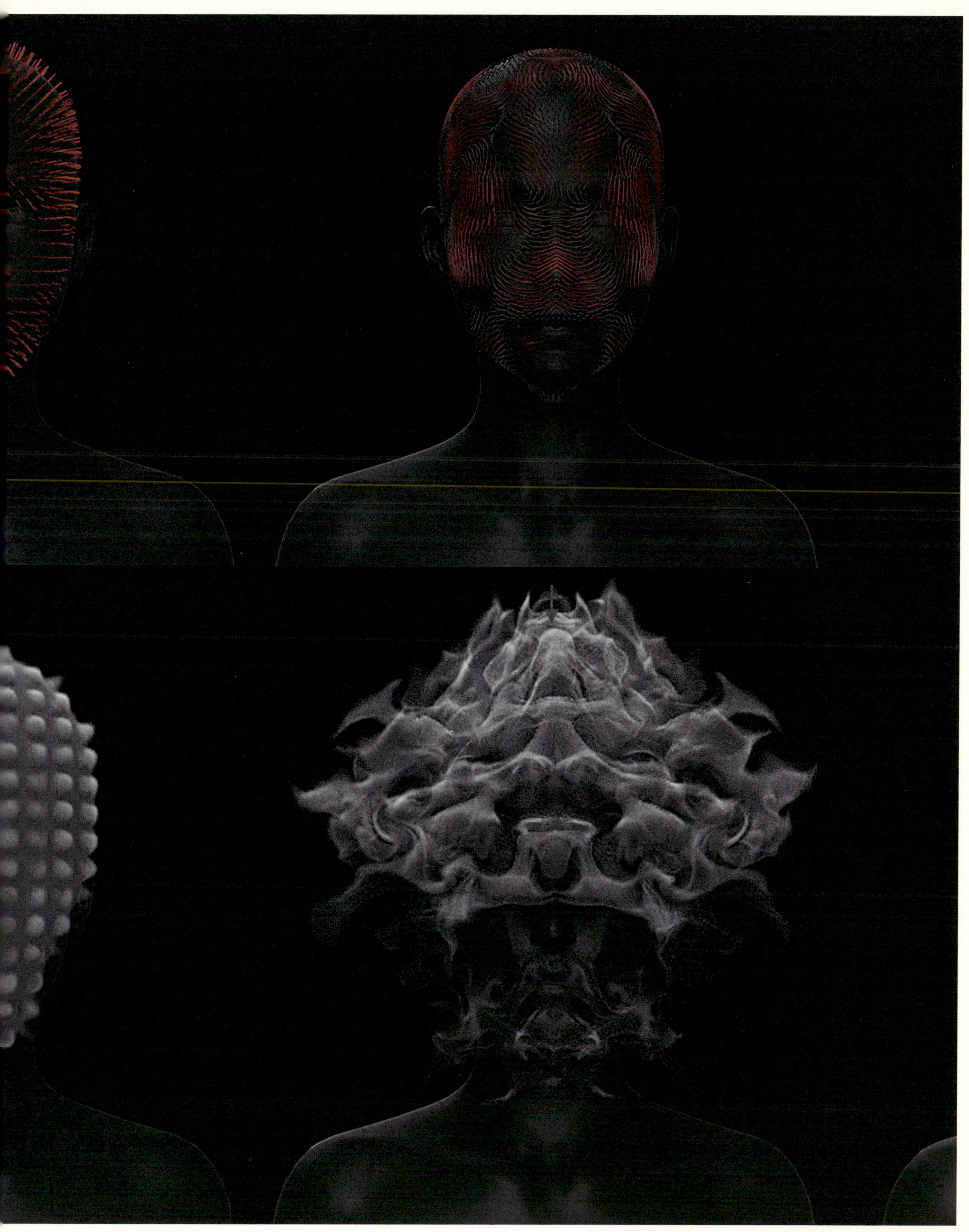

VESPERS I

Neri Oxman and
The Mediated Matter Group
Vespers I. 2018
Photopolymers
Dimensions variable
Produced by Stratasys Ltd.
Research team: Christoph Bader, Dominik Kolb, James C. Weaver, Neri Oxman
Collaborators and contributors: Gal Begun; Boris Belocon; Naomi Kaempfer; MIT Media Lab; Danielle van Zadelhoff; Wyss Institute, Harvard University

Created in collaboration with Stratasys Ltd. for the New Ancient Collection.

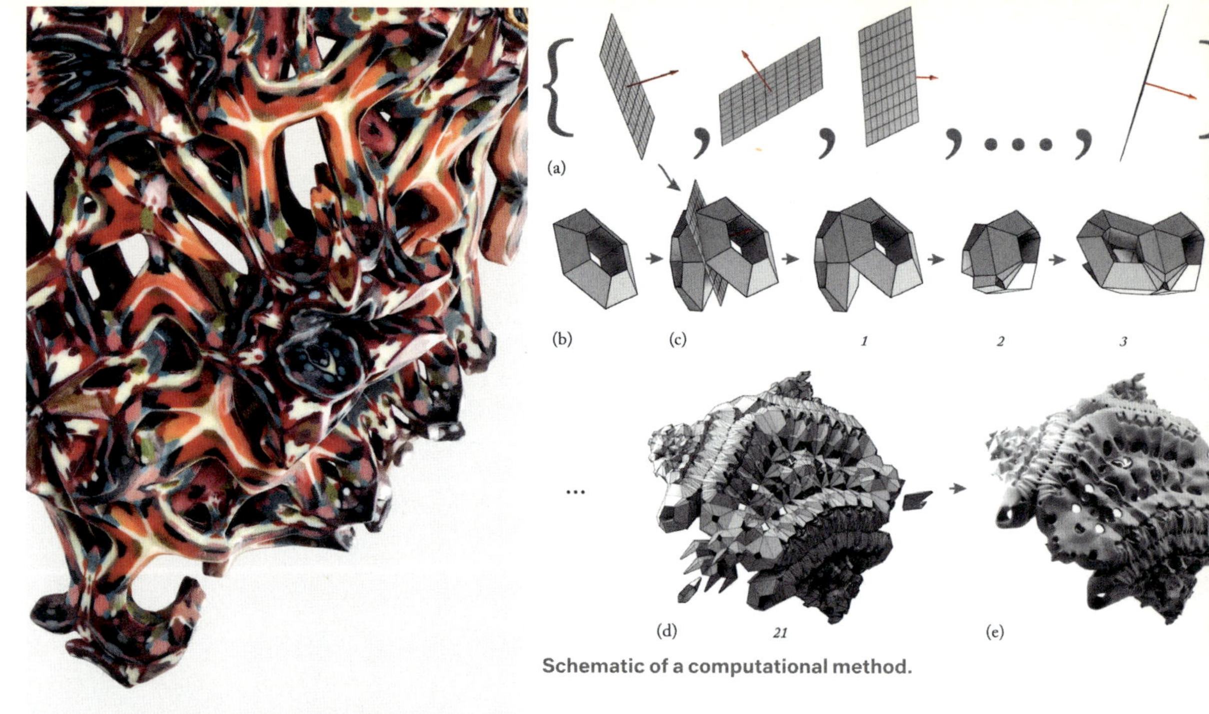

Schematic of a computational method.

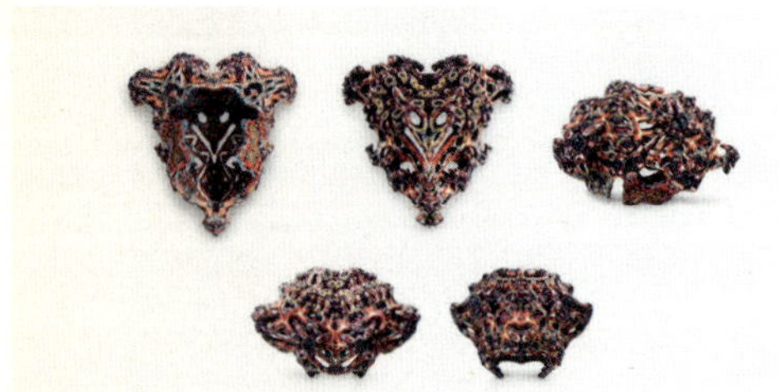
Mask 1.

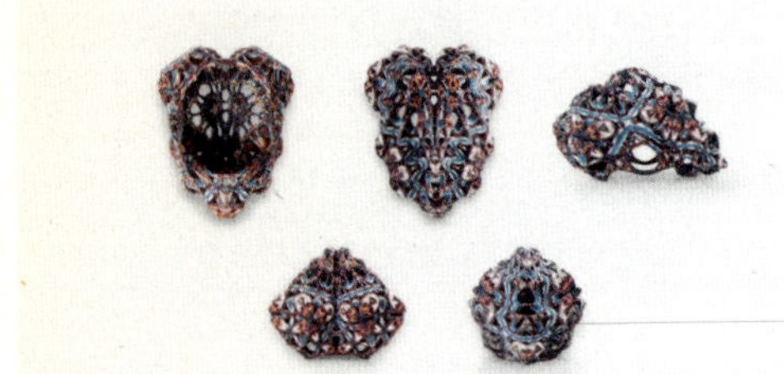
Mask 2.

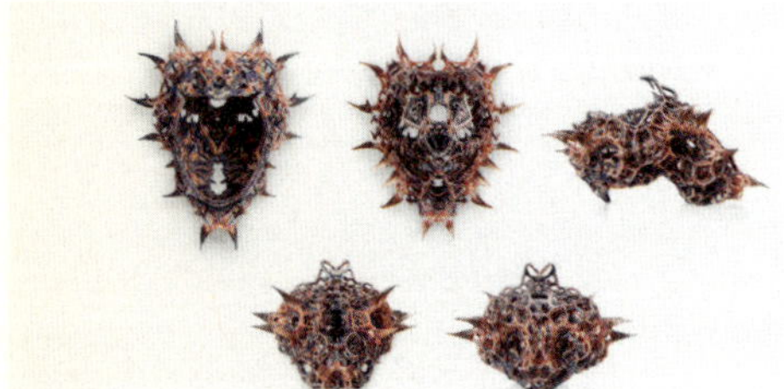
Mask 3.

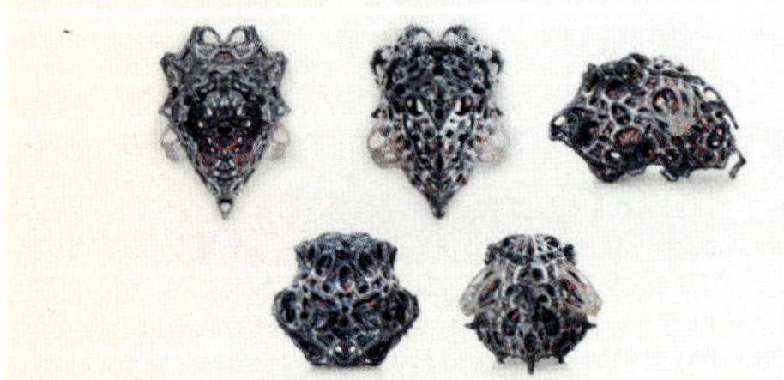
Mask 4.

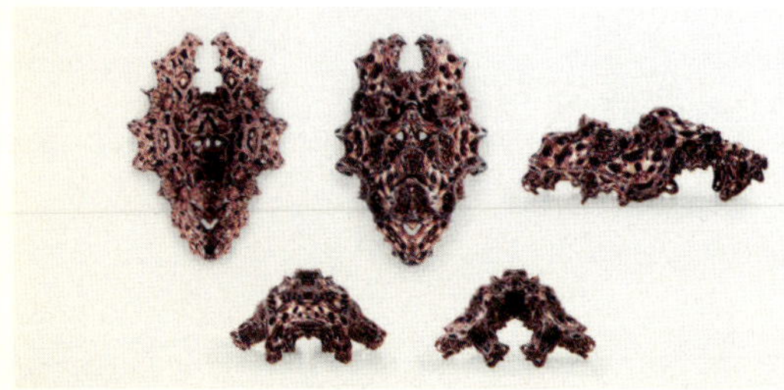
Mask 5.

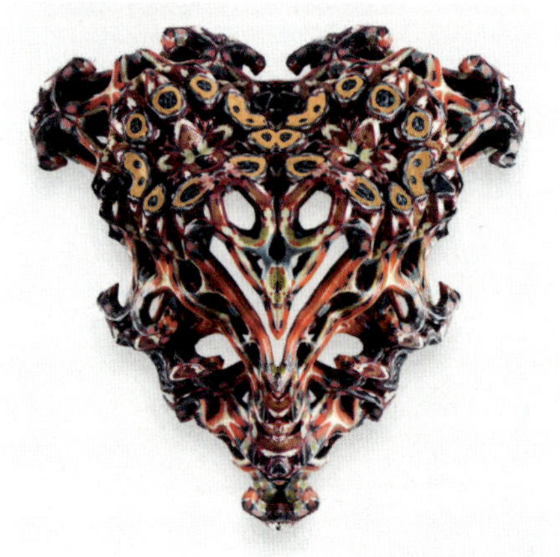

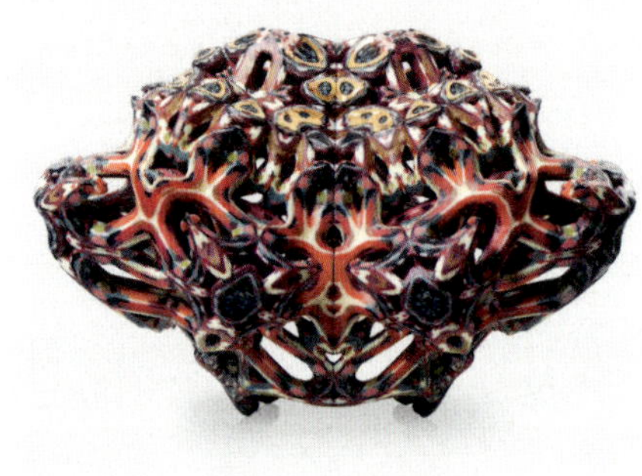
Mask 1.

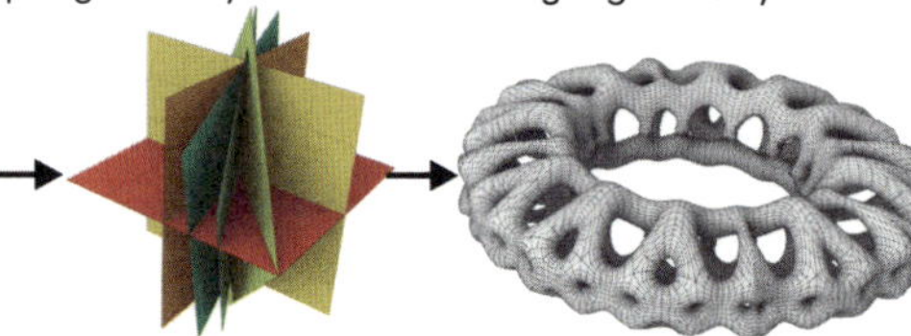

Schematic of a generated shape adapted to a given target geometry using a novel generative design method.

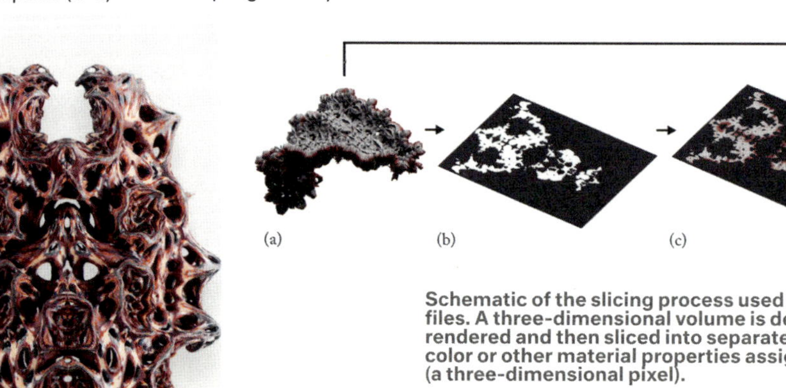

Schematic of the slicing process used to create print files. A three-dimensional volume is designed and rendered and then sliced into separate files with color or other material properties assigned per voxel (a three-dimensional pixel).

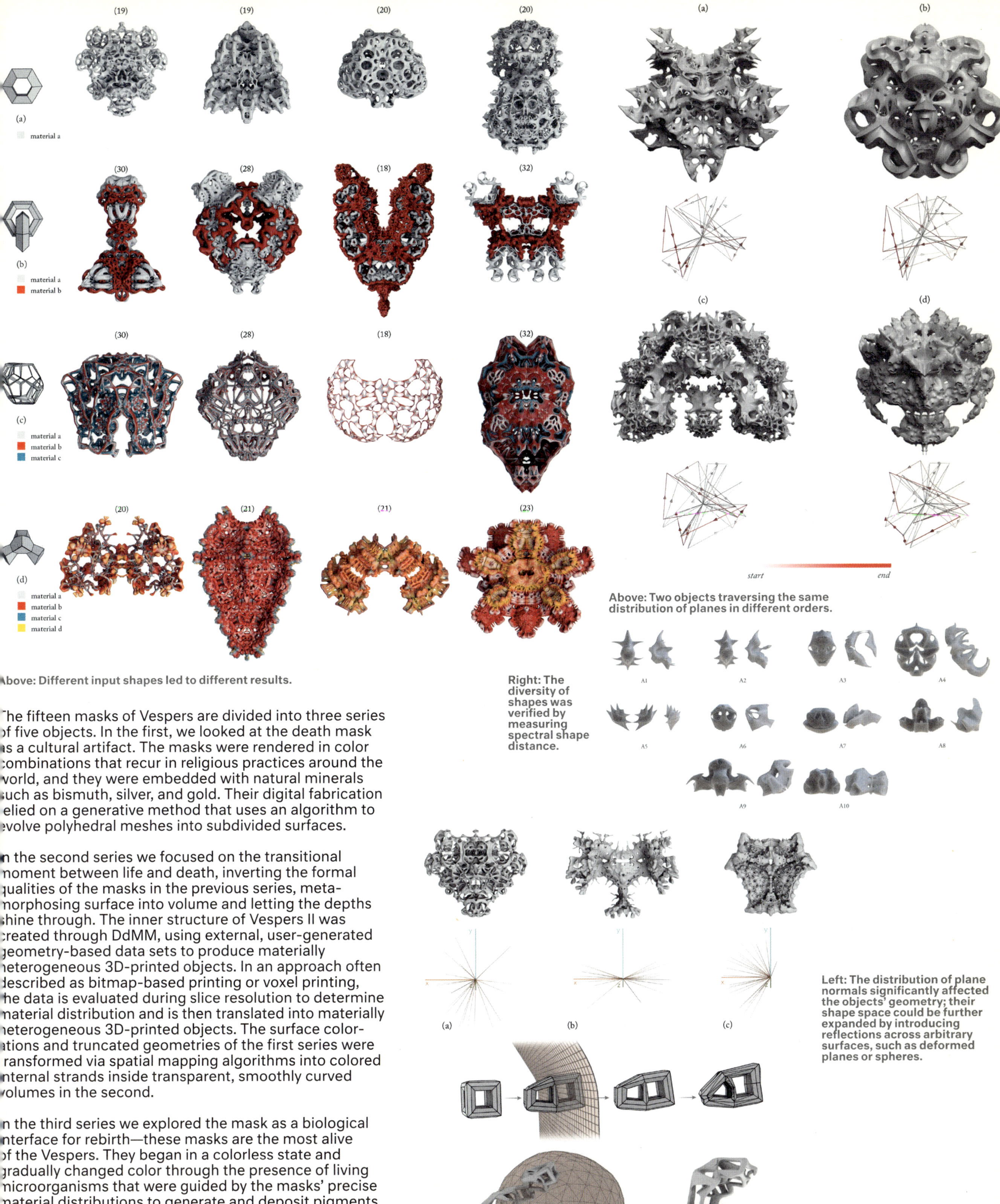

Above: Two objects traversing the same distribution of planes in different orders.

Right: The diversity of shapes was verified by measuring spectral shape distance.

Left: The distribution of plane normals significantly affected the objects' geometry; their shape space could be further expanded by introducing reflections across arbitrary surfaces, such as deformed planes or spheres.

Above: Different input shapes led to different results.

The fifteen masks of Vespers are divided into three series of five objects. In the first, we looked at the death mask as a cultural artifact. The masks were rendered in color combinations that recur in religious practices around the world, and they were embedded with natural minerals such as bismuth, silver, and gold. Their digital fabrication relied on a generative method that uses an algorithm to evolve polyhedral meshes into subdivided surfaces.

In the second series we focused on the transitional moment between life and death, inverting the formal qualities of the masks in the previous series, metamorphosing surface into volume and letting the depths shine through. The inner structure of Vespers II was created through DdMM, using external, user-generated geometry-based data sets to produce materially heterogeneous 3D-printed objects. In an approach often described as bitmap-based printing or voxel printing, the data is evaluated during slice resolution to determine material distribution and is then translated into materially heterogeneous 3D-printed objects. The surface colorations and truncated geometries of the first series were transformed via spatial mapping algorithms into colored internal strands inside transparent, smoothly curved volumes in the second.

In the third series we explored the mask as a biological interface for rebirth—these masks are the most alive of the Vespers. They began in a colorless state and gradually changed color through the presence of living microorganisms that were guided by the masks' precise material distributions to generate and deposit pigments according to the spatial logic of the masks in the second series. The organisms produced pigments that emulated the colors we used in Vespers I, and the nonliving materials in the masks of Vespers III provided a habitat for the living ones.

VESPERS II

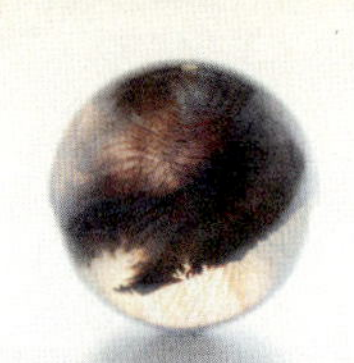
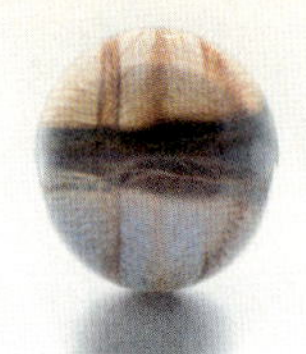
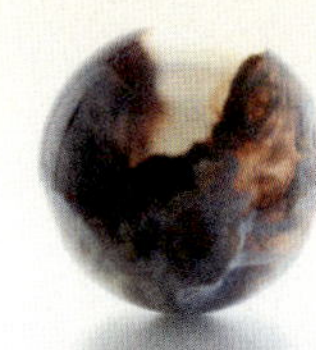
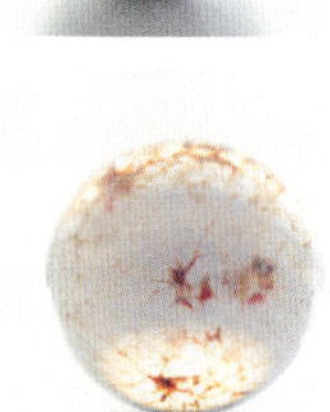

Objects generated from volumetric high-resolution data, leveraging generative material modeling during printing.

Neri Oxman and
The Mediated Matter Group
Vespers II. 2018
Photopolymers
Dimensions variable
Produced by Stratasys Ltd.
Research team: Christoph Bader, Dominik Kolb, James C. Weaver, Neri Oxman
Collaborators and contributors:
Gal Begun; Boris Belocon; Ahmed Hosny; Naomi Kaempfer; MIT Media Lab; Danielle van Zadelhoff; Wyss Institute, Harvard University

Created in collaboration with Stratasys Ltd. for the New Ancient Collection

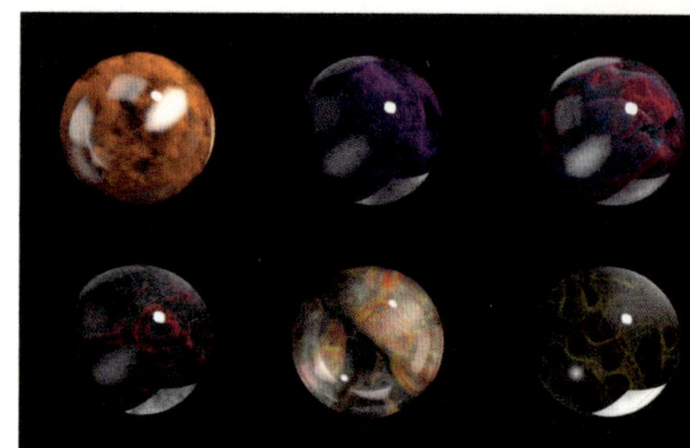

Various examples of material distribution.

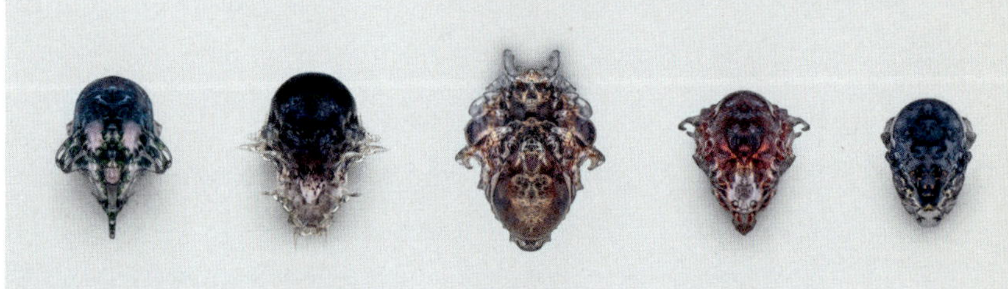

The Vespers II series.

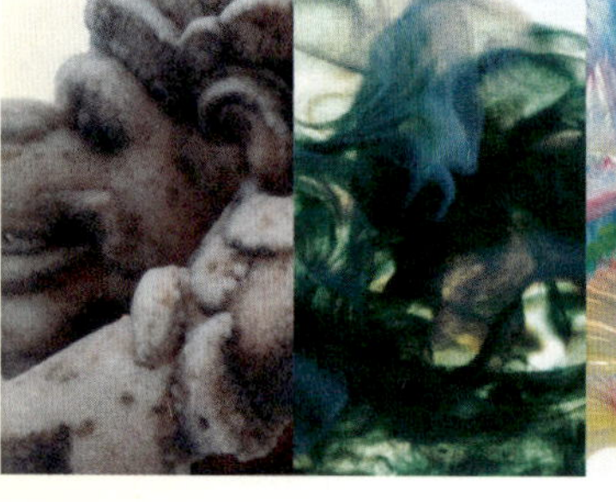
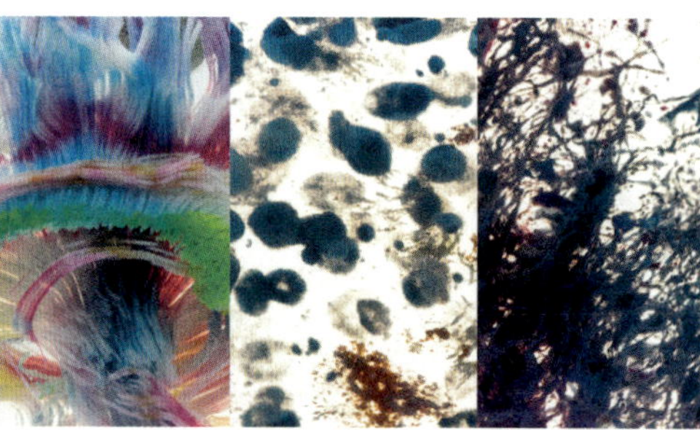
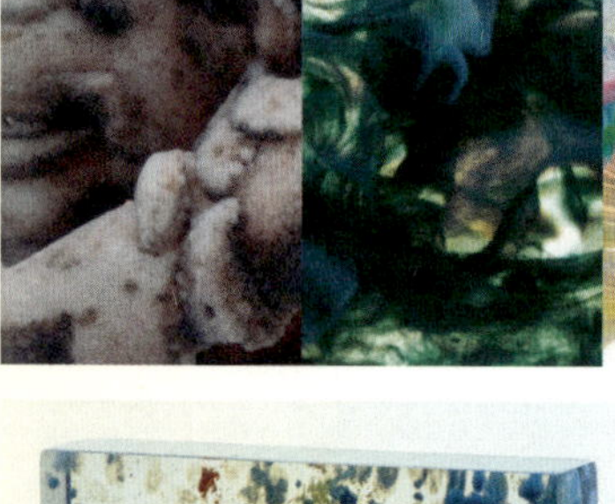

Details of different 3D-printed data sets.

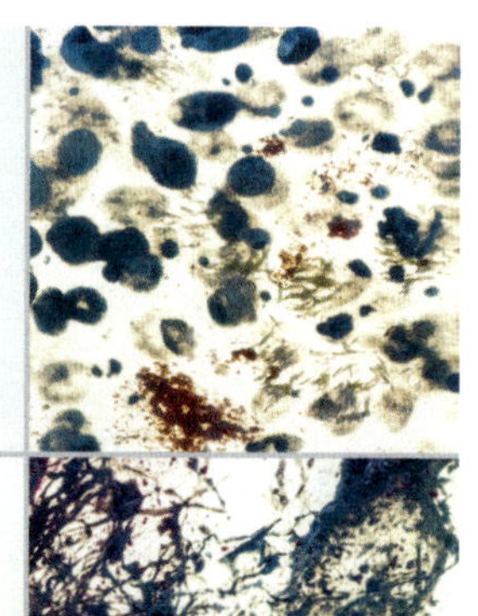
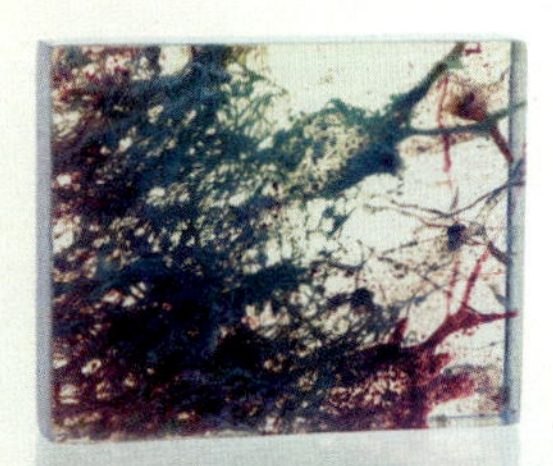

3D-printed models of image-based data, collected from living human lung tissue (top) and from biopsied mouse hippocampus (bottom).

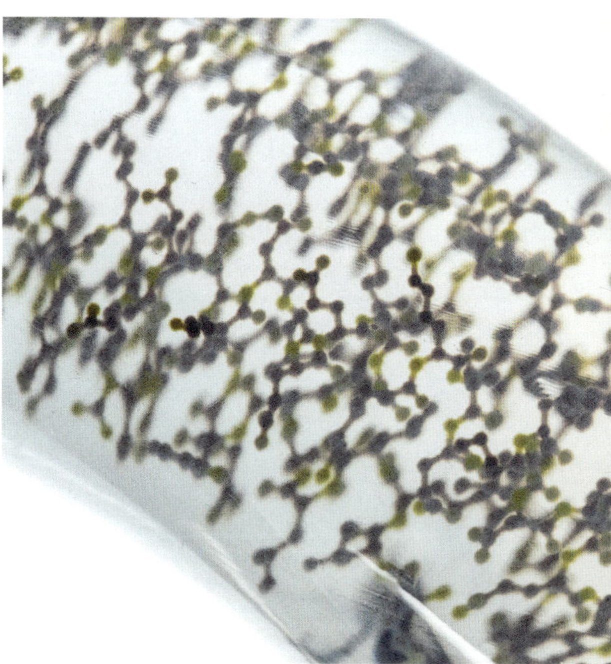

Object generated from a point cloud data set of a scanned artwork.

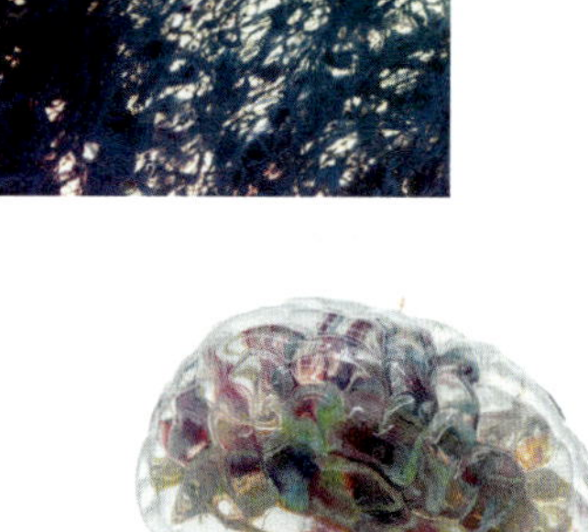
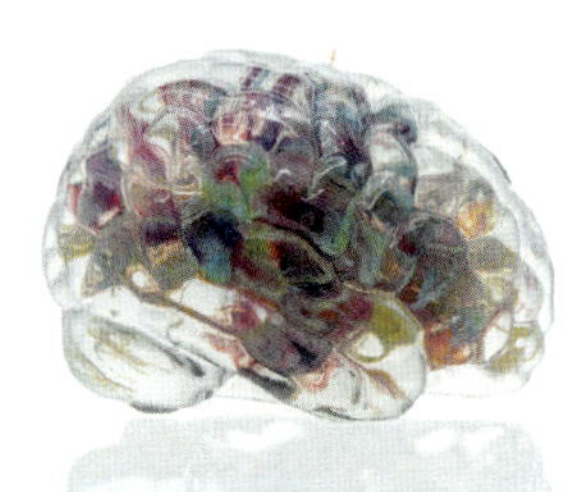
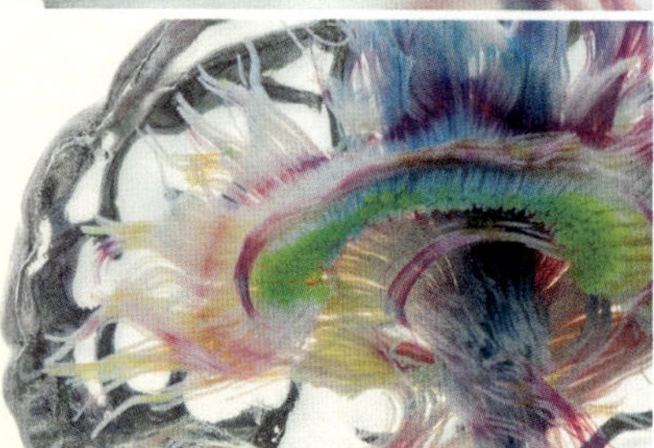
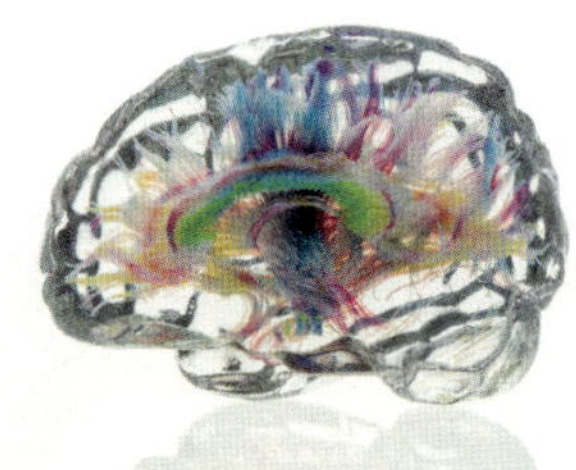

Postprocessing the 3D-printed object for optical clarity demonstrated the high resolution of the modeling method.

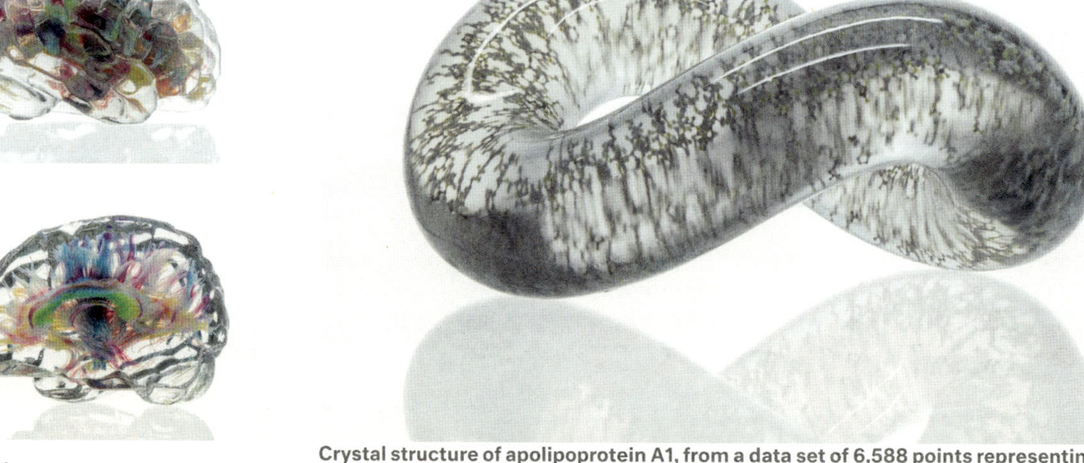

Crystal structure of apolipoprotein A1, from a data set of 6,588 points representing each atom and 13,392 line segments representing interatomic bonds.

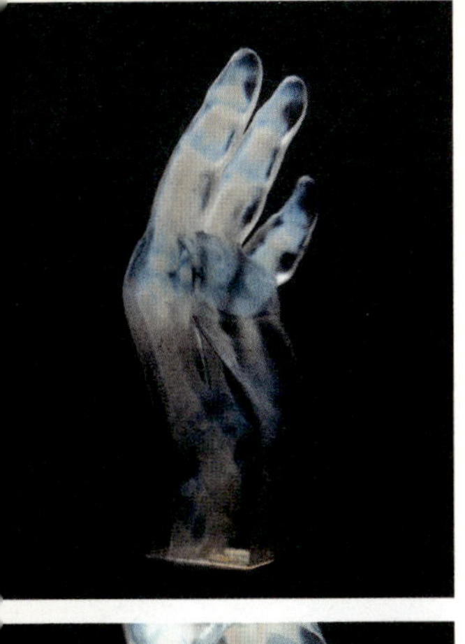

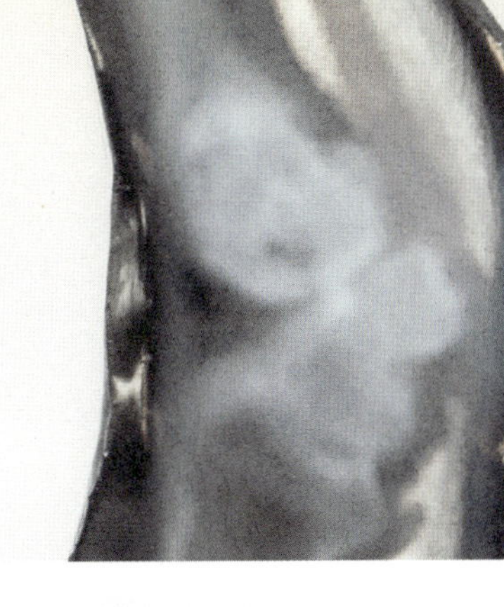

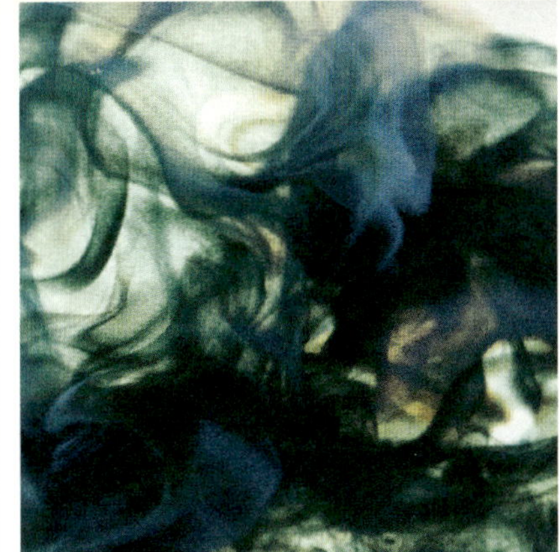

3D-printed objects generated from volumetric data.

Computational fluid simulation of the chaotic mixing of white and green fluids in a transparent volume (right).

CT scan of an arthritic hand, with transparency adjusted according to differences in bone-mineral density: white areas representing the highest density, transparent regions representing skin and soft tissue, and a gradient of semitransparent areas representing muscles, tendons, and lower-density bone (center and left, top and bottom).

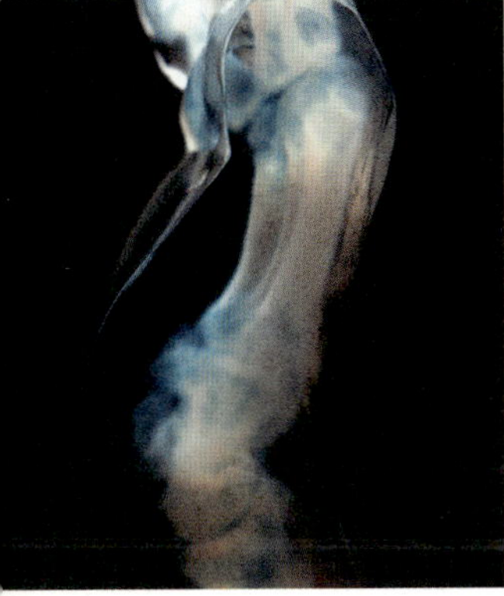

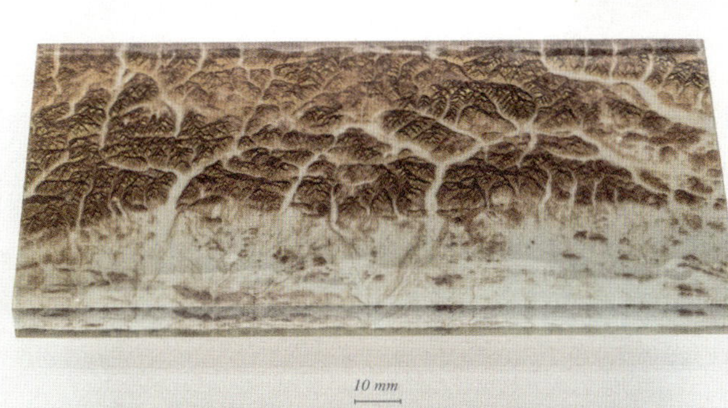

Elevation map of the Brooks Range in Northern Alaska, using information from the Elevation Derivatives for National Application database. Each voxel was color coded to reflect the degree of slope, with beige indicating flat areas and dark brown indicating steep inclines.

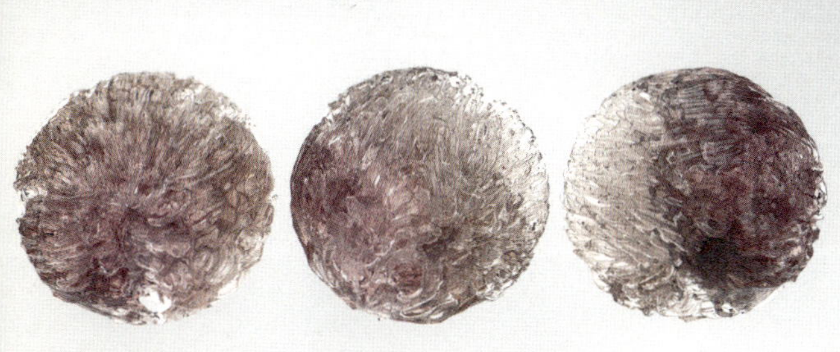

Early prototypes demonstrating various material distributions.

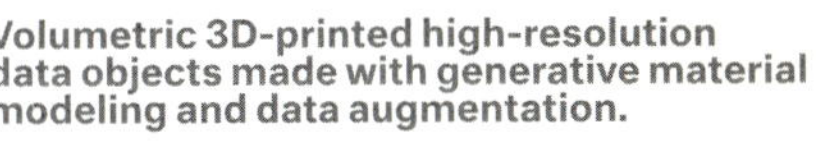

Volumetric 3D-printed high-resolution data objects made with generative material modeling and data augmentation.

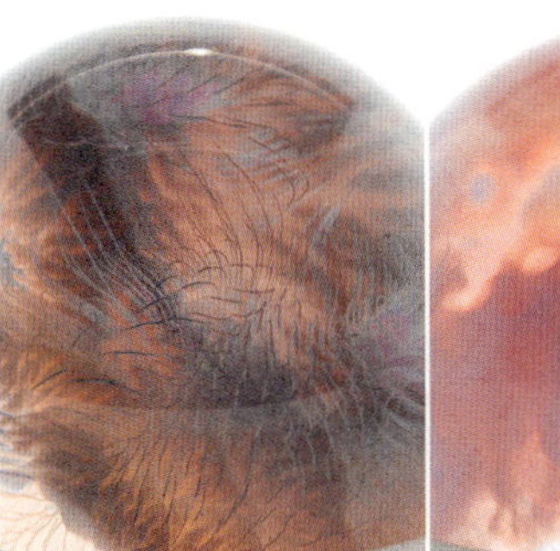

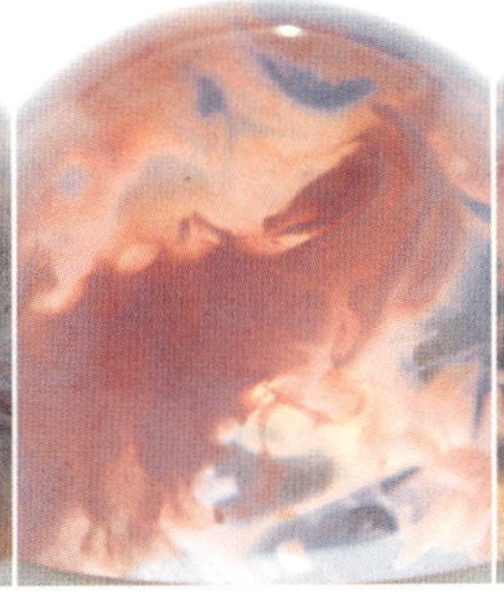

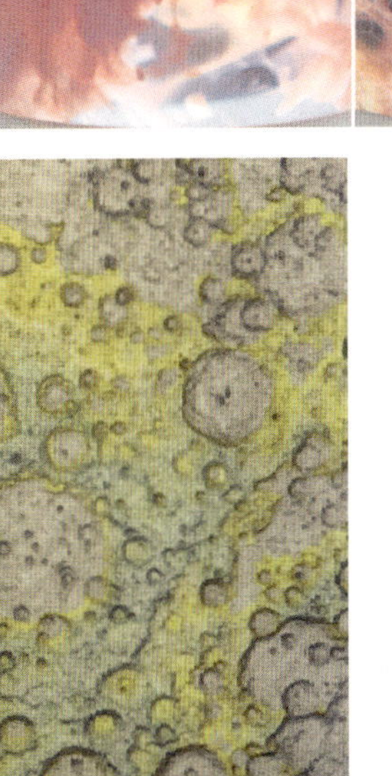

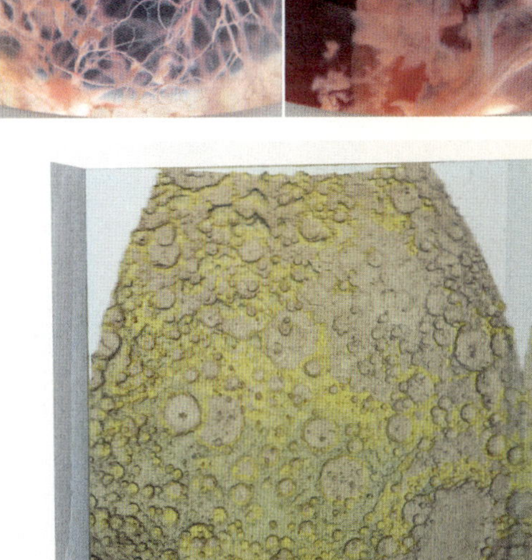

Objects generated from a digital elevation model of the moon captured by NASA's Lunar Reconnaissance Orbiter, consisting of 21 million points.

Mask 1 (detail).

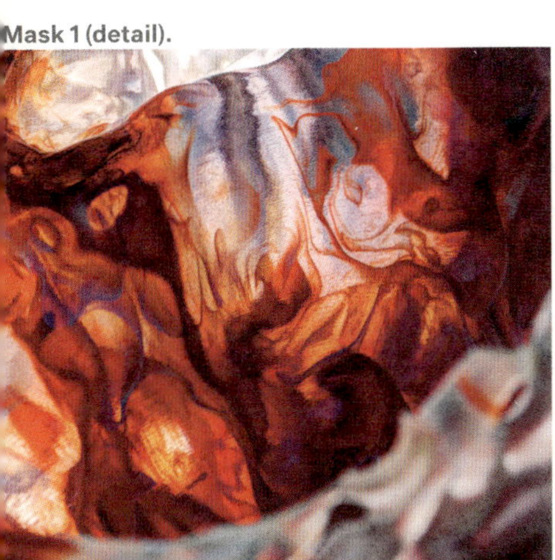

Mask 2 (detail).

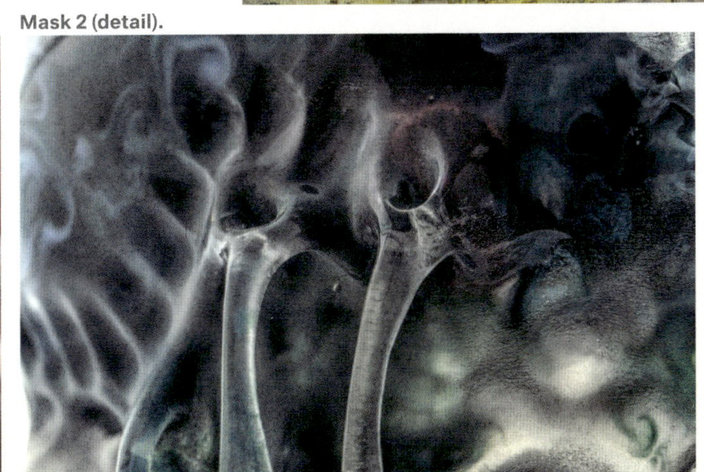

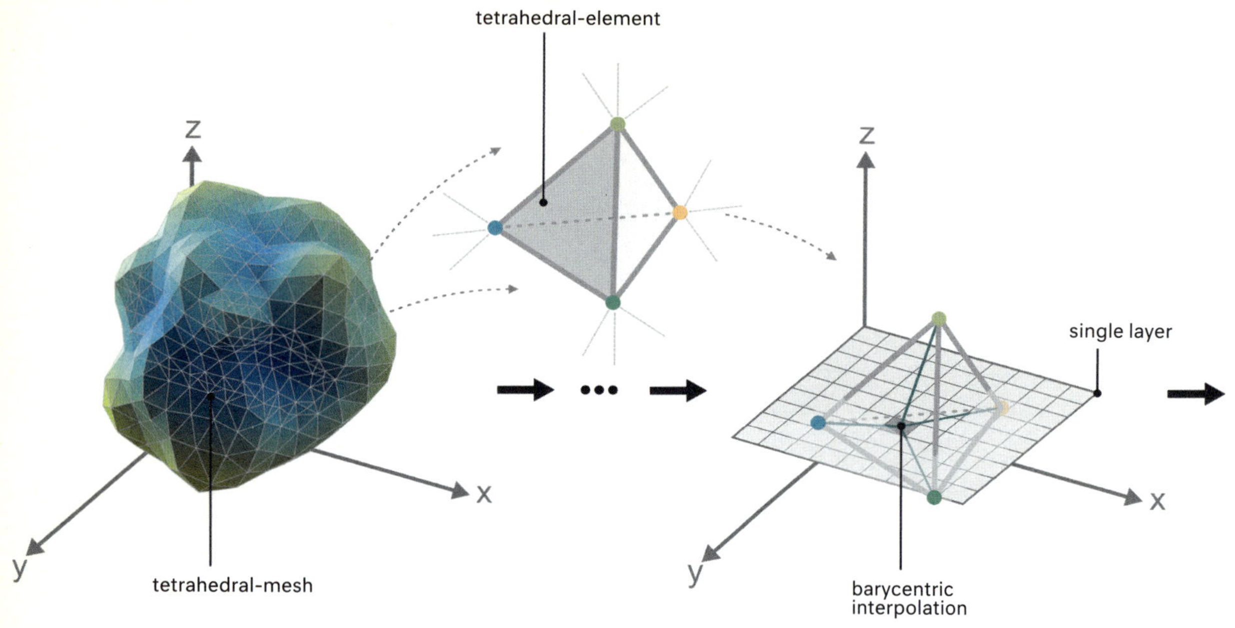

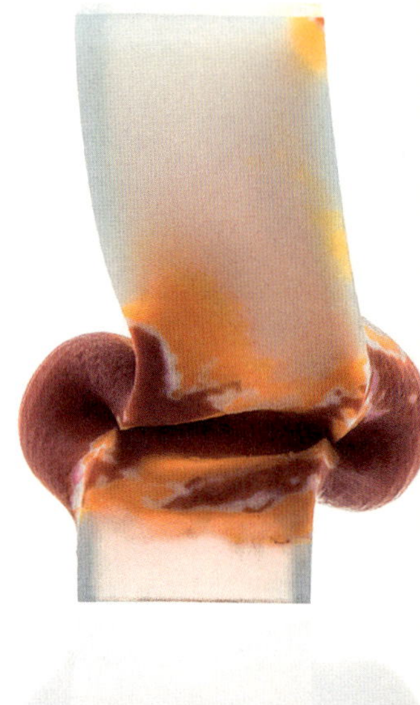

Schematic of the conversion of tetrahedral meshes into printable models. The mesh was the result of a simulated deformation of a compressed square tube using finite element methods. Red opaque areas indicate high deformation, and translucent white areas indicate low deformation.

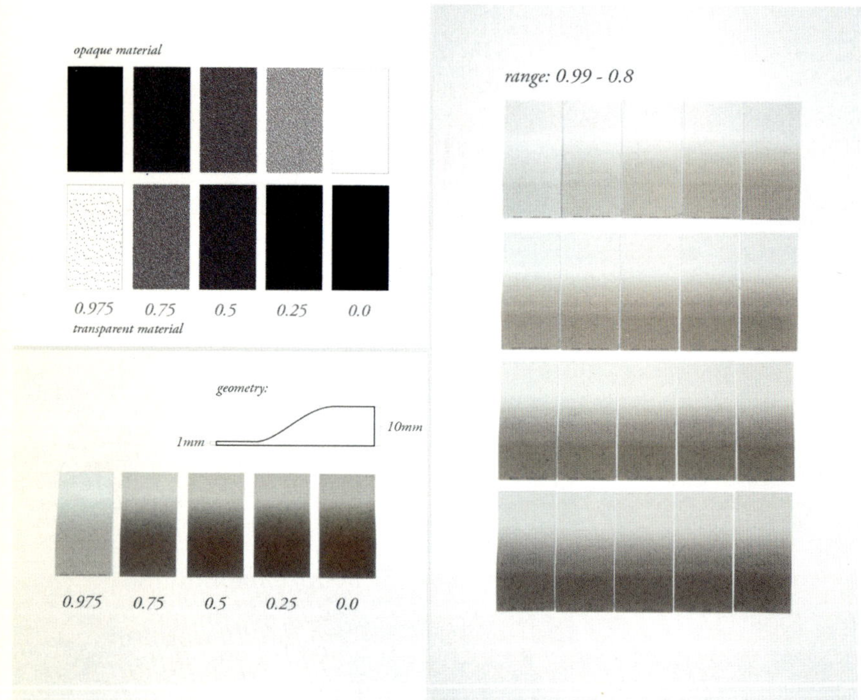

Diagram of a typical single layer of different material mixing ratios achieved through material dithering (the conversion of continuous materials into 3D-printable ones, top left), with the corresponding printed objects (bottom left and right). Variations in optical quality were tuned through transparent-opaque resin mixing ratios.

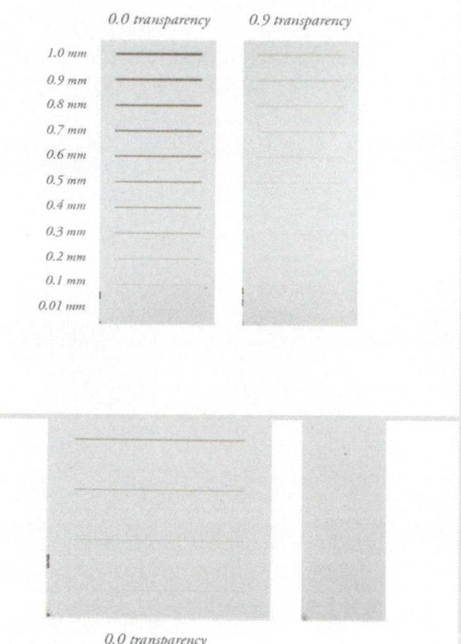

Studies of the effect of mixing ratios and element sizes on the perceived transparency of printed objects.

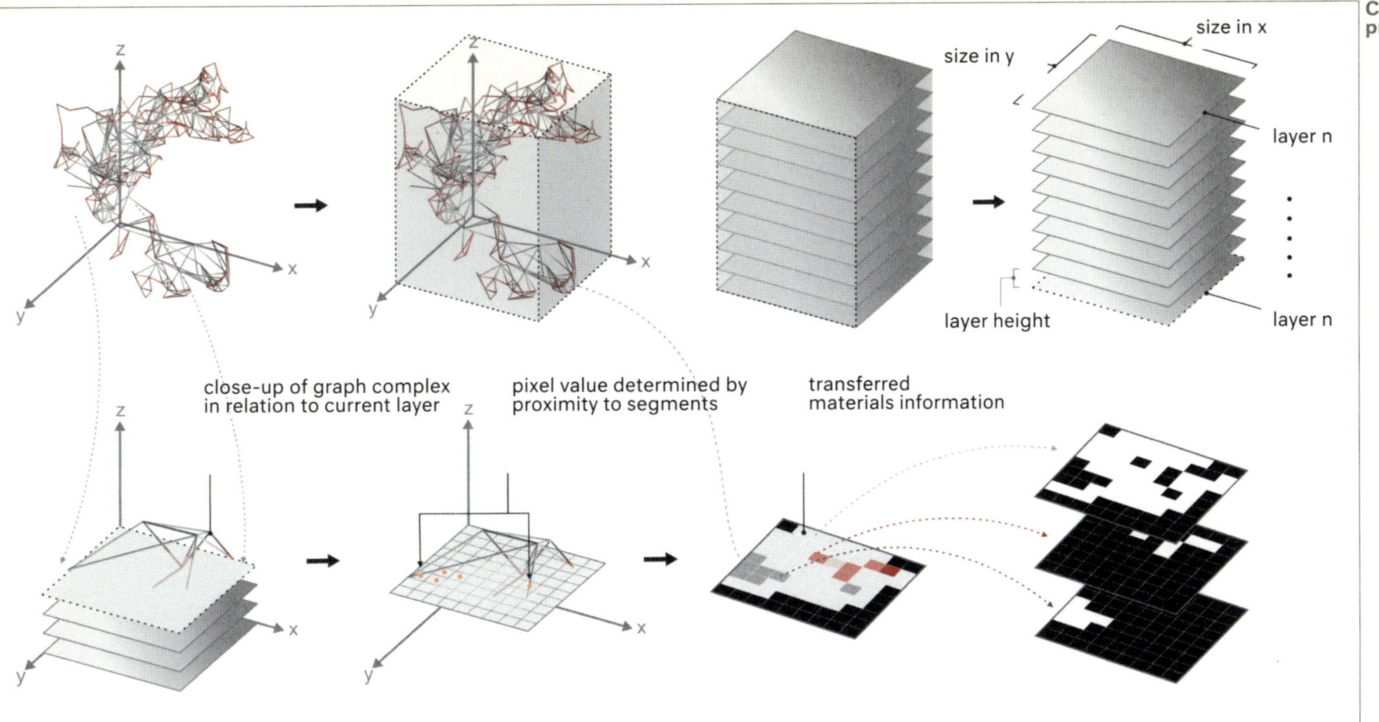

Curve and graph data-processing workflows.

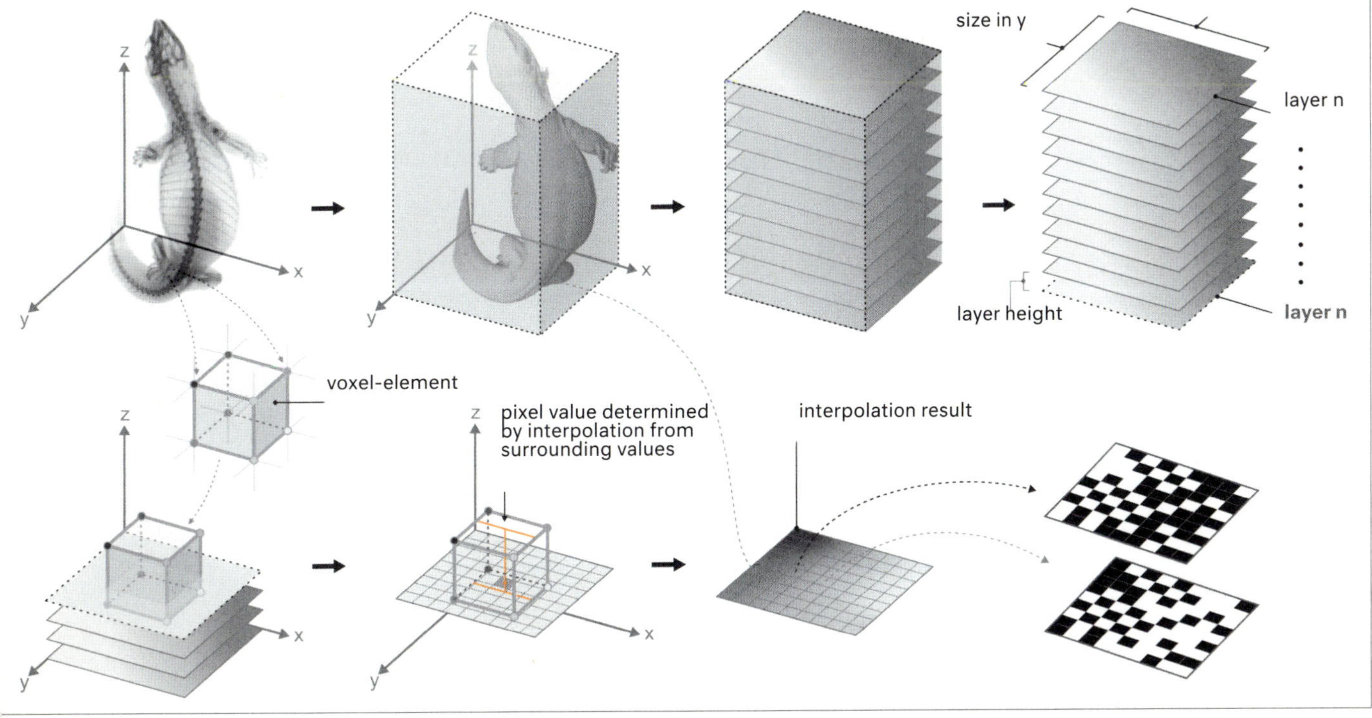

Volumetric data-processing workflows.

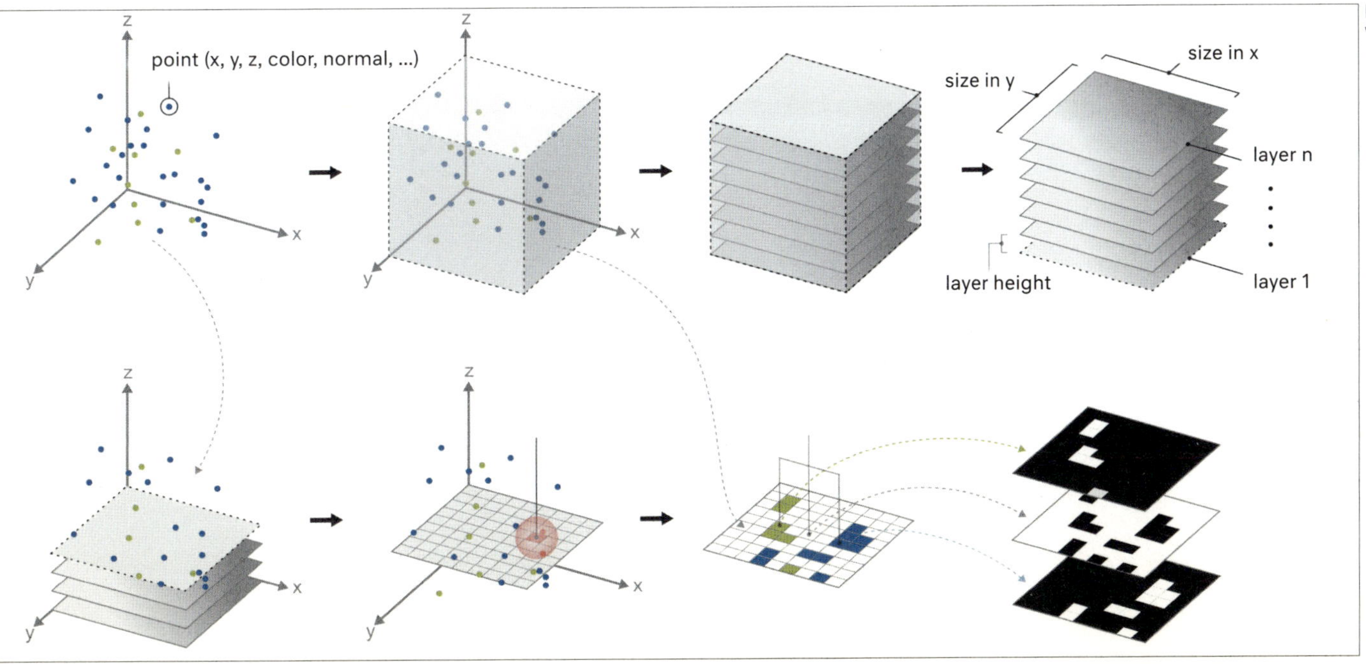

Point cloud data-processing workflows.

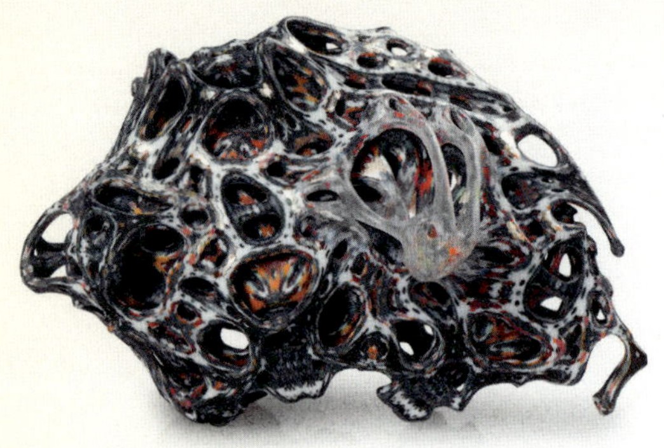
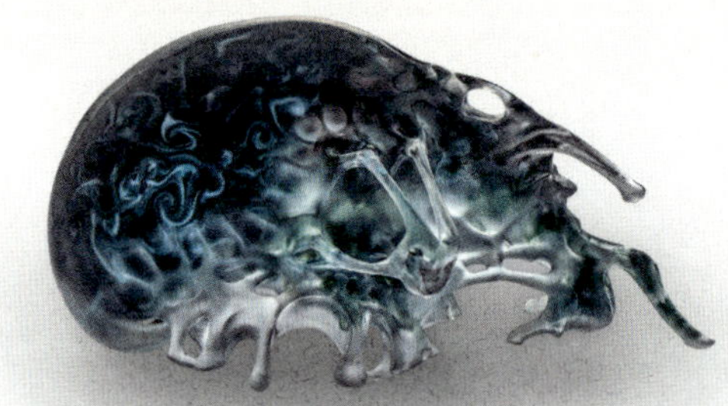
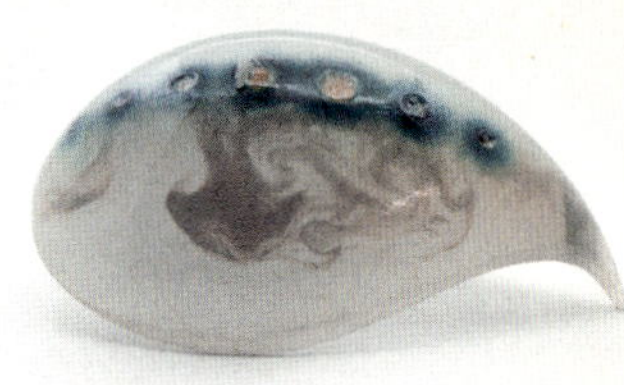
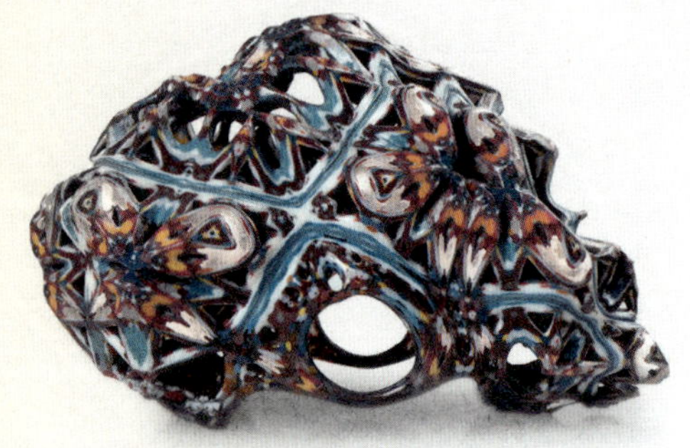
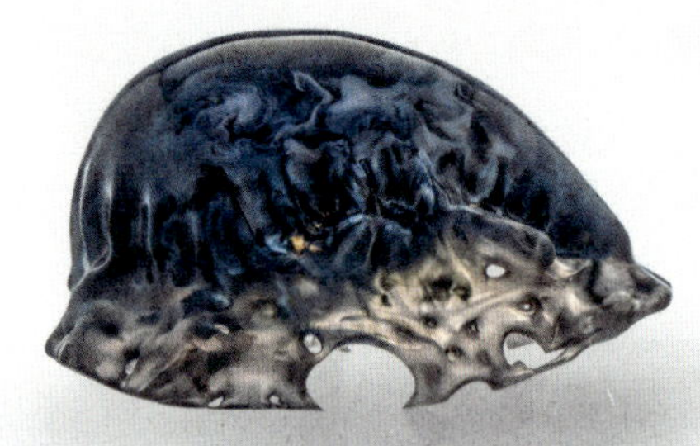
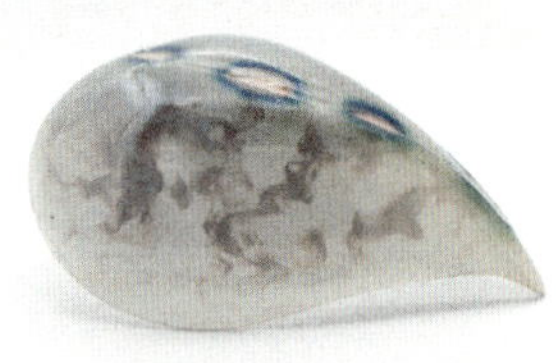
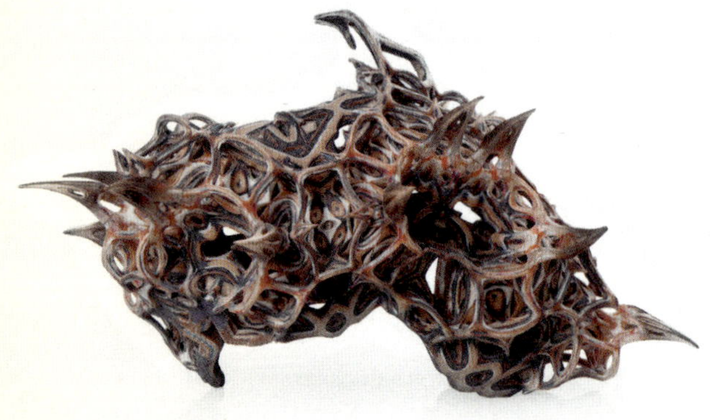
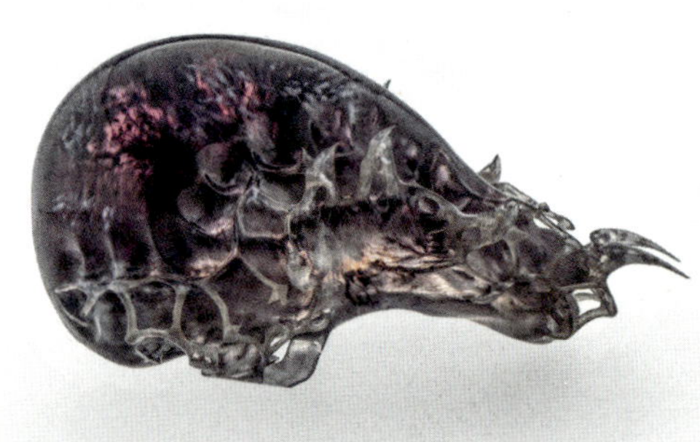
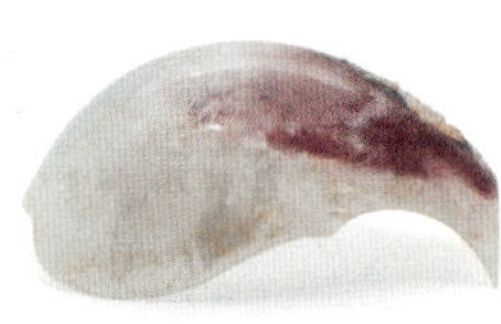
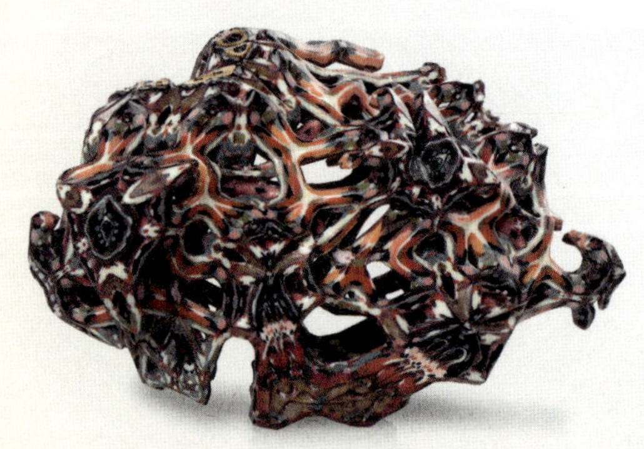
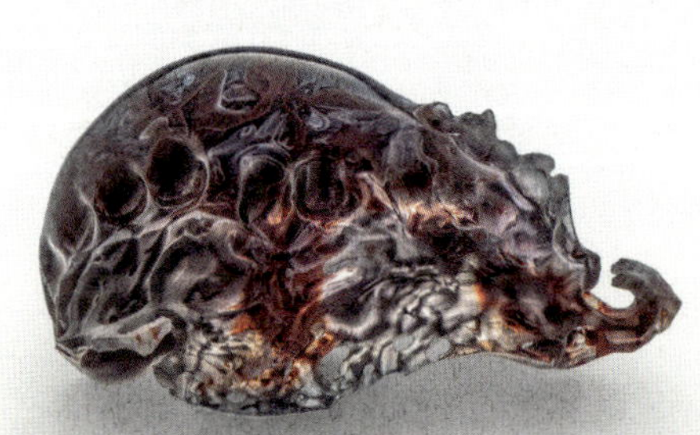
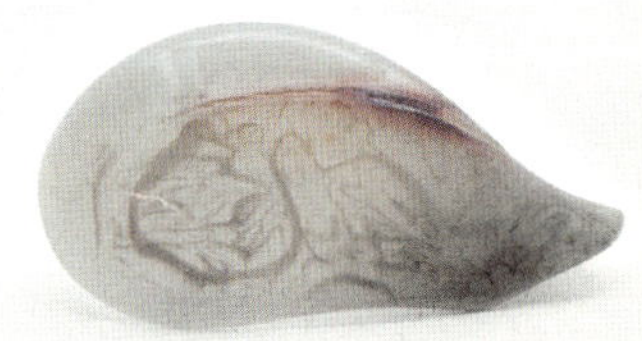

Vespers
Top to bottom: Masks 4, 2, 3, and 1 in all three series.

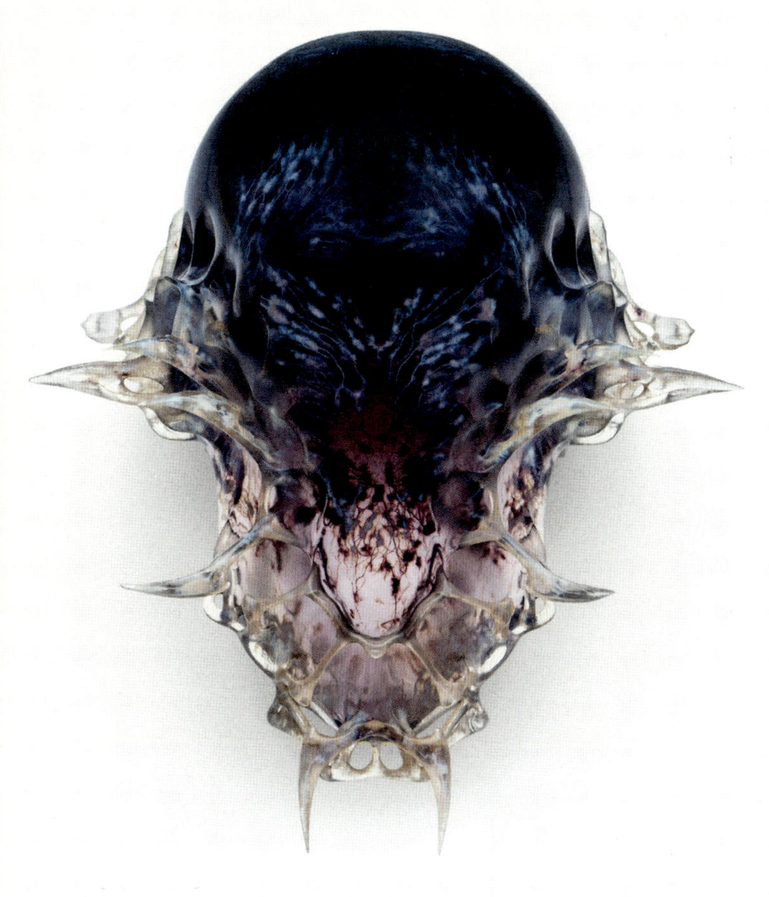

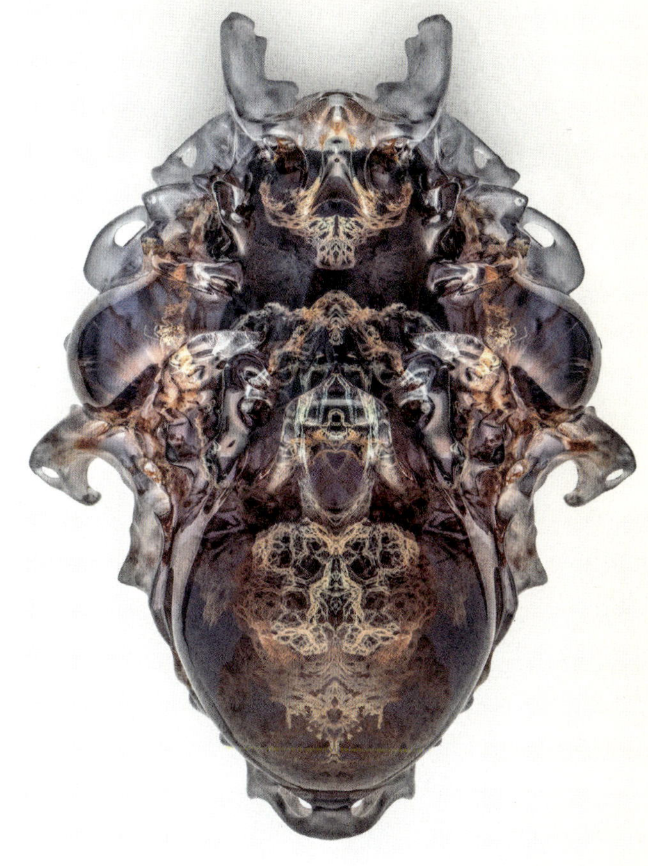

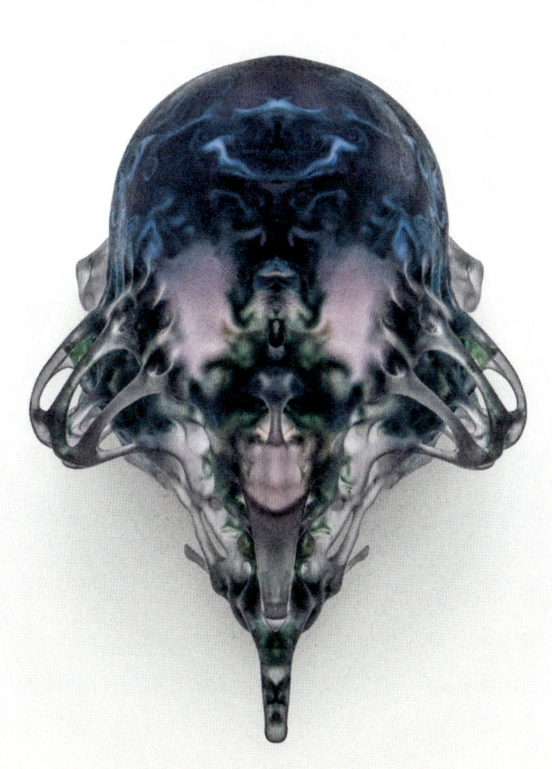

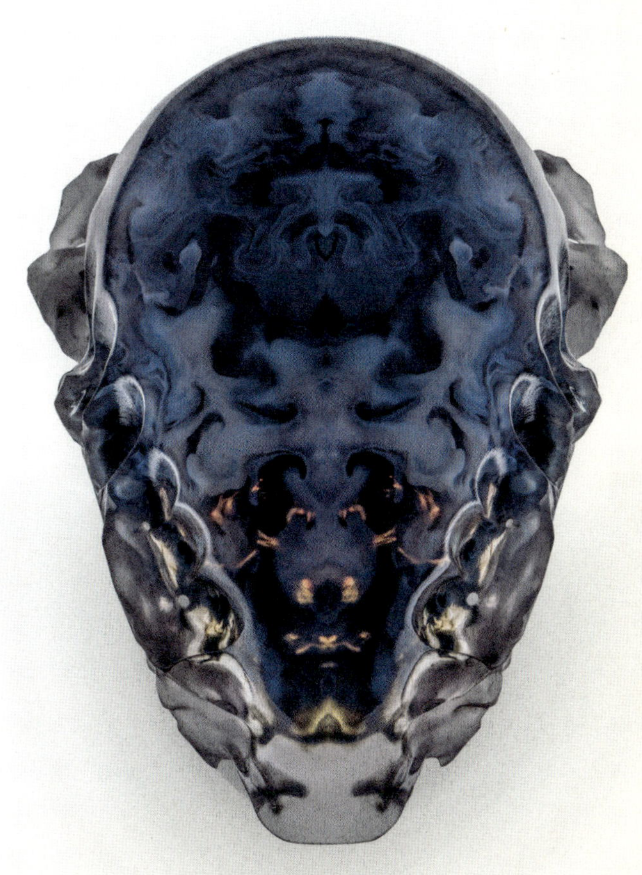

Vespers II
Clockwise, from upper left: Masks 3, 5, 2, 4.

Vespers II
3D-printed objects made with generative material modeling and data augmentation.

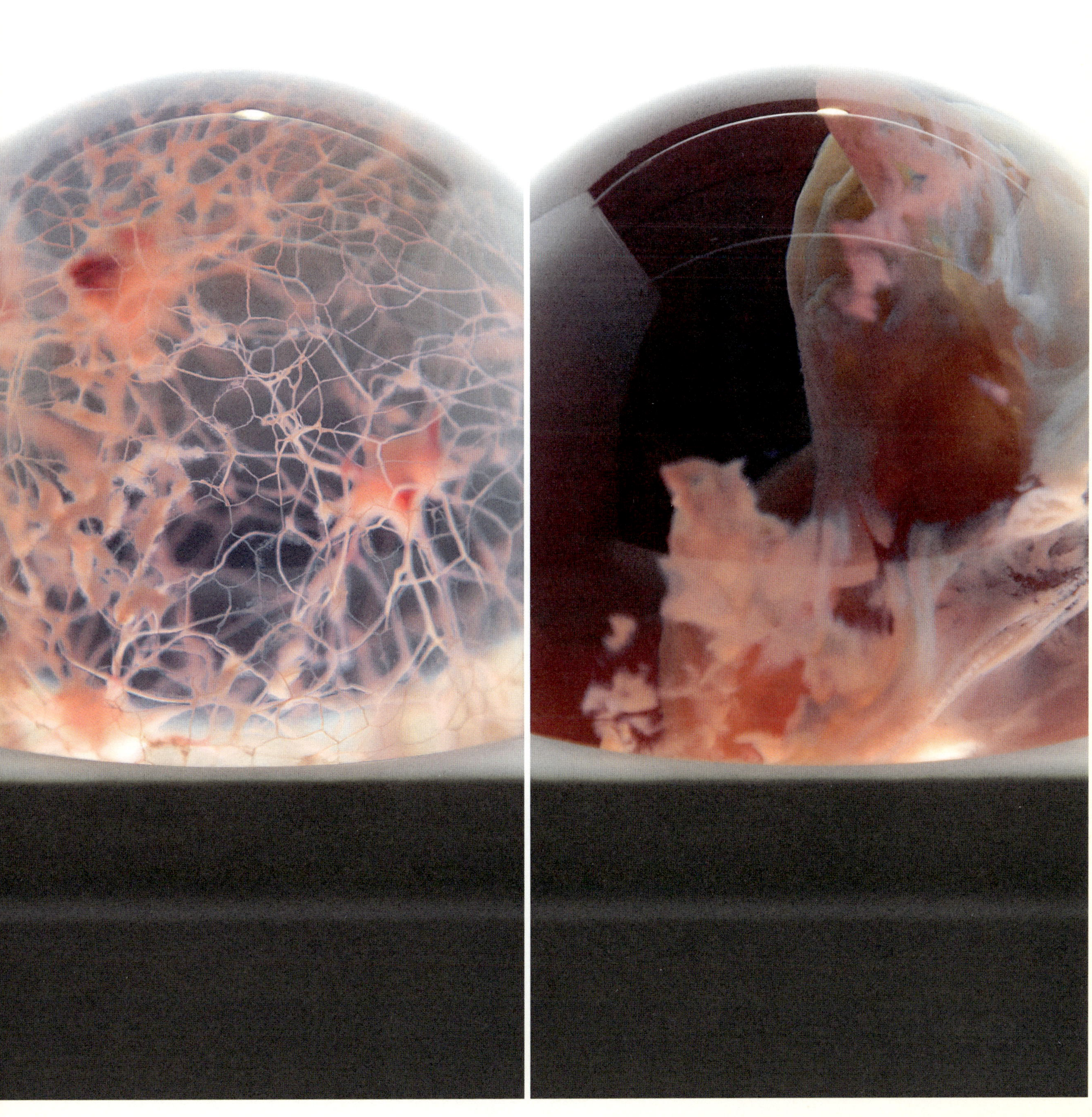

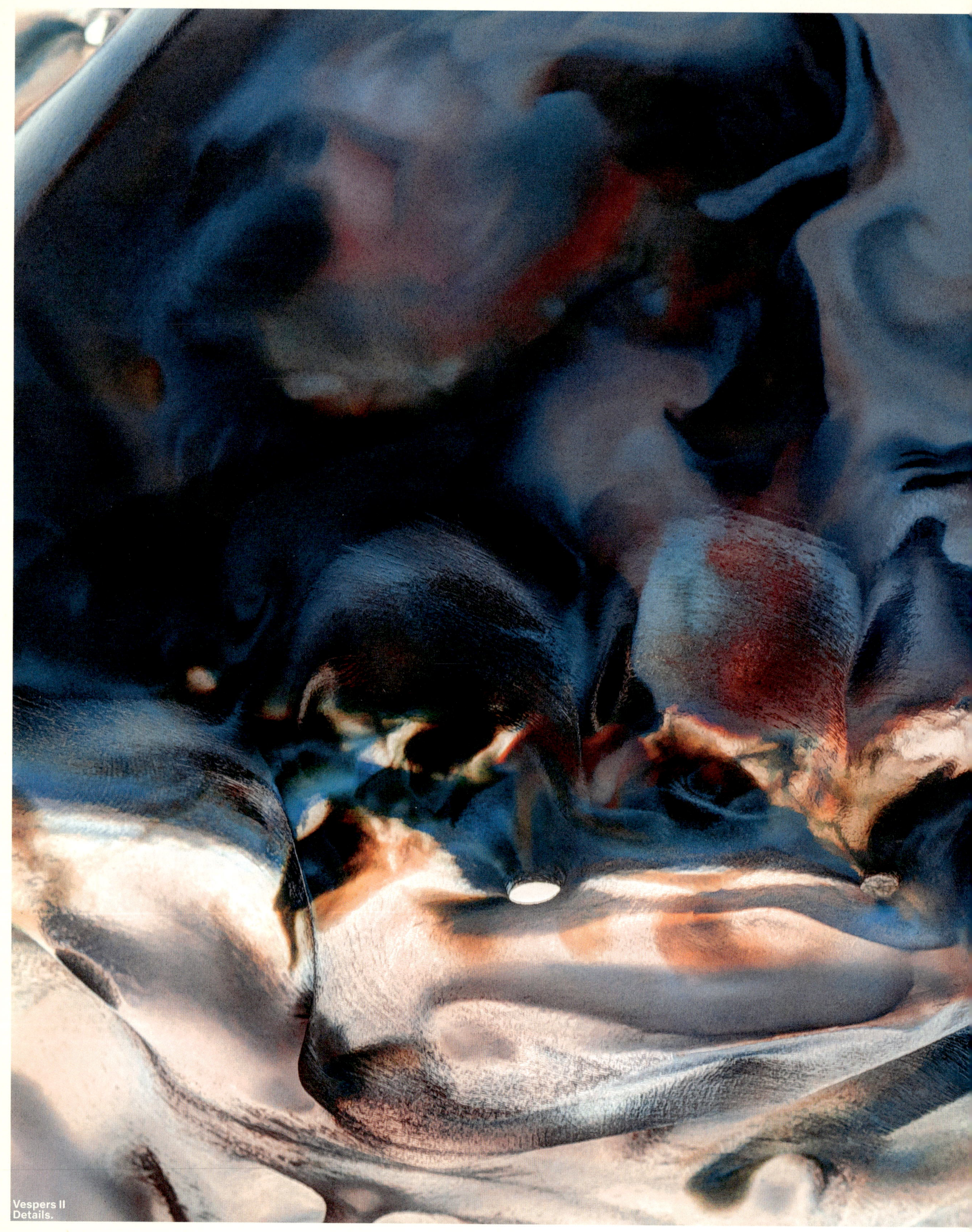

Vespers II
Details.

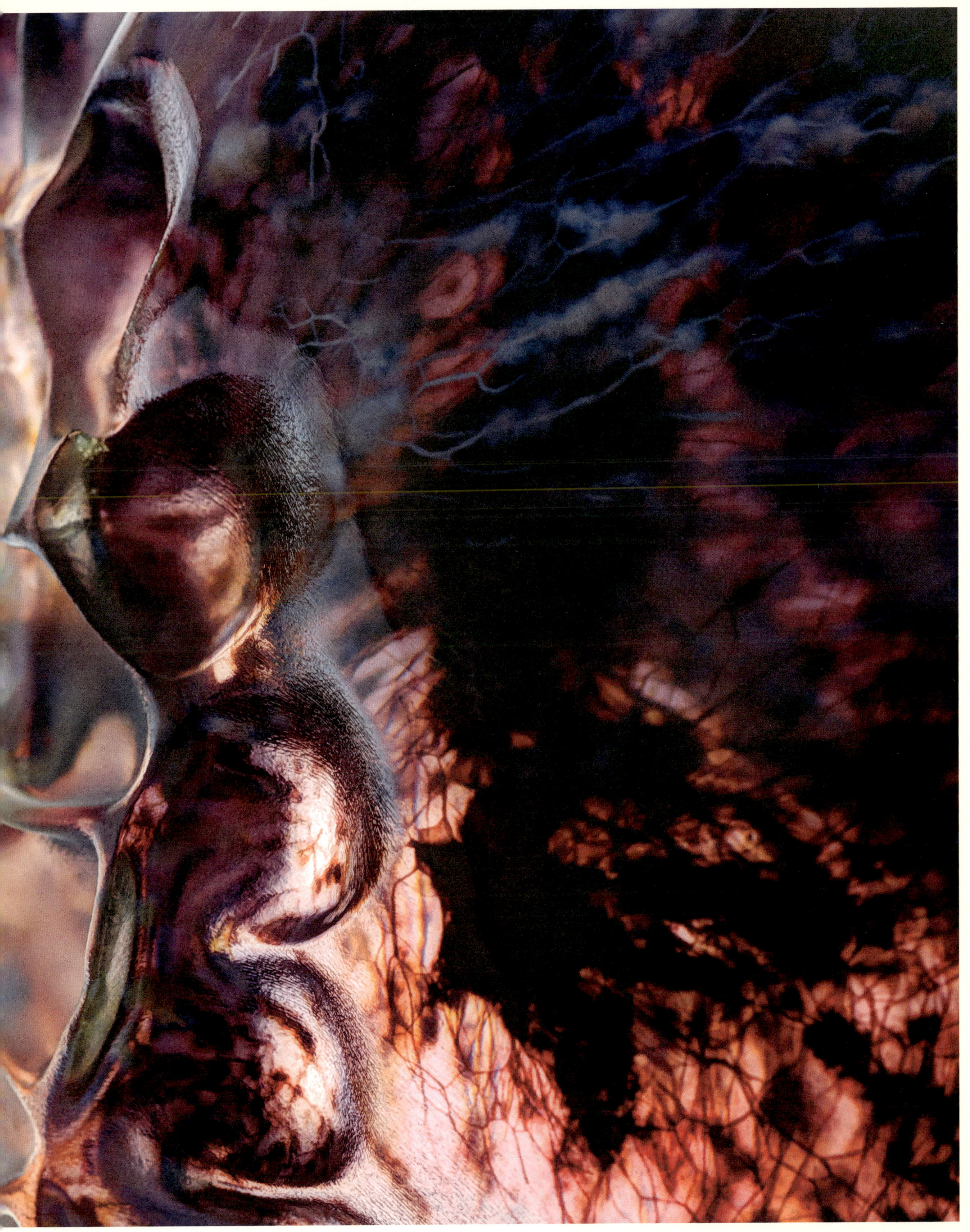

VESPERS III

Neri Oxman and
The Mediated Matter Group
Vespers III. 2018
Photopolymers, *Escherichia coli* bacteria, hydrogel, nutrient media, and chemical signal
Dimensions variable
Produced in part by Stratasys Ltd.
Research team: Christoph Bader, Rachel Soo Hoo Smith, Dominik Kolb, Sunanda Sharma, João Costa, James C. Weaver, Neri Oxman
Collaborators and contributors: Gal Begun; Boris Belocon; Kelly Egorova; Jeremy Flower; Ahmed Hosny; Noah Jakimo; Naomi Kaempfer; Wendy Salmon; Tzu-Chieh Tang; MIT Environmental Health and Safety; Media Lab Facilities; MIT Center for Bits and Atoms; MIT Media Lab; Danielle van Zadelhoff; Wyss Institute, Harvard University

Created in collaboration with Stratasys Ltd. for the New Ancient Collection

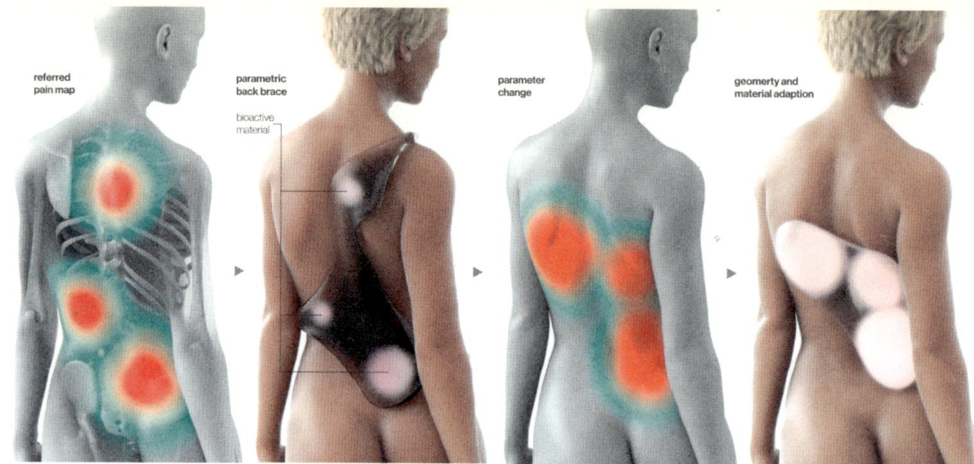

A potential application for the hybrid livin materials (HLMs) developed for the Vespers III series: biomedical wearables and prostheses, such as a devic containing cells that create topical analgesics on demand.

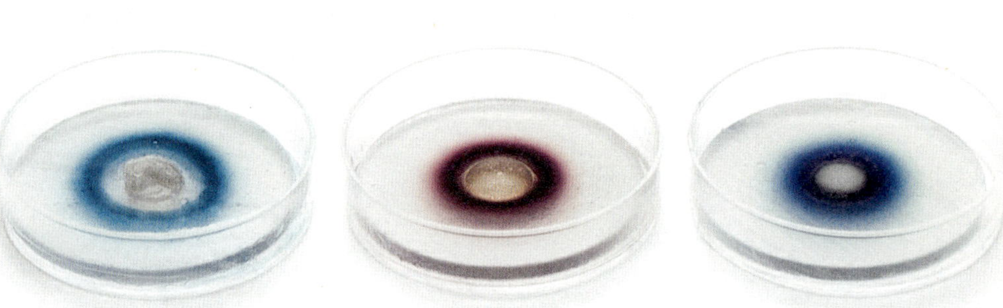

Prototype HLM disks demonstrating different patterns and colors created by bacteria.

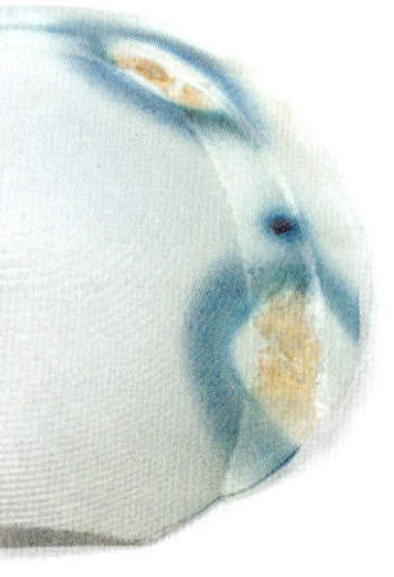

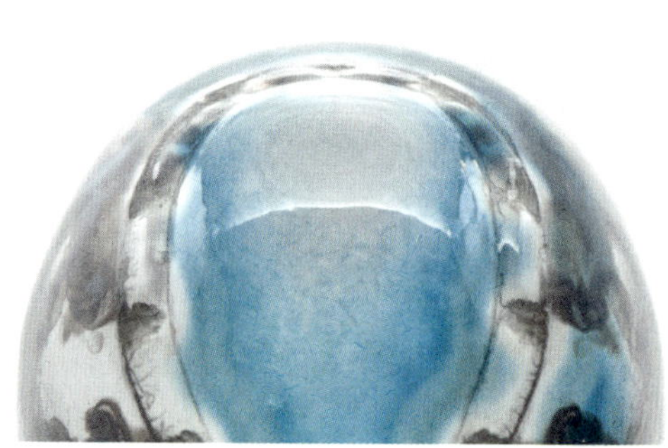

HLM prototypes containing and templating genetically engineered bacteria.

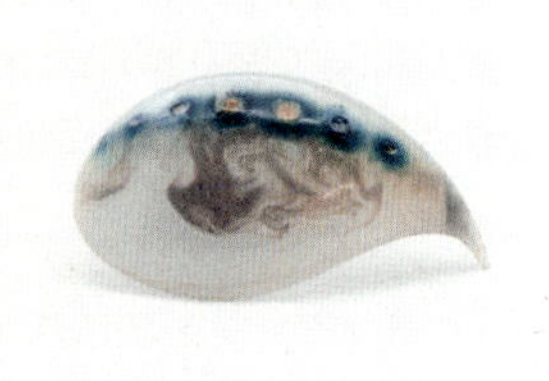

Mask 4.

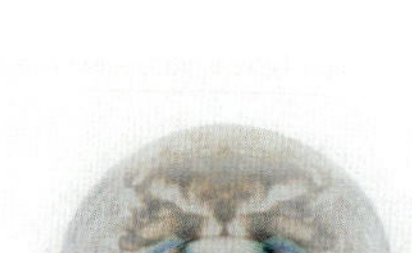

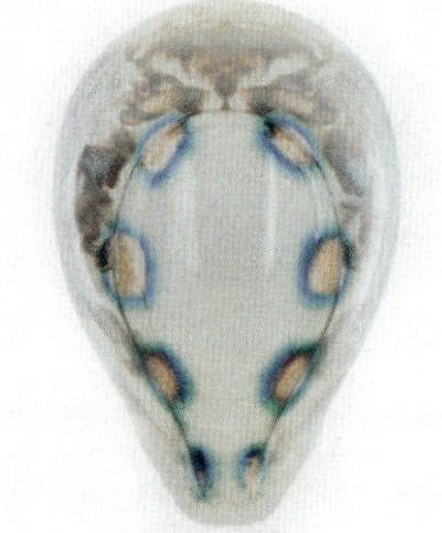

Mask 2.

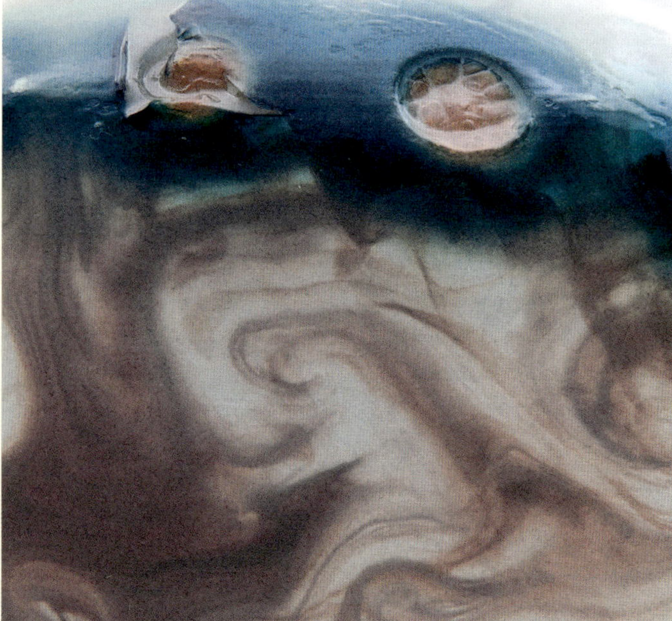

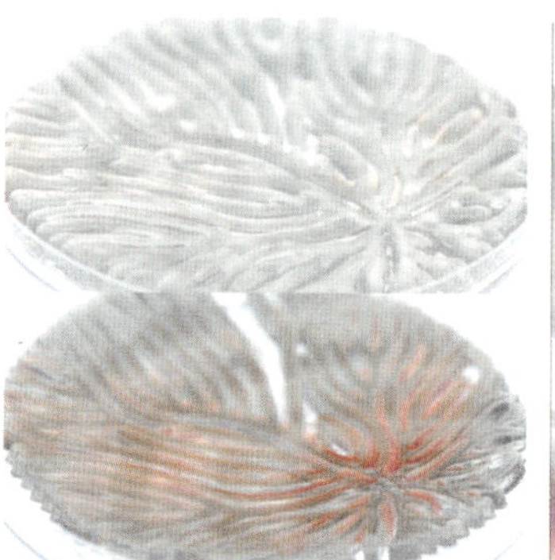

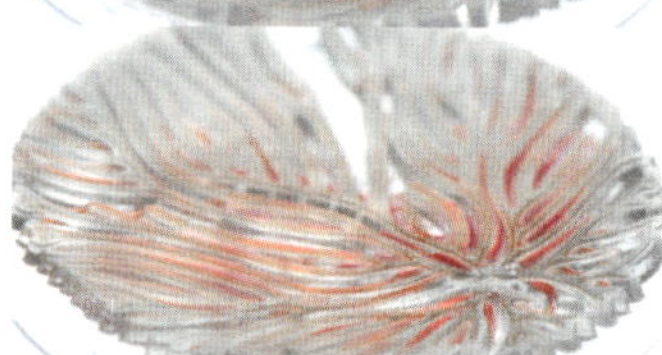

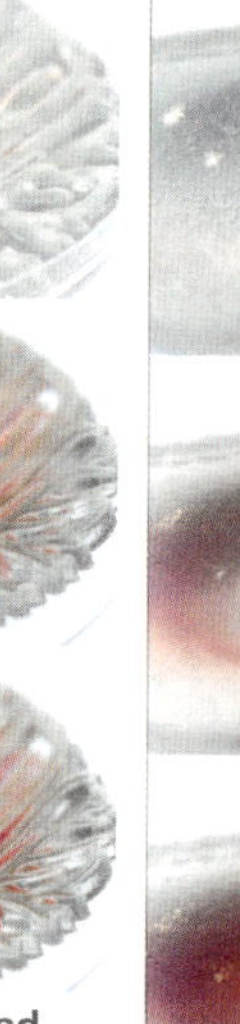

Leaf-shaped prototype imaged over time, showing gradual production of red pigment.

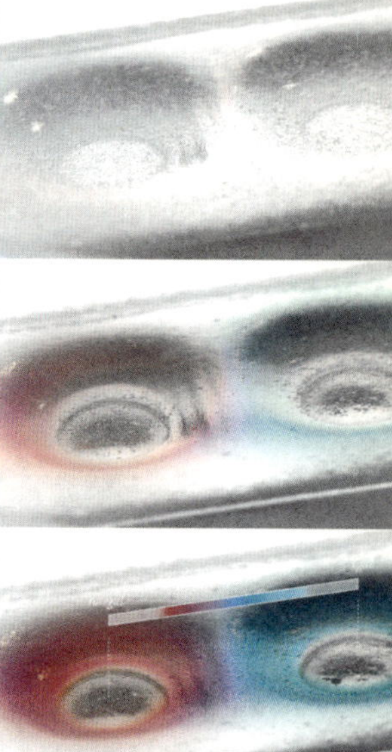

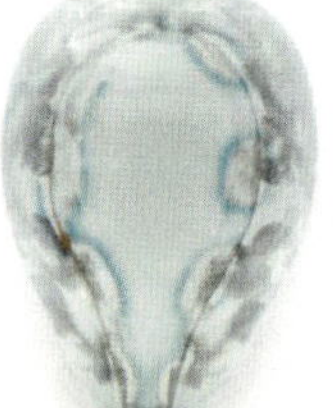

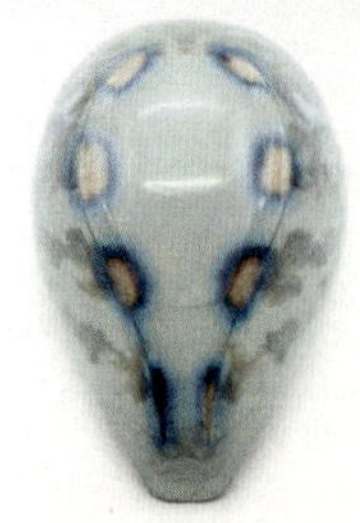

print file version_1 → *print file version_2* → *print file version_3*

Printed iterations of a single mask, demonstrating the relationship between material distribution and bacteria-mediated pigmentation.

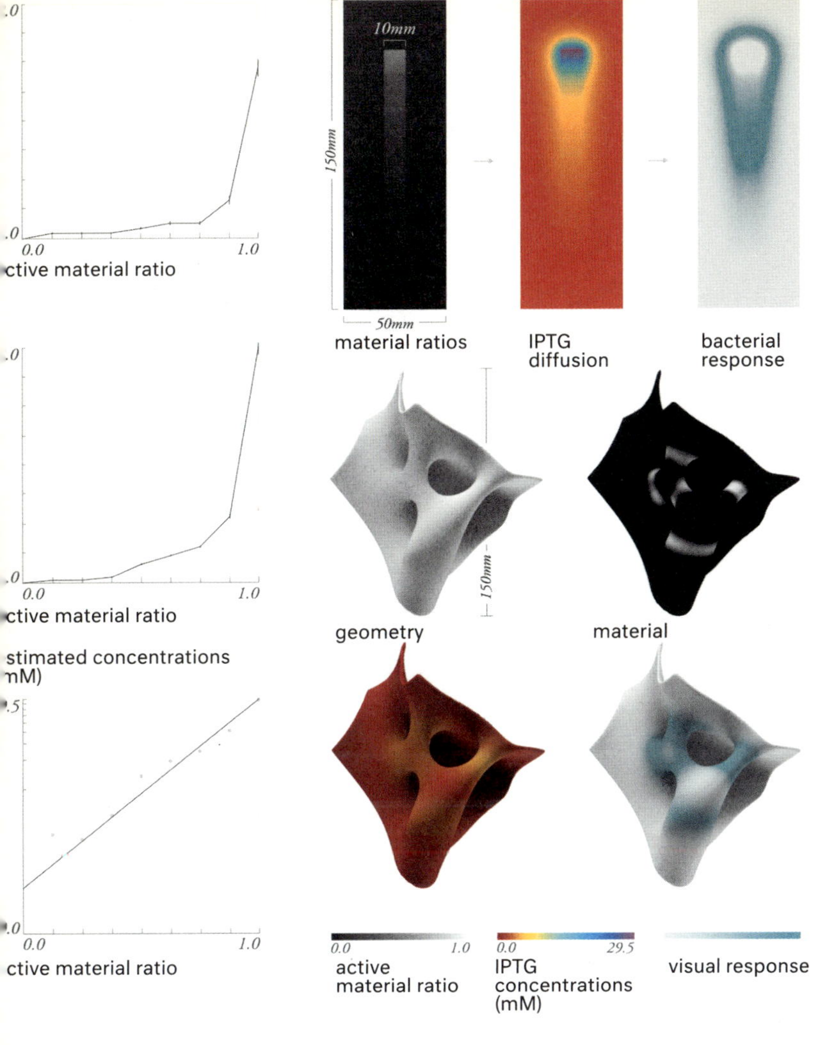

he computational design rocess involved the imulation of chemical iffusion and bacterial esponse on two- and three-imensional geometries.

eft: Prototype with two pigments esponding to chemical signals.

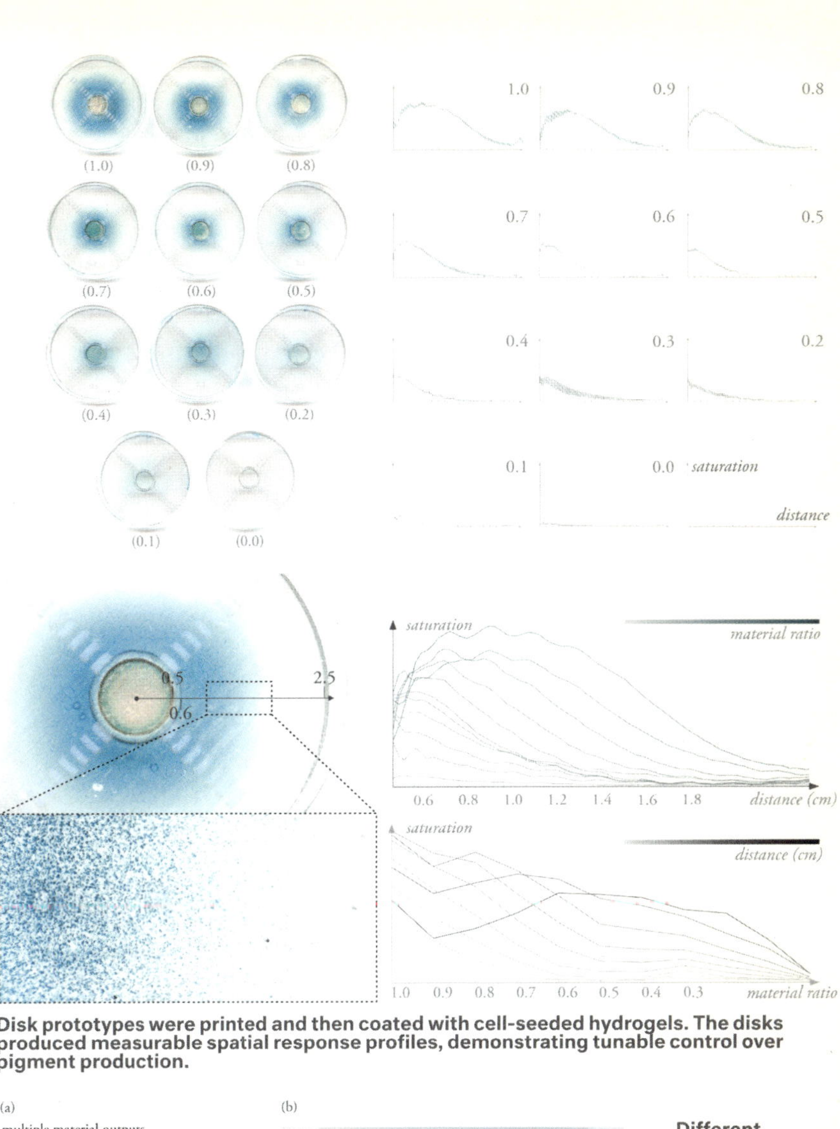

Disk prototypes were printed and then coated with cell-seeded hydrogels. The disks produced measurable spatial response profiles, demonstrating tunable control over pigment production.

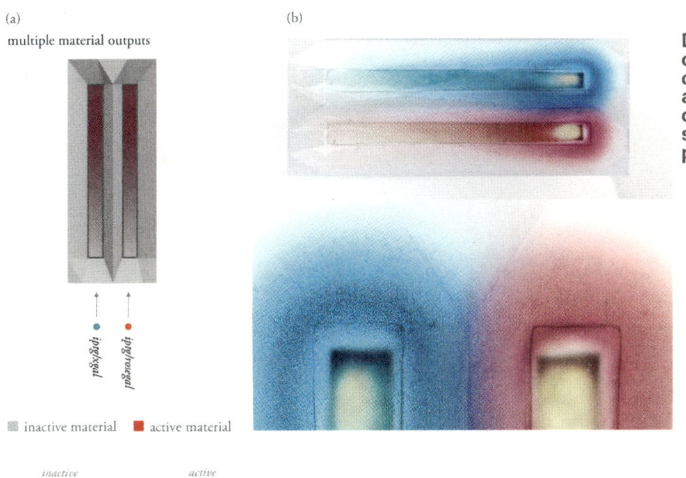

Different combinations of chemical signals and engineered cells produced spatially specific pigments.

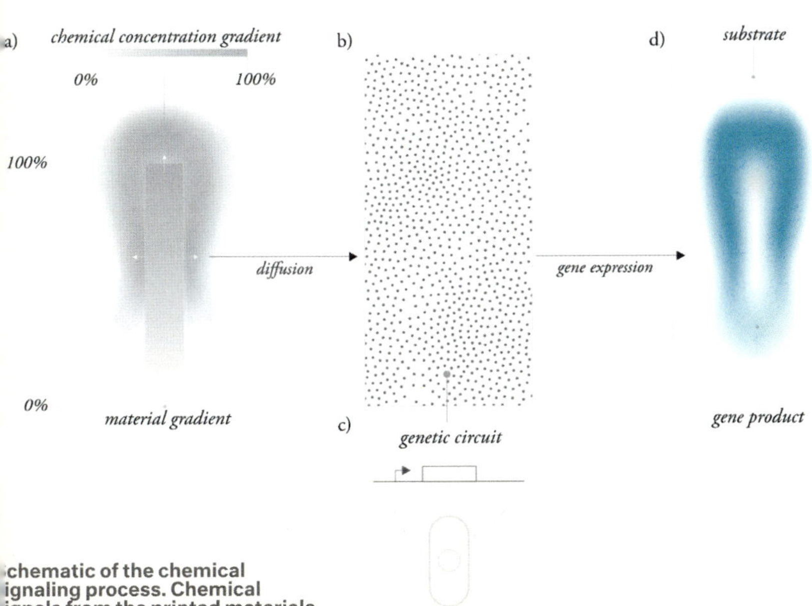

chematic of the chemical ignaling process. Chemical ignals from the printed materials vere sensed by genetically ngineered bacteria, which led o the expression of proteins and hen to pigment formation.

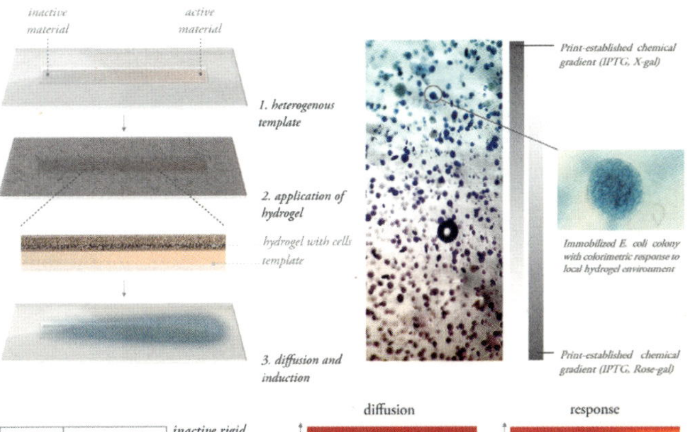

Each mask in Vespers III was computationally designed with a specific geometry and chemical-material distribution. The masks were 3D printed and coated with a cell-seeded hydrogel; the cells responded to the chemical signals and produced pigments across the surface of the mask.

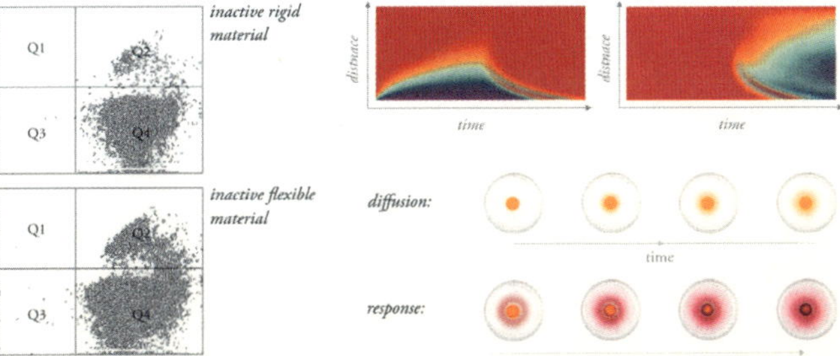

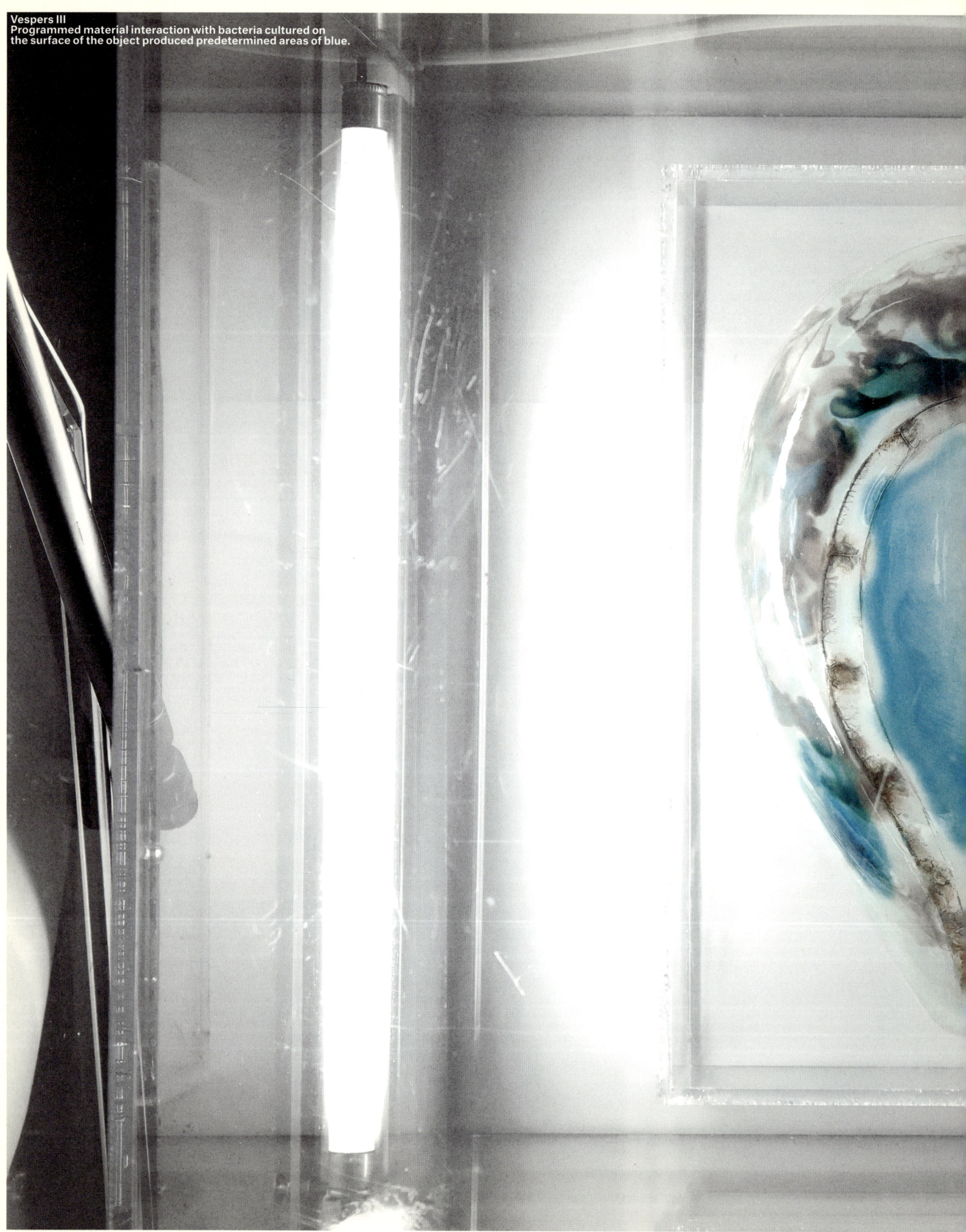
Vespers III
Programmed material interaction with bacteria cultured on the surface of the object produced predetermined areas of blue.

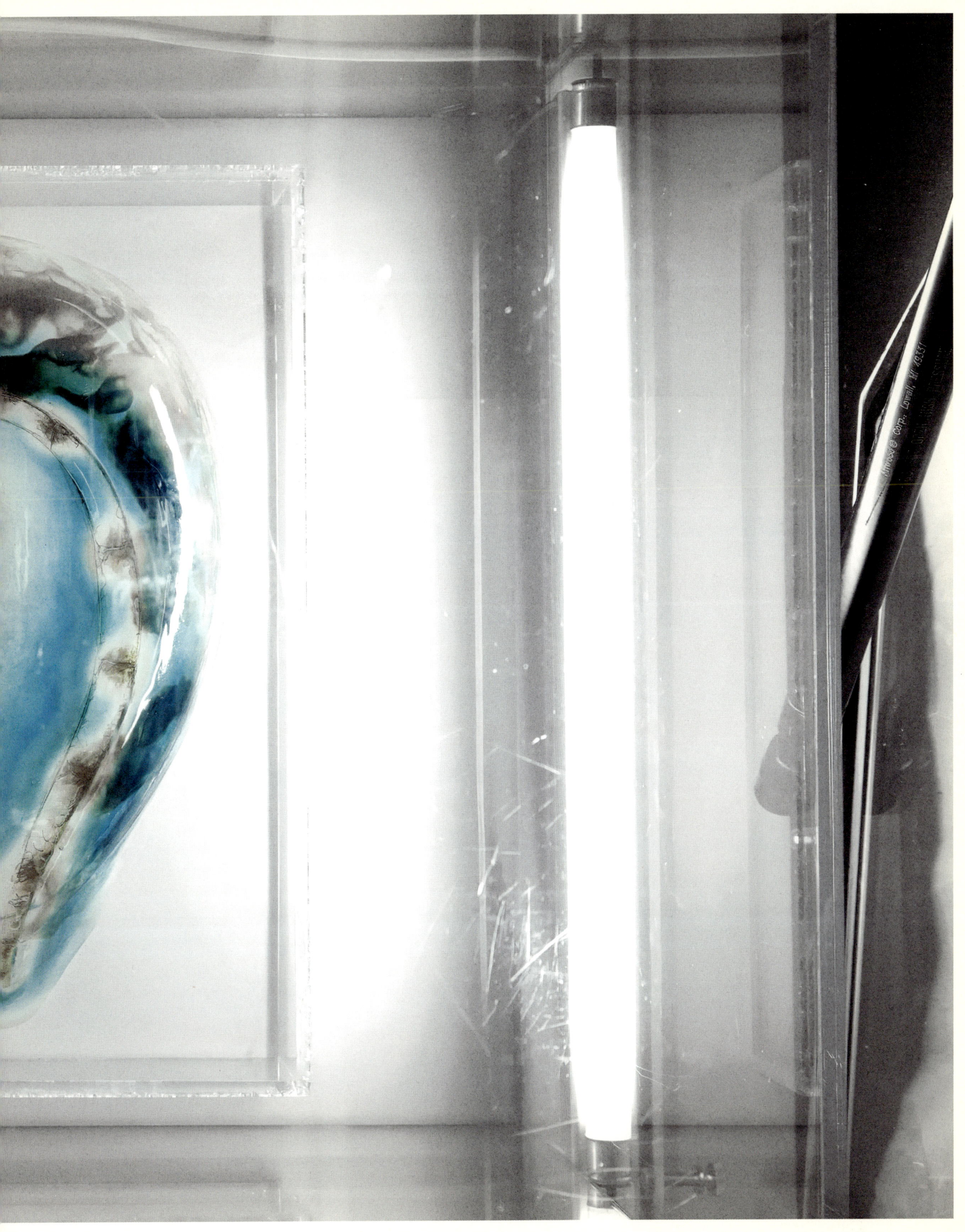
Corp.: Lowell, MI 49331

Vespers III
Mask 1.

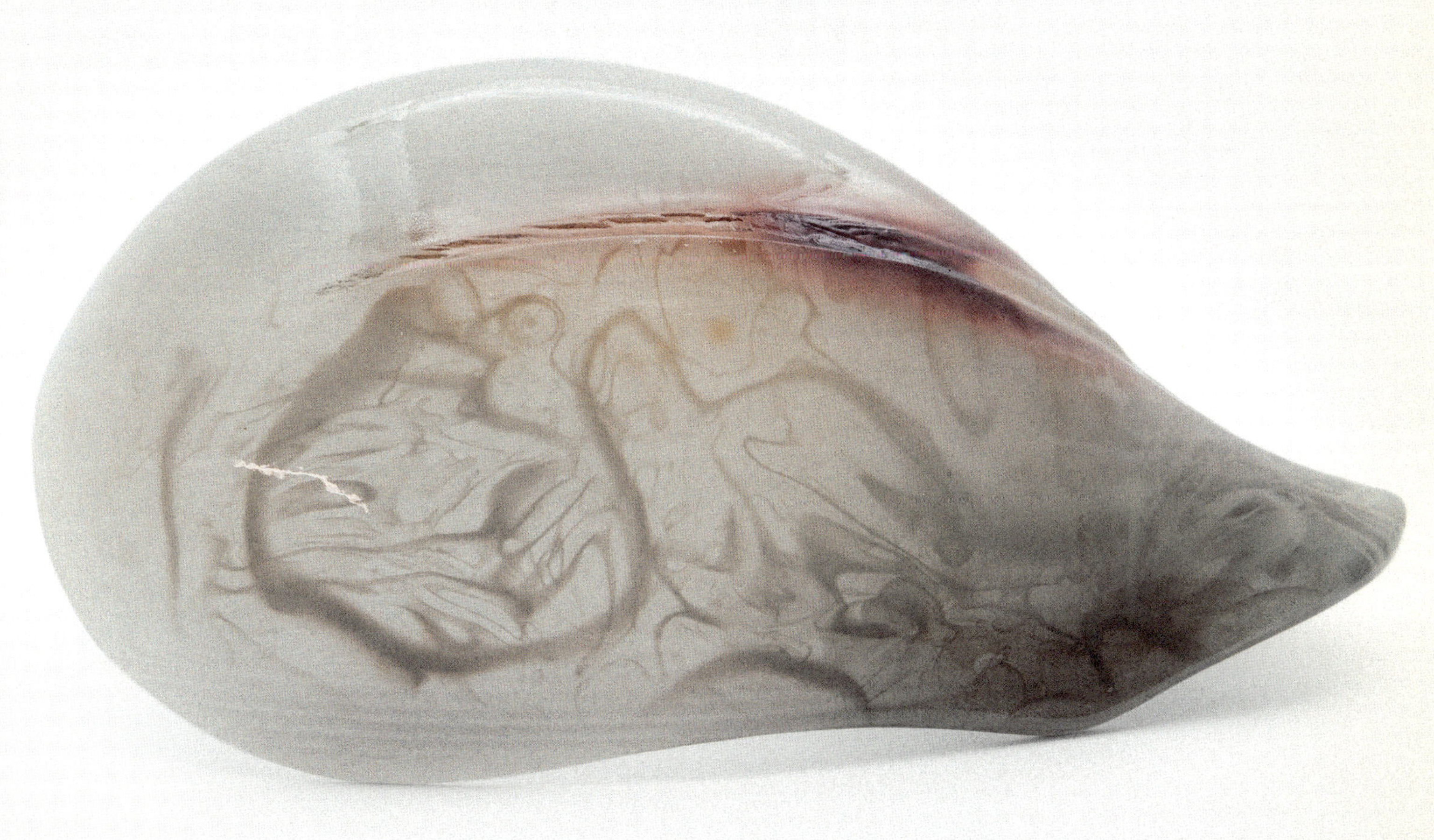

TOTEMS

Neri Oxman and
The Mediated Matter Group
Totems. 2019
Photopolymer resin, aluminum, steel, and melanins from bird feathers, squid ink, mushroom enzyme, L-tyrosine, and alkaline water
3D-printed block: 5 ⅞ × 5 ⅞ × 19 5⁄16 in. (15 × 15 × 49 cm)
Produced by Stratasys Ltd.
An MIT Media Lab project
Research team: Sunanda Sharma, Christoph Bader, Rachel Soo Hoo Smith, Felix Kraemer, João Costa, Joseph H. Kennedy Jr., Neri Oxman
Undergraduate researchers: Joseph Faraguna, Sara L. Wilson, Sangita Vasikaran
Collaborators and contributors: Natalia Casas; Eric de Broche des Combes, Luxigon; Kelly Egorova; Osvaldo Golijov; Gianluca di Ioia; Dechuan Meng; Hans Martin Pech; Christopher Voigt; Susan Williams; Nitzan Zilberman; Stratasys Ltd.; Front Inc.; Bodino
Acknowledgments: NOE LLC, Design Indaba, Robert Wood Johnson Foundation, Estée Lauder, GETTYLAB, XXII Triennale di Milano, 2019

Commissioned for *Broken Nature: Design Takes On Human Survival*, XXII Triennale di Milano, 2019

Earth's biodiversity has come under unprecedented threat. One of the main goals of our research has been to find and develop materials and chemical substances that can sustain and enhance the survival of all species. Beginning with natural materials produced by organisms that have survived climate change thus far, we have been able to draw on a palette of materials born of adaptability and resilience.

Colors darkened as L-tyrosine, a common amino acid, was converted into a melanin-like material through the action of the enzyme tyrosinase.

Melanin is one such substance. As a biomarker of evolution, it is a ubiquitous pigment that defines the color of skin, fur, hair, and eyes, and is found in different forms: in the blue of peacock feathers, the ochre of butterfly wings, and the brown skin of a ripening pear. Its presence in millions of organisms, across multiple phylogenetic kingdoms at different stages of evolution, represents unity in the variety of life on Earth. As a substance resistant to degradation and a heterogeneous pigment, melanin functions differently in different organisms: defending against ultraviolet radiation and oxidative stress, harvesting energy, binding metals, and providing thermal regulation. Fossils more than 160 million years old contain nearly the same type of melanin that can now be chemically synthesized in a laboratory, making it a substance both ancient and modern. Melanin is a critical element for health and today is considered the key to human survival.

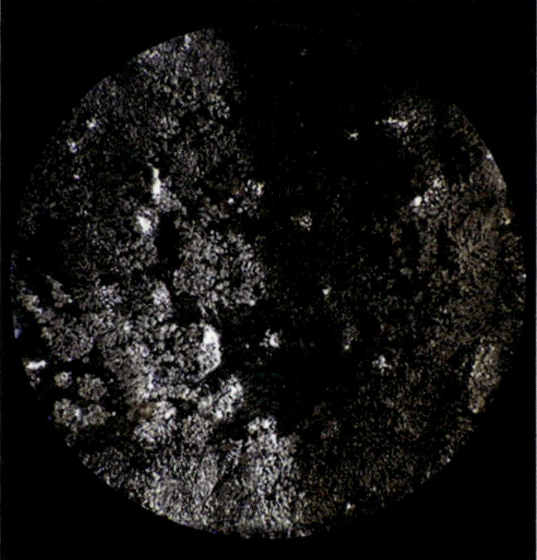

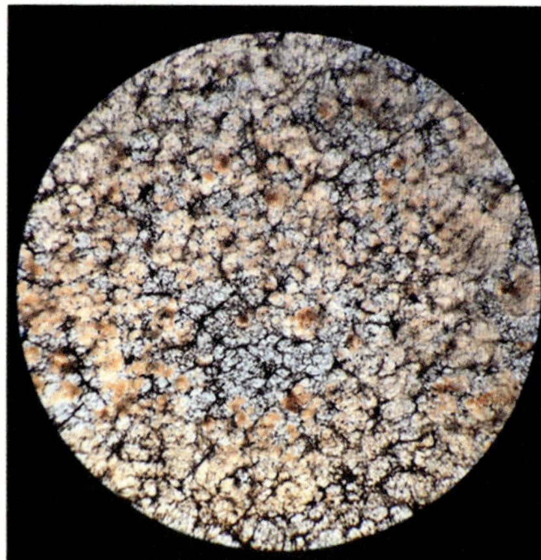

Optical microscope images taken using reflected and transmitted illumination show the product of an enzymatic reaction of tyrosinase.

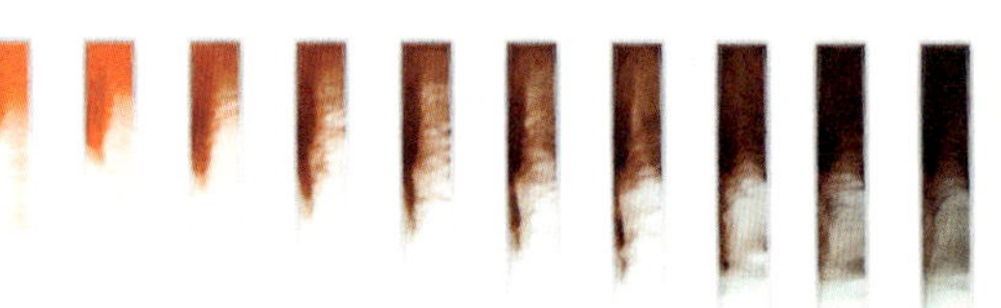

Time-lapse photographs of the chemical reaction between tyrosinase and L-tyrosine. The colors, movement, and patterns changed over time as more and more of the substrate was converted, generating flow within a rectangular 3D-printed dish.

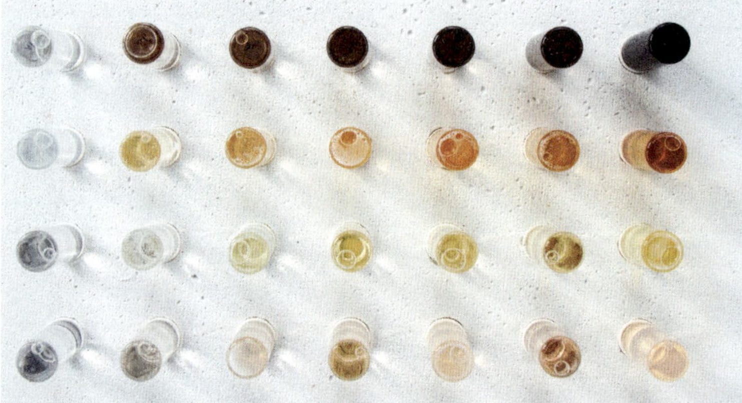

Library of melanins and melanin-like materials.

With Totems we experimented with programming melanin's interaction across scales and species, in order to understand, explain, and predict how it and other derivative pigments could be generated on demand, and how the environmental and human factors involved in its creation and form might be tuned or even reversed.

Petri dishes containing melanin made by engineered bacteria responding to tuned chemical signals.

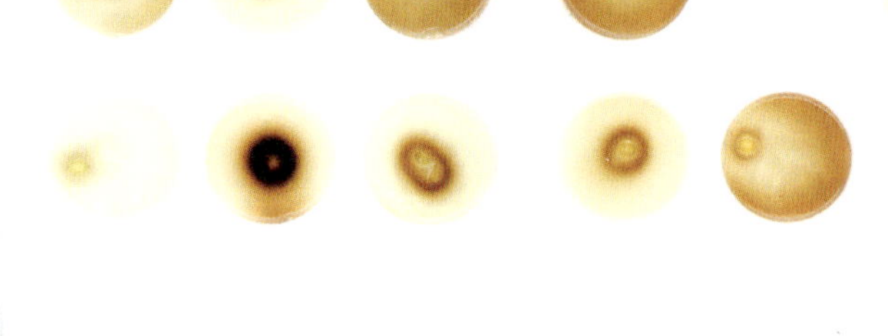

Partial library of melanin-like pigments created through synthesis and extraction processes. Each vial contained a different color, each absorbing light from different parts of the spectrum.

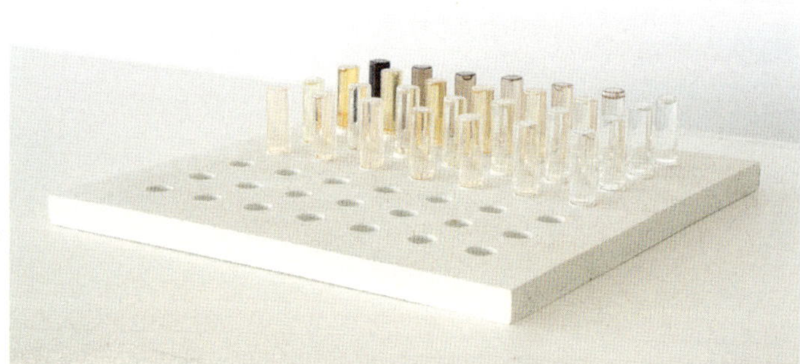

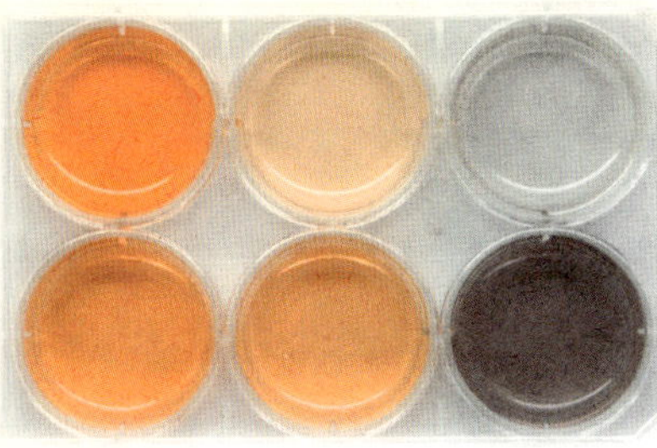

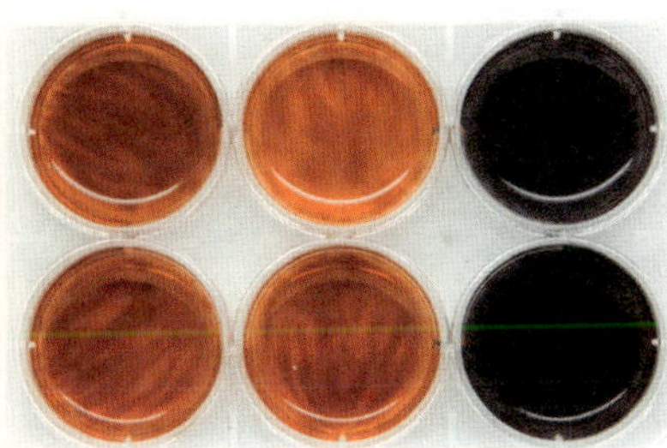

Progression of an enzymatic reaction over six hours, with earliest stages at the top. The four wells at the left of each tray contained only the enzyme and substrate; the two at the right included copper sulfate.

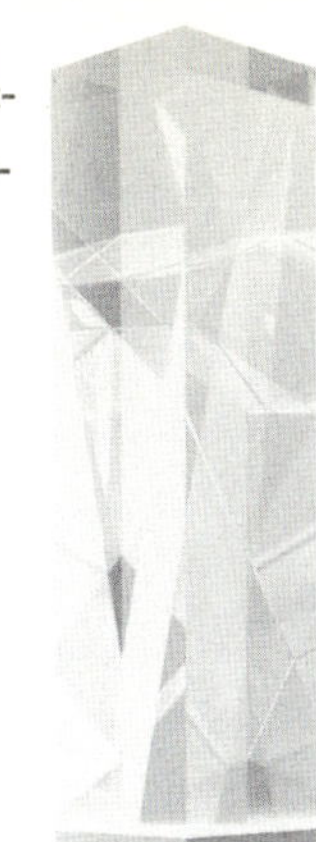

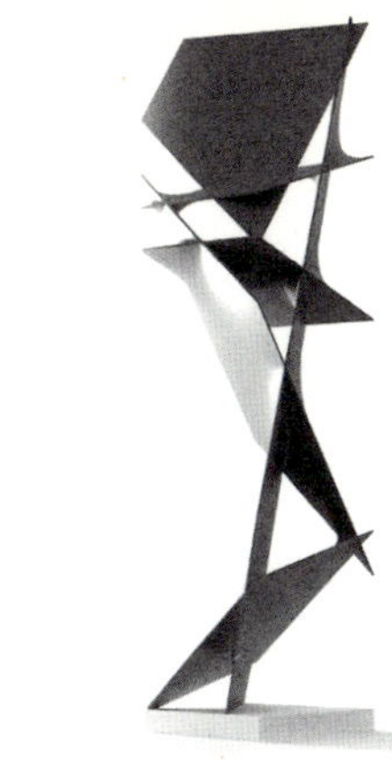

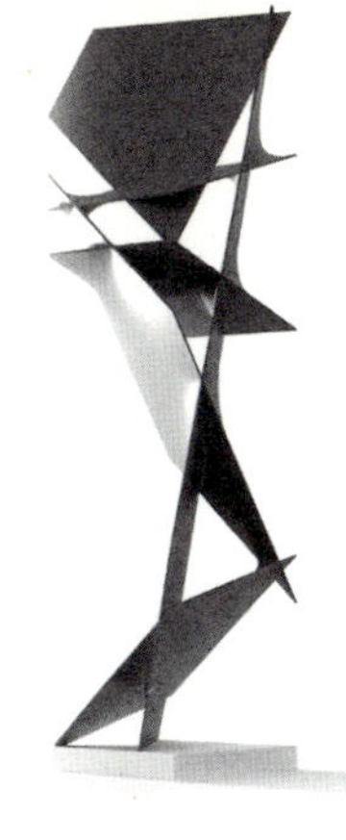

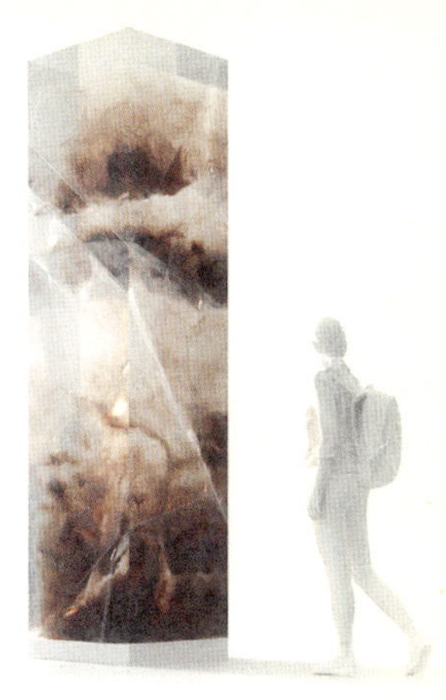

Early renderings of a column containing melanin, demonstrating the influence of geometry on an enzymatic reaction.

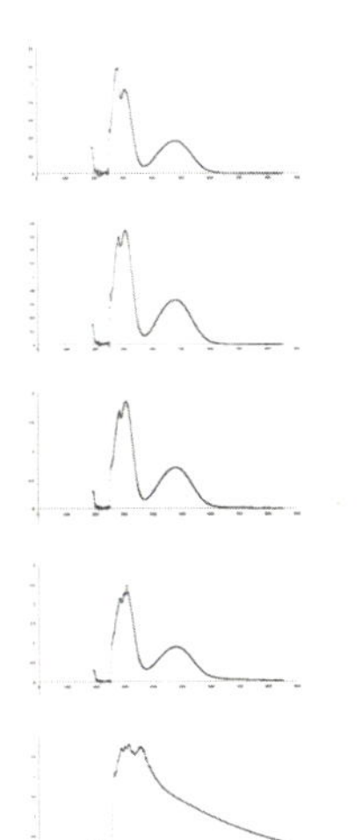

The absorption spectra of electromagnetic radiation transmitted through synthesized pigments suspended in water (left); structures for chemicals created through the reaction process (right).

We created melanin by producing a reaction with an active chemical—the enzyme tyrosinase, which, when extracted from a mushroom, converts the amino acid tyrosine (a protein building block) into melanin. The pigments can then be deployed in liquid or powder form to create inks that we printed in two and three dimensions. We also extracted pigment from bird feathers and cuttlefish, and with them we began our experiments in additive manufacturing in a series of spherical objects featuring computationally grown channels filled with melanin. Our goal, however, was to bring the project to an architectural scale, and for *Broken Nature: Design Takes On Human Survival* at the 2019 Milan Triennale, we created a column—a chemical totem—with 3D-printed channels running throughout it, containing liquid melanin.

The benefits of using melanin in architectural construction remain unexplored but intriguing: Could it produce optical changes in a building's facade depending on the time of day and season? Could it equip a building to be a responsive greenhouse? Totems is currently designed to contain melanin for UV protection; in the future we hope to develop ways of printing semipermeable architectural skins.

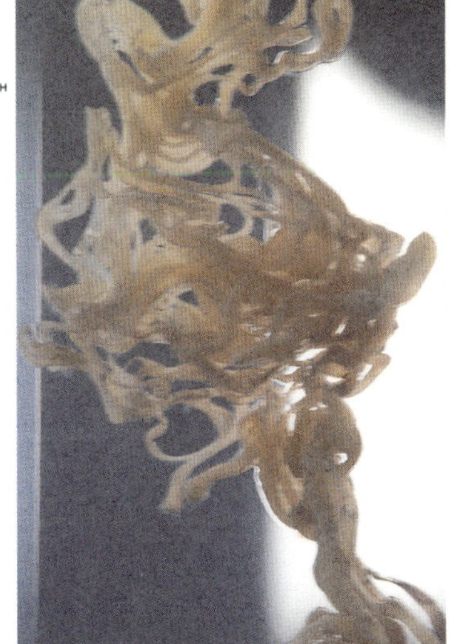

Detail of a 3D-printed element containing six distinct channels filled with liquid melanin. A liquid printing technique was used to create it.

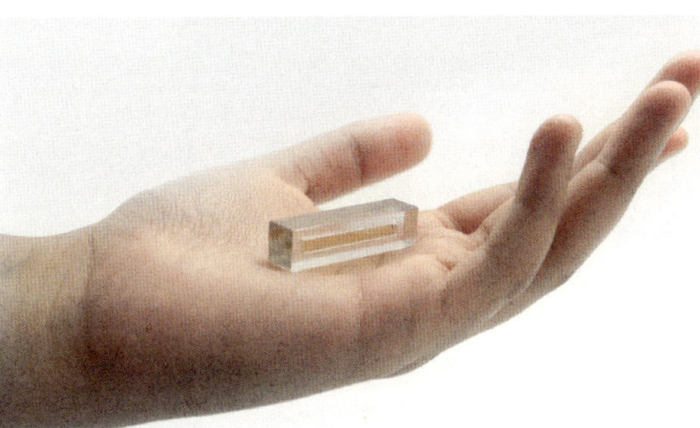

3D-printed prototype from a digital fabrication process in which biochemical solutions were printed as liquid resins.

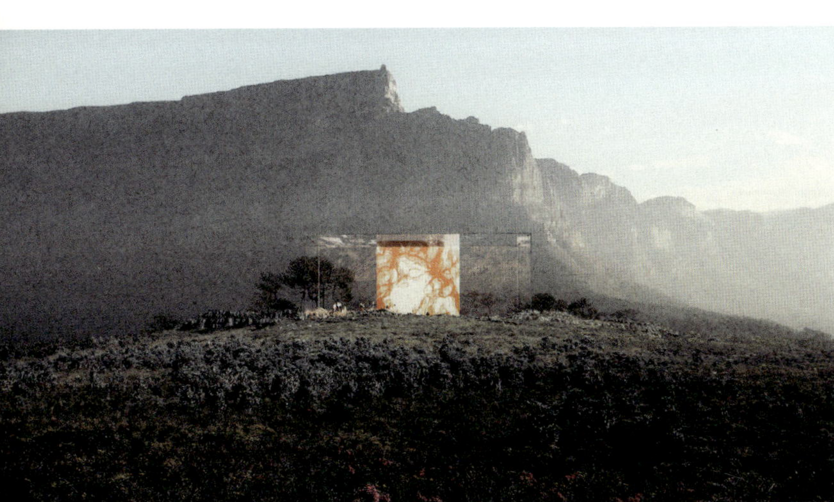

Proposal for an environmentally responsive glass structure. The walls are infused with melanin synthesized from organisms native to Table Mountain, in Cape Town, South Africa, the project's site.

It is of course as a societal construct rather as than a biological element that melanin is most widely understood; it continues to determine the destiny of individuals and whole racial groups to an enormous degree. Attention to the duality of biology and culture is critical for designers who have taken up the engineering of living systems. What will be the biological and cultural implications, now that melanin can be engineered?

Totems is the first iteration of a long-term project initiated in 2018 under the guidance of Ravi Naidoo at the annual Design Indaba conference in Cape Town, South Africa. By investigating the practical possibilities of melanin in the fields of architecture and design, we hope to investigate our own relationship with biology through the use of materials that both sustain and complicate our existence.

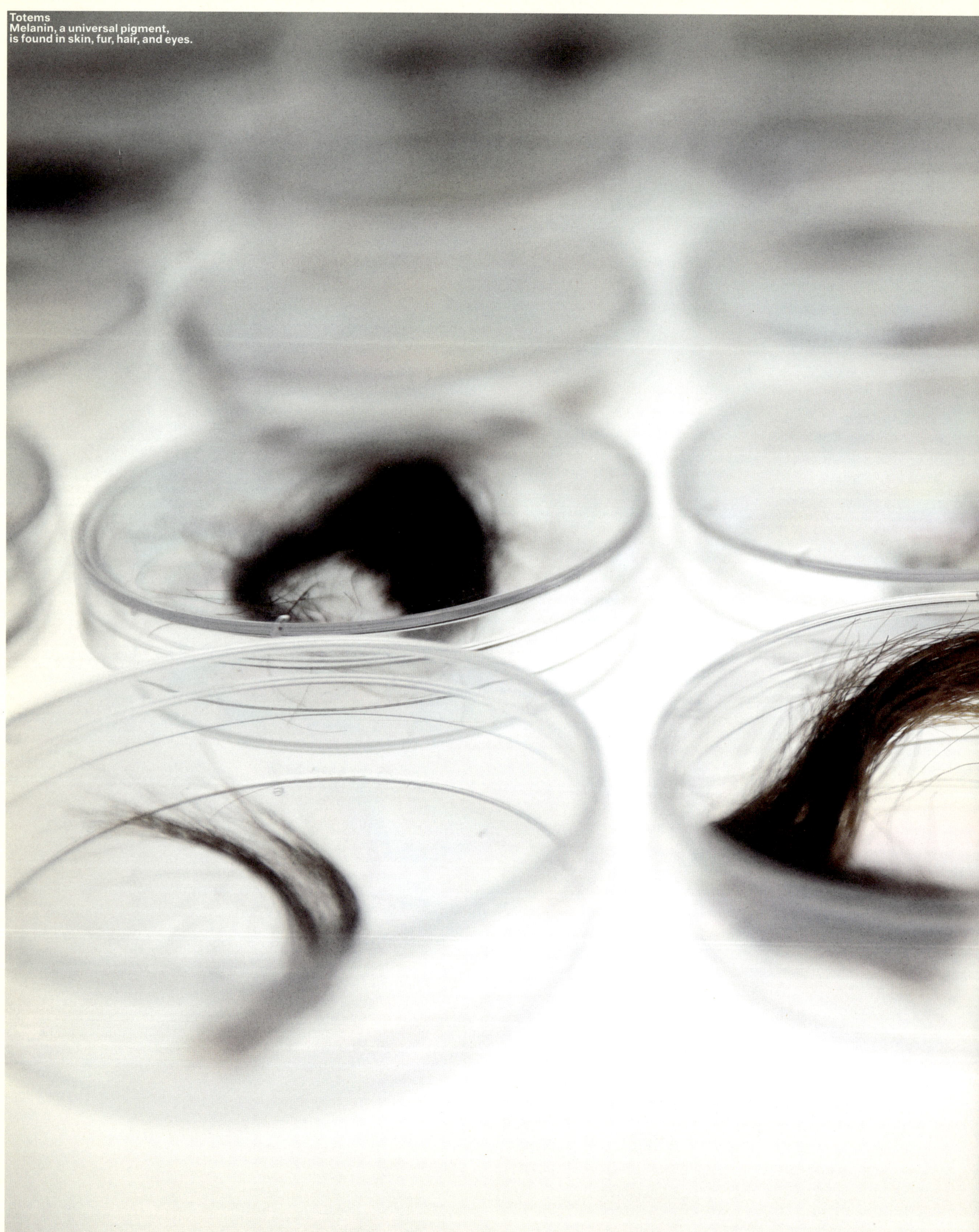
Totems
Melanin, a universal pigment, is found in skin, fur, hair, and eyes.

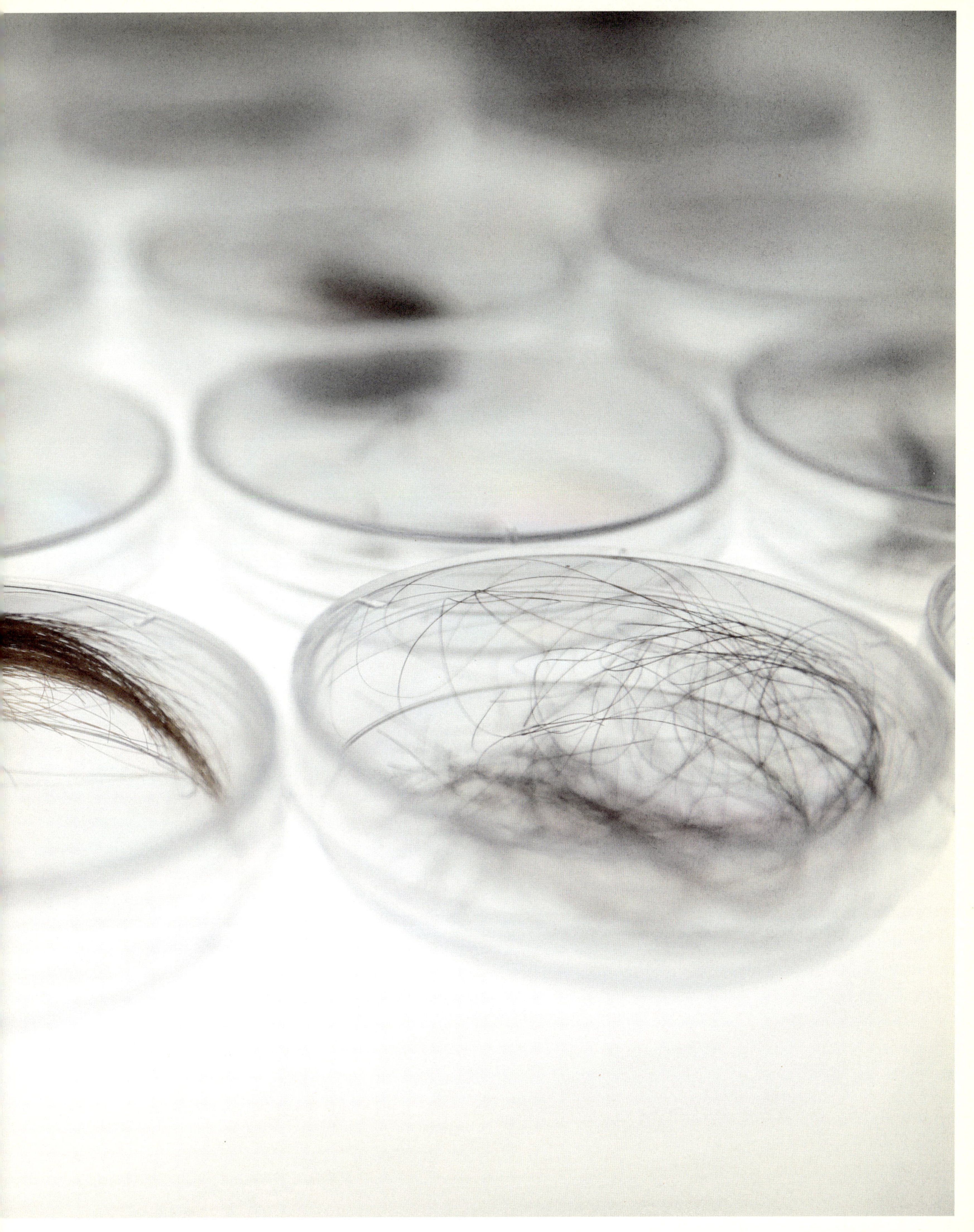

Totems
Melanins and melanin-like materials were created through synthesis and extraction processes to create a subset of the many colors visible in nature.

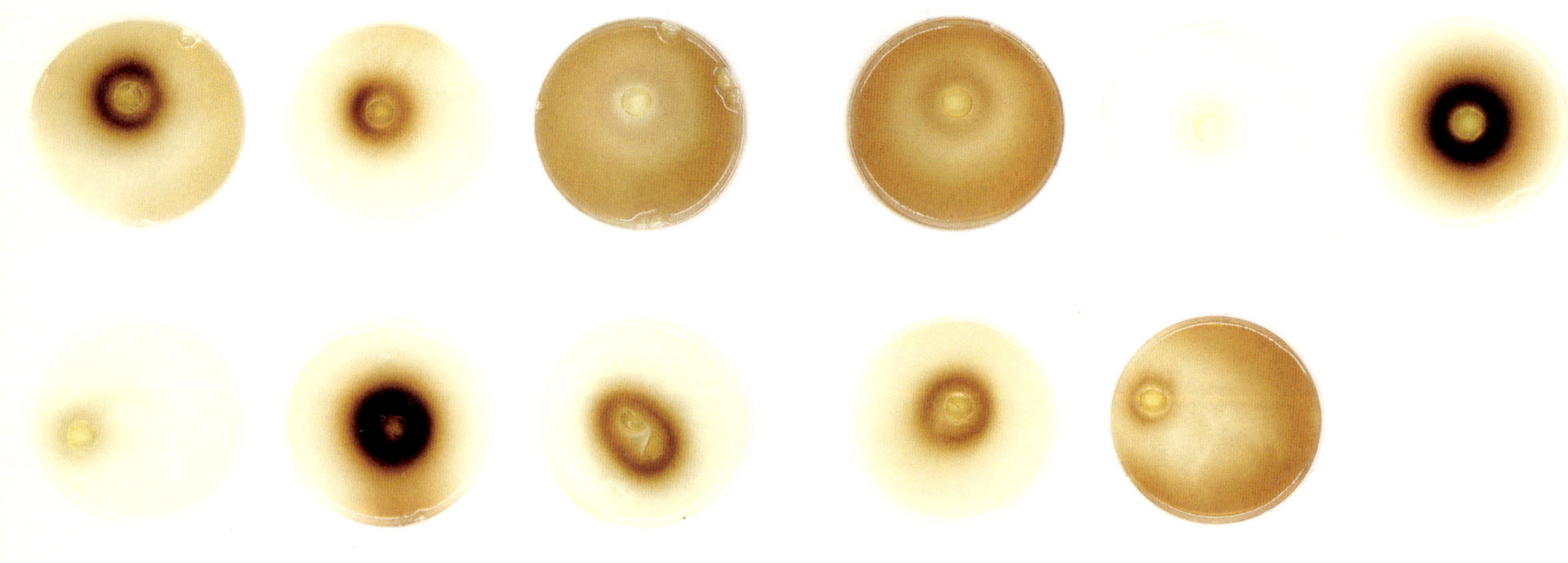

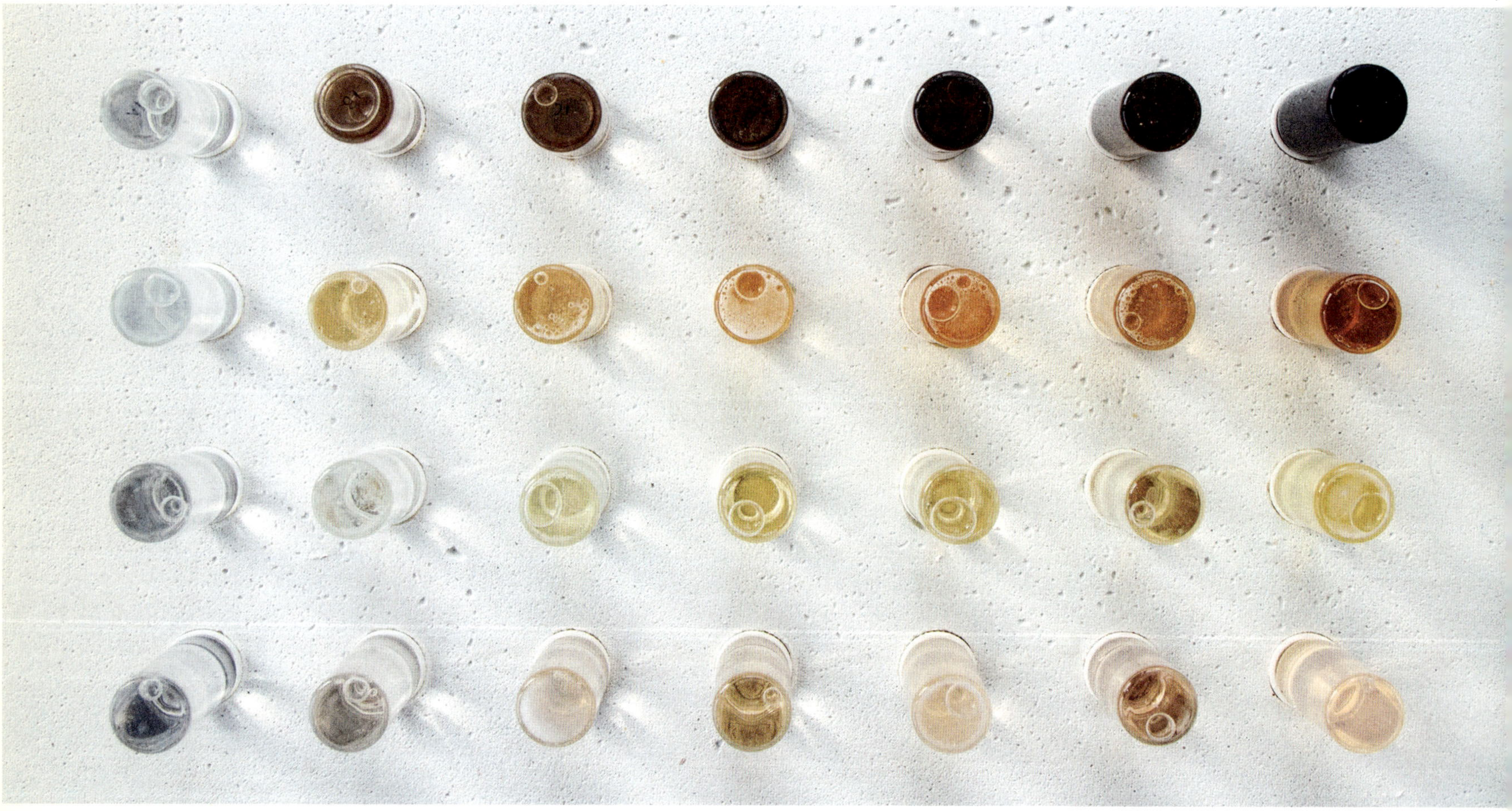

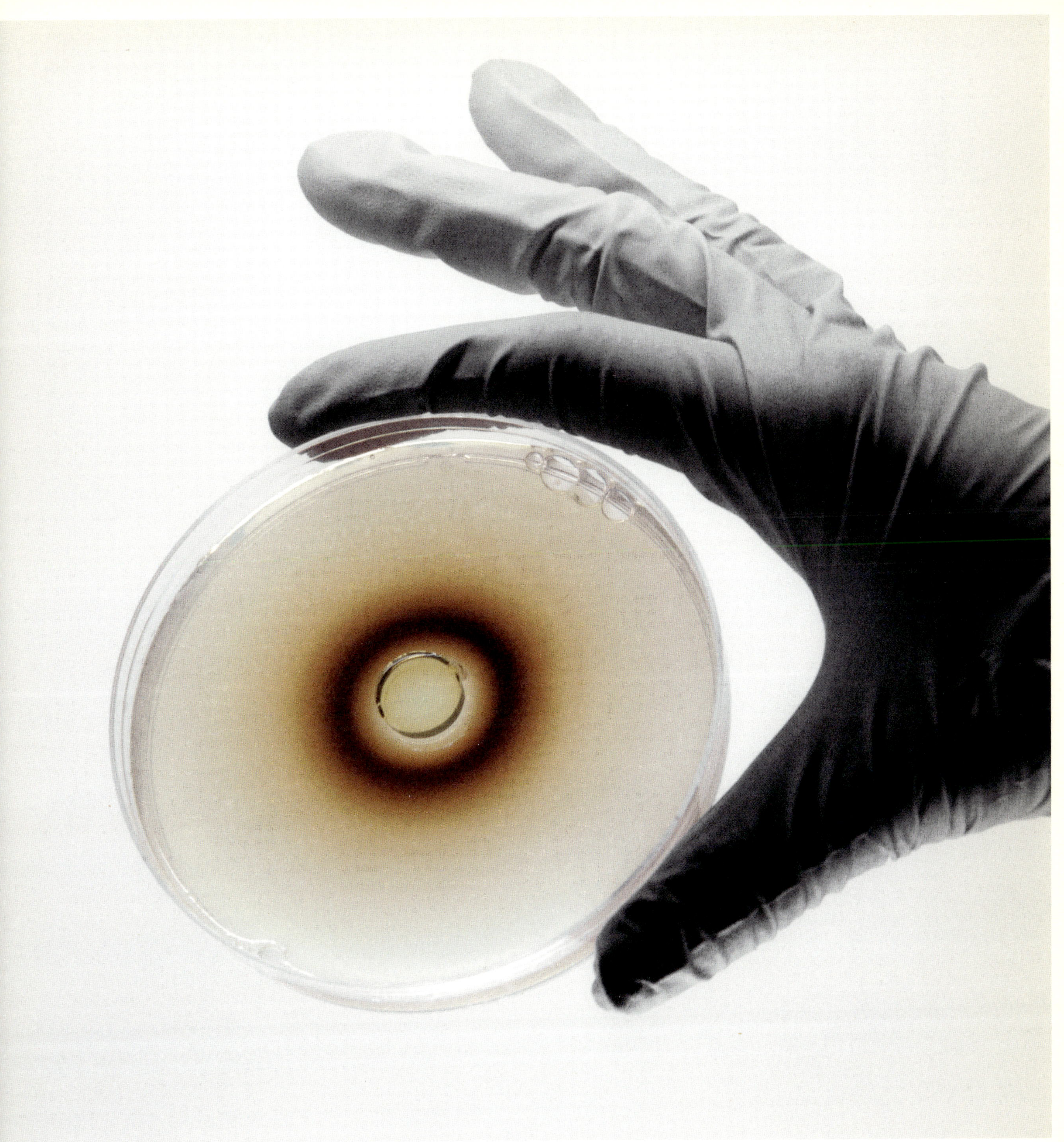

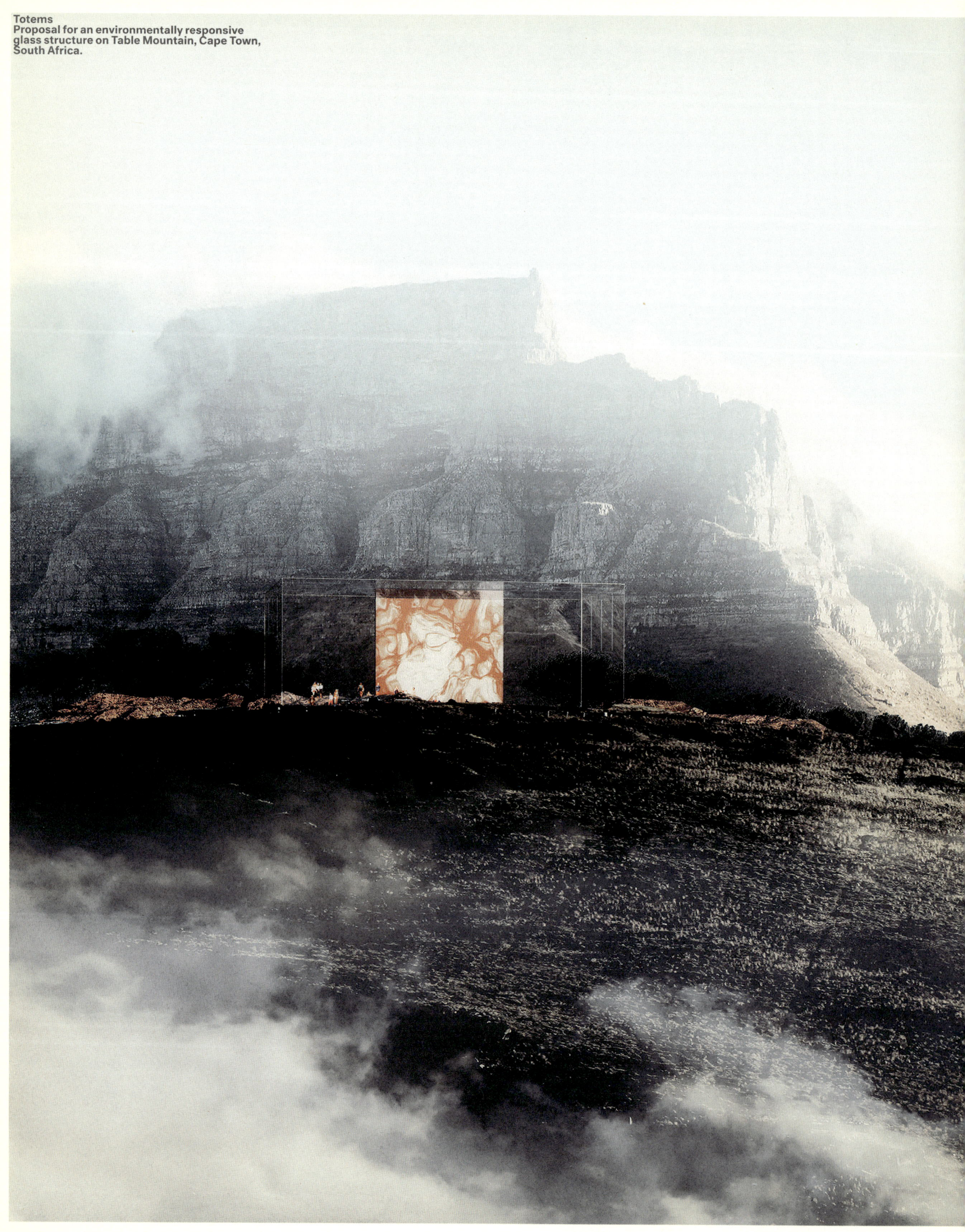

Totems
Proposal for an environmentally responsive glass structure on Table Mountain, Cape Town, South Africa.

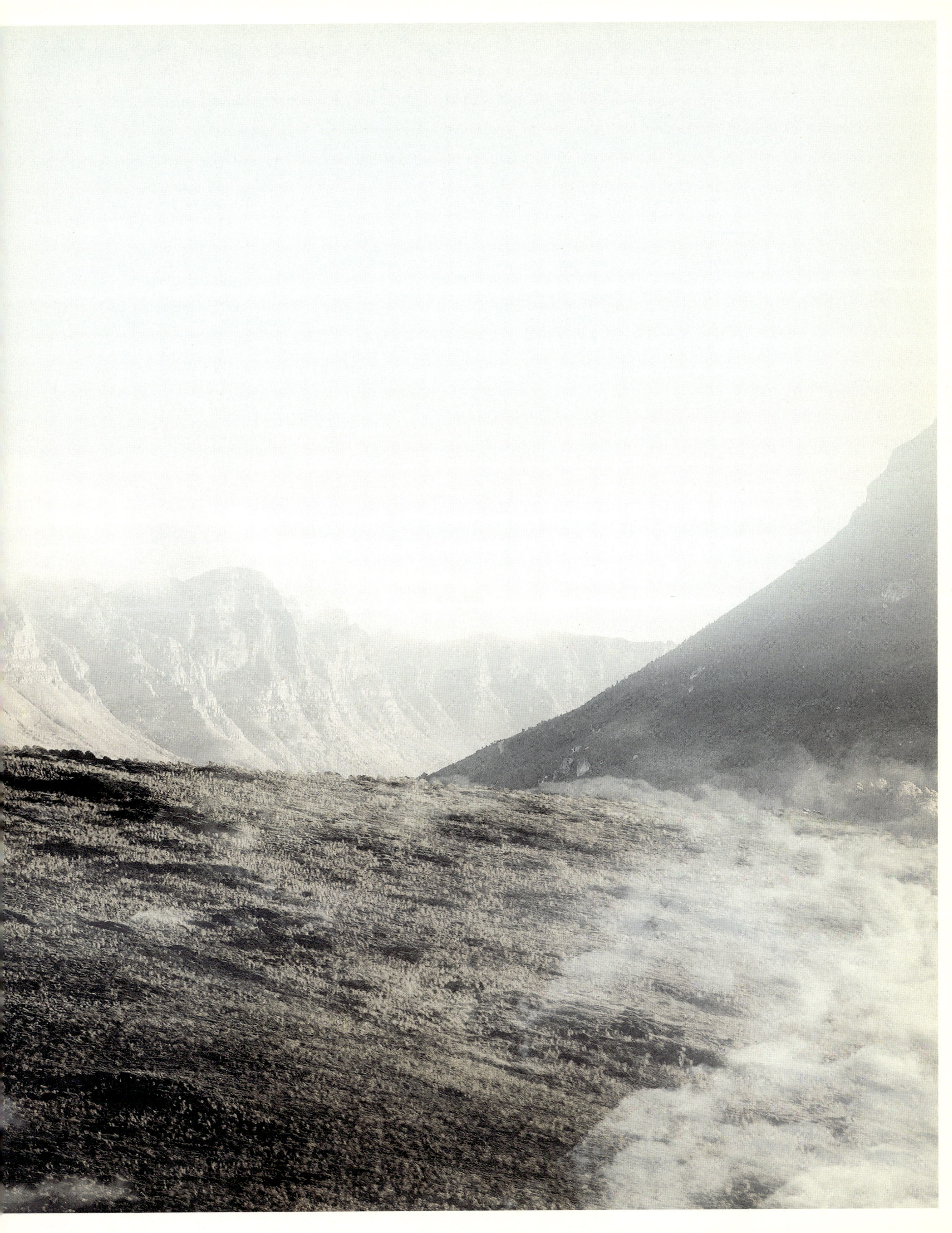

Selected Bibliography

Papers and articles

In chronological order, by project

General

Oxman, Neri. "Towards a Material Ecology." In *ACADIA 2012: Synthetic Digital Ecologies*, ed. Jason Kelly Johnson, Mark Cabrinha, and Kyle Steinfeld, 19–20. Proceedings of the 32nd Annual Conference of the Association for Computer-Aided Design in Architecture, California College of the Arts, San Francisco, October 18–21, 2012. New York: ACADIA, 2012.

———. "Material Ecology." In *Theories of the Digital in Architecture*, ed. Rivka Oxman and Robert Oxman, 321–28. London: Routledge, 2013.

Oxman, Neri, Christine Ortiz, Fabio Gramazio, and Matthias Kohler. "Material Ecology." In "Material Ecology: Design and Computational Issues," ed. Oxman, Ortiz, and Gramazio, 1–2. Special issue, *Computer-Aided Design* 60 (March 2015).

Oxman, Neri. "Age of Entanglement." *Journal of Design and Science*, no. 1 (January 2016).

Materialecology

Oxman, Neri. "Get Real: Towards Performance-Driven Computational Geometry." *International Journal of Architectural Computing* 5, no. 4 (December 2007): 663–84.

———. "Oublier Domino: On the Evolution of Architectural Theory from Spatial to Performance-Based Programming." In *What Matter(s)?*, ed. Kostas Terzidis, 393–402. Proceedings of the First International Conference on Critical Digital, Harvard University Graduate School of Design, Cambridge, Mass., April 18–19, 2008.

———. "Rapid Gestalt(en)." In *High Tech Solutions and Concepts*, ed. Rudolf Meyer, 61–69. Proceedings of Euro-uRapid 2008: International User's Conference on Rapid Prototyping & Rapid Tooling & Rapid Manufacturing, Berlin, September 23–24, 2008. Magdeburg, Germany: Fraunhofer Allianz Rapid Prototyping, 2008.

———. "Material-Based Design Computation: Tiling Behavior." In *ACADIA 09: reForm(); Building a Better Tomorrow*, ed. Tristan D'Estrée Sterk, Russell Loveridge, and Douglas Pancoast, 122–27. Proceedings of the 29th Annual Conference of the Association for Computer-Aided Design in Architecture, Chicago, October 22–25, 2009.

———. "Structuring Materiality: Design Fabrication of Heterogeneous Materials." In "The New Structuralism: Design, Engineering, and Architectural Technologies," ed. Rivka Oxman and Robert Oxman. Special issue, *Architectural Design* 80, no. 4 (July–August 2010): 78–85.

———. "Finite Element Synthesis." In *Innovative Developments in Virtual and Physical Prototyping*, ed. Paulo Jorge Bártolo et al., 719–24. Proceedings of the 5th International Conference on Advanced Research in Virtual and Rapid Prototyping, Leiria, Portugal, September 28–October 1, 2011. Boca Raton, Fla.: CRC Press, 2011.

Oxman, Neri, Steven Keating, and Elizabeth Tsai. "Functionally Graded Rapid Prototyping." In *Innovative Developments in Virtual and Physical Prototyping*, ed. Paulo Jorge Bártolo et al., 483–90. Proceedings of the 5th International Conference on Advanced Research in Virtual and Rapid Prototyping, Leiria, Portugal, September 28–October 1, 2011. Boca Raton, Fla.: CRC Press, 2011.

Duro-Royo, Jorge, Katia Zolotovsky, Laia Mogas-Soldevila, Swati Varshney, Neri Oxman, Mary C. Boyce, and Christine Ortiz. "MetaMesh: A Hierarchical Computational Model for Design and Fabrication of Biomimetic Armor Surfaces." In "Material Ecology: Design and Computational Issues," ed. Oxman, Ortiz, and Fabio Gramazio. Special issue, *Computer-Aided Design* 60 (March 2015): 14–27.

Aguahoja

Mogas-Soldevila, Laia, Jorge Duro-Royo, and Neri Oxman. "Water-Based Robotic Fabrication: Large-Scale Additive Manufacturing of Functionally Graded Hydrogel Composites via Multichamber Extrusion." *3D Printing and Additive Manufacturing* 1, no. 3 (September 2014): 141–51.

Mogas-Soldevila, Laia, and Neri Oxman. "Water-Based Engineering and Fabrication: Large-Scale Additive Manufacturing of Biomaterials." In *Adaptive Architecture and Programmable Matter: Next Generation Building Skins and Systems from Nano to Macro*, ed. Jenny Sabin, Maria Paz Gutierrez, and Christian Santangelo, 46–53. Proceedings of the Materials Research Society (MRS) Spring Meeting, San Francisco, April 6–10, 2015.

Mogas-Soldevila, Laia, Jorge Duro-Royo, Daniel Lizardo, Markus Kayser, Winston Patrick, Sunanda Sharma, Steven Keating, John Klein, Chikara Inamura, and Neri Oxman. "Designing the Ocean Pavilion: Biomaterial Templating of Structural, Manufacturing, and Environmental Performance." In *Future Visions*. Proceedings of IASS2015 Annual International Symposium, Amsterdam, August 17–20, 2015.

Mogas-Soldevila, Laia, Jorge Duro-Royo, and Neri Oxman. "Form Follows Flow: A Material-Driven Computational Workflow for Digital Fabrication of Large-Scale Hierarchically Structured Objects." In *Acadia 2015: Computational Ecologies*, ed. Lonn Combs and Chris Perry, 185–93. Proceedings of the Association for Computer-Aided Design in Architecture 2015 International Conference (ACADIA), Cincinnati, October 19–25, 2015.

Mogas-Soldevila, Laia, Jorge Duro-Royo, and Neri Oxman. "Flow-Based Fabrication: An Integrated Computational Workflow for Design and Digital Additive Manufacturing of Multifunctional Heterogeneously Structured Objects." *Computer-Aided Design* 69 (December 2015): 143–54.

Duro-Royo, Jorge, Joshua Van Zak, Andrea S. Ling, Yen-Ju (Tim) Tai, and Neri Oxman. "Parametric Chemistry: Reverse Engineering Biomaterial Composites for Additive Manufacturing of Bio-Cement Structures across Scales." In *Challenges for Technology Innovation: An Agenda for the Future*, ed. Fernando Moreira da Silva et al., 217–23. Proceedings of the International Conference on Sustainable Smart Manufacturing (S2M 2016), Lisbon, October 20–22, 2016. Boca Raton, Fla.: CRC Press, 2017.

Tai, Yen-Ju (Tim), Christoph Bader, Andrea S. Ling, Jean Disset, Barrak Darweesh, Jorge Duro-Royo, Joshua Van Zak, Nic Hogan, and Neri Oxman. "Designing (for) Decay: Parametric Material Distribution for Hierarchical Dissociation of Water-Based Biopolymer Composites." In *Creativity in Structural Design*. Proceedings of the Annual IASS Symposium, Boston, July 16–20, 2018.

Lee, Nic, Ramon Weber, Joseph H. Kennedy, Sunanda Sharma, Jorge Duro-Royo, and Neri Oxman. "Continuous Gradation of Multi-Material Biopolymer Hydrogels." Paper presented at Materials Research Society (MRS) Fall Meeting, Boston, December 1–6, 2019.

Lee, Nic, Ramon Weber, Joseph H. Kennedy, Jorge Duro-Royo, and Neri Oxman. "Functionally Graded Architectural Biocomposites." Paper to be delivered at the FABRICATE conference, University College London, April 2–3, 2020.

Silk Pavilion

Tsai, Elizabeth, Michal Firstenberg, Jared Laucks, Benjamin Lehnert, Yoav Sterman, and Neri Oxman. "CNSilk: Spider-Silk Inspired Robotic Fabrication." In *Rob/Arch 2012: Robotic Fabrication in Architecture, Art, and Design*, ed. Sigrid Brell-Çokcan and Johannes Braumann, 160–66. Proceedings of Rob/Arch 2012: Robotic Fabrication in Architecture, Art, and Design, Vienna, December 17–18, 2012. Vienna: Springer, 2013.

Oxman, Neri, Jared Laucks, Markus Kayser, Carlos David Gonzalez Uribe, and Jorge Duro-Royo. "Biological Computation for Digital Design and Fabrication: A Biologically Informed Finite Element Approach to Structural Performance and Material Optimization of Robotically Deposited Fiber Structures." In *Computation and Performance*, ed. Rudi Stouffs and Sevil Sariyildiz, 1–10. Proceedings of the 31st International Conference on Education and Research in Computer-Aided Architectural Design in Europe, Faculty of Architecture, Delft University of Technology, the Netherlands, September 18–20, 2013.

———. "Silk Pavilion: A Case Study in Fibre-Based Digital Fabrication." In *FABRICATE: Negotiating Design & Making*, ed. Fabio Gramazio, Matthias Kohler, and Silke Langenberg, 248–55. Proceedings of the FABRICATE conference, ETH, Zurich, February 14–15, 2014. Zurich: gta Verlag, 2014; London: UCL Press, 2017.

Oxman, Neri, Jorge Duro-Royo, Steven Keating, Benjamin Peters, and Elizabeth Tsai. "Towards Swarm Printing." In "Made by Robots: Challenging Architecture at the Large Scale," ed. Fabio Gramazio and Matthias Kohler. Special issue, *Architectural Design* 84, no. 3 (May–June 2014): 108–15.

Oxman, Neri, Jared Laucks, Markus Kayser, Elizabeth Tsai, and Michal Firstenberg. "Freeform 3D Printing: Towards a Sustainable Approach to Additive Manufacturing." In *Green Design, Materials and Manufacturing Processes*, ed. Helena Bártolo et al., 479–84. Proceedings of the 2nd International Conference on Sustainable Intelligent Manufacturing, Lisbon, June 26–29, 2014. Boca Raton, Fla.: CRC Press, 2013.

Duro-Royo, Jorge, and Neri Oxman. "Towards Fabrication Information Modeling (FIM): Four Case Models to Derive Designs Informed by Multi-Scale Trans-Disciplinary Data." In *Adaptive Architecture and Programmable Matter: Next Generation Building Skins and Systems from Nano to Macro*, ed. Jenny Sabin, Maria Paz Gutierrez, ıd Christian Santangelo, 38–45. ·roceedings of the Materials Research ociety (MRS) Spring Meeting, San ·rancisco, April 6–10, 2015.

osta, João, Christoph Bader, unanda Sharma, Jessica Xu, and eri Oxman. "Integrated Design nd Digital Fabrication of Bio- omeomorphic Structures across cales." In *Creativity in Structural esign*. Proceedings of the Annual ASS Symposium, Boston, July 5–20, 2018.

lass

lein, John, Michael Stern, Markus ayser, Chikara Inamura, Giorgia ranchin, Shreya Dave, James Veaver, Peter Houk, Paolo Colombo, nd Neri Oxman. "Additive Manu- ıcturing of Optically Transparent lass." *3D Printing and Additive Ianufacturing* 2, no. 3 (September 015): 92–105.

run, Pierre-Thomas, Chikara ıamura, Daniel Lizardo, Giorgia ranchin, Michael Stern, Peter ouk, and Neri Oxman. "The Molten lass Sewing Machine." *Philo- ophical Transactions of the Royal ociety A: Manufacturing, Physical nd Engineering Sciences* 375, o. 2,093 (2016): 1–12.

ıamura, Chikara, Michael Stern, aniel Lizardo, Peter Houk, and eri Oxman. "High-Fidelity Additive Ianufacturing of Transparent lass Structures." In *Creativity in tructural Design*. Proceedings of he Annual IASS Symposium, oston, July 16–20, 2018.

——. "Additive Manufacturing of ransparent Glass Structures." *3D rinting and Additive Manufacturing* , no. 4 (December 2018): 268–83.

maginary Beings

orges, Jorge Luis. *The Book of maginary Beings*. Trans. Norman homas di Giovanni. New York: utton, 1969.

)xman, Neri, and Jesse Louis osenberg. "Material-Based Design omputation: An Inquiry into igital Simulation of Physical Iaterial Properties as Design enerators." *International Journal f Architectural Computing*, no. 5 January 2007): 26–44.

)xman, Neri. "Variable Property apid Prototyping." *Virtual and hysical Prototyping* 6, no. 1 June 2011): 3–31.

)xman, Neri, Elizabeth Tsai, and Iichal Firstenberg. "Digital nisotropy: A Variable Elasticity apid Prototyping Platform." *Virtual nd Physical Prototyping* 7, no. 4 September 2012): 261–74.

Doubrovski, Eugeni L., Elizabeth Tsai, Daniel Dikovsky, Jo M. Geradts, Hugh Herr, and Neri Oxman. "Voxel-Based Fabrication through Material Property Mapping: A Design Method for Bitmap Printing." In "Material Ecology: Design and Computational Issues." Special issue, *Computer-Aided Design* 60 (March 2015): 3–13.

Vespers/Lazarus

Bader, Christoph, Dominik Kolb, James C. Weaver, and Neri Oxman. "Data-Driven Material Modeling with Functional Advection for 3D-Printing of Materially Hetero- geneous Objects." *3D Printing and Additive Manufacturing* 3, no. 2 (June 2016): 71–79.

Bader, Christoph, and Neri Oxman. "Recursive Symmetries for Geometrically Complex and Materially Heterogeneous Additive Manufacturing." *Computer-Aided Design* 81 (December 2016): 39–47.

McEoin, Ewan, Simon Maidment, Megan Patty, and Pip Wallis, eds. *NGV Triennial 2017*. Melbourne: National Gallery of Victoria, 2017.

Bader, Christoph, Dominik Kolb, James C. Weaver, Sunanda Sharma, Ahmed Hosny, João Costa, and Neri Oxman. "Making Data Matter: Voxel Printing for the Digital Fabrication of Data across Scales and Domains." *Science Advances* 4, no. 5 (May 2018): eaas8652.

Smith, Rachel Soo Hoo, Christoph Bader, Sunanda Sharma, Dominik Kolb, Tzu-Chieh Tang, Ahmed Hosny, Felix Moser, James Weaver, Christopher A. Voigt, and Neri Oxman. "Hybrid Living Materials: Digital Design and Fabrication of 3D Multi-Material Structures with Programmable Biohybrid Surfaces." *Advanced Functional Materials*. Published ahead of print, December 18, 2019. onlinelibrary .wiley.com/doi/10.1002/adfm .201907401.

Totems

Oxman, Neri, and The Mediated Matter Group. "Totems." In *Broken Nature: Design Takes On Human Survival*, ed. Paola Antonelli and Ala Tannir, 62. New York: Rizzoli Electa, 2019.

Sharma, Sunanda, Bianca Datta, V. Michael Bove, and Neri Oxman. "Synthesizing Tunable Artificial Color: A Combined Approach." Paper presented at Materials Research Society (MRS) Fall Meeting, Boston, December 1–6, 2019.

Patents by Neri Oxman and The Mediated Matter Group
In chronological order

Keating, Steven, and Neri Oxman. "Methods and Apparatus for Computer-Assisted Spray Foam Fabrication." U.S. Patent 9,566,742 B2, issued February 14, 2017 (MIT 15527T).

———. "Jamming Methods and Apparatus." U.S. Patent 9,764,220 B2, issued September 17, 2017 (MIT 14894T).

Peters, Benjamin, and Neri Oxman. "Methods and Apparatus for Actuated Fabricator." U.S. Patent 9,764,378 B2, issued September 19, 2017 (MIT 15533T).

Klein, John, Giorgia Franchin, Michael Stern, Markus Kayser, Chikara Inamura, Shreya Dave, Neri Oxman, and Peter Houk. "Methods and Apparatus for Additive Manu- facturing of Glass." U.S. Patent 9,896,368 B2, issued February 20, 2018 (MIT 17063TK C1).

Inamura, Chikara, Daniel Lizardo, Michael Stern, Peter Houk, Tal Achituv, and Neri Oxman. "Methods and Apparatus for Additive Manufacturing with Molten Glass." U.S. Patent 9,919,510 B2, issued March 20, 2018 (MIT18655T).

Peters, Benjamin, and Neri Oxman. "Methods and Apparatus for Actuated Fabricator." U.S. Patent 10,189,076 B2, issued January 29, 2019 (MIT 15533T C1).

Keating, Steven, and Neri Oxman. "Methods and Apparatus for Computer-Assisted Spray Foam Fabrication." U.S. Patent 10,189,187 B2, issued January 29, 2019 (MIT 15527T C1).

Bader, Christoph, Dominik Kolb, James Weaver, and Neri Oxman. "Methods and Apparatus for 3D Printing of Point Cloud Data." U.S. Patent 10,259,164, issued April 16, 2019 (MIT 18871JT).

Klein, John, Giorgia Franchin, Michael Stern, Markus Kayser, Chikara Inamura, Shreya Dave, Neri Oxman, and Peter Houk. "Methods and Apparatus for Additive Manufacturing of Glass." U.S. Patent 10,266,442, issued April 23, 2019 (MIT 17063TK C2).

Duro-Royo, Jorge, Laia Mogas- Soldevila, and Neri Oxman. "Methods and Apparatus for Additive Manufacturing along User- Specified Toolpaths." U.S. Patent 10,286,606, issued May 14, 2019 (MIT 17388T).

Peters, Benjamin, and Neri Oxman. "Methods and Apparatus for Actuated Fabricator." U.S. Patent 10,391,550 B2, issued August 27, 2019 (MIT 15533T C2).

Inamura, Chikara, Daniel Lizardo, Michael Stern, Peter Houk, Tal Achituv, and Neri Oxman. "Methods and Apparatus for Additive Manufacturing with Molten Glass." U.S. Patent 10,464,305, issued November 5, 2019 (MIT 18655T C1).

Further reading
In alphabetical order, by subject

Nature and design

Alberti, Marina. *Cities That Think Like Planets: Complexity, Resilience, and Innovation in Hybrid Ecosystems*. Seattle: University of Washington Press, 2016.

Antonelli, Paola, and Ala Tannir, eds. *Broken Nature: Design Takes On Human Survival*. New York: Rizzoli Electa, 2019.

Benyus, Janine M. *Biomimicry*. New York: HarperCollins, 2002.

Brownell, Blaine, and Marc Swackhamer, eds. *Hypernatural: Architecture's New Relationship with Nature*. New York: Princeton Architectural Press, 2015.

Clément, Gilles. *"The Planetary Garden" and Other Writings*. Trans. Sandra Morris. Philadelphia: University of Pennsylvania Press, 2015.

Crist, Eileen, and H. Bruce Rinker, eds. *Gaia in Turmoil: Climate Change, Biodepletion, and Earth Ethics in an Age of Crisis*. Cambridge, Mass.: MIT Press, 2010.

Demos, T. J. *Decolonizing Nature*. Berlin: Sternberg Press, 2016.

Graham, James, Caitlin Blanchfield, Alissa Anderson, Jordan Carver, and Jacob Moore. *Climates: Architecture and the Planetary Imaginary*. New York: Columbia Books on Architecture and the City; Zurich: Lars Müller, 2016.

Haraway, Donna. *Staying with the Trouble: Making Kin in the Chthulucene*. Durham, N.C.: Duke University Press, 2016.

Kolbert, Elizabeth. *The Sixth Extinction: An Unnatural History*. New York: Henry Holt, 2014.

Latour, Bruno. *Facing Gaia: Eight Lectures on the New Climatic Regime*. Trans. Catherine Porter. Cambridge: Polity Press, 2017.

Lipps, Andrea, Matilda McQuaid, Caitlin Condell, and Gène Bertrand, eds. *Nature: Collaborations in Design*. New York: Cooper Hewitt, Smithsonian Design Museum, 2019.

Lovelock, James. *Gaia: A New Look at Life on Earth.* Oxford: Oxford University Press, 1979.

Morton, Oliver. *The Planet Remade: How Geoengineering Could Change the World.* Princeton, N.J.: Princeton University Press; London: Granta Books, 2015.

Morton, Timothy. *Hyperobjects: Philosophy and Ecology after the End of the World.* New York: Columbia University Press, 2013.

Pearce, Peter. *Structure in Nature Is a Strategy for Design.* Cambridge, Mass.: MIT Press, 1980.

Pollan, Michael. *Second Nature: A Gardener's Education.* New York: Grove Press, 1991.

Sachs, Angeli, ed. *Nature Design: From Inspiration to Innovation.* Baden, Switzerland: Lars Müller, 2007.

Vogel, Steven. *Comparative Biomechanics: Life's Physical World.* Princeton, N.J.: Princeton University Press, 2013.

Organic design, biodesign, and synthetic biology

Agapakis, Christina M. "Designing Synthetic Biology." *ACS Synthetic Biology* 3, no. 3 (March 2014): 121–28.

Antonelli, Paola, ed. *Design and the Elastic Mind.* New York: The Museum of Modern Art, 2008.

Benjamin, David. *Now We See Now: Architecture and Research by The Living.* New York: Monacelli Press, 2018.

Church, George, and Ed Regis. *Regenesis: How Synthetic Biology Will Reinvent Nature and Ourselves.* New York: Basic Books, 2014.

Cogdell, Christina. *Toward a Living Architecture? Complexism and Biology in Generative Design.* Minneapolis: University of Minnesota Press, 2019.

Costa, Beatriz da, and Kavita Philip. *Tactical Biopolitics: Art, Activism, and Technoscience.* Cambridge, Mass.: MIT Press, 2010.

Ginsberg, Alexandra Daisy, Jane Calvert, Pablo Schyfter, Alistair Elfick, and Andrew D. Endy. *Synthetic Aesthetics: Investigating Synthetic Biology's Designs on Nature.* Cambridge, Mass.: MIT Press, 2014.

Kac, Eduardo. *Signs of Life: Bio Art and Beyond.* Cambridge, Mass.: MIT Press, 2009.

Kellert, Stephen R., Judith Heerwagen, and Martin Mador. *Biophilic Design: The Theory, Science and Practice of Bringing Buildings to Life.* Hoboken, N.J.: John Wiley & Sons, 2008.

Myers, William. *Bio Design: Nature Science Creativity.* New York: The Museum of Modern Art, 2012.

Pandilovski, Melentie, ed. *Art in the Biotech Era.* Adelaide: Experimental Art Foundation, 2008.

Roosth, Sophia. *Synthetic: How Life Got Made.* Chicago: University of Chicago Press, 2017.

Sabin, Jenny E., and Peter Lloyd Jones. *LabStudio: Design Research between Architecture and Biology.* New York: Routledge, 2018.

Terranova, Charissa N., and Meredith Tromble, eds. *The Routledge Companion to Biology in Art and Architecture.* New York: Routledge, 2017.

Wilson, Stephen. *Art and Science Now.* London: Thames & Hudson, 2013.

Design, computation, performance, and optimization

Aranda, Benjamin, and Chris Lasch, eds. *Tooling.* Pamphlet Architecture 27. New York: Princeton Architectural Press, 2006.

Banham, Reyner. *The Architecture of the Well-Tempered Environment.* London: Architectural Press; Chicago: University of Chicago Press, 1969.

Bottazzi, Roberto. *Digital Architecture beyond Computers: Fragments of a Cultural History of Computational Design.* London: Bloomsbury Visual Arts, 2018.

Frazer, John. *An Evolutionary Architecture.* London: Architectural Association, 1995.

Hensel, Michael. *Performance-Oriented Architecture: Rethinking Architectural Design and the Built Environment.* Chichester, U.K.: John Wiley & Sons, 2013.

Kolarevic, Branko, and Ali Malkawi, eds. *Performative Architecture: Beyond Instrumentality.* London: Routledge, 2005.

Leach, Neil. *Designing for a Digital World.* Chichester, U.K.: Wiley-Academy, 2002.

Leach, Neil, and Philip F. Yuan. *Computational Design.* Tongji, China: Tongji University Press, 2018.

Lynn, Greg, ed. *Archaeology of the Digital.* Berlin: Sternberg Press; Montreal: Canadian Centre for Architecture, 2013.

Menges, Achim, and Sean Ahlquist. *Computational Design Thinking.* Chichester, U.K.: John Wiley & Sons, 2011.

Negroponte, Nicholas. *The Architecture Machine: Toward a More Human Environment.* Cambridge, Mass.: MIT Press, 1970.

Oxman, Neri. "Per Formative: Towards a Post Materialist Paradigm in Architecture." In "Taboo," ed. John Capen Brough, Seher Erdogan, and Parsa Khalili. Special issue, *Perspecta: The Yale Architectural Journal,* no. 43 (2010): 19–30.

Rid, Thomas. *The Rise of the Machines: A Cybernetic History.* New York: W. W. Norton, 2016.

Materials and digital fabrication

Boucher, Marie-Pier, Stefan Helmreich, Leila W. Kinney, Skylar Tibbits, Rebecca Uchill, and Evan Ziporyn, eds. *Being Material.* Cambridge, Mass.: MIT Press, 2019.

Cruz, Marcos, and Steve Pike, eds. *Neoplasmatic Design.* London: John Wiley & Sons, 2008.

Glynn, Ruairi, and Bob Sheil, eds. *Fabricate: Making Digital Architecture.* Cambridge, Ontario: Riverside Architectural Press, 2011; London: UCL Press, 2017.

Gramazio, Fabio, and Matthias Kohler. *Digital Materiality in Architecture.* Baden, Switzerland: Lars Müller, 2008.

Hays, Stephanie G., William G. Patrick, Marika Ziesack, Neri Oxman, and Pamela A. Silver. "Better Together: Engineering and Application of Microbial Symbioses." *Current Opinion in Biotechnology* 36 (December 2015): 40–49.

Kolarevic, Branko, ed. *Architecture in the Digital Age: Design and Manufacturing.* New York: Spon Press, 2003.

Kolarevic, Branko, and Kevin R. Klinger, eds. *Manufacturing Material Effects: Rethinking Design and Making in Architecture.* New York: Routledge, 2008.

Menges, Achim, ed. "Material Synthesis: Fusing the Physical and the Computational." Special issue, *Architectural Design* 85, no. 5 (November 2015).

Picon, Antoine. *Digital Culture in Architecture: An Introduction for the Design Professions.* Basel: Birkhäuser, 2010.

Sachs, Emanuel, Michael Cima, James Cornie, David Brancazio, Jim Bredt, Alain Curodeau, Tailin Fan, Satbir Khanuja, Alan Lauder, John Lee, and Steve Michaels. "Three-Dimensional Printing: The Physics and Implications of Additive Manufacturing." *CIRP Annals: Manufacturing Technology* 42, no. 1 (1993): 257–60.

Sheil, Bob, ed. *Manufacturing the Bespoke: Making and Prototyping Architecture.* Chichester, U.K.: John Wiley & Sons, 2012.

Tibbits, Skylar, ed. *3D Printing and Additive Manufacturing* 3, no. 2 (June 2016).

Oxman, Neri. *Active Matter.* Cambridge, Mass.: MIT Press, 2017.

Oxman, Neri. *Self-Assembly Lab: Experiments in Active Matter.* New York: Routledge, 2017.

Vincent, Julian. *Structural Biomaterials.* London: Macmillan Education, 1982.

Vollrath, Fritz. "Strength and Structure of Spiders' Silks." *Journal of Biotechnology* 74 (2000): 67–83.

hotograph credits

ndividual works appearing in this book ay be protected by copyright in the nited States and other countries and ay not be reproduced without the ermission of the rights holders. In eproducing the images contained in is publication, the Museum obtained e permission of the rights holders henever possible. Should the Museum ave been unable to locate a rights older, notwithstanding good-faith fforts, it requests that any contact formation concerning such rights olders be forwarded, so that they ay be contacted for future editions.

ages 2, 7, 10, 11: Courtesy Neri Oxman nd The Mediated Matter Group; pages , 8: Courtesy Neri Oxman, photograph y Yoram Reshef; pages 178–79: Courtesy eri Oxman and The Mediated Matter roup, photograph by Yoram Reshef; ages 182–83: Courtesy Neri Oxman

he Natural Evolution of Architecture

ig. 1: © 2008 The Museum of Modern rt, New York. Photograph by Jonathan uzikar; figs. 2–4, 11–14, 23, 40–45: ourtesy Neri Oxman; figs. 5–10: Courtesy eri Oxman and The Mediated Matter roup; figs. 15, 16: © 2015 Tomás araceno. Courtesy the artist; Andersen's, openhagen; Ruth Benzacar, Buenos ires; Tanya Bonakdar Gallery, New York/ os Angeles; Pinksummer contem-orary art, Genoa; Esther Schipper, erlin. Photographs by Studio Tomás araceno (fig. 15) and Brett Moen (fig. 6); fig. 17: © Agnieszka Kurant. ourtesy the artist and Tanya Bonakdar allery, New York / Los Angeles; figs. 8–22: © 2009 Dunne & Raby; fig. 24: © ichael Moran; fig. 25: © 2000-01 The issue Culture & Art Project (Oron atts, Ionat Zurr, Guy Ben Ary); fig. 26: 2008 The Musuem of Modern Art, ew York. Photograph by Martin Seck; g. 27: © 2019 Diller + Scofidio; fig. 28: 2015 Walter Pichler; fig. 29: © 2015 aimund Abraham; fig. 32: © 1965 rançois Dallegret; fig. 33 © 1970 assachusetts Institute of Technology, y permission of The MIT Press; fig. 34 Peter Eisenman, fonds Canadian entre for Architecture; figs. 35, 36: © 019 and courtesy David Benjamin; fig. 7: © 2019 and courtesy Self-Assembly ab, MIT + Invena; figs. 38–39 © 2008 he Museum of Modern Art, New York. hotographs by Pablo Enriquez; gs. 46–48 Courtesy Stewart Brand

Limbs of Nature

fig. 1: © The Museum of Modern Art, New York. Department of Architecture and Design Study Center. Photograph by George Barrows; fig. 2: © 2019 Karl Blossfeldt / Artists Rights Society (ARS), New York; fig. 4: Richard Nickel Archive, Ryerson and Burnham Archives, The Art Institute of Chicago. Digital File #201006_101116-007; fig. 5: Embryological House records, Canadian Centre for Architecture, Montreal; fig. 6: Courtesy Carl Solway Gallery, Cincinnati; fig. 7: © J. Paul Getty Trust. Getty Research Institute, Los Angeles (2004.R.10); fig. 8: Courtesy The Estate of R. Buckminster Fuller; fig. 9: saai | Archiv für Architektur und Ingenieurbau am KIT | Werkarchiv Frei Otto; fig. 10: Photograph by Central Press/Hulton Archive/Getty Images; figs. 11, 12: Archigram Archives 2019; fig. 13: © The Museum of Modern Art, New York

Armour/MetaMesh

All images courtesy Neri Oxman

Raycounting

All images courtesy Neri Oxman
Page 52, top left (silk-coated nylon structure) and page 53, bottom right (acrylic-based polymer structure): photographs by Mikey Siegel

Cartesian Wax

All images courtesy Neri Oxman.
All photographs by Mikey Siegel except as noted
Page 58, center right (multiple tiles): © The Museum of Modern Art, New York. Photograph by Jonathan Muzikar

Monocoque/Beast

All images courtesy Neri Oxman, except as noted
Pages 64–65 (single prototypes): photographs by Mikey Siegel; page 65, bottom (three prototypes): © The Museum of Modern Art, New York. Photograph by Thomas Griesel; page 67 (Beast prototype and detail): photographs by Yoram Reshef

Aguahoja I and II; Silk Pavilion I and II

All images courtesy Neri Oxman and The Mediated Matter Group

Glass I

All images courtesy Neri Oxman and The Mediated Matter Group except as noted
Page 117, bottom right (G3DP depositing molten glass, three images): photographs by Steven Keating; page 118, top right (thermal image): courtesy Sadie Forbes; page 118, bottom left (three-point bending test): photograph by Kyle Hounsell; page 119, upper left and center left (table installation seen from above and glass object illuminated from above): photographs by John Werner; page 119, center middle and lower right (object illuminated from above and table installation): photographs by Andy Ryan; page 119, bottom right (scanning electron microscope): image by James C. Weaver

Glass II

All images courtesy Neri Oxman and The Mediated Matter Group
Page 128, center right (MIT Media Lab installation): photograph by Andy Ryan; page 128 (Tokyo installation) and page 129, center (experiments for Glass III): photographs by Daniel Lizardo; pages 128, bottom left, and 130–31 (Milan installation): photographs by Lexus; page 129, top right (Milan installation): photograph by Johnathan Williams

Imaginary Beings

All images courtesy Neri Oxman except as noted. All photographs by Yoram Reshef except as noted
Page 135, bottom right (3D printer, two images): courtesy Neri Oxman and The Mediated Matter Group, no photographer; page 134 (Minotaur Head with Lamella): image by by Turlif Vilbrandt (Symvol, Uformia)

Lazarus

All images courtesy Neri Oxman and The Mediated Matter Group
Page 142, center left (front view), page 143, bottom center (humans wearing Lazarus), and pages 144–45: photographs by Yoram Reshef; page 142, bottom (children holding Lazarus): photographs by Danielle van Zadelhoff

Vespers I

All images courtesy Neri Oxman and The Mediated Matter Group.
All photographs by Yoram Reshef

Vespers II

All images courtesy Neri Oxman and The Mediated Matter Group
Page 151, bottom center and left (details), pages 154–55, and pages 157–58: photographs by Yoram Reshef

Vespers III

All images courtesy Neri Oxman and The Mediated Matter Group

Totems

All images courtesy Neri Oxman and The Mediated Matter Group

Works in the collection of The Museum of Modern Art, New York

Committee on Architecture and Design Funds
Imaginary Beings: Arachne, Daphne, Doppelgänger, Gravida, Medusa, Minotaur Head with Lamella, Minotaur Head with Sutures

Gift of The Contemporary Arts Council of The Museum of Modern Art
Raycounting, Cartesian Wax, Monocoque, Subterrain

The Modern Women's Fund
Glass I, Glass II

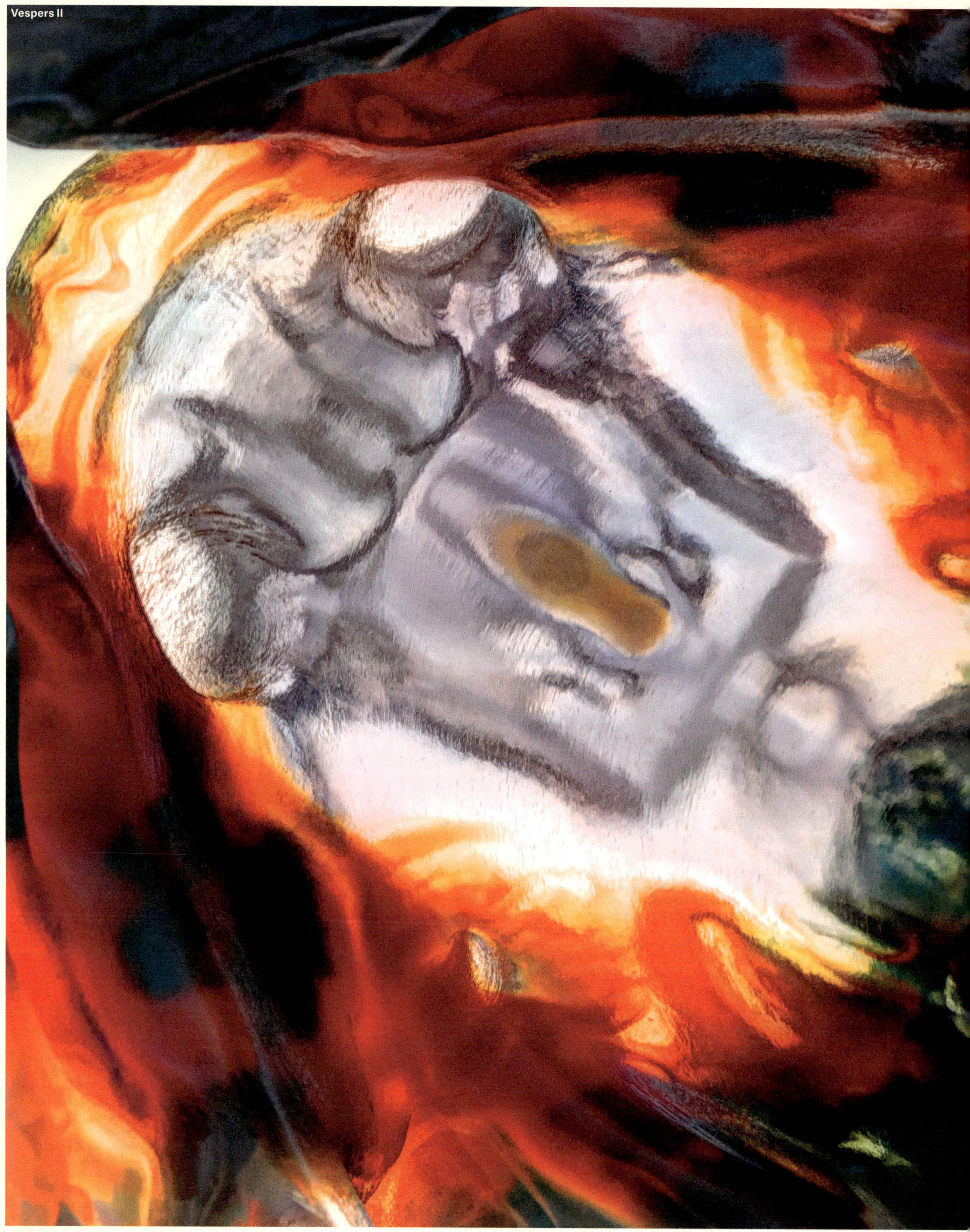

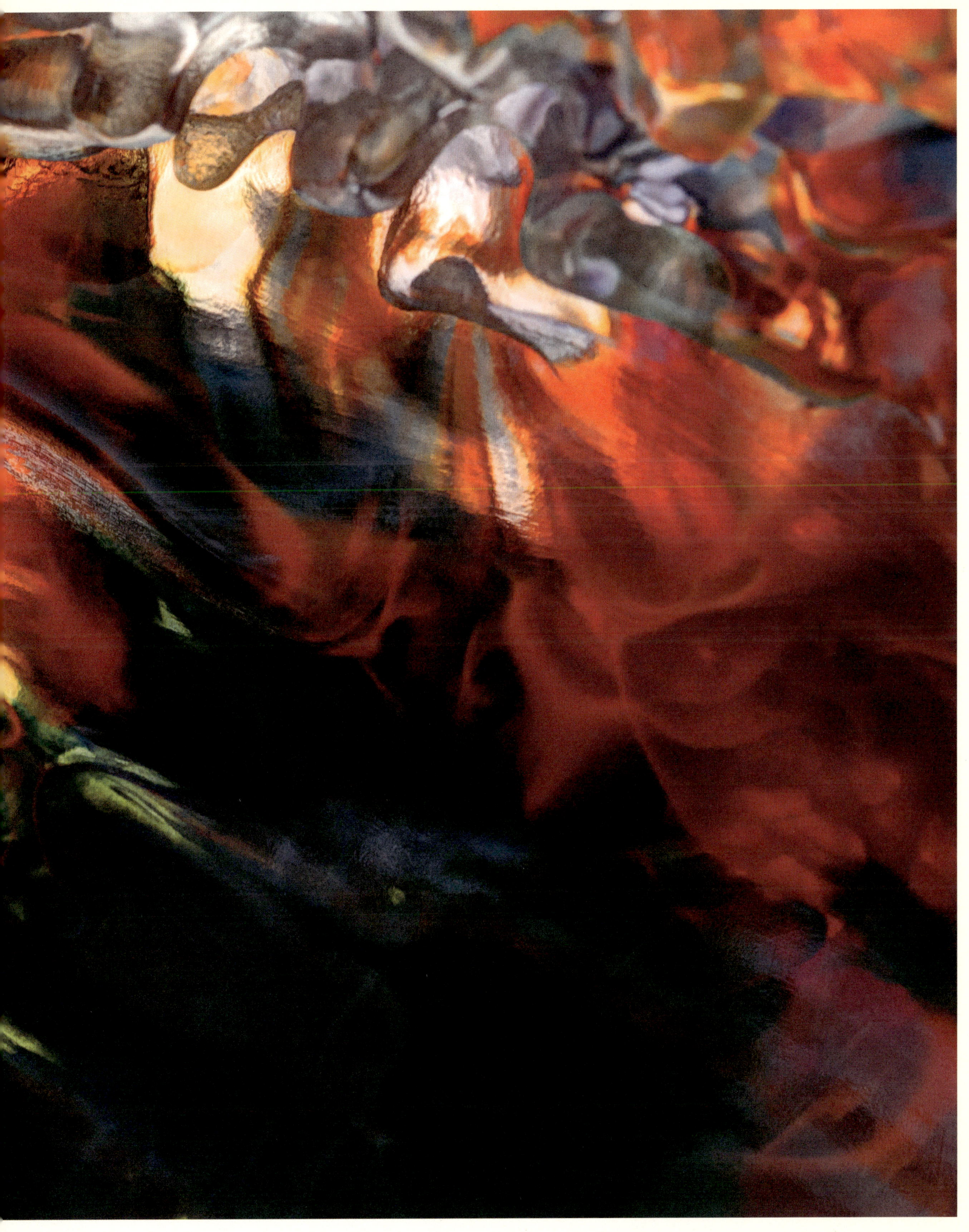

Acknowledgments

Although the planning of *Neri Oxman: Material Ecology* officially began in November 2018, the exhibition has been years in the making. I have been plotting a project of this nature ever since I met Neri, in 2007, and was electrified by the vision in her work, the solidity of her science, the elegance of her delivery, and the generosity of her intellect. I had found a kindred spirit, a friend, and a comrade in arms. This book and exhibition have been an opportunities for both of us to take stock of the ways in which we and our fields of research have evolved and the roles of architecture and design have been recalibrated. It has also been an opportunity to understand what it means to be a human being fully engaged with the complex ecosystems of nature. I thank her for giving me the chance to go on this adventure with her and for allowing me to curate her work without pressure or interference. Those of you who work as curators know how rare an event this is, and how it can happen only with the most centered and unselfish of artists.

Together with my ally, Anna Burckhardt, and on behalf of The Museum of Modern Art, I also wish to thank the team of researchers at MIT Media Lab's Mediated Matter Group (Christoph Bader, João Costa, Jean Disset, Jorge Duro-Royo, Joseph Kennedy, Felix Kraemer, Nic Lee, Ren Ri, Sunanda Sharma, Rachel Soo Hoo Smith, Michael Stern, Ramon Weber, and Susan Williams) and their collaborators, who patiently listened to our inquiries, generously accommodated our multiple requests, and hosted us at their lab several times for deep-immersion sessions. Their tireless work on Silk Pavilion II and on all other projects have made the exhibition and publication as memorable as they are, and for that we are truly grateful. Special thanks go to Kelly Egorova and Nitzan Zilberman, both of them great production managers who expertly took charge of logistics, and Tilman Bayer, who managed the photo archive with calmness and skill. Natalia Casas, former Administrative Assistant, and Becca Bly, Administrative Assistant, kept everyone in check and on schedule. At the MIT Media Lab, I wish to thank Claudia Robaina, Program Manager, Director's Fellows: and Alexandra Kahn, Senior Press Liaison. Jason Balesta, Counsel in the Office of the General Counsel at MIT, efficiently considered loan and image requests.

A number of collaborators provided key contributions to this project. The entire team at Front Inc. and Bodino were instrumental in the design, engineering, and production of Silk Pavilion II and Totems. At Front we wish to thank in particular Paolo Bortolazzo, Associate, and Marc Simmons, Principal, who listened to our suggestions and worked tirelessly to make this magnificent project a reality. Naomi Kaempfer, Creative Director for Art Fashion and Design at Stratasys, Neri's trusted collaborator for many years—and my coconspirator since 2006, when we first managed to include 3D-printed design objects in MoMA's collection—answered questions and always made herself available, no matter how urgent the request. I thank Bree Rodrigues, the sole external lender to the exhibition, for generously agreeing to part with Armour for a few months. The team at Sibling Rivalry (Mikon Van Gastel, Shelby Ross, Gabe Darling, Natasha Nussbaum, and Justin Zurrow) created a beautiful VR experience devoted to the Vespers masks; although we were not able to include it in the exhibition for want of space, it is available for viewing (at www.anddesignvr.com)—and we recommend it. Another unrealized project is a children's book, which we hope will happen in the future: we would like to thank Maria Popova, Kelsey Jubin, Michelle Millar Fisher, and Maya Bradford for their advice. I'd also like to single out a few friends and colleagues who have lent their valuable time, connections, expertise, and advice: Sunny Bates, Adam Bly, Oron Catts, Marshall Ganz, Jamer Hunt, Sylvia Lavin, Janna Levin, Ravi Naidoo, Ala Tannir, and Kevin Slavin. Last but not least, I would like to acknowledge Neri's partner, Bill Ackman, and mine, Larry Carty, for their steadfast belief in our work—individual and joint—and for their generosity with their time and attention whenever we have needed to bounce ideas or resolve dilemmas.

All of my patient colleagues at The Museum of Modern Art deserve particular acknowledgment for their indispensable contributions. Director Glenn D. Lowry's vision and confidence have been crucial at every twist and turn of this project. For this, and for much more, I am deeply grateful. I owe a debt of gratitude also to Peter Reed, Senior Deputy Director, Curatorial Affairs, and James Gara, Chief Operating Officer, for their guidance. In the Department of Exhibition Administration and Planning, I thank the indefatigable, unflappable, and generous Ramona Bannayan, Senior Deputy Director, Exhibitions and Collections, and Erik Patton, Director, and Jennifer Cohen, Associate Director, Exhibition Planning and Administration, for their command of this complex undertaking, which involved live silkworms, biodegradable substances, and brand-new, untested materials. I salute Margaret Aldredge, Exhibition Manager, and Rachel Kim, Senior Exhibition Manager, for overseeing logistics, contracts, and budgets; and Jessica Nielsen, Associate Registrar, and Steven Wheeler, Registrar, for masterfully coordinating shipments and keeping track of all the objects. My thanks also to Roger Griffith, Sculpture Conservator; Chris McGlinchey, Scientist; and Joy Bolser, Fellow, who kept organisms, polymers, biopolymers, and other assorted matter healthy; experimented with new structures in 3D printed glass; and researched the particularities of silk.

Material Ecology is the first exhibition to use the entire ground-floor space, newly expanded for MoMA's reopening in 2019. Its installation was expertly designed by Lana Hum, Director, and LJ McNerney, Assistant Production Manager, Department of Exhibition Design and Production, with the assistance of Benjamin Akhavan, whose scale objects for the exhibition model deserve a display of their own. Lana led MoMA's outstanding in-house carpentry and operations crew (Matthew Cox, Senior Production Manager; James Majewski, Project Manager, Housekeeping; Jason Fry, Lead Carpenter; J ohn Wood, Chris DaSilva, and Craig Anderson, Carpenters; Bryan Reyna, Foreman, Paint Shop; James Allgeier, Painter; Sean Brown, Foreman, Lighting Studio; Andrew Tedeschi and Raymond Martin, Mechanics) to produce a great exhibition. Peter Perez, Foreman, Frame Shop, and Jeroen Potjer, Senior Structural Engineer at Arup, provided invaluable guidance. Tom Krueger, Assistant Manager, Art Handling and Preparation, led the incomparable art-handling squad and was fundamental in coordinating all aspects of the installation. Thank you also to Aaron Harrow, AV Design Manager; Tal Marks, Manager; Mike Gibbons, AV Exhibitions Foreperson; Travis Kray, AV Technician; and the rest of their team in the Audiovisual Department for doing justice to MMG's beautiful videos.

This book came to life under the guidance of Christopher Hudson, Publisher, Department of Publications, who had faith in the project from its conception; and Don McMahon, Editorial Director, who provided invaluable feedback. Emily Hall, Editor, was steadfast in shaping the complex technical texts included in this volume with the meticulous assistance of Oriana Tang, Intern. Hannah Kim, Business and Marketing Manager, and aomi Falk, Rights Coordinator, advised n contracts and image permissions. My eep appreciation and respect go to the erson who made this book (and every ther book I have conceived in twenty-six ears at MoMA) happen: Marc Sapir, roduction Director, who not only master-ninded it, guided us, and tolerated my elays but also moved mountains so that ve could work with the best book esigner in the world—Irma Boom. Irma, s is her habit, dove into Neri's work and ersonality, and surprised both Neri and ne with an interpretation that made erfect sense and showed the grit and nergy of the process behind Neri's rojects. And after that, she waded earlessly through hundreds of images, rom technical drawings to stylized etails, to produce an inspired visual ynthesis. I still cannot believe that I was ucky enough to work with her for a econd time (the first was *Design and the lastic Mind* in 2008). The catalogue also enefited greatly from the essay ontributed by Hadas Steiner, Associate rofessor, State University of New York at uffalo, as well as the texts provided by Jeri and MMG. I would also like to thank ennifer Liese for editing my essay and naking it a thousand times better.

)n MoMA's Creative Team, I would like to onvey appreciation to Leah Dickerman,)irector, Editorial and Content Strategy, nd Prudence Pfeiffer, Managing Editor, or understanding the importance of Jeri's work and suggesting that Neri vrite the set of principles that became a entral aspect of my introductory essay. Rob Giampietro, Director of Design; Claire Corey, Production Manager; Elle Kim, Senior Art Director; David Klein, Senior Graphic Designer; and Olya Domoradova, Senior Graphic Designer, reated and implemented the exhibition's nmistakable graphic identity. Rebecca Stokes, Director of Marketing Campaigns, nd Wendy Olson, Marketing Manager, long with their team, made sure that he exhibition would be widely seen. My colleagues in the Department of Education merit a special shout-out. Wendy Woon, Deputy Director of Education; Pablo Helguera, Director of Adult and Academic Education; Elizabeth Margulies, Director, Family Programs and Resources; Adelia Gregory, Associate Educator, Public Programs and Gallery nitiatives; and Leticia Gutierrez, Associate Educator, Gallery and Learning Programs, supported the show from the beginning and prioritized it in their public programs. Sara Bodinson, Director, Jenna Madison, Assistant Director, nterpretation and Digital Learning, and Jackie Neudorf, Assistant Editor, helped shape the exhibition labels.

In the Department of Communications, my heartfelt gratitude goes to Amanda Hicks, Director, for her invaluable guidance, and to Stephanie Katsias, Publicist, who coordinated the exhibition's media coverage with confidence and composure. Meagan Johnson, Director of Institutional Giving and Development Operations; Jessica Smith, Assistant Director of Institutional Giving; and Anna Vallifuoco, Manager of Institutional Giving, Global Partnerships, worked hard to underwrite this ambitious project, while Patty Lipshutz, General Counsel, and Nancy Adelson, Deputy General Counsel, handled the many legal aspects related to the exhibition and were always available to advise me in time of need.

To my colleagues in the Department of Architecture and Design I offer my heartfelt appreciation—especially since some of them have had to listen to me insist on the urgency of this exhibition for an entire decade! I am deeply grateful to Martino Stierli, Philip Johnson Chief Curator of Architecture and Design, who believed in the project and supported it in every way possible: to Juliet Kinchin, Curator, and Sean Anderson, Associate Curator, who helped me in many curatorial and noncuratorial ways; and to Paul Galloway, Collections Specialist; Emma Presler, Department Manager; and Pamela Popeson, Preparator. My thanks to Nadine Dosa, former Department Assistant, who elegantly coordinated everything from payments to meetings. Anna and I were blessed with the generous help of several talented interns, including Ariana Kalliga, Anna Talley, and Ruby Tian. Mathilde Loiseau deserves a special mention for providing key research for this book, assistance in several phases of the exhibition design, and more.

In closing, I would like to recognize Anna Burckhardt, the colleague with whom I was lucky to share this unique experience. Her brilliant mind, eye, and hand have graced all aspects of this book and exhibition, and her ability to stay calm and focused through thick and thin have made her an invaluable partner. I thank her with all my heart.

Paola Antonelli
The Museum of Modern Art, New York

The Mediated Matter Group at the MIT Media Lab

Neri Oxman, Director
Christoph Bader, Tilman Bayer, Becca Bly, João Costa, Jean Disset, Jorge Duro-Royo, Kelly Egorova, Joseph H. Kennedy Jr., Felix Kraemer, Nic Lee, Ren Ri, Sunanda Sharma, Rachel Soo Hoo Smith, Michael Stern, Ramon Weber, Susan Williams, Nitzan Zilberman

Former members
Tal Achituv, Levi Cai, Natalia Casas, Barrak Darweesh, Sara Falcone, Giorgia Franchin, Carlos David Gonzalez Uribe, Chikara Inamura, Nassia Inglessis, Viirj Kan, Markus Kayser, Steven Keating, John Klein, Dominik Kolb, Julian Leland, Andrea Ling, Daniel Lizardo, Laia Mogas-Soldevila, Benjamin Peters, William Patrick, Yen-Ju (Tim) Tai, Owen Trueblood, Joshua Van Zak, Tomer Weller

Project collaborators
Mary Ann Babula; Matthew Bradford; Gal Begun; Boris Belocon; Prof. Katia Bertoldi; Davide Biasetto, Il Brolo Società Agricola S.r.l.; David J. Benyosef; Silvia Cappellozza, Council for Agricultural Research and Agricultural Economy Analysis (CREA-AA); Prof. W. Craig Carter; Allen Chen; Prof. George Church; Rachel Diaz-Granados; Joseph Faraguna; Prof. Javier G. Fernandez; Yi Gong; Danielle Grey-Stewart; Aury Hay; Stephanie Hays; Joe Hicklin, MathWorks; Ahmed Hosny; Peter Houk; Brian Huang; Ava Iranmanesh; Noah Jakimo; Naomi Kaempfer; Neils La White; Prof. Tim Lu; Dr. Anne Madden, North Carolina State University; Daniel Maher; Dechuan Meng; Philip Norwood; Jessica O'Keefe; Prof. Fiorenzo Omenetto, Tufts University; Johannes Overvelde, Bas; Keren Oxman; Hans Martin Pech; Robert Philips; Dan Robertson; Alessio Saviane, CREA-AA; Sol Schade, Advanced Functional Fabrics of America (AFFOA); Susan Shapiro; Prof. Pamela Silver; Eléonore Tham; Iris van Herpen; Christopher Voigt; James C. Weaver; Forrest Whitcher; Dr. Noah Wilson-Rich and The Best Bees Company; Sara Wilson; Amelia Wong; Wyss Institute, Harvard University

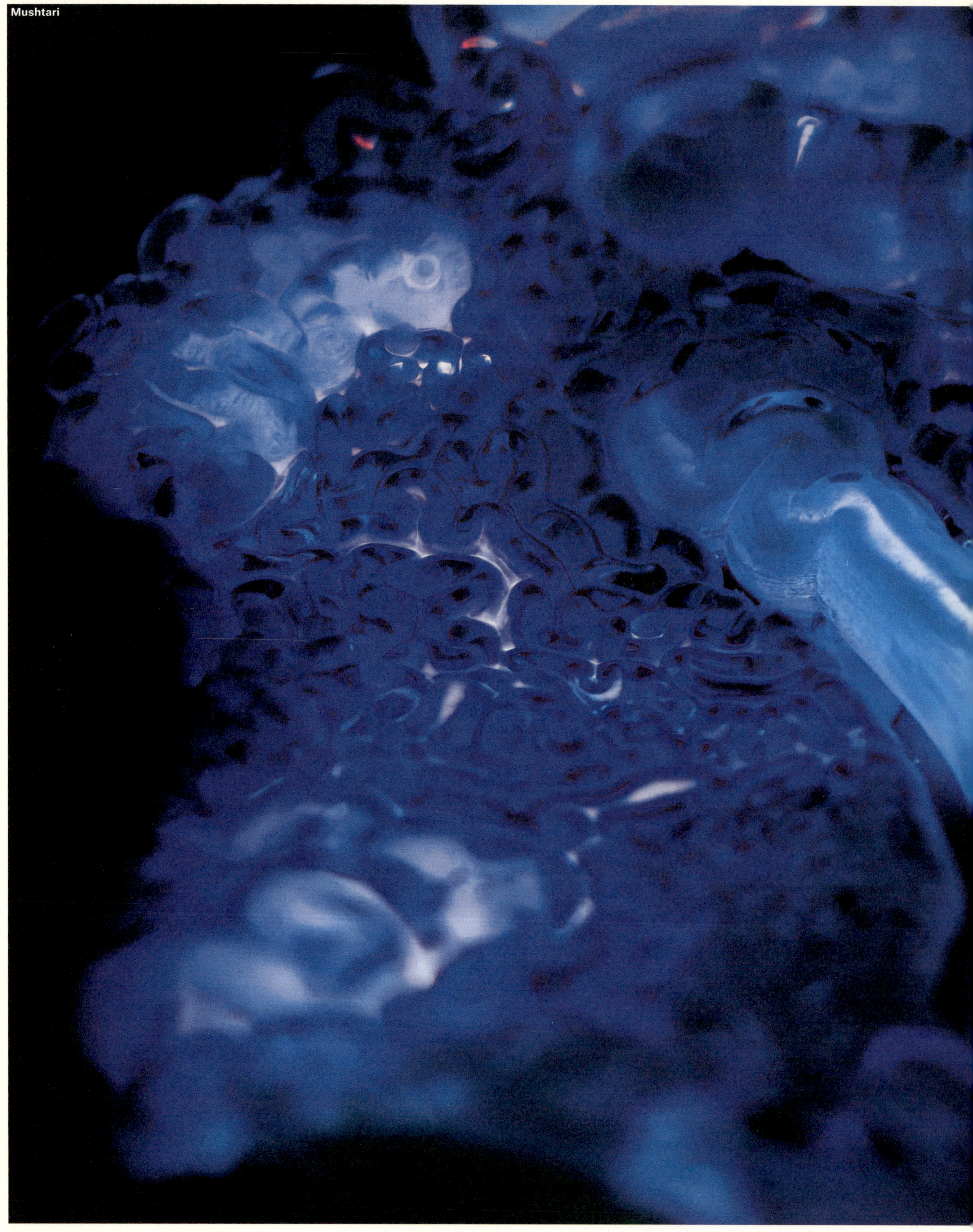
Mushtari

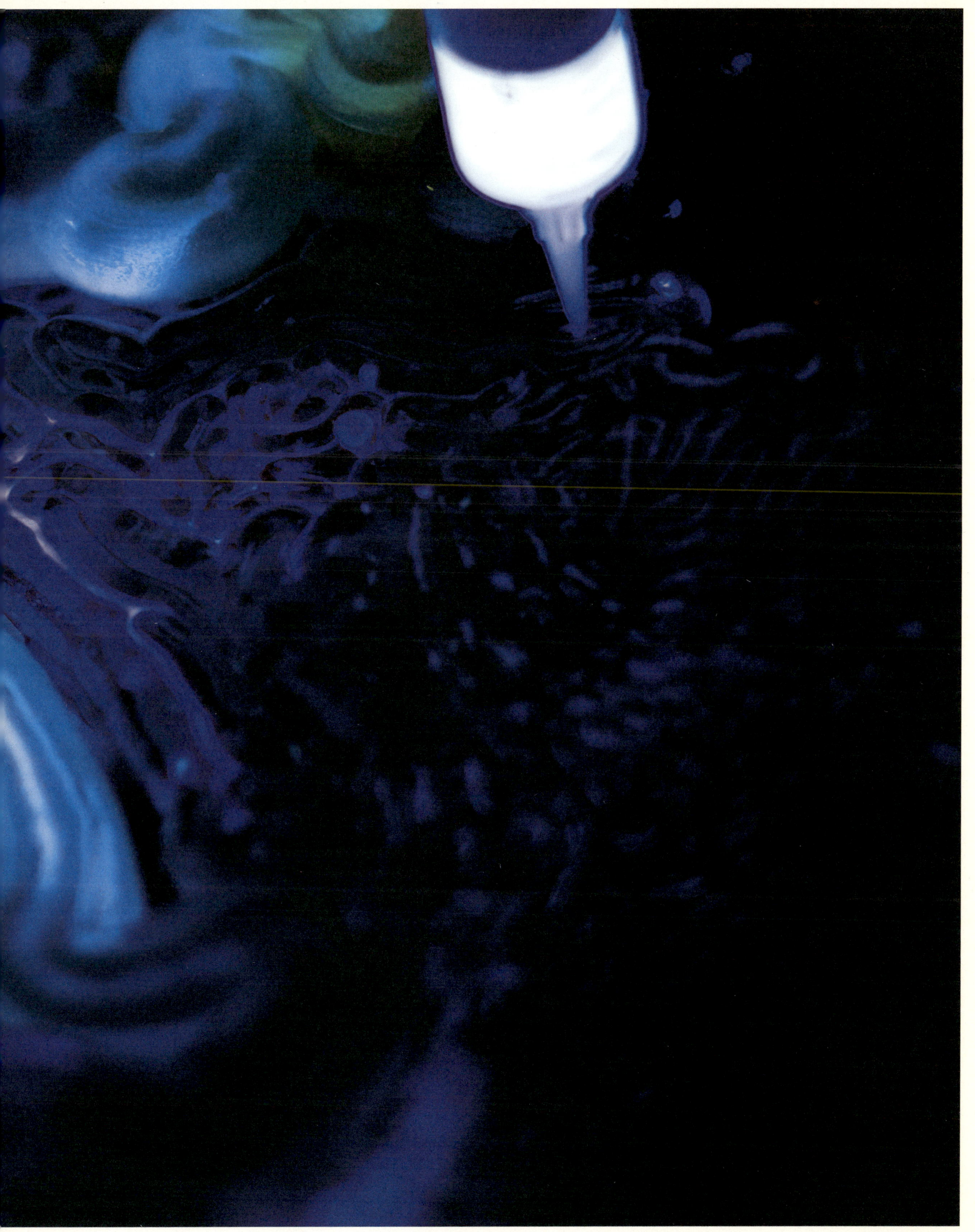

Trustees of The Museum of Modern Art

Leon D. Black
Chairman

Ronnie Heyman
President

Sid R. Bass
Mimi Haas
Marlene Hess
Maja Oeri
Richard E. Salomon
Vice Chairmen

Glenn D. Lowry
Director

Richard E. Salomon
Treasurer

James Gara
Assistant Treasurer

Patty Lipshutz
Secretary

Ronald S. Lauder
Honorary Chairman

Robert B. Menschel
Chairman Emeritus

Jerry I. Speyer
Chairman Emeritus

Agnes Gund
President Emerita

Marie-Josée Kravis
President Emerita

Wallis Annenberg*
Lin Arison**
Sarah Arison
Sid R. Bass
Lawrence B. Benenson
Leon D. Black
David Booth
Eli Broad*
Clarissa Alcock Bronfman
Patricia Phelps de Cisneros
Steven Cohen
Edith Cooper
Douglas S. Cramer*
Paula Crown
David Dechman
Anne Dias Griffin
Glenn Dubin
Lonti Ebers
Joel S. Ehrenkranz
John Elkann
Laurence D. Fink
H.R.H. Duke Franz of Bavaria**
Glenn Fuhrman
Kathleen Fuld
Howard Gardner*
Maurice R. Greenberg**
Agnes Gund
Mimi Haas
Marlene Hess
Ronnie Heyman
AC Hudgins
Barbara Jakobson
Jill Kraus
Marie-Josée Kravis
June Noble Larkin*
Ronald S. Lauder
Wynton Marsalis**
Robert B. Menschel
Khalil Gibran Muhammad
Philip S. Niarchos
James G. Niven
Peter Norton
Daniel S. Och
Maja Oeri
Eyal Ofer
Michael S. Ovitz
Emily Rauh Pulitzer
David Rockefeller, Jr.*
Sharon Percy Rockefeller
Lord Rogers of Riverside**
Richard E. Salomon
Ted Sann**
Anna Marie Shapiro
Lila Silverman**
Anna Deavere Smith
Jerry I. Speyer
Jon Stryker
Daniel Sundheim
Tony Tamer
Steve Tananbaum
Yoshio Taniguchi**
Jeanne C. Thayer*
Alice M. Tisch
Edgar Wachenheim III
Gary Winnick
Xin Zhang

Ex Officio

Glenn D. Lowry
Director

Bill de Blasio
Mayor of the City of New York

Corey Johnson
Speaker of the Council of the City of New York

Scott M. Stringer
Comptroller of the City of New York

Ann Fensterstock and Tom Osborne
Co-Chairmen of The Contemporary Arts Council

Alvin Hall and Nancy L. Lane
Co-Chairmen of The Friends of Education

* Life Trustee
** Honorary Trustee